Guidebook to organic synthesis

Guidebook to organic synthesis

Raymond K. Mackie and David M. Smith

Department of Chemistry, University of St Andrews

Longman London and New York

Longman Group Limited
Longman House
Burnt Mill, Harlow, Essex, UK

*Published in the United States of America
by Longman Inc., New York*

© Longman Group Limited 1982

First published 1982

British Library Cataloguing in Publication Data

Mackie, Raymond Keith
  Guidebook to organic synthesis.
  1. Chemistry. Organic-Synthesis
  I. Title   II. Smith, David Macdonald
  547'.2   QD262   80-41904
  ISBN 0-582-45592-8

Printed in Great Britain by The Pitman Press Ltd, Bath

Contents

Foreword

Synthesis, now more than ever, is a vital and exciting part of organic chemistry. This Guidebook is for the benefit of the student faced with an ever increasing multitude of synthetic methods, routes, approaches and concepts; it is designed therefore to smooth the way by picking out the essentials. The authors' treatment is unusual because they have opted to bring together the traditional and the new by combining the approach based on synthetic methods with that involving more recent synthon-disconnection ideas. In addition they encourage readers to gain experience in the latter approach by including worked examples throughout the text.

Because this was not to be an advanced or comprehensive treatise, the problem of what to omit must have been daunting, but the authors have drawn on their long experience of University teaching to produce a readable summary of the essentials of organic synthesis which will be a valuable supplement for students and for their teachers.

J. I. G. Cadogan

Preface

This book is intended for students who are beginning a detailed study of organic synthesis. It is assumed that they will already have completed a course in elementary organic chemistry which includes the reactions of simple functional groups and the basic concepts of reaction mechanism and of stereochemistry. We have particularly in mind students in the third and fourth years of a Scottish Honours Chemistry course, and their counterparts elsewhere in the world.

Organic synthesis has, rightly, commanded a good deal of attention from research chemists in recent times, and as the expansion and diversification of the subject continues, the student is confronted by a more and more bewildering array of synthetic methods. It thus becomes more and more difficult for the student, especially at an early stage in his career, to take a firm grasp of the subject and to gain an appreciation of its underlying principles. It is all too easy, as we know from experience, for synthetic strategy to be lost in this veritable jungle of exotic new methods and reagents. Or, to use another botanical metaphor, it is all too easy not to see the wood for the trees.

This does not set out, therefore, to be an advanced treatise, and still less to provide a compendium of the latest synthetic methods. More senior students who require these are amply provided for elsewhere. The choice of material for inclusion has not been an easy one, and it inevitably reflects to some extent our own personal interests; but we hope that it provides a satisfactory blend of old and new, and we are sorry if any reader is aggrieved by the omission of a 'favourite' reaction.

The book falls into several main sections. At the outset (Ch. 2) we provide an outline of functional group chemistry, most (if not all) of which we assume to be merely revision of material already learnt. Chapters 3–7 are then concerned with the formation of carbon–carbon bonds and related processes by which a molecular framework is constructed. In the third section (Chs. 8–10) we return to functional group interconversion (reduction, oxidation, and the use of protective groups). There then follow three short chapters outlining some synthetic uses of reagents containing boron, phosphorus, and silicon, and in the final chapter we discuss a few syntheses, of widely differing types, which are taken directly from the original literature.

With regard to reaction mechanisms, we have made extensive cross-reference throughout the text to Dr Peter Sykes' highly successful *A Guidebook to Mechanism in Organic Chemistry*; the page numbers refer to the fifth edition (Longman, 1981). We have also made extensive use of illustrative examples, and we have quoted product yields in every case where yields have been recorded in the literature, since we believe that yield is an important consideration in the choice of a synthetic method.

At an early stage in the text (Ch. 3) we have introduced the synthon-disconnection approach to synthetic design: this is the approach first advocated (or so we believe) by Professor E. J. Corey in the 1960s and developed with such outstanding success by Dr Stuart Warren in his programme-type book of problems, *Designing Organic Syntheses* (Wiley, 1978). The term *synthon* is used by chemists nowadays in two entirely different senses, but we have adhered to Corey's original definition, i.e. that a synthon is 'a structural unit within a molecule which is related to possible synthetic operations'.

All of our colleagues in Organic Chemistry at St Andrews, and some in other universities, have helped us in one way or another with the preparation of this book, and we gladly acknowledge our indebtedness to them. We are particularly grateful to three friends, Professor Hamish Wood (Strathclyde) and Drs Derek Leaver and Hamish McNab (Edinburgh), who willingly agreed to read the complete typescript and have thus helped us to improve and clarify the original version. We also thank our good friend and former leader, Professor John Cadogan, for initiating the writing of this book, for contributing the Foreword, and for much advice and encouragement throughout the period of writing. Above all, however, we must thank our typist, Mrs Wilma Pogorzelec, whose typing skill and knowledge of chemical terminology made the transition from pencilled manuscript to finished typescript a particularly painless one for the authors.

St Andrews   R.K.M.
August 1980   D.M.S.

Important notice

This book is not intended to be a 'recipe book' for the experimentalist. The reactions cited here should be regarded merely as an illustration of the general principles; readers seeking to use these reactions in practice are *strongly advised* to refer first to the original literature for experimental details.

Readers are also reminded of the various hazards entailed in organic chemical reactions, and of the consequent need for proper safety precautions. The fire hazard associated with many common solvents is well-known, but particular care must also be taken where a compound is liable to be explosive (e.g. azides, diazo-compounds): corrosive (e.g. phenols), skin-irritant, toxic (e.g. methylating agents) or carcinogenic (e.g. benzene, N-nitroso compounds, and certain arylamines). Further details of such hazards, and safety precautions, are to be found in practical handbooks, e.g. in Vogel's *Textbook of Practical Organic Chemistry* (Fourth Edition, Longman, 1978).

Abbreviations and other trivial names

DCC N,N'-Dicyclohexylcarbodi-imide,

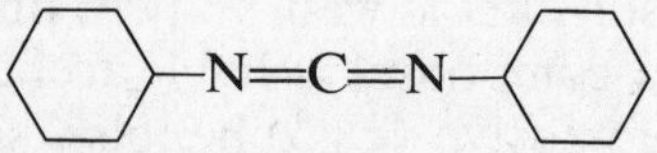

Diglyme   Diethyleneglycol dimethyl ether,
                    $CH_3O(CH_2)_2O(CH_2)_2OCH_3$

Digol   Diethylene glycol, $HO(CH_2)_2O(CH_2)_2OH$

DMF N,N-Dimethylformamide, $(CH_3)_2NCHO$

HMPA   Hexamethylphosphoramide, $[(CH_3)_2N]_3PO$

(also HMPT)   (triamide)

LDA   Lithium di-isopropylamide, $[(CH_3)_2CH]_2N^-Li^+$

NBS N-Bromosuccinimide,

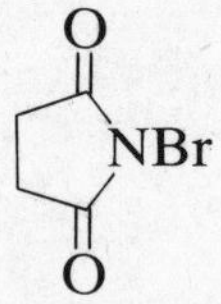

Ph   Phenyl,

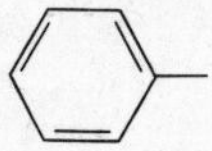

Sulpholan   Tetramethylene sulphone,

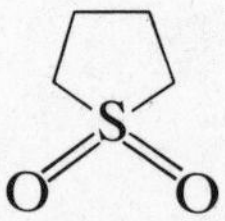

THF   Tetrahydrofuran,

Ts   Toluene-p-sulphonyl, CH_3

(also tosyl)

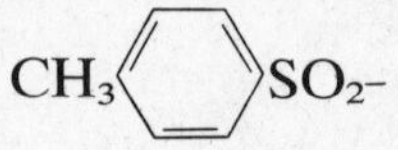

1   Introduction

Of all the principal constituent parts of present-day organic chemistry, synthesis is the one with perhaps the longest history. The ideas of functionality and stereochemistry, for example, have their origins in the second half of the nineteenth century, and the concepts of bonding and reaction mechanism, as we know them today, undoubtedly belong to the present century. Synthesis, however, has constituted an important part of organic chemistry from the very beginnings of the subject, and thus has a history stretching back over many centuries. It has to be admitted, however, than most of the early work was fragmentary in character, depending as it did on starting materials isolated from natural sources in doubtful states of purity; the *development* of organic synthesis on a systematic basis belongs to the nineteenth century, even if its *origins* are much earlier.

In more recent times, the growth of organic synthesis has kept pace with the growth of organic chemistry as a whole. As understanding of structural and theoretical chemistry has increased, and as experimental methods have been developed and refined, the chemist has been able to set himself more and more ambitious synthetic objectives. These lead in turn to the discovery of new reactions and to the perfection of new experimental methods, and thence to new synthetic targets; and so on. Thus present-day organic synthesis often appears to the student as a vast assembly of factual information without much by way of structure or rationale.

Since our own student days, in the mid-1950s, the teaching of functional group chemistry has been revolutionised, and in most cases greatly simplified, by the use of reaction mechanism. The corresponding revolution in the teaching of synthesis has, unaccountably, taken somewhat longer, but is now well under way.

The fundamental ideas which lie behind this revolution are neither complicated nor new. They consist in recognising that a covalent bond is formed, in the vast majority of synthetically useful processes, by the interaction of an electrophilic and a nucleophilic atom: and in recognising the various structural units (called **synthons**) which go to make up a given synthetic target molecule. These ideas have been familiar to synthetic chemists for decades, but have rarely been included in undergraduate textbooks (or, we suspect, in lecture courses).

In 1835, the German chemist, Friedrich Wöhler, who was one of the pioneers of organic synthesis, wrote a letter to his mentor, the great Jöns Jacob Berzelius, which included the following often-quoted remarks[1]:

> Organic chemistry just now is enough to drive one mad. It gives me the impression of a primeval tropical forest, full of the most remarkable things; a monstrous and boundless thicket, with no way of escape, into which one may well dread to enter.

If any reader of this Introduction feels like that about organic *synthesis*, let him read on. This Guidebook has been written especially for him and those who share his view. It may not lead him right through the forest, but we hope that it will at least provide a reliable pathway, over solid ground, as far as the first clearing.

Note

1.   A. Findlay, *A Hundred Years of Chemistry* (2nd edn), Duckworth, London, 1947, p. 21.

2 Functionalisation and interconversion of functional groups

Important aspects of synthesis are the introduction of functional groups into a molecule and interconversion of functional groups. We shall show that in some instances it is relatively easy to functionalise certain positions whereas in other situations functionalisation is impossible and the desired product can only be obtained by a series of interconversions of functional groups.

In this chapter, we shall attempt to bring together, in outline only, a variety of reactions which the successful synthetic chemist will require to have at his command. Further details of the reactions mentioned in this chapter will be found in standard works on organic chemistry and Sykes describes the mechanisms of many of the reactions.

2.1 Functionalisation of alkanes

The unreactivity of alkanes towards electrophilic and nucleophilic reagents will be familiar to the reader. Alkanes are, however, reactive in radical reactions, particularly halogenation. Such reactions are nevertheless of limited synthetic use, due to the difficulties encountered in attempts to control them.

Because of the higher reactivity of Cl· than Br·, chlorination tends to be less selective than bromination and, indeed, 2-bromo-2-methylpropane is almost exclusively formed when isobutane reacts with bromine at 300° whereas chlorination results in a $2:1$ mixture of 1-chloro- and 2-chloro-2-methylpropane.

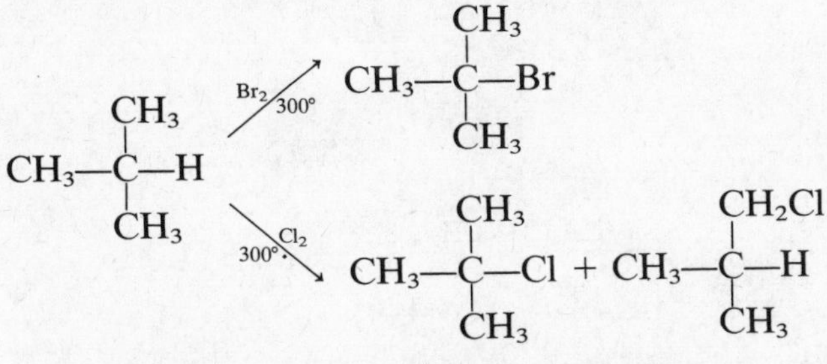

On the other hand, rearrangements are encountered in the intermediate radicals with less frequency than in the corresponding carbonium ions. Thus, only 1-chloro-2,2-dimethylpropane results when

2,2-dimethylpropane is chlorinated:

$$(CH_3)_4C \xrightarrow[\text{u.v.}]{Cl_2} (CH_3)_3CCH_2Cl \text{ [not } (CH_3)_2CClCH_2CH_3]$$

2.2   Functionalisation of alkenes

Unlike alkanes, alkenes contain two sites at which functionalisation can be carried out with a high degree of specificity. These are (a) at the C=C double bond and (b) at the carbon adjacent to the double bond – the *allylic* position.

The chemistry of alkenes is largely concerned with reactions of electrophiles with the double bond. The mechanism of these reactions is discussed by Sykes (pp. 175–91) and will not be discussed in this chapter. It is, however, necessary to recall that addition of electrophiles to unsymmetrical alkenes proceeds through the more stable carbonium ion,[1] resulting in the product in which the more positive moiety of the reagent has become attached to the **less** substituted alkene carbon (the 'Markownikoff' product). Scheme 2.1 summarises addition reactions involving propene. Further discussion of certain of these reactions will be found in later chapters: catalytic hydrogenation (Ch. 8), oxidation (Ch. 9) and hydroboration (Ch. 11). Some comment on the synthetic utility of the remaining reactions outlined in scheme 2.1 will now be considered.

Strong acids, e.g. HCl, HBr, HI, H$_2$SO$_4$, and CF$_3$CO$_2$H, add to alkenes as indicated in scheme 2.1, but weaker acids, e.g. CH$_3$CO$_2$H

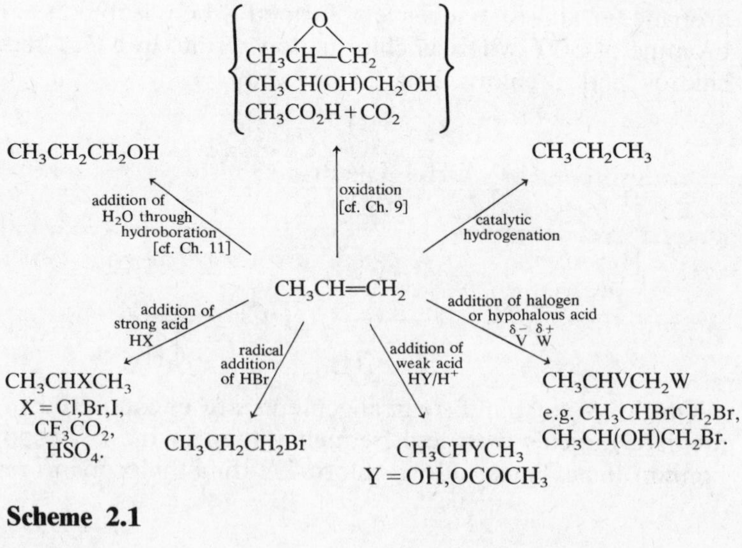

Scheme 2.1

and H_2O, require acid catalysis. The products of such reactions are alkyl halides, alkyl sulphates, alkyl trifluoroacetates, alkyl acetates and alcohols respectively. In all cases, 'Markownikoff' addition is observed except in the case of HBr: in this case, unless the alkene is rigorously purified so that peroxide impurities are excluded, 'anti-Markownikoff' addition is observed. This is due to the fact that, in presence of peroxide, a radical mechanism is followed (Sykes, pp. 307–10).

The orientation of addition in this case is governed by the relative stabilities of possible carbon radicals formed in the *addition* step. The result is that, for addition to simple alkenes, the *bromine* atom becomes attached to the **less** substituted carbon. This should be compared with the situation pertaining in ionic addition of hydrogen halides to alkenes when the *hydrogen* becomes attached to the **less** substituted carbon. Thus in this case the structure of the alkyl bromide formed by reaction of an alkane with hydrogen bromide can be determined by the conditions employed in the reaction.

$$CH_3CH_2CH_2Br \xleftarrow[\text{+peroxide}]{\text{HBr}} CH_3CH{=}CH_2 \xrightarrow[\text{peroxide-free}]{\text{HBr}} CH_3CHBrCH_3$$

It may also be noted from scheme 2.1 that either of the two possible alcohols may be obtained by addition of H—OH to the double bond. Further discussion of this topic will be found in Chapter 11.

Addition of mixed halides, e.g. ICl, and of hypohalous acids is also governed by the Markownikoff Rule, i.e. the more positive end of the

$$CH_3CH{=}CH_2 + \overset{\delta-\ \delta+}{V W} \longrightarrow CH_3CHVCH_2W$$

dipolar molecule becomes attached to the less substituted carbon:

$$CH_3CH(OH)CH_2Br \xleftarrow{\text{HOBr}} CH_3CH{=}CH_2 \xrightarrow{\text{ICl}} CH_3CHClCH_2I$$

The intermediate in reactions involving halogens and hypohalous acids is a halonium ion (1), reaction of which with a nucleophile leads to a *trans* addition product. In the case of addition of hypohalous acid,

(1)

the *trans*-halohydrin formed can be converted, by treatment with base, into an oxiran (epoxide):

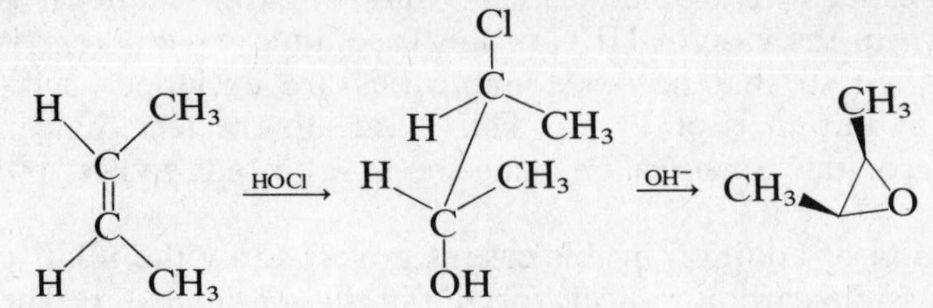

$$(Z) \text{ or } cis \qquad threo \qquad cis$$

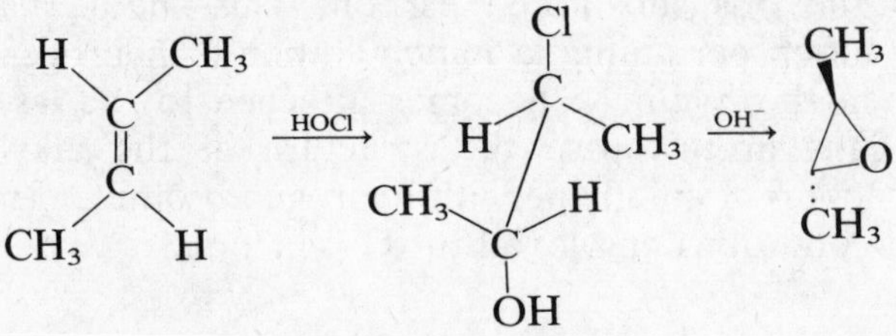

$$(E) \text{ or } trans \qquad erythro \qquad trans$$

An alternative means by which alkenes may be functionalised is reaction at the allylic position. Carbon–hydrogen bonds adjacent to the carbon–carbon double bond, the *allylic hydrogens*, are susceptible to oxidation and to halogenation. Although the majority of these halogenation reactions are free-radical processes, ionic reactions can also take place.

The most commonly used reagent for bromination is *N*-bromo-succinimide (NBS) and, since the reaction involves an intermediate allyl radical, a mixture of bromides can be expected:

$$RCH_2CH=CH_2 \xrightarrow[\text{(PhCO}_2)_2]{\text{NBS}} R\dot{C}HCH=CH_2 \longleftrightarrow RCH=CHCH_2^{\cdot}$$

$$\downarrow \text{NBS}$$

$$RCHBrCH=CH_2 + RCH=CHCH_2Br$$
$$(Z \text{ and } E \text{ isomers})$$

However, in simple cases such as cyclohexene, a good yield of the bromoalkene is obtained.

The introduction of oxygenated functional groups at allylic positions will be discussed in Chapter 9.

2.3 Functionalisation of alkynes

Most of the chemistry of alkynes is concerned with their reactivity towards electrophiles. As in the case of alkenes considered in the previous section, reactions with halogens, hydrogen halides and acids are synthetically useful. Hydrogenation of alkynes is also of considerable significance and will be discussed in Chapter 8. In addition, a terminal acetylene is a weak acid and the anion derived from it is of importance in carbon–carbon bond forming reactions (cf. sections 3.4.2.iii and 4.3).

Reactions of bromine with an alkyne results in the formation of a *trans*-dibromide and addition of lithium bromide to the reaction mixture increases the yield of the product. Reaction with hydrogen halide is of greater complexity, often following a *cis* stereochemistry. However, when the triple bond is not conjugated with an aromatic ring, the *trans* isomer predominates. Also, the addition of solvent may be a competing reaction, but this can be suppressed by carrying out the reaction in presence of a quaternary ammonium halide. These complications reduce the synthetic utility of the reaction:

$$C_2H_5C\equiv CC_2H_5 \xrightarrow[CH_3CO_2H]{HCl}$$

$$\underset{(40-72\,\%)}{\underset{Cl}{\overset{C_2H_5}{\diagdown}}C=C\underset{C_2H_5}{\overset{H}{\diagup}}}$$

$$+ \quad \underset{(<1\,\%)}{\underset{H}{\overset{C_2H_5}{\diagdown}}C=C\underset{Cl}{\overset{C_2H_5}{\diagup}}} \quad + \quad [C_2H_5CH=CC_2H_5] \quad \underset{OCOCH_3}{|}$$

$$\Big\downarrow {H_2O}$$

$$\underset{(28-60\,\%)}{C_3H_7COC_2H_5}$$

$$C_2H_5C\equiv CC_2H_5 \xrightarrow[CH_3CO_2H]{\overset{HCl/}{(CH_3)_4N^+Cl^-}}$$

$$\underset{(91-97\,\%)}{\underset{Cl}{\overset{C_2H_5}{\diagdown}}C=C\underset{C_2H_5}{\overset{H}{\diagup}}}$$

$$+ \quad \underset{(<0.2\,\%)}{\underset{H}{\overset{C_2H_5}{\diagdown}}C=C\underset{Cl}{\overset{C_2H_5}{\diagup}}} \quad + \quad \underset{(3-9\,\%)}{C_3H_7COC_2H_5}$$

$$PhC\equiv CCH_3 \xrightarrow[CH_3CO_2H]{HCl}$$

$$\underset{(67-78\,\%)}{\underset{Cl}{\overset{Ph}{\diagdown}}C=C\underset{H}{\overset{CH_3}{\diagup}}} \quad + \quad \underset{(11-19\,\%)}{\underset{Cl}{\overset{Ph}{\diagdown}}C=C\underset{CH_3}{\overset{H}{\diagup}}} \quad + \quad \underset{(10-13\,\%)}{PhCOC_2H_5}$$

Addition of water and of carboxylic acids to alkynes is catalysed by mercuric oxide. In the former case the product from a terminal alkyne is a methyl ketone and in the latter an enol ester:

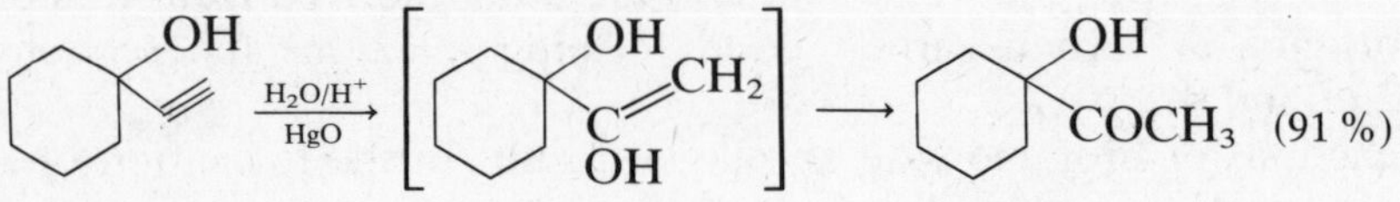

$$CH\equiv CH \xrightarrow[\text{ClCH}_2\text{CO}_2\text{H}]{\text{HgO}} CH_2=CHOCOCH_2Cl \quad (49\%)$$

The commonly used synthetic procedures are shown in scheme 2.2.

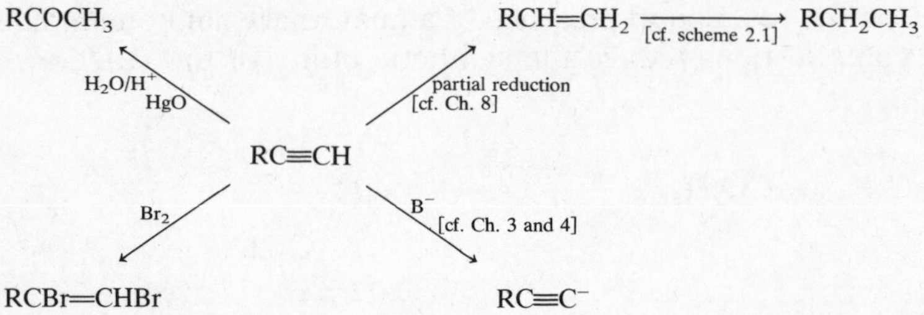

Scheme 2.2

2.4 Functionalisation of aromatic hydrocarbons

2.4.1 Substitution at a ring position

The characteristic reaction of benzene is an electrophilic addition-elimination reaction, the overall effect of which is substitution. This is the most widely used procedure for the introduction of functional groups on to the benzene ring. Scheme 2.3 outlines some of the more important reactions.

Some brief comments on the synthetic utility of these reactions is appropriate here but for a more detailed account of electrophilic aromatic substitution the reader is directed to one of the monographs on the topic and to Sykes, pp. 129–48.

Friedel–Crafts **alkylation** leads to polyalkylation in most cases, since the product alkylbenzene is more reactive towards electrophiles than is benzene. Hence an indirect synthesis, via acylation and reduction, is often desirable. Cyclopropane, alkenes, and alcohols may be used in place of alkyl halides in the alkylation reaction. The Vilsmeier and Gattermann aldehyde syntheses may be regarded as extensions of the Friedel–Crafts acylation reactions (cf. also section 5.4.2).

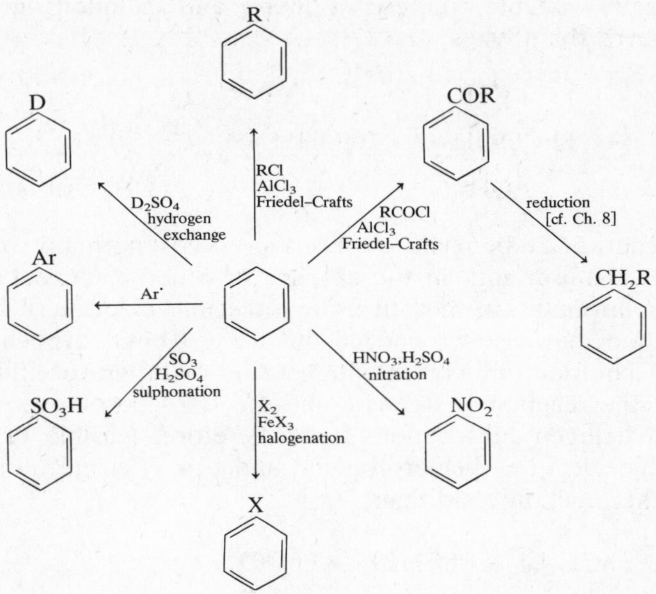

Scheme 2.3

Direct **halogenation** of benzene by molecular halogen catalysed by a Lewis acid is restricted to chlorination and bromination. Iodine is not sufficiently reactive to iodinate benzene, but toluene can be iodinated using iodine monochloride and zinc chloride. Fluorination is carried out by indirect methods, e.g. from diazonium salts to be described later in this chapter.

Sulphonation is an easily reversible reaction and this makes the sulphonic acid group a useful blocking group in synthesis.

Arylation of benzene can be carried out by free radical reactions involving diaroyl peroxides or *N*-nitrosoacetanilides, by the Gomberg reaction involving the alkaline decomposition of arenediazonium salts in benzene, or, perhaps most simply, by reaction with a primary arylamine and an alkyl nitrite.

2.4.2 Reaction in the side chain

Alkylbenzenes can be functionalised either in the side chain or in the ring. The latter will be discussed shortly. The side chain is susceptible to attack by radicals and also to oxidation at the position adjacent to the ring (the *benzylic* position). The oxidation of a methyl group involves three levels of oxidation: -CH$_2$OH, -CHO, and -CO$_2$H (cf. sections 9.2.2 and 9.2.3).

The benzylic position is also susceptible to autoxidation and the commercially valuable synthesis of phenol and acetone from cumene makes use of this (Sykes, p. 127):

$$PhCH \begin{matrix} CH_3 \\ | \\ | \\ CH_3 \end{matrix} \xrightarrow{O_2} PhCOOH \begin{matrix} CH_3 \\ | \\ | \\ CH_3 \end{matrix} \xrightarrow{H^+} PhOH + \underset{CH_3 \quad CH_3}{\overset{O}{\underset{\|}{C}}}$$

Halogenation at benzylic positions proceeds normally by a free radical mechanism and, in the absence of other reactive functional groups, is normally carried out using molecular chlorine or bromine. Chlorination may also be carried out using t-butyl hypochlorite or sulphuryl chloride, and bromination using N-bromosuccinimide. In all cases, the reaction is stepwise and the steps become slower with increased halogen substitution. It is, therefore, feasible to prepare benzyl chloride, α,α-dichlorotoluene and α,α,α-trichlorotoluene by varying the reaction conditions:

$$PhCH_3 \rightarrow PhCH_2Cl \rightarrow PhCHCl_2 \rightarrow PhCCl_3$$

2.5   Functionalisation of substituted benzene derivatives

Substituted benzene derivatives undergo electrophilic and free radical substitution reactions analogous to those described previously for benzene. However, in **electrophilic substitution,** the substituents already present in the ring direct an incoming electrophile into certain position(s) and affect the rate of substitution to such an extent that certain reactions (e.g. alkylation of nitrobenzene) cannot be carried out and others not possible with benzene can take place (e.g. reaction

Table 2.1 Orientation and rate of electrophilic substitution of substituted benzenes

Substituent	Orientation of electrophilic substitution	Rate of substitution relative to that of benzene
alkyl or aryl	o-, p-	faster
—OH, —OR	o-, p-	faster
—NH$_2$, —NHR, —NR$_2$	o-, p-	faster
halogen	o-, p-	similar or slower
$\diagdown$C=O	m-	slower
—C≡N	m-	slower
—NO$_2$	m-	slower
—SO$_3$H	m-	slower
—CF$_3$	m-	slower

of sodium phenoxide with diazonium salts). The mechanism of electrophilic substitution and the effect of various functional groups on orientation and rate of substitution is described by Sykes (pp. 149–61). A simplified general guide to these effects is given in table 2.1.

Two points are worth noting at this stage. Firstly, when more than one substituent is already on the benzene ring, the most strongly electron-donating group controls the position of further substitution. Secondly, in order to minimize possible substitution at nitrogen, aromatic amines are usually converted into acetanilides before substitution is carried out. This also serves to reduce the reactivity of the ring towards electrophilic substitution. Below are given some examples which may help the reader to understand the application of the rules:

(a)
$$\text{OH} \xrightarrow{\text{fuming HNO}_3} \text{OH with } O_2N, NO_2, NO_2$$

Mononitration takes place with dilute nitric acid, indicating that phenol is much more reactive than benzene. The hydroxyl group is *o*-/*p*- directing.

(b)
$$\text{NH}_2 \xrightarrow{\text{Br}_2, \text{H}_2\text{O}} \text{NH}_2 \text{ with Br, Br, Br}$$

No Lewis acid catalyst is required and the reaction cannot be stopped at the mono- or the di-bromo stage. The amino group is *o*-/*p*-directing and controls the orientation of addition rather than the more weakly *o*-/*p*-directing bromine. Monobromination can be affected by way of acetanilide:

$$\text{NHCOCH}_3 \xrightarrow[\text{acetic acid}]{\text{Br}_2} \text{NHCOCH}_3 \text{ (Br)} + \left[\text{NHCOCH}_3 \text{ Br} \right]$$

minor product

$$\xrightarrow[\text{OH}^-, \text{H}_2\text{O}]{\text{H}^+ \text{ or}} \text{NH}_2 \text{ (Br)}$$

(c)
$$\text{Cl, NO}_2 \xrightarrow[\text{fuming HNO}_3]{\text{conc. H}_2\text{SO}_4} \text{Cl, NO}_2, NO_2$$

Much more vigorous conditions are required for this reaction since both substituents retard nitration. The orientation is governed by the o-/p-directing chlorine.

The directional effects in **free radical substitution** reactions are much less pronounced, and it is normal to expect all three isomers from, for example, phenylation of a monosubstituted benzene:

(62 %)   (10 %)   (28 %)

Nucleophilic substitution is accelerated by electron-withdrawing substituents, e.g. NO_2. However, a leaving group such as halogen is also required and the reaction is not considered at this point.

2.6 Functionalisation of simple heterocyclic compounds

In the space available, it is only possible to deal in outline with some of the more important reactions of simple heterocyclic compounds and, for further detail on them and on reactions of more complex heterocyclic compounds, the reader is directed to more comprehensive texts.

Pyridine is a weak mono-acidic base which possesses a considerable degree of aromatic character. In reacts, for example, with methyl iodide to form quaternary salts which on heating rearrange to 2- and 4-methylpyridinium iodides. Molecular orbital calculations indicate that C-3 is the carbon having the highest electron density but, even at this position, the electron density is much lower than that of benzene. Electrophilic substitution, therefore, requires forcing conditions and the reactions which may be useful are summarized in scheme 2.4. Free radical phenylation results in the formation of a mixture of all three monophenylpyridines. Nucleophilic substitution results in substitution mainly at the 2-position and these reactions are summarised in scheme 2.5.

Pyridine-N-oxide, prepared most readily by treatment of pyridine with a peracid such as peracetic or perbenzoic acid, is a much weaker base than pyridine. It is readily nitrated in the 4-position and the directive effect of the N-oxide is such that 4-substitution takes place in most cases except those with a hydroxy or dimethylamino group in the 2-position. If the 4-position is blocked, nitration usually fails. Direct halogenation and sulphonation do not proceed readily and, as

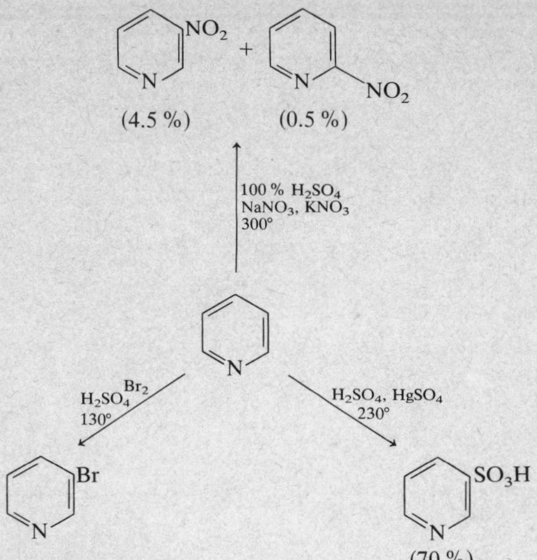

Scheme 2.4

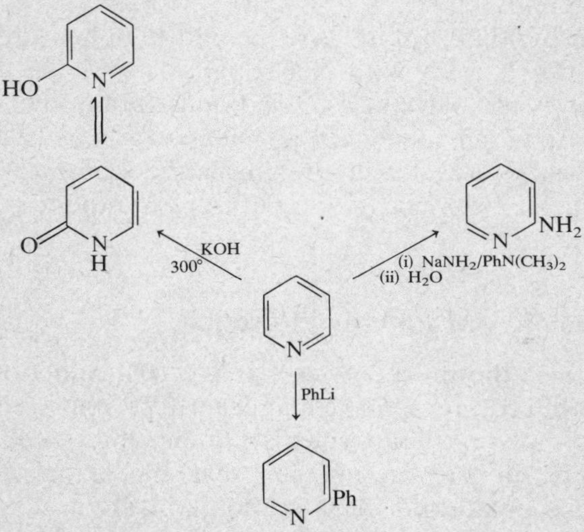

Scheme 2.5

with pyridine itself, the Friedel–Crafts reaction fails. Pyridine-*N*-oxide is converted into 2-acetoxypyridine by acetic anhydride, and on heating with bromine and acetic anhydride-sodium acetate 3,5-dibromopyridine-*N*-oxide is formed. Chlorination at the 4-position with deoxygenation can be carried out using phosphorus penta-chloride. Deoxygenation of *N*-oxides is readily carried out using, for example, phosphorus trichloride. These reactions are summarized in scheme 2.6.

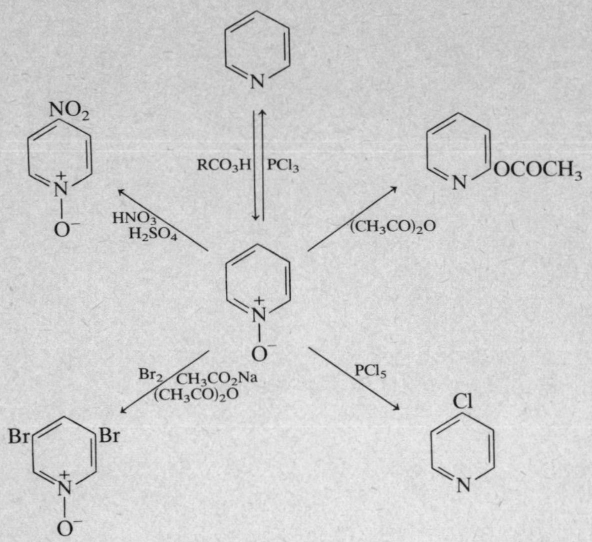

Scheme 2.6

In contrast to pyridine, furan, pyrrole, and thiophen are electron-rich molecules which react with electrophiles mainly in the 2- and 5-positions. However, under acidic conditions furan, and to a lesser extent pyrrole, are polymerised. Direct halogenation of furan, pyrrole and thiophen usually results in the formation of polyhalogenated products. Scheme 2.7 summarises some useful synthetic reactions of these compounds.

2.7 Interconversion of functional groups

As we have seen in the preceding sections, certain functional groups are readily introduced in a specific manner whilst others are not. It is now the job of the synthetic chemist to be able to interconvert functional groups in such a manner that the remainder of the molecule remains unaffected. This section will attempt to show, in outline only, how specific functional groups can be interconverted.

2.7.1 Transformation of the hydroxyl group

Alcohols are weak bases which are capable of reacting as nucleophiles. Reaction of alcohols with acid chlorides or anhydrides results in the formation of *esters*. In most cases the reaction is promoted by the addition of a tertiary base. The alkoxide ion is a stronger nucleophile which can react with alkyl halides, sulphonates and sulphates to form *ethers*. However, elimination competes with substitution in reactions involving secondary and tertiary halides.

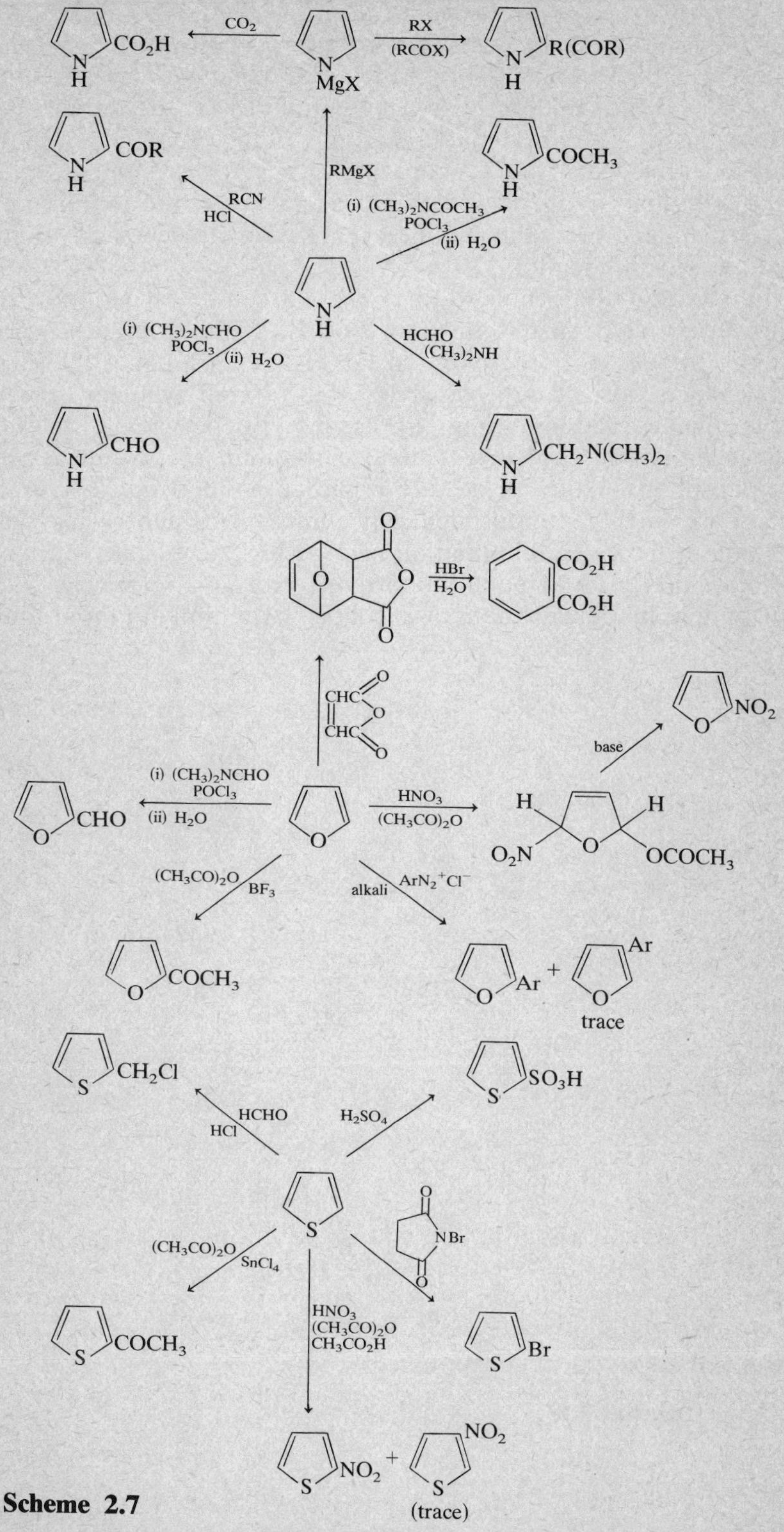

Scheme 2.7

Pyrrole reactions:

- CO_2 → pyrrole-CO_2H
- RX (RCOX) → pyrrole-R(COR)
- RMgX → pyrrole-MgX
- RCN, HCl → pyrrole-COR
- (i) $(CH_3)_2NCOCH_3$, $POCl_3$; (ii) H_2O → pyrrole-$COCH_3$
- (i) $(CH_3)_2NCHO$, $POCl_3$; (ii) H_2O → pyrrole-CHO
- HCHO, $(CH_3)_2NH$ → pyrrole-$CH_2N(CH_3)_2$

Furan reactions:

- HBr, H_2O → benzene-CO_2H, CO_2H
- base → furan-NO_2
- (i) $(CH_3)_2NCHO$, $POCl_3$; (ii) H_2O → furan-CHO
- HNO_3, $(CH_3CO)_2O$ → O_2N / furan / $OCOCH_3$
- $(CH_3CO)_2O$, BF_3 → furan-$COCH_3$
- alkali, $ArN_2^+Cl^-$ → furan-Ar + furan-Ar (trace)

Thiophene reactions:

- HCHO, HCl → thiophene-CH_2Cl
- H_2SO_4 → thiophene-SO_3H
- $(CH_3CO)_2O$, $SnCl_4$ → thiophene-$COCH_3$
- N-Br (succinimide) → thiophene-Br
- HNO_3, $(CH_3CO)_2O$, CH_3CO_2H → thiophene-NO_2 + thiophene-NO_2 (trace)

Alkyl halides may be prepared from alcohols using reagents such as thionyl chloride for chlorides, constant boiling hydrobromic acid or phosphorus tribromide for bromides, and iodine with red phosphorus for iodides. Mild conditions must be employed for the preparation of tertiary halides to prevent elimination taking place, e.g. t-butanol shaken with concentrated hydrochloric acid gives t-butyl chloride. Some additional reagents which are useful when the more common reagents induce rearrangement, racemisation or decomposition will be discussed in Chapter 12.

Dehydration of alcohols to form *alkenes* can be carried out using a wide variety of Brønsted and Lewis acids. With strong acids, acyclic alcohols appear to be dehydrated largely by an E1 mechanism, and the products derived are usually of the Saytzeff type (i.e. the most stable alkene predominating; Sykes, p. 250) perhaps with skeletal rearrangement of the intermediate carbonium ion. Some reagents, e.g. phosphorus oxychloride, are regarded as inducing dehydrations which are highly stereospecifically *trans*, consistent with an E2 mechanism. Since E1 elimination may be less stereospecific than E2, choice of reagent may be an important factor in determining product distribution in dehydration of alcohols. Attempted preparation of

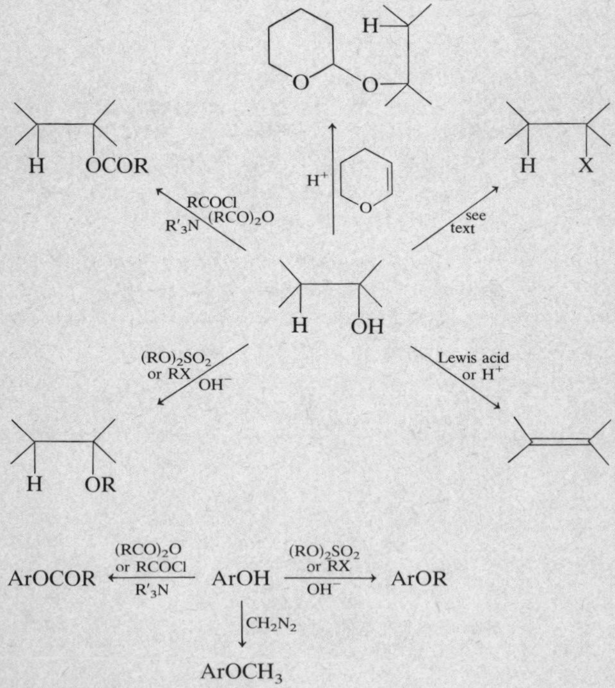

Scheme 2.8

tertiary alcohols by, for example, the Grignard reaction (cf. section 4.1.2) often results in spontaneous dehydration to the alkene.

Alcohols add to 2,3-dihydropyran under acidic conditions to give mixed *acetals* which are used to protect hydroxyl groups (cf. section 10.2.1). The reactions of alcohols with carbonyl and carboxyl groups are discussed later (cf. sections 2.7.5 and 2.7.6).

Phenols can be alkylated and acylated by ways similar to those used for alcohols. Aryl methyl ethers are often prepared by reaction of the phenol with diazomethane. The preparation of aryl halides from phenols is of little preparative significance.

The main transformations using alcohols and phenols are shown in scheme 2.8.

2.7.2 Transformation of the amino group

The amino group is basic and reacts as a nucleophile with alkyl halides, giving rise to secondary and tertiary amines and to quaternary ammonium salts. Acid chlorides and anhydrides give *amides* (section 2.7.6). The sulphonamide derived from reaction of a primary amine with a sulphonyl chloride has an acidic hydrogen which may be removed to produce a strongly nucleophilic species:

$$ArNH_2 + TsCl \longrightarrow ArNHTs \xrightarrow{NaOH} Ar\bar{N}Ts \xrightarrow{RCl} ArN(R)Ts$$

$$[Ts = p\text{-}CH_3C_6H_4SO_2^-]$$

For aliphatic amines, reaction of a primary amine with nitrous acid is of little preparative significance due to the formation of a complex mixture of products, except in cases where elimination reactions cannot take place. However, it has recently been shown that primary aliphatic amines can be transformed into a wide variety of products by converting the amino group into a better leaving group such as 2,4,6-triphenylpyridine which can be displaced by a range of nucleophiles. Examples of these transformations include the following:

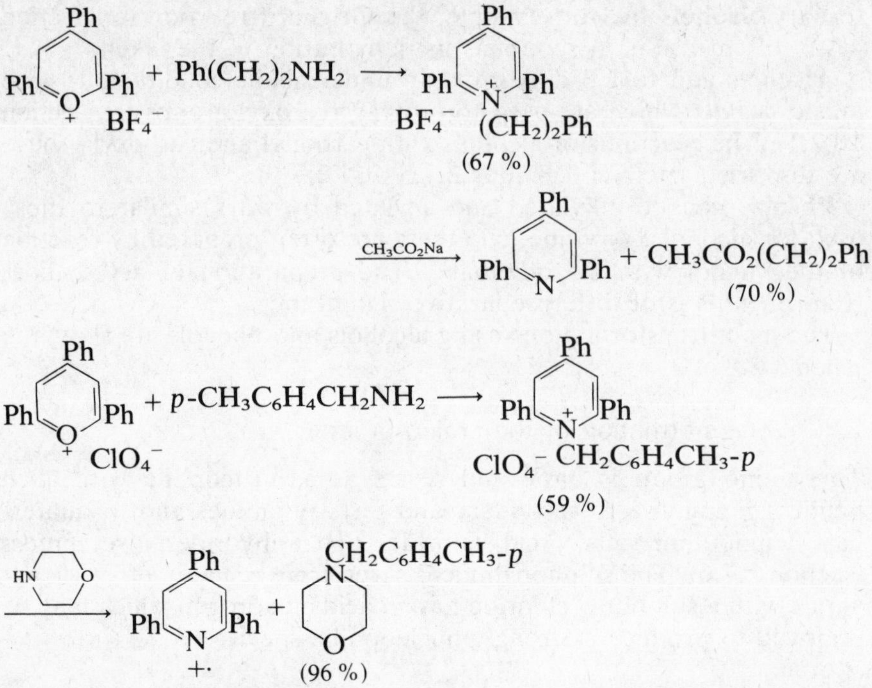

Treatment of secondary amines with nitrous acid results in the formation of N-nitroso compounds which can be reduced to *N,N*-disubstituted hydrazines. The reaction of tertiary aliphatic amines is complex and of no preparative importance.

The reaction of primary aromatic amines with nitrous acid is of considerable significance. The *diazonium salts* so formed can undergo a wide variety of transformations which are of preparative use. These together with other reactions involving the amino-group are collected in scheme 2.9. The reactions of amino groups with carbonyl compounds will be considered later (Ch. 6 and 7).

2.7.3 Transformation of halogeno-compounds

A halogen, in addition to providing a good leaving group in nucleophilic substitution reactions, withdraws electrons from the adjacent carbon atom. Hence, alkyl halides participate in a wide variety of nucleophilic substitution reactions. Reactions with alcohols and with amines have already been mentioned, and reactions with thiolate anions, cyanide ions, anions derived from acetylenes, and other carbanions are all valuable (cf. sections 3.3.1 and Ch. 4 and 5). Elimination reactions, may, however, complicate the situation, especially in the case of secondary halides, and for many tertiary halides only elimination products are obtained.

$$RNR'_2 \xrightarrow{R'Cl} R\overset{+}{N}R'_3 \; Cl^-$$

$$\Big\uparrow R'Cl$$

RNHCOR′

$$RNHR' \xrightarrow{HONO} RR'NNO$$

$$\downarrow \text{reduction}$$

$$RR'NNH_2$$

$$RNH_2 \xrightarrow[\text{(R'CO)}_2\text{O}]{\text{R'COCl or}} RNHCOR'$$

$$RNH_2 \xrightarrow{R'Cl} RNHR'$$

$$RNH_2 \xrightarrow{ArSO_2Cl} RNHSO_2Ar$$

$$RNH_2 \xrightarrow{\text{see text}} RX$$

C-nitroso compound

$$\Big\uparrow HONO$$

$$ArNR_2 \xrightarrow{RCl} Ar\overset{+}{N}R_3 \; Cl^-$$

$$\Big\uparrow RCl$$

ArNHCOR

$$ArNHR \xrightarrow{HONO} \overset{Ar}{\underset{R}{\diagdown}} NNO$$

$$\downarrow \text{reduction}$$

$$\overset{Ar}{\underset{R}{\diagdown}} NNH_2$$

$$ArNH_2 \xrightarrow[\text{(RCO)}_2\text{O}]{\text{RCOCl or}} ArNHCOR$$

$$ArNH_2 \xrightarrow{RCl} ArNHR$$

$$ArNH_2 \xrightarrow{Ar'SO_2Cl} ArNHSO_2Ar'$$

$$ArNH_2 \xrightarrow[\text{H}_2\text{SO}_4]{\text{NaNO}_2} ArN_2{}^+ HSO_4{}^-$$

PhAr

$$ArN_2{}^+ HSO_4{}^- \xrightarrow[\text{PhH}]{OH^-} PhAr$$

$$ArN_2{}^+ HSO_4{}^- \xrightarrow[\text{Cu}_2\text{(CN)}_2]{KCN} ArCN$$

$$ArN_2{}^+ HSO_4{}^- \xrightarrow[\text{OH}^-]{PhOH} ArN_2 \!\!\left\langle \bigcirc \right\rangle\!\! OH$$

$$ArN_2{}^+ HSO_4{}^- \xrightarrow[\text{aqueous solution}]{\text{boil}} ArOH$$

$$ArN_2{}^+ HSO_4{}^- \xrightarrow{H_3PO_2} ArH$$

$$ArN_2{}^+ HSO_4{}^- \xrightarrow[\text{HX}]{Cu_2X_2} ArX \quad [X = Cl,\, Br]$$

$$ArN_2{}^+ HSO_4{}^- \xrightarrow{KI} ArI$$

$$ArN_2{}^+ HSO_4{}^- \xrightarrow{NaBF_4} ArN_2{}^+ BF_4{}^-$$

$$ArN_2{}^+ BF_4{}^- \xrightarrow{NaNO_2} ArNO_2$$

$$ArN_2{}^+ BF_4{}^- \xrightarrow{\text{pyrolysis}} ArF$$

Scheme 2.9

Alkyl halides may be hydrolysed to alcohols using sodium hydroxide, but in the case of most secondary and tertiary halides elimination is a competitive reaction. Elimination is favoured in the case of a strong base reacting in a non-polar solvent at high temperature. Base catalysed elimination from secondary and tertiary halides normally obeys the Saytzeff Rule.

Alkyl halides react with certain metals to form metal alkyls. Of particular synthetic importance are alkyl-lithium derivatives and Grignard reagents, RMgX. These reagents are strong bases and their synthetic utility will be discussed in Chapter 4. A summary of the reactions of alkyl halides is given in scheme 2.10.

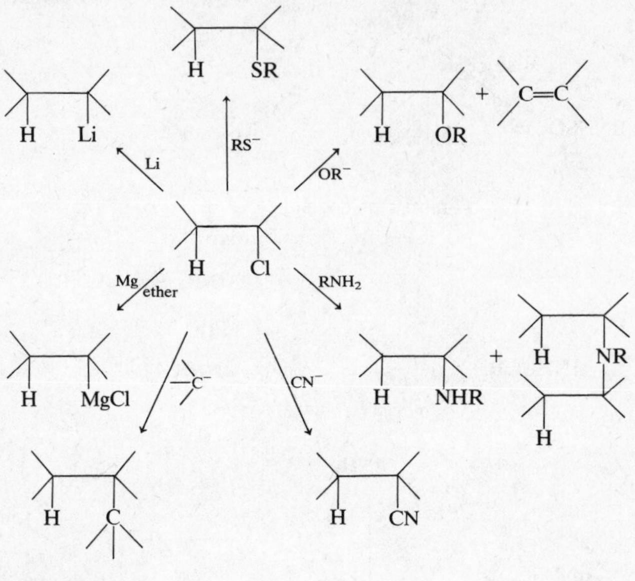

Scheme 2.10

Aryl halides are less reactive towards nucleophiles than alkyl halides, except in cases where there are sufficient electron-withdrawing substituents in positions *ortho-* and/or *para-* to the halogen. 2- and 4-halogenopyridines are also susceptible to nucleophilic attack. Aryl halides form aryl-lithiums and Grignard reagents.

2.7.4   Transformation of nitro-compounds

Aliphatic nitro-compounds are of lesser synthetic importance than are aromatic nitro-compounds. However, a stable carbanion can be formed on the carbon adjacent to the nitro-group and such carbanions can be used in many of the reactions to be described in Chapters 3 and 5.

Due to the ease of formation of aryl nitro-compounds, they are of great importance for introducing a nitrogen-containing function on to the aromatic ring. Reduction with a wide variety of reagents (e.g. Sn/HCl, Fe/HCl, Raney Ni/H$_2$, Raney Ni/N$_2$H$_4$) causes conversion to the amino group whose synthetic versatility has just been discussed. Reduction to hydroxylamines, azo compounds, and *N,N'*-disubstituted hydrazines is also possible (see scheme 2.11) depending on the reagent chosen.

$$ArN_2^+ \longleftarrow ArNH_2 \xrightarrow{H_2SO_5} ArNO$$

[Scheme 2.9]

ArNO₂ ⟷ ArNH₂ (CF₃CO₃H, see text)

$$ArNO_2 \xrightarrow[Zn]{NH_4Cl} ArNHOH \quad \xrightarrow{K_2Cr_2O_7} \quad ArNH_2$$

$$ArN=NAr \xleftarrow[NaOH]{As_2O_3}$$
$$\overset{+}{\underset{O^-}{}}$$

ArNO₂ — Zn, NaOH (aqueous), Zn NaOH/methanol → ArN=NAr

ArNHNHAr ← Zn, NaOH(aqueous)

Scheme 2.11

2.7.5 Transformation of aldehydes and ketones

Oxidation (Ch. 9) and reduction (Ch. 8) of these compounds will be dealt with later, as will their reactions with carbon nucleophiles (Ch. 3 and 5). Aldehydes and ketones react reversibly under acidic conditions with alcohols to give firstly hemi-acetals and hemi-ketals and then acetals and ketals:

$$\underset{\text{(ROH + }C=O)}{ROH + C=O} \underset{H_2O}{\overset{H^+}{\rightleftharpoons}} \underset{\text{(RO, HO, C)}}{} \underset{H_2O,-ROH}{\overset{ROH, H^+}{\rightleftharpoons}} \underset{\text{(RO, RO, C)}}{}$$

The acetals and ketals derived by reaction of aldehydes and ketones with ethylene glycol are used to protect the carbonyl group during reactions carried out under neutral or alkaline conditions. The analogous dithioketals are used in a conversion of carbonyl groups into methylene groups. The reaction requires, however, a large excess of Raney nickel:

$$C=O + \underset{CH_2SH}{\overset{CH_2SH}{|}} \xrightarrow{H^+} C\underset{S}{\overset{S}{<}}\underset{CH_2}{\overset{CH_2}{|}} \xrightarrow[nickel]{Raney} CH_2$$

2.7.6 Transformation of acids and acid derivatives

Carboxylic acids are converted by acid-catalysed reaction with alcohols into *esters*. For methyl esters another convenient method is the use of diazomethane. For more complex esters, reaction of the alcohol with the acid chloride or with the anhydride may be more satisfactory. Many of the procedures used for amide formation will also serve in esterification. Another method of ester formation is the reaction of an alkyl halide with the silver salt of the carboxylic acid.

Acid chlorides are usually prepared by reaction of the acid with thionyl chloride. They are converted into *anhydrides* by reaction with the sodium salt of the acid. Reaction of acid chlorides with diazomethane results in the formation of diazoketones which are converted by treatment with moist silver oxide into the carboxylic acid containing an additional methylene group. Reduction of acid chlorides is considered in section 8.4.4.

Amides can be prepared by reaction of ammonia or the appropriate amine with anhydrides, esters or acid chlorides. Alternative methods of amide formation, used widely in peptide syntheses, will be discussed in Chapter 14. Primary amides can be dehydrated to nitriles which can also be prepared by reaction of alkyl halides with potassium cyanide. A useful synthetic reaction of amides is their conversion into *amines* on treatment with bromine and alkali (the Hofmann reaction). Alternative procedures for converting acids and

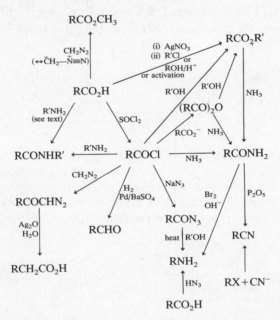

Scheme 2.12

their derivatives into amines are the thermal degradation of acid azides in alcoholic solvents (the Curtius reaction) and the treatment of carboxylic acids with hydrazoic acid (the Schmidt reaction).

The interconversions described in this section are summarised in scheme 2.12. Carbon–carbon bond forming reactions involving acid derivatives will be discussed in Chapters 3–5.

Note

1. The systematic name for trico-ordinate carbocations, R_3C^+, is **carbenium** ion. This nomenclature distinguishes such species from the pentacoordinate ions, R_5C^+, which are correctly named carbonium ions. However, the term carbenium ion is not yet widely used and in this book we, in common with Sykes, have used the traditional term carbonium ion for R_3C^+.

3 Formation of carbon–carbon bonds: the principles

It is surely obvious that an essential part of most organic syntheses is the construction of the carbon skeleton of the desired end-product. It is true that for some small molecules containing, say, up to six or seven carbon atoms, it may be possible to achieve their synthesis from readily available starting materials merely by functionalisation and/or functional group interconversions. This also holds for derivatives of simple ring systems, e.g. benzene, cyclohexane, pyridine, etc., and for molecules which are simply related to abundant naturally-occurring compounds like glucose or cholesterol or penicillins. These, however, represent exceptions to the general rule.

3.1 General strategy

The construction of the molecular framework for any given target compound is, however, not merely a matter of joining together the requisite number of carbon atoms in the right way: attention must be paid to the position of functional groups in the end-product.

For example, suppose one were asked to devise a synthesis for compound A (1), below.[1] It has a straight chain of twenty-one carbon atoms, with a Z- (or cis-) double bond between C_6 and C_7 and a ketonic carbonyl group on C_{11}.

$$\begin{array}{c}\underset{CH_3(CH_2)_4}{\overset{H}{\diagdown}}C=C\underset{\underset{(1)}{(CH_2)_3\overset{O}{\overset{\|}{C}}(CH_2)_9CH_3}}{\overset{H}{\diagup}}\end{array}$$

Compound A

How, then, does one set about such a synthesis? Straight-chain C_{21} compounds are not readily available, and so the chain has to be built up from smaller units. But which smaller units? a C_{10} and a C_{11} compound? or three C_7 compounds? or seven C_3 compounds? Does it matter?

The answer to this last question is quite clear: of course it matters. And it matters for two main reasons:

(i) Firstly, as a rule, **the fewer the number of steps in a synthesis, the better**. Few organic reactions proceed in anything approaching quantitative yield, 70–80 % being normally regarded as highly satisfactory. So even with a 70% yield at each stage, a three-stage synthesis gives an overall yield of only 34 %, and a five-stage synthesis gives only 17 %.

(ii) Secondly, it matters because of the functional groups in the end-product. If the end-product had been the C_{21} alkane (2), it might have been equally satisfactory to use a C_{10} and a C_{11} compound, or a C_{12} and a C_9 compound, or a C_{16} and a C_5 compound, or any other suitable pair. But the alkane is no good at all as an intermediate on the way to compound A, because there is no obvious way of inserting the functional groups into the alkane at the appropriate positions. So (and this is another general rule) **the necessary functionality must be built into the carbon skeleton as the latter is being assembled**.

$$CH_3(CH_2)_{19}CH_3$$

(2)

This need to build in functionality imposes severe restrictions on the number of ways in which one can construct the C_{21} chain in the synthesis of compound A. In the course of this chapter and the chapters which follow, it will become clear that the principal reactions leading to carbon-carbon bond formation are those in which either (i) both of the carbons to be joined initially bear functional groups, or (ii) one of the carbons initially bears a functional group and the other is directly adjacent to a functional group. Very often, as we shall see, the result of such a reaction is to leave a functional group in the product either at the point of joining of the fragments, or one carbon atom away from the joint.

So one now has a few clues about possible synthetic approaches to compound A. With regard to the left-hand 'end', one might consider using a C_5 compound and try to form the 5,6-bond, or a C_7 compound and try to form the 7,8-bond. One might even try to use a C_6 compound to make the double bond directly. At the other 'end', the 10,11- or 11,12-bond, flanking the functionalised carbon, ought to be the easiest to form.

Compound A may thus be 'dissected' in a number of ways, as indicated in the diagram below. Whether any of these can be developed into a practicable synthetic method, we shall see in due course.

$$CH_3(CH_2)_3CH_2 \dashv CH \doteq CH \dashv CH_2 - CH_2 - CH_2 \dashv \overset{\overset{O}{\parallel}}{C} \dashv CH_2(CH_2)_8CH_3$$
$$\quad\quad\quad 5 \quad\quad 6 \quad 7 \quad\quad 8 \quad\quad 9 \quad\quad 10 \quad\quad 11 \quad 12$$

——— C$_5$ —— +	——— C$_6$	+	—— C$_{10}$	
——— C$_5$ —— +	——— C$_5$	+	—— C$_{11}$	
——— C$_6$ ——	+ — C$_5$	+	—— C$_{10}$	Compound A
——— C$_6$ ——	+ — C$_4$	+	—— C$_{11}$	? →
——— C$_7$ ——	+ — C$_4$	+	—— C$_{10}$	
——— C$_7$ ——	+ — C$_3$	+	—— C$_{11}$	

3.2   Disconnections and synthons

In its simplest terms, a carbon–carbon bond may be defined as the sharing of a pair of electrons between the carbon atoms. There are two ways in which such a bond may be formed: in the first, each carbon atom contributes one electron to the shared pair, and in the second, one of the carbons provides both electrons for the shared pair. These possibilities may be represented schematically as follows:

$$\ce{>C^{.} \quad ^{.}C< \longrightarrow >C-C<} \tag{3.1}$$

$$\ce{>C:\quad \overset{+}{C}< \longrightarrow >C-C<} \tag{3.2}$$

The first of these processes (3.1) is, of course, a radical reaction. In the simplified form above, the combination of two radicals is shown, but other variants are possible, such as (3.3), the addition of a radical to a double bond:

$$\ce{>C^{.} \quad C=C< \longrightarrow >C-C-C<} \tag{3.3}$$

(This is, of course, an important step in the polymerisation of alkenes.) The second process is the more familiar type of laboratory reaction in which a nucleophile reacts with an electrophile. These are represented above by a carbanion and a carbonium ion respectively, but such a reaction is an extreme case, and more usual variants include the following (3.4–3.6):

$$\ce{-\overset{-}{C}:\quad \overset{\delta+}{C}\overset{\delta-}{X} \longrightarrow C-C \quad X ;} \tag{3.4}$$

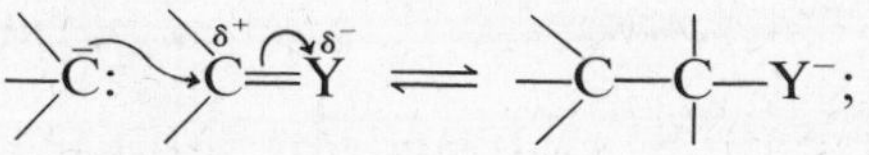

$$\overset{\backslash}{\underset{/}{C}}{:}\quad \overset{\backslash}{\underset{/}{C}}{\overset{\delta^+}{=}}\overset{\delta^-}{Y} \;\rightleftharpoons\; \overset{\backslash}{\underset{/}{C}}-\overset{|}{\underset{|}{C}}-Y^{-}; \tag{3.5}$$

$$-\overset{\backslash}{\underset{/}{C}}\overset{Z^{\delta^+}}{\overset{\delta^-}{\underset{}{}}}\;\overset{\delta^+}{\underset{/}{C}}\overset{\delta^-}{\underset{}{X}}\;\longrightarrow\; \overset{+}{Z}\quad \overset{|}{\underset{/|}{C}}-\overset{|}{\underset{|\backslash}{C}}\quad X^{-};\,\text{etc.} \tag{3.6}$$

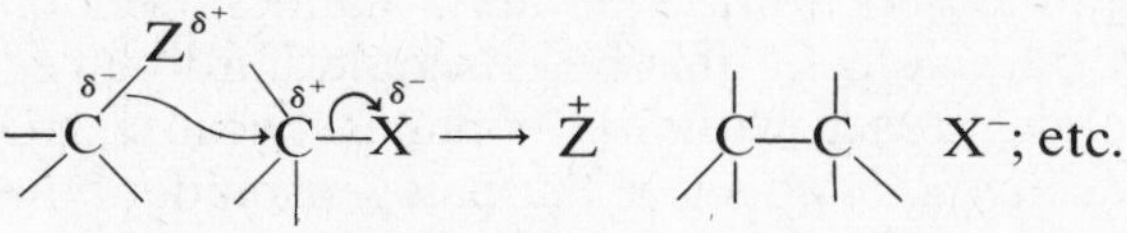

In summary, therefore:

The formation of a carbon–carbon single bond implies the interaction of two carbon radicals, or the reaction of a nucleophilic carbon species with an electrophilic carbon species.

We may put this statement into a 'shorthand' form, using the mathematical symbol ⇒ in place of the word 'implies':

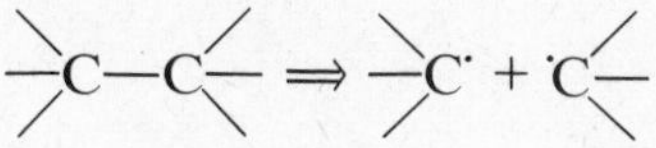

$$\overset{\backslash}{\underset{/}{C}}-\overset{/}{\underset{\backslash}{C}}{\Rightarrow}\;\overset{\backslash}{\underset{/}{C}}{\cdot}\;+\;{\cdot}\overset{/}{\underset{\backslash}{C}} \tag{3.7}$$

or

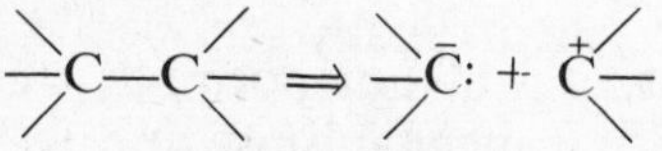

$$\overset{\backslash}{\underset{/}{C}}-\overset{/}{\underset{\backslash}{C}}{\Rightarrow}\;\overset{\backslash}{\underset{/}{\bar{C}}}{:}\;+\;\overset{+}{\underset{\backslash}{C}}\overset{/}{} \tag{3.8}$$

It is important to remember the distinction between the 'implies' symbol, ⇒, and the reaction arrow, →. The processes (3.7) and (3.8) do not represent synthetic reactions: in fact they are the precise opposites of such reactions, (3.7) being the opposite of reaction (3.1), and (3.8) being the opposite of reaction (3.2). Processes such as (3.7) and (3.8) are called **disconnections**, and the products of disconnections [e.g. the fragments on the right-hand side of (3.7) and (3.8)] are called **synthons**. The usefulness of the disconnection/synthon approach will become clearer in due course (cf. sections 4.4 and 5.5).

3.3 Electrophilic carbon species

Although there are several important radical reactions which lead to carbon–carbon bond formation (e.g. arylation: section 2.4.1) the vast majority of useful laboratory methods for joining two carbon atoms are electrophile-nucleophile interactions, as expressed in reaction (3.2) or one of its variants such as (3.4–3.6). So far in this chapter we have dealt with these processes only in very general terms; we now consider in detail the functional groups and other features in a molecule which confer electrophilic or nucleophilic character on one (or more) of its carbon atoms.

3.3.1   Alkylating agents

One of the first general reactions learned by most students of organic chemistry is the nucleophilic substitution of alkyl halides (section 2.7.3, scheme 2.10), and the reader of this book should hardly need reminding that alkyl halides react with nucleophiles because the electron-withdrawing inductive ($-I$) effect of the halogen renders the halogen-bearing carbon electron-deficient, i.e. electrophilic (cf. Sykes, pp. 21–2). The mechanistic and stereochemical aspects of these reactions, especially the distinction between the unimolecular, stepwise S_N1 process (3.9) and the bimolecular, concerted S_N2 process (3.10) should be familiar to most readers (cf. Sykes, Ch. 4).

$$\overset{\delta+}{\underset{}{C}}-\overset{\delta-}{X} \rightleftharpoons \overset{}{C^+} \quad X^- \xrightarrow{\ Nu^-\ } -C-Nu + X^- \qquad (3.9)$$

$$Nu^- \quad \overset{\delta+}{\underset{}{C}}-\overset{\delta-}{X} \rightleftharpoons \left[\overset{\delta-}{Nu} \cdots \overset{|}{C} \cdots \overset{\delta-}{X} \right] \longrightarrow Nu-\overset{|}{C} \quad X^- \qquad (3.10)$$

For the purposes of this book, however, it will not normally be necessary to distinguish between S_N1 and S_N2 mechanisms, and we shall represent the reaction of an alkyl halide with a nucleophile simply as

$$Nu^{\curvearrowright} R{-}X \longrightarrow Nu-R \quad X^-$$

irrespective of the detailed mechanism. Since the nucleophile becomes attached to the alkyl group, it is said to have been **alkylated** and the process is known as **alkylation**.

In the examples above, X has been used to represent a halogen. However, alkyl halides are not the only useful alkylating agents: as long as the C–X bond is sufficiently polarised, and as long as X^- is a stable anion which is neither a strong nucleophile nor a strong base,[2] (i.e. the anion of a strong acid) nucleophilic substitution may occur. The most common alternatives to alkyl halides are the alkyl esters of sulphonic acids, especially methanesulphonates ('mesylates', 3) and toluene-*p*-sulphonates ('tosylates', 4). For methylations, dimethyl sulphate (5) is also frequently used, and one of the most powerful methylating agents available is methyl fluorosulphonate ('magic methyl', 6), although it has the disadvantage of being extremely toxic.

$R-OSO_2CH_3$ $R-OSO_2\!\!\overset{\text{(ring)}}{\bigcirc}\!\!CH_3$ $CH_3-OSO_2O-CH_3$ CH_3-OSO_2F

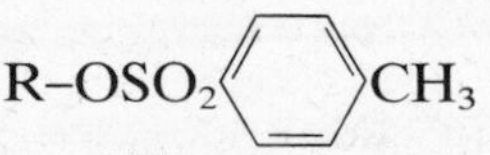

(R–OMs) (R–OTs) (5) (6)

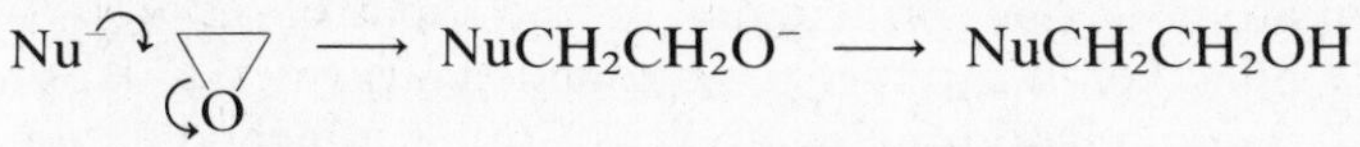

(3) (4)

$Nu \overset{\curvearrowright}{\underset{\underset{O}{\overset{\triangledown}{}}}{}} \longrightarrow NuCH_2CH_2O^- \longrightarrow NuCH_2CH_2OH$ (3.11)

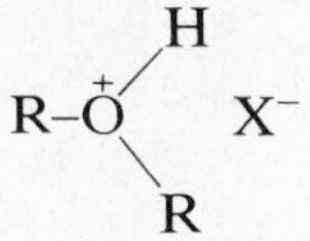

(7)

$R-\overset{+}{O}H_2 \;\; X^-$ $R-\overset{\overset{H}{|}}{\underset{\underset{R}{|}}{\overset{+}{O}}} \;\; X^-$ $R-N_2^+ \;\; X^-$ $R-\overset{\overset{R}{|}}{\underset{\underset{R}{|}}{\overset{+}{O}}} \;\; BF_4^-$

(8) (9) (10) (11)

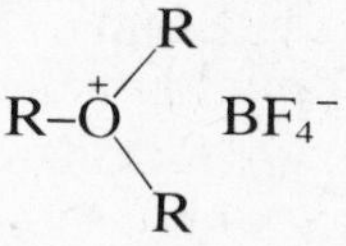

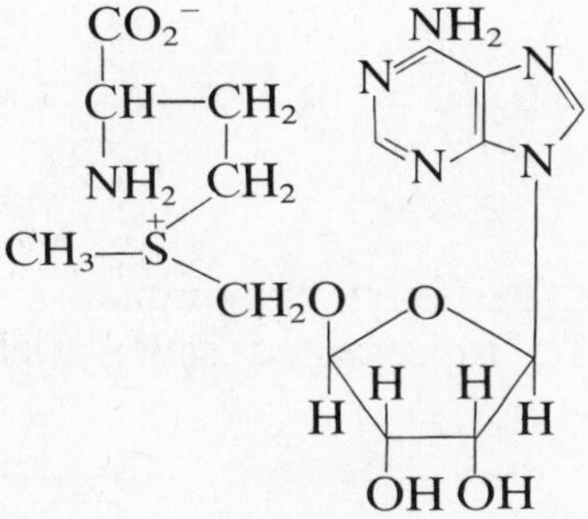

(12)

Carboxylic esters (R–OCOR′) are not effective alkylating agents because reactions with nucleophiles occur preferentially at the car-bonyl group (cf. section 3.3.2). Alcohols and ethers (and their sulphur-containing counterparts) are also ineffective alkylating agents, because -OH, -OR, -SH, and -SR are poor **leaving groups** (i.e. they do not meet the criteria of the previous paragraph). Oxirans (epoxides), e.g. (7), can function as alkylating agents despite having no good leaving group, because the reaction with the nucleophile involves ring-opening and hence relief of strain (3.11).

Protonated alcohols (8) and ethers (9), and also diazonium salts (10), are possible sources of carbonium ions, and should thus be potential alkylating agents, but their formation normally requires acidic conditions under which carbon nucleophiles would be proto-nated and hence be rendered inactive. However, trialkyloxonium salts (11), e.g. trimethyl- and triethyloxonium fluoroborates, are powerful alkylating agents, and it is of interest to note that a trialkylsulphonium compound, S-adenosylmethionine (12), is one of Nature's principal methylating agents.

3.3.2 Carbonyl compounds

The reader should already be familiar with the electrophilic nature of carbonyl compounds, and the consequent reactivity of such molecules towards nucleophiles (cf. Sykes, Ch. 8). The electrophilicity of the carbonyl carbon atom is due principally to the electron-accepting **mesomeric** (-M) effect of the oxygen, although it also depends on the electron-donating or -withdrawing ability of other attached atoms or groups. The order of reactivity in carbonyl compounds is as follows:

$$R-\overset{\overset{\displaystyle O}{\|}}{C}-\overset{+}{N}R_3 \ \ X^- > R-\overset{\overset{\displaystyle O}{\|}}{C}-Cl > R-\overset{\overset{\displaystyle O}{\|}}{C}-OCOR > R-\overset{\overset{\displaystyle O}{\|}}{C}-H >$$

$$R-\overset{\overset{\displaystyle O}{\|}}{C}-alkyl > R-\overset{\overset{\displaystyle O}{\|}}{C}-aryl > R-\overset{\overset{\displaystyle O}{\|}}{C}-OR' > R-\overset{\overset{\displaystyle O}{\|}}{C}-NR'_2 > R-\overset{\overset{\displaystyle O}{\|}}{C}-O^-.$$

The general reaction of carbonyl compounds, $R-\overset{\overset{\displaystyle O}{\|}}{C}-X$, with carbon nucleophiles may be represented schematically as follows (3.12):

$$\tag{3.12}$$

The initially formed anion (13) may then undergo further reaction in one of three ways.

(i) If X is a leaving group (i.e. forms a stable anion), it may be eliminated as X^-. The net result (3.13) is substitution of the group X by the carbon nucleophile; the nucleophile becomes attached to an acyl group, and is thus said to have undergone **acylation**.

$$\tag{3.13}$$

(ii) If X is not a leaving group, the anion (13) is likely to pick up a proton from the reaction medium, either immediately (if the reaction is conducted in a protic solvent) or during the isolation procedure (the 'work-up'). In such cases the net result is **addition** to the carbonyl group (3.14).

$$\underset{(13)}{\overset{O^-}{\underset{R}{\overset{|}{>}}}C-C\overset{\diagup}{\underset{\diagdown X}{}}} \xrightarrow[\substack{\text{or } H^+ \text{ (in} \\ \text{work-up)}}]{H-Y \text{ (solvent)}} \underset{(14)}{\overset{OH}{\underset{R}{\overset{|}{>}}}C-C\overset{\diagup}{\underset{\diagdown X}{}}} \tag{3.14}$$

(iii) If X is not a leaving group, and the adduct (14) also contains an acidic hydrogen adjacent to the hydroxyl group, elimination of water may follow the nucleophilic addition (3.15). This addition-elimination sequence is called a **condensation** reaction.[3]

$$(13) \longrightarrow \underset{\underset{(14)}{H \quad R \ X}}{\overset{OH}{\overset{|}{-}C-C\overset{|}{\diagdown}}} \xrightarrow{-H_2O} \overset{X}{>}C=C\overset{\diagup}{\underset{\diagdown R}{}} \tag{3.15}$$

All of these reactions will be considered in more detail in Chapters 4, 5 and 7.

3.3.3 Electrophilic carbon–nitrogen reagents

One would expect compounds containing carbon-nitrogen double bonds to resemble carbonyl compounds in their reactions with nucleophiles. Imino-compounds do indeed react with nucleophiles (3.16) in accord with this expectation, although such reactions are in general much less useful than the corresponding processes involving carbonyl compounds. On the other hand, if the nitrogen is positively charged the carbon becomes highly electrophilic, and nucleophilic addition to iminium salts (3.17) is the key to important synthetic procedures such as the Mannich and Vilsmeier–Haack–Arnold reactions (sections 5.4.3 and 5.4.2).

$$\underset{R \quad X}{>}C\overset{R'N}{\underset{C}{\overset{\|}{\diagdown}}} \longrightarrow \underset{R}{>}C-\overset{R'N^-}{\underset{|}{C}}-X \longrightarrow \underset{R}{>}C-\overset{R'NH}{\underset{|}{C}}-X, \text{ etc.} \tag{3.16}$$

$$\underset{R \quad X}{>}C\overset{\overset{+}{N}R'_2}{\underset{C}{\overset{\|}{\diagdown}}} \longrightarrow \underset{R \quad X}{>}C-\overset{NR'_2}{\underset{|}{C}} \tag{3.17}$$

Cyano-compounds also behave as electrophiles (3.18), nucleophilic addition to such compounds giving anions of imines. The imines themselves are frequently not stable, and undergo hydrolysis to carbonyl compounds (cf. Sykes, pp. 238–9):

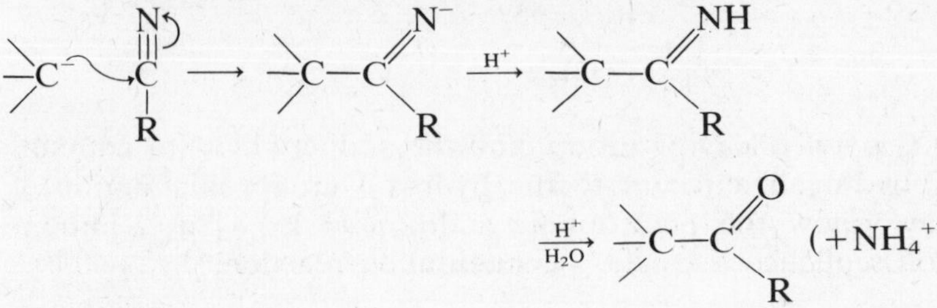

$$\xrightarrow[H_2O]{H^+} \quad \text{(+NH}_4^+) \quad (3.18)$$

3.3.4 Electrophilic alkenes

One does not expect an alkene, which is an electron-rich species, to function as an electrophile: indeed, one is accustomed to alkenes being *nucleophilic* and reacting with electrophiles. However, if the carbon atom, one position removed from the double bond, is electrophilic, then nucleophilic attack may occur not only at the electrophilic carbon but at the 'far' end of the double bond as in (3.19) and (3.20) (cf. Sykes, pp. 108–9 and 195–9).

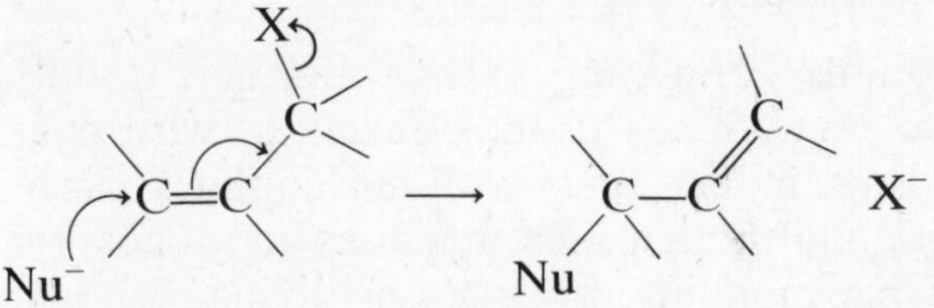

$$(3.19)$$

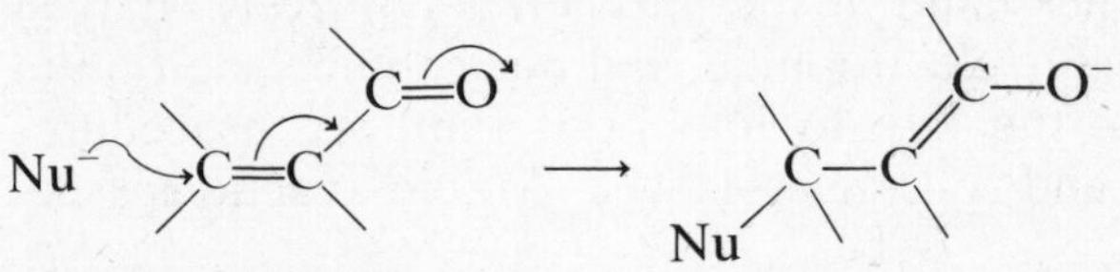

$$(3.20)$$

3.3.5 Carbenes

These neutral, electron-deficient species $\left(:C\underset{Y}{\overset{X}{\diagdown}}\right)$ which are highly reactive electrophiles (cf. Sykes, pp. 259–61), are of synthetic interest principally for their reactions with alkenes (section 7.2.3) and with electron-rich aromatic molecules (section 5.4.2).

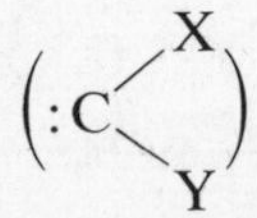

3.4 Nucleophilic carbon species

In our original schematic representation of the electrophile-nucleophile interaction [section 3.2: reaction (3.2)], the nucleophile was represented by a carbanion, and so we now consider the molecular features which promote the formation of these and related nucleophilic species.

3.4.1 Grignard and related organometallic reagents

Most readers will already be familiar with Grignard reagents, RMgX, where R is an alkyl (or aryl) group and X is a halogen (usually bromine or iodine). These are undoubtedly the most widely used of the organometallic reagents in nucleophile-electrophile reactions: they are simply made from alkyl (or even aryl) halides and magnesium in a dry ether solvent, and are stable in this solution, although they are rapidly decomposed by oxygen and by water and other protic solvents (cf. below). The exact structure of Grignard reagents and the exact mechanisms by which they react with electrophiles, are matters of some dispute (cf. Sykes, pp. 217–18), but these need not concern us here, since we are primarily concerned in this book with the *products* of their reactions. For synthetic purposes, they may be regarded as having the structure RMgX, which is polarised $\overset{\delta^-}{R}-\overset{\delta^+}{MgX}$ or even $\overset{-}{R}\overset{+}{MgX}$: they behave as carbanions and are adequately represented by the synthon R^-. They suffer from the disadvantage, however, of being very strong bases, and will abstract even feebly acidic protons, e.g. from water, alcohols, or even primary and secondary amines:

$$R - MgX + H - Y \rightarrow R - H + YMgX \ (Y = OH, OR', NHR', NR'_2)$$

The corresponding organozinc reagents are very seldom used nowadays: they are less reactive than Grignard reagents, and are allegedly more difficult to handle. The lower reactivity, however, is utilised in the **Reformatsky reaction** (section 4.2.2).

Dialkylcadmium reagents (R_2Cd), which are preparable from Grignard reagents and cadmium chloride, are also less reactive than Grignard reagents (the metal is less electropositive than magnesium), but are sometimes used by virtue of their greater selectivity towards electrophiles. Alkyllithium and aryllithium compounds, on the other hand, are more reactive, and even less selective, than the corresponding Grignard reagents.

One group of organometallic reagents which may function as useful carbon nucleophiles and which exhibit an unusual degree of selectivity are organocopper reagents. These are of two types, both derived from copper(I) halides and alkyllithium compounds (3.21–3.22).

$$2RLi + Cu_2X_2 \rightarrow 2RCu + 2LiX \tag{3.21}$$

$$4RLi + Cu_2I_2 \rightarrow 2R_2CuLi + 2LiI \tag{3.22}$$

Of these, the first (the simple alkylcopper) is sparingly soluble in organic solvents unless it is complexed using ligands such as trialkyl-phosphines, but the second (called a **lithium dialkylcuprate**) is soluble in ethers and is hence of immediate use as a reagent.

The synthetic uses of all these organometallic reagents are described in Chapter 4.

3.4.2   Stabilised carbanions

The vast majority of stabilised carbanions are produced by heterolysis of a C–H bond (3.23):

$$\overset{\diagdown}{\underset{\diagup}{C}}\!\!-\!\!H \;\rightleftharpoons\; \overset{\diagdown}{\underset{\diagup}{C}}{}^{-} \; \overset{+}{H} \tag{3.23}$$

In most cases, however, this type of ionisation does not occur spontaneously to any significant extent, since it is rare for a hydrogen bonded to carbon to be strongly acidic. A base is therefore used in order to promote the heterolysis (3.24–3.25):

$$\overset{\diagdown}{\underset{\diagup}{C}}\!\!-\!\!H + \bar{B} \;\rightleftharpoons\; \overset{\diagdown}{\underset{\diagup}{C}}{}^{-} + H\!\!-\!\!B \tag{3.24}$$

$$\overset{\diagdown}{\underset{\diagup}{C}}\!\!-\!\!H + \ddot{B} \;\rightleftharpoons\; \overset{\diagdown}{\underset{\diagup}{C}}{}^{-} + H\!\!-\!\!\overset{+}{B} \tag{3.25}$$

It is important to remember, of course, that reactions (3.24) and (3.25) are equilibria, since carbanions are themselves basic and can recapture protons. It follows, therefore, that if deprotonation of the $\geqslant$C–H compound is to be complete (or effectively complete), B^- or $\ddot{B}$ must be a much stronger base than the carbanion. Or, to put it another way, the more strongly acidic the $\geqslant$C–H compound, the weaker the base required for complete deprotonation. Also it must be remembered that **for complete deprotonation a molar equivalent of base is required**. These points may at first sight appear trivial, but they are in fact very important in determining the outcome of some carbanion reactions (sections 5.1 and 5.2).

The structural features which enhance the acidity of $\geqslant$C–H compounds and which stabilise carbanions are described by Sykes (pp. 265–9), and so only a summary is given here.

(i) The best stabilisation of the carbanion is achieved when the anionic centre is adjacent to an electron-accepting (-M) group, such as carbonyl, cyano, nitro, or sulphonyl. Stabilisation results from the delocalisation of the negative charge:

$$\overset{\displaystyle O}{\underset{\displaystyle}{>\!\!\bar{C}\!-\!\overset{\|}{C}\!<}} \longleftrightarrow \overset{\displaystyle O^-}{\underset{\displaystyle}{>\!\!C\!=\!\overset{|}{C}\!<}} \; ; \quad \overset{\displaystyle O}{\underset{\displaystyle O^-}{>\!\!\bar{C}\!-\!\overset{+}{N}\!\!<}} \longleftrightarrow \overset{\displaystyle O^-}{\underset{\displaystyle O^-}{>\!\!C\!=\!\overset{+}{N}\!\!<}} \quad , \text{etc.}$$

Of these -M groups, the order of stabilising effect is $NO_2 > CO > SO_2 \simeq CN$; among carbonyl groups, the order is as expected, e.g. aldehyde > ketone > ester. If the anionic centre is flanked by two -M groups, additional delocalisation of the charge is possible, the stability of the anion is considerably increased, and its basicity is correspondingly decreased; if three -M groups flank the anion, it is scarcely basic at all.

(ii) When the anionic centre is adjacent to an inductively electron-withdrawing (-I) group, stabilisation of the carbanion results, although not surprisingly this type of stabilisation is less effective than that involving a -M group. Two or more -I substituents make a moderately stable carbanion, e.g. $^-CF_3$ or $\bar{C}H(SR)_2$, as do -I groups carrying a positive charge, e.g.

$$>\!\!\overset{-}{\underset{}{C}}\!-\!\overset{+}{P}R_3 \quad \text{or} \quad >\!\!\overset{-}{\underset{}{C}}\!-\!\overset{+}{\underset{}{S}}R_2 \quad \text{or} \quad >\!\!\overset{-}{\underset{}{C}}\!-\!\overset{+}{\underset{\displaystyle\underset{\|}{O^-}}{S}}\!-\!R.$$

For this type of stabilisation, it is noteworthy that (fluorine apart) the most effective stabilising groups are those in which the atom next to the anionic centre is one from the second row of the periodic table (in particular phosphorus or sulphur). These atoms have unoccupied $3d$ orbitals which, in principle, may overlap with the $2p$ orbital of carbon which contains the lone pair and hence exert a type of mesomeric stabilisation of the negative charge. The extent of such overlap, and hence the degree of mesomeric stabilisation of such carbanions, is a matter of current debate, but such arguments need not concern us in this book, where reaction products take precedence over the finer points of mechanism.

(iii) When the anionic centre resides on a triply bonded carbon atom, a degree of stabilisation is conferred on the carbanion. Alk-1-ynes, although by no means strong acids, are nevertheless much stronger acids than alkanes, and are thus deprotonated easily by alkyl carbanions (Grignard and similar reagents) and also by amide ion:

$$R'C\!\equiv\!\overset{\frown}{C}\!-\!H \; \overset{\frown}{\,}\bar{N}H_2 \longrightarrow R'C\!\equiv\!C^- + NH_3$$

The alkynyl carbanion is stabilised, relative to an alkyl carbanion, by virtue of the high s character of the orbital containing the unshared electron pair. Hydrogen cyanide is considerably more acidic than alkynes, and cyanide ion is much more stabilised than alkynyl ions: this enhanced stability presumably results from the polarisation of the π-bond system, which depletes the carbon of electrons, and thus reduces the availability of the lone pair for bonding.

(iv) A carbanion is greatly stabilised if the lone pair of electrons which is responsible for the negative charge forms part of an aromatic system. This is a relatively uncommon situation, but it explains, for example, the high stability of the anion derived from cyclopentadiene:

(a 6π-electron system)

3.4.3 Alkenes, arenes, and heteroarenes

The student of organic chemistry learns at a very early stage that alkenes react with electrophiles (by an addition process); that arenes (benzene, naphthalene, etc.) do likewise (by addition-elimination); and, at a later stage perhaps, that heteroaromatic compounds (furan, thiophen, pyridine, indole, etc.) react more or less in the same way as arenes. So alkenes, arenes, and heteroarenes must be considered as nucleophilic carbon species.

The principal reactions of these classes of compound are summarised in Chapter 2 (sections 2.2 and 2.4–2.6), and it must be obvious that among these there are very few which involve carbon–carbon bond formation. For alkenes, none at at all are listed. For benzene and its heterocyclic analogues, only the Friedel–Crafts reaction has wide generality, and even that does not apply to ring systems with reduced nucleophilicity, such as nitrobenzene and pyridine.

The reader who has studied the chemistry of simple benzene derivatives will recall that an electron-donating (+M) substituent greatly enhances the reactivity of a benzene ring towards electrophiles (cf. Sykes, pp. 152–4). The effect is pronounced in arylamines and phenols, and even more so in phenoxide ions; reaction occurs o- and p- to the substituent, e.g.

, etc. (E⁺ = an electrophile)

Reaction may also occur, of course, at the substituent and not on the ring:

$$R_2\ddot{N}{-}\bigcirc \xrightarrow{\ E^+\ } R_2\overset{+}{N}{-}\bigcirc \;;\quad \overset{O^-}{\bigcirc} \xrightarrow{\ E^+\ } \overset{O{-}E}{\bigcirc}$$

The reactions of phenols and arylamines with electrophiles include several useful carbon–carbon bond-forming reactions (cf. sections 5.4.2 and 5.4.3).

Electron-donating substituents have a similar activating effect on simple alkenes. So **enols** (15), **enolate** ions (16), and **enamines** (17) react with electrophiles as follows (although, as above, reaction may also occur at oxygen or nitrogen):

$$E^+\!\!\curvearrowright\!\! \underset{(15)}{C{=}C\overset{\displaystyle \ddot{O}H}{}} \longrightarrow E{-}\overset{|}{C}{-}\overset{\displaystyle \overset{+}{O}H}{C} \tag{3.26}$$

$$\Big\downarrow {-H^+}$$

$$E^+\!\!\curvearrowright\!\! \underset{(16)}{C{=}C\overset{\displaystyle O^-}{}} \longrightarrow E{-}\overset{|}{C}{-}\overset{\displaystyle O}{C} \tag{3.27}$$

$$E^+\!\!\curvearrowright\!\! \underset{(17)}{C{=}C\overset{\displaystyle \ddot{N}R_2}{}} \longrightarrow E{-}\overset{|}{C}{-}\overset{\displaystyle \overset{+}{N}R_2}{C} \tag{3.28}$$

These are three very important reactions, as will be seen later (in Ch. 5). Reaction (3.27) should already be familiar, because (16) is nothing more than the alternative canonical form for the carbanion

$$\overset{|}{C}{-}C\overset{\displaystyle O}{}$$. Compound (15) is tautomeric with a carbonyl com-

pound, and (3.26) should perhaps be rewritten as:

$$H{-}\overset{|}{C}{-}C\overset{\displaystyle O}{} \rightleftharpoons \;C{=}C\overset{\displaystyle OH}{} \xrightarrow{\ E^+\ } E{-}\overset{|}{C}{-}C\overset{\displaystyle \overset{+}{O}H}{} \left[\xrightarrow{\ -H^+\ } E{-}\overset{|}{C}{-}C\overset{\displaystyle O}{}\right]$$

In this expanded form, the importance of the reaction becomes clear: it shows that **enolisable carbonyl compounds can react as nucleophiles even in the absence of a (carbanion-forming) base.**

A similar expansion of reaction (3.28) shows that the enamine reaction is no more than a variant of (3.26), at least from the synthetic viewpoint: it permits electrophilic substitution at the α-position of an enolisable carbonyl compound using a base no stronger than a secondary amine:

$$H-\underset{|}{\overset{|}{C}}-C{\overset{O}{\diagup}} \xrightarrow[\underset{(trace)}{H^+}]{R_2NH} H-\underset{|}{\overset{|}{C}}-\underset{\diagdown}{\overset{NR_2}{C}}-OH \xrightarrow[-H_2O]{H^+} \underset{\diagup}{\overset{\diagdown}{C}}=\underset{\diagdown}{\overset{NR_2}{C}}$$

$$\xrightarrow{E^+} E-\underset{|}{\overset{|}{C}}-\underset{\diagdown}{\overset{\overset{+}{N}R_2}{C}} \xrightarrow{H_2O} E-\underset{|}{\overset{|}{C}}-C{\overset{O}{\diagup}} + HNR_2$$

The reactions of enols, enolates, and enamines will be considered further in Chapter 5.

Notes

1.   Compound A is an example of a class of compounds known as *insect pheromones*. This particular compound (Z-heneicos-6-en-11-one) is produced by the female of the Douglas fir tussock moth in order to attract a mate, and the synthetic material is thus of value as a lure to trap male moths and hence limit the breeding process of what is regarded by the forester as a pest.

2.   For the distinction between nucleophilicity and basicity, see Sykes, p. 95.

3.   The term *condensation* is one for which few present day text-books of organic chemistry seem prepared to offer a definition. In common practice the term is used loosely to describe reactions of various kinds: thus, for example, the Claisen ester 'condensation' and Dieckmann 'condensation' are in reality acylations of esters (sections 5.2.2 and 7.1.1); the aldol 'condensation', the benzoin 'condensation', and the Michael 'condensation' are three different types of addition reaction (sections 5.2.4.1; 5.6.2, Note 10; and 5.1.5) leading to three very different types of product; and the Claisen–Schmidt and Knoevenagel–Doebner 'condensations' are addition-eliminations of the general type (3.15). For the purposes of this book, where product types are the main concern, we have chosen to avoid possible confusion by using the term *condensation* to refer only to reactions of this last type. Such reactions will be considered further in sections 5.1.4; 5.2.4; and 7.1.

4   Formation of carbon–carbon bonds: reactions of organometallic compounds

So far we have dealt with carbon–carbon bond-forming reactions only in general terms. We now turn to consider some of the most important of these processes, and we begin in this chapter with what, on paper at least, are the simplest: the reactions involving organometallic species.

4.1   Grignard reagents and electrophiles

As already noted (section 3.4.1) a Grignard reagent, RMgX, in ethereal solution (usually diethyl ether or tetrahydrofuran) acts as a source of unstabilised carbanions, i.e. R^-. R is commonly alkyl or aryl, but may also be alkenyl or alkynyl (see section 4.1.5).

4.1.1   Alkylation

Grignard reagents undergo alkylation to give alkanes:

$$XMg\!-\!\overset{\curvearrowright}{R} \quad R'\!-\!\overset{\curvearrowleft}{Y} \longrightarrow R\!-\!R' + MgXY \qquad (4.1)$$

Thus, for example,

$$CH_3(CH_2)_{11}MgI + CH_3(CH_2)_{11}I \longrightarrow CH_3(CH_2)_{22}CH_3 \quad (63\,\%)$$

$$PhCH_2MgCl + CH_3(CH_2)_3OTs \longrightarrow Ph(CH_2)_4CH_3 \qquad (61\,\%)$$

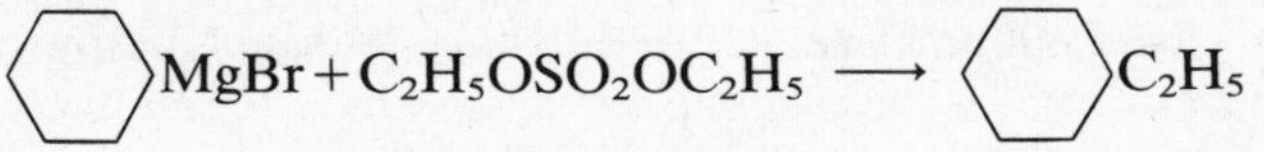

$$\text{MgBr} + C_2H_5OSO_2OC_2H_5 \longrightarrow \text{C}_2H_5 \qquad (80\,\%)$$

$$PhMgBr + \text{ClCH}_2\text{—Cl} \longrightarrow PhCH_2\text{—Cl} \qquad (88\,\%)$$

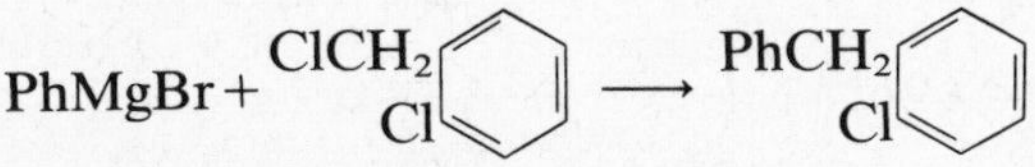

[Ts = toluene-*p*-sulphonyl]

Although good yields are obtained in the above cases, these are exceptions to the general rule: alkylations of Grignard reagents frequently proceed in very low yield (especially when the leaving group Y is a halogen), because of the intervention of side-reactions (e.g. elimination of hydrogen halide, or redox processes giving radicals). One generally useful alkylation, however, is the reaction with oxiran (ethylene oxide):

$$XMg{-}R \quad \overset{}{\underset{}{\bigtriangleup}}O \longrightarrow R{-}CH_2CH_2{-}O^- \overset{+}{M}gX \overset{H^+}{\longrightarrow} RCH_2CH_2OH \quad (4.2)$$

For example,

$$PhMgCl + \overset{}{\underset{}{\bigtriangleup}}O \overset{THE}{\longrightarrow} PhCH_2CH_2OH \quad (88\%)$$

4.1.2 Reactions with carbonyl compounds

These are by far the most useful reactions of Grignard compounds. With aldehydes and ketones, the reaction is addition: formaldehyde is converted into primary alcohols, other aldehydes into secondary alcohols, and ketones into tertiary alcohols:

$$XMg{-}R \quad \overset{R^1}{\underset{R^2}{{>}C{=}O}} \longrightarrow \overset{R^1}{\underset{R^2}{R{-}C{-}O^- \overset{+}{M}gX}} \overset{H^+}{\longrightarrow} \overset{R^1}{\underset{R^2}{R{-}C{-}OH}} \quad (4.3)$$

Thus,

$$\bigcirc\!\!\!\!\rangle MgCl + CH_2O \longrightarrow \bigcirc\!\!\!\!\rangle CH_2OH \quad (65\%)$$

$$(CH_3)_2CHMgBr + CH_3CHO \longrightarrow (CH_3)_2CH\overset{OH}{\underset{|}{C}}HCH_3 \quad (52\%)$$

$$CH_3(CH_2)_3MgBr + CH_3COCH_3 \longrightarrow CH_3(CH_2)_3\overset{OH}{\underset{|}{C}}(CH_3)_2 \quad (90\%)$$

With acyl halides, anhydrides, and esters, the first stage of the reaction is acylation, giving a ketone. This may then react with a second molecule of the Grignard reagent, the final product being a tertiary alcohol:

$$XMg{-}R \quad \overset{R^1}{\underset{Y}{{>}C{=}O}} \longrightarrow \overset{R^1}{\underset{Y}{R{-}C{-}O^- \overset{+}{M}gX}} \longrightarrow \overset{R^1}{\underset{R}{{>}C{=}O}}$$

$$\overset{RMgX}{\longrightarrow} \overset{R^1}{\underset{R}{R{-}C{-}OH}} \quad (4.4)$$

Of these reactions, those with esters are the most reliable and give the best yields:

$$2CH_3MgI + PhCOCl$$

$$\begin{array}{c} OH \\ | \\ PhC(CH_3)_2 \quad (40\text{–}50\%) \end{array}$$

$$2CH_3MgI + (PhCO)_2O$$

$$2CH_3MgI + CH_3(CH_2)_2CO_2C_2H_5 \longrightarrow CH_3(CH_2)_2\overset{\displaystyle OH}{\underset{\displaystyle |}{C}}(CH_3)_2 \quad (88\%)$$

$$3CH_3CH_2MgBr + CO(OC_2H_5)_2 \longrightarrow \left[CH_3CH_2\overset{\displaystyle \bar{O}\overset{+}{M}gBr}{\underset{\displaystyle OC_2H_5}{C}}OC_2H_5 \right.$$

$$\longrightarrow CH_3CH_2\overset{\displaystyle O}{\overset{\|}{C}}OC_2H_5 \longrightarrow CH_3CH_2\overset{\displaystyle \bar{O}\overset{+}{M}gBr}{\underset{\displaystyle OC_2H_5}{-C-}}CH_2CH_3$$

$$\longrightarrow CH_3CH_2\overset{\displaystyle O}{\overset{\|}{C}}CH_2CH_3 \longrightarrow \left. (CH_3CH_2)_3COH \quad (86\%) \right]$$

Formate esters, of course, give secondary alcohols [$R^1 = H$ in reaction (4.4)], e.g.

$$CH_3(CH_2)_3MgBr + HCO_2C_2H_5 \longrightarrow [CH_3(CH_2)_3CHO]$$

$$\longrightarrow CH_3(CH_2)_3\overset{\displaystyle OH}{\underset{\displaystyle |}{C}}H(CH_2)_3CH_3 \quad (83\%)$$

It is seldom possible to stop these reactions at the half-way stage, but carbonyl compounds have occasionally been isolated under special conditions, e.g. when the acyl compound is used in excess, when the reaction temperature is low, or when the product is sterically hindered:

$$CH_3(CH_2)_3MgCl + (CH_3CO)_2O \text{ (excess)} \xrightarrow{-80°} CH_3(CH_2)_3COCH_3 \quad (80\%)$$

$$(CH_3)_3CCH_2MgCl + (CH_3)_3CCOCl \longrightarrow (CH_3)_3CCH_2\overset{\displaystyle O}{\overset{\|}{-C-}}C(CH_3)_3$$

$$(87\%)$$

With carbon dioxide, Grignard reagents undergo carboxylation, giving carboxylate ions:

$$XMg{\frown}R \quad \overset{O}{\underset{O}{\overset{\|}{C}}} \longrightarrow R-\overset{O}{\overset{\|}{C}}\diagdown_{O^-\overset{+}{M}gX} \quad \xrightarrow{H^+} \quad R-\overset{O}{\overset{\|}{C}}\diagdown_{OH} \qquad (4.5)$$

eg.

(70 %)

The reader may well wonder why, in this case, the carbonyl compound does not apparently react with a second molecule of the Grignard reagent. Carboxylate ions, contrary to popular belief, are not entirely unreactive towards Grignard reagents. Indeed, formate ions react very readily, giving aldehydes, e.g.

$$C_2H_5MgBr + HCO_2H \xrightarrow[0°]{THF} HCO_2^-\overset{+}{M}gBr + C_2H_6$$

$$CH_3(CH_2)_5MgBr + HCO_2^-\overset{+}{M}gBr \xrightarrow[\substack{0°-20°,\\40\ min}]{THF}$$

$$CH_3(CH_2)_5\overset{\overset{\displaystyle \bar{O}\overset{+}{M}gBr}{|}}{\underset{\underset{\displaystyle H}{|}}{C}}{-}\bar{O}\overset{+}{M}gBr \xrightarrow[H_2O]{H^+} CH_3(CH_2)_5CHO \quad (75\ \%)$$

and other carboxylates also react, although more slowly, e.g.

$$(CH_3)_2CHCH_2MgBr + CH_3CO_2^-Na^+$$

$$\xrightarrow[\substack{room\ temperature,\\24\ hours}]{ether} (CH_3)_2CHCH_2COCH_3 \quad (ca.\ 25\ \%)$$

Carboxylation, however, is normally carried out using a large excess of carbon dioxide, and is much faster than addition to the carboxylate ion in any case; thus the latter is not normally an important side-reaction.

With primary and secondary amides, as with carboxylic acids, the principal action of the Grignard reagent is to remove the (acidic) proton from the nitrogen or oxygen:

$$XMg{\frown}R \quad {\frown}H{\frown}O_2C-R' \longrightarrow RH + R'CO_2^-\overset{+}{M}gX$$

$$XMg{\frown}R \quad {\frown}H{\frown}NHCOR' \longrightarrow RH + R'CO\bar{N}H\ \overset{+}{M}gX$$

The reaction with tertiary amides, however, constitutes an interesting synthesis of carbonyl compounds:

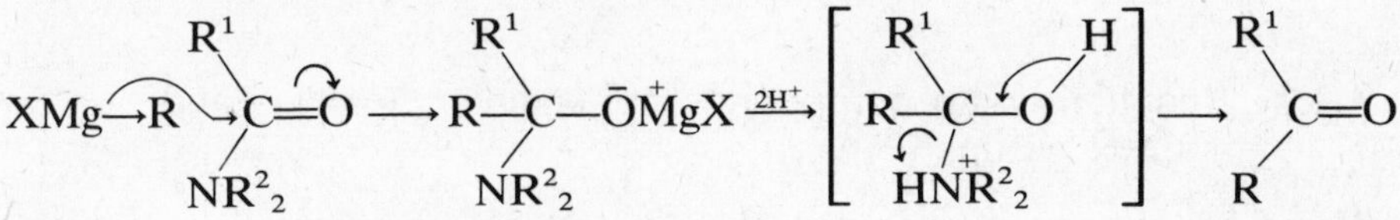

$$
\underset{NR^2_2}{XMg{\to}R{\to}\underset{}{C}{=}O} \longrightarrow \underset{NR^2_2}{R{-}\underset{}{C}{-}\bar{O}\overset{+}{M}gX} \xrightarrow{2H^+} \left[\underset{H\overset{+}{N}R^2_2}{R{-}\underset{}{C}{-}O} \right] \longrightarrow \underset{R}{\overset{R^1}{C}{=}O}
$$

$$
(1) \qquad\qquad\qquad\qquad\qquad\qquad (4.6)
$$

[In this reaction the initial adduct (1) does not collapse directly to the carbonyl compound because NR^2_2 is a very poor leaving group. $\overset{+}{N}HR^2_2$, however, produced by protonation during the work-up, is by contrast an excellent leaving group.]

Examples of this include:

$$CH_3(CH_2)_3C{\equiv}CMgBr + HCON(CH_3)_2 \longrightarrow CH_3(CH_2)_3C{\equiv}CCHO$$

$$(51\%)$$

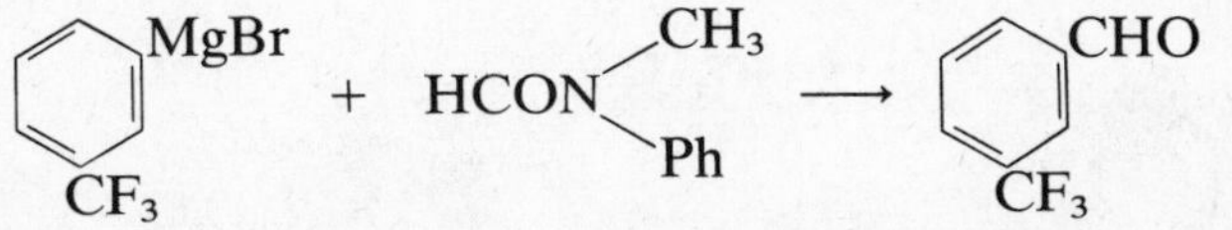

$$(90\%)$$

The above has not been developed extensively as a route to aldehydes, despite the ready availability of reagents such as dimethyl-formamide. The other classical Grignard synthesis of aldehydes, using orthoformate esters (trialkoxymethanes), is equally simple and often gives better yields:

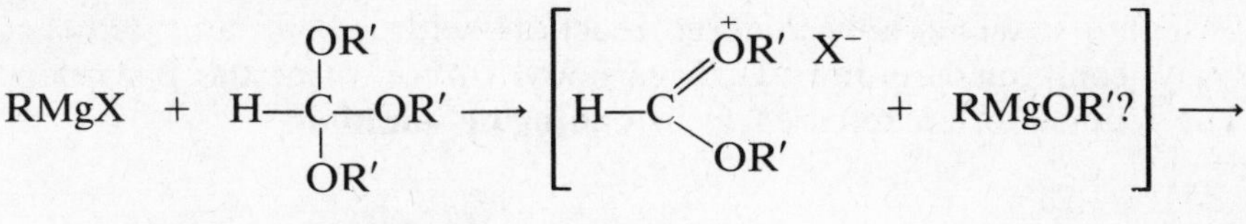

$$
RMgX + H{-}\underset{OR'}{\overset{OR'}{C}}{-}OR' \longrightarrow \left[H{-}\underset{OR'}{\overset{\overset{+}{O}R'\ X^-}{C}} + RMgOR'? \right] \longrightarrow
$$

$$(R' = CH_3\ \text{or}\ C_2H_5)$$

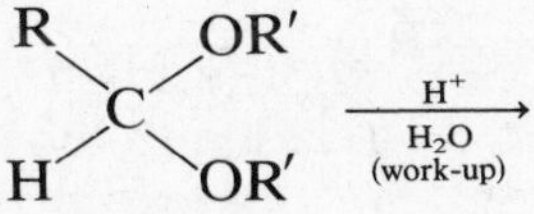

$$
\underset{H}{\overset{R}{C}}\underset{OR'}{\overset{OR'}{}} \xrightarrow[\substack{H_2O \\ (\text{work-up})}]{H^+} RCHO \qquad\qquad (4.7)
$$

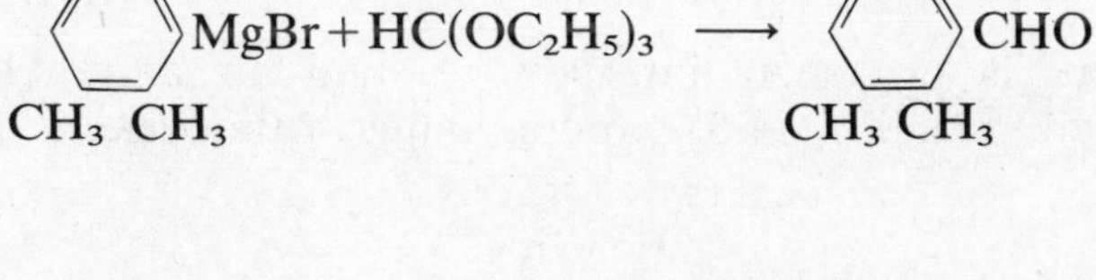

e.g.

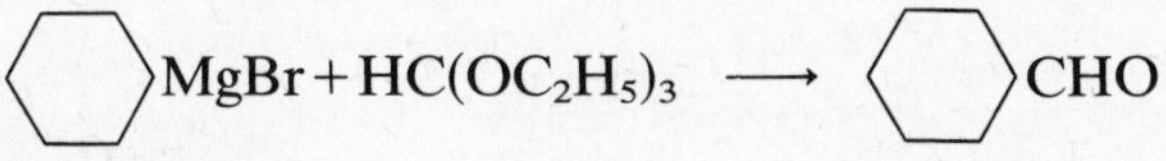

The recent method using formate ions (cf. p. 42) also provides a simple alternative.

4.1.3 Reactions with compounds containing $\diagup\!\!\!C{=}N{-}$ and $-C{\equiv}N$ groups

These follow the expected pathways, as the undernoted examples show. The reaction with cyano-compounds constitutes a useful general route to ketones:

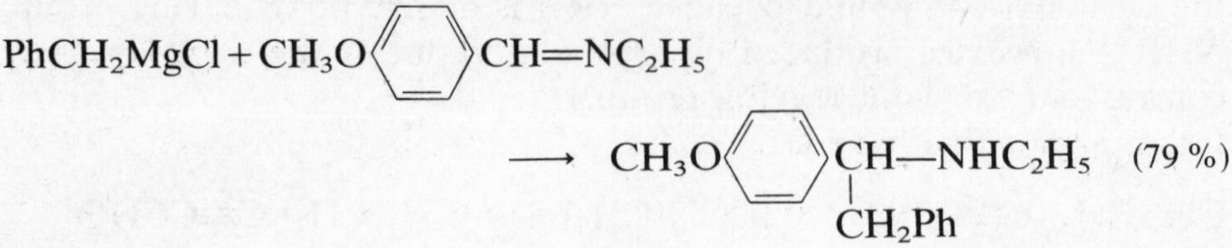

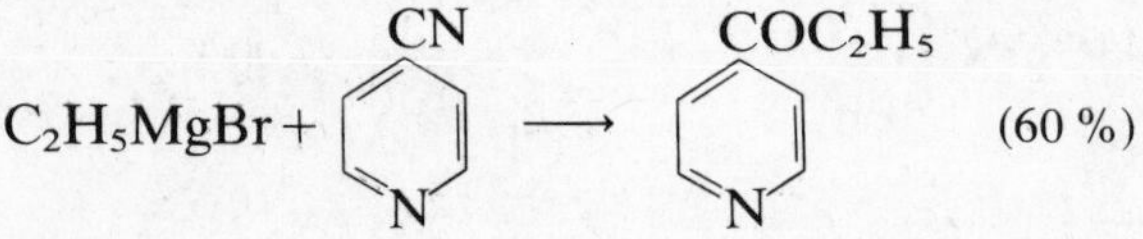

$PhCH_2MgCl + CH_3O\!\!\diagtext\!\!CH{=}NC_2H_5$

$$\longrightarrow \quad CH_3O\!\!\diagtext\!\!\underset{\underset{CH_2Ph}{|}}{CH}{-}NHC_2H_5 \quad (79\,\%)$$

$C_2H_5MgBr + \underset{N}{\overset{CN}{\bigcirc}} \longrightarrow \underset{N}{\overset{COC_2H_5}{\bigcirc}} \quad (60\,\%)$

4.1.4 Reactions with $\diagup\!\!\!C{=}\overset{|}{C}{-}\overset{|}{C}{=}O$ and related systems

Grignard reagents may undergo reactions with α,β-unsaturated carbonyl compounds either at the carbonyl carbon or at the β-carbon. The latter is often referred to as **conjugate addition**.

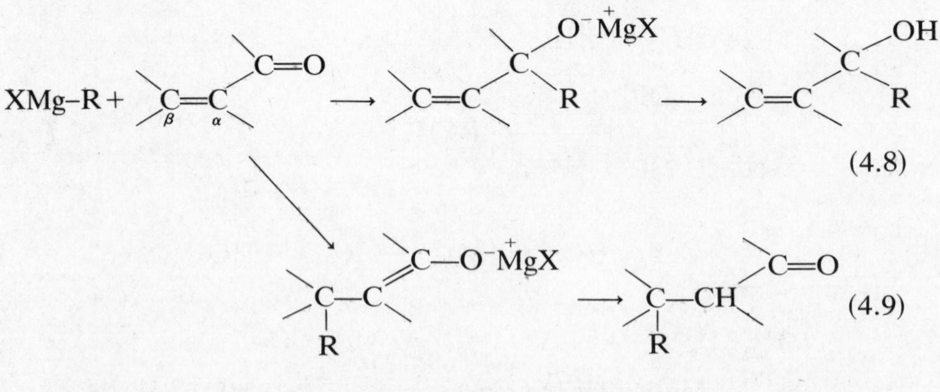

$$(4.8)$$

$$(4.9)$$

Where there is no steric interference, and an alkyl Grignard reagent is used, reaction (4.8) predominates; otherwise, conjugate addition may also occur, and this is not therefore a reliable general method:

$$CH_3MgCl + CH_3CH{=}CHCHO \longrightarrow CH_3CH{=}CHCHCH_3 \quad (81\%)$$
$$\overset{\displaystyle OH}{\phantom{CH_3CH{=}CHCH}|}$$

PhMgBr + [cyclohexene with COPh] $\longrightarrow$ [cyclohexane with COPh and Ph] (quantitative)

PhMgBr + [cyclohexene with CO$_2$CH$_3$] $\longrightarrow$ do. (82 %)

$$\left[\text{but PhMgBr} + \text{[cyclohexene with CN]} \longrightarrow \text{[cyclohexene with COPh]} \quad \text{only} \quad (40\%) \right]$$

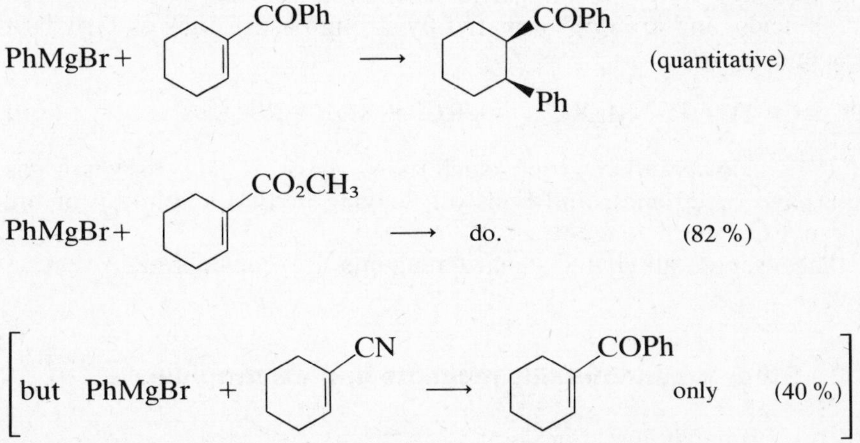

Reaction of Grignard reagents with 'π-deficient' heterocycles such as pyridine, quinoline and isoquinoline generally occur at a position adjacent to the heteroatom:

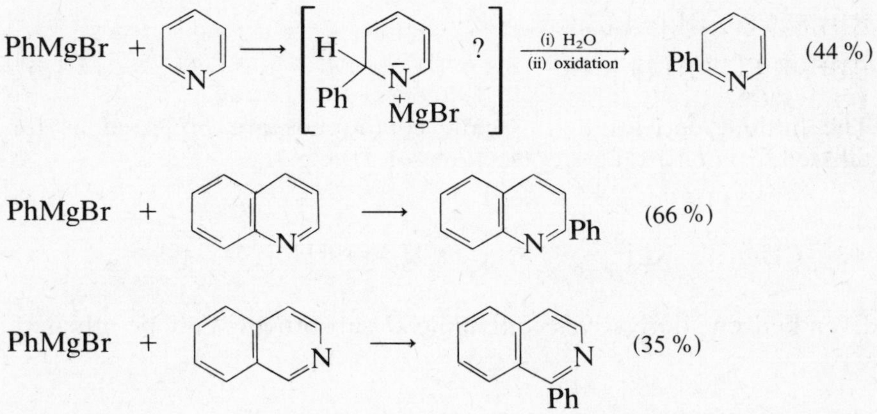

PhMgBr + [pyridine] $\longrightarrow$ $\left[\text{[dihydropyridine intermediate with H, Ph, N}^-\text{, MgBr}^+\text{]} \ ? \right]$ $\xrightarrow[\text{(ii) oxidation}]{\text{(i) H}_2\text{O}}$ Ph[pyridine] (44 %)

PhMgBr + [quinoline] $\longrightarrow$ [quinoline with Ph at 2-position] (66 %)

PhMgBr + [isoquinoline] $\longrightarrow$ [isoquinoline with Ph] (35 %)

4.1.5 Alkenyl and alkynyl Grignard reagents

Halogenoalkenes, in which the halogen is attached directly to one of the doubly bonded carbons, are unreactive halides (cf. Sykes, p. 85), and they form Grignard reagents only with difficulty. Longer times and higher temperatures are required than for alkyl and aryl halides: e.g. tetrahydrofuran (b.p. 65°) is normally used as solvent in place of diethyl ether, and several hours may be necessary for complete reaction with magnesium.

Alkynyl Grignard reagents ($RC \equiv CMgX$), however, are relatively easy to prepare, and are not prepared from halogenoalkynes at all. As we have noted in an earlier section (3.4.2, iii), alk-1-ynes are weak acids, and are deprotonated by strong bases – such as Grignard reagents:

$$RC \equiv C-H \quad R'-MgX \quad \longrightarrow \quad RC \equiv CMgX + R'H \qquad (4.10)$$

If R' is a lower alkyl group, such as $-CH_3$ or $-C_2H_5$, $R'H$ is a gas (methane or ethane), and boils off, leaving behind a solution of the alkynyl Grignard reagent.

The uses of alkynyl Grignard reagents are considered in section 4.3.1.

4.2 Other organometallic reagents and electrophiles

4.2.1   Organolithium reagents

These are usually prepared either from the appropriate halide and metallic lithium, or (especially for the less reactive halides) by halogen–metal exchange, i.e. reaction of the halide with a pre-formed alkyl-lithium:

$$RBr + 2Li \rightarrow RLi + LiBr \qquad (4.11)$$

$$RBr + R'Li \rightarrow RLi + R'Br \qquad (4.12)[1]$$

The lithium derivatives of acidic compounds are prepared as for alkynyl Grignard reagents (section 4.1.5), e.g.

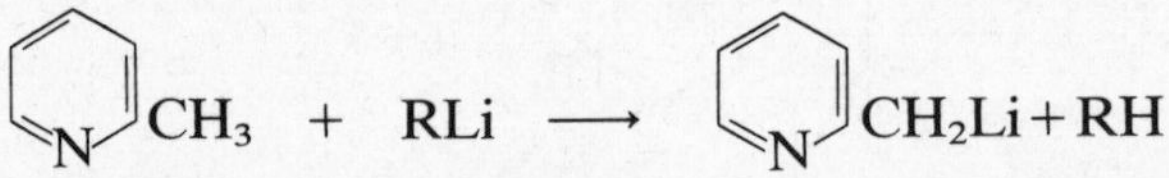

Even benzene derivatives containing -I substituents may be lithiated, e.g.

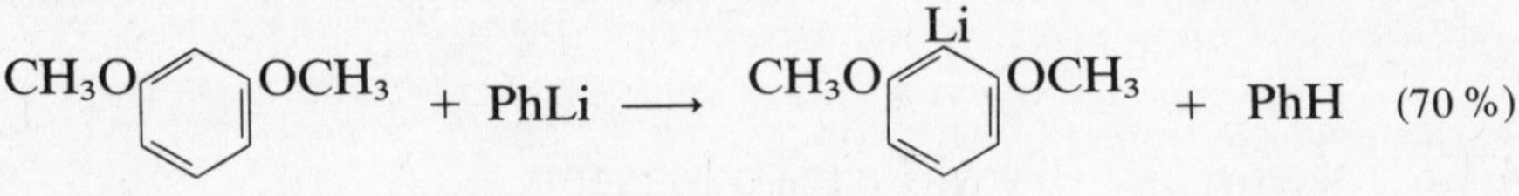

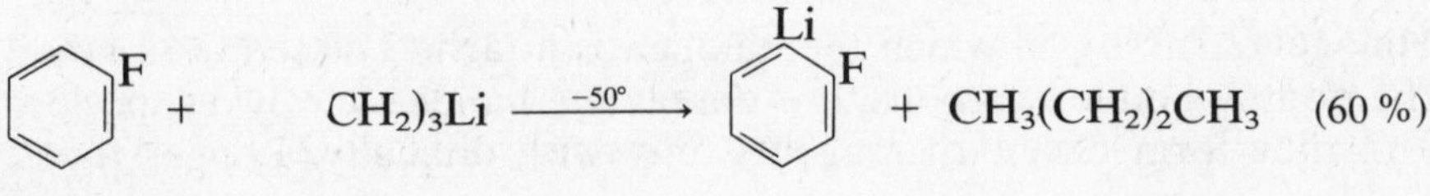

[The o-halogenophenyl–lithiums decompose at higher temperatures giving benzyne (cf. section 7.2.1)].

Organolithium reagents are more strongly nucleophilic than the corresponding Grignard reagents. They undergo all the addition reactions of the latter, in certain cases more efficiently: for example, the conversion of carboxylate ions into ketones, the conversion of tertiary amides into aldehydes and ketones, and addition to $>C=N-$ bonds:

$$PhLi + PhCO_2^-Li^+ \xrightarrow[\text{5–6 h}]{\text{ether}} PhCOPh \quad (70\,\%)$$

$$CH_3Li + PhCH=CHCO_2^-Li^+ \xrightarrow[\text{2 h}]{\text{ether}} PhCH=CHCOCH_3 \quad (69\,\%)$$

$$CH_3(CH_2)_9Li + HCON(CH_3)_2 \longrightarrow CH_3(CH_2)_9CHO \quad (60\,\%)$$

$$CH_3(CH_2)_9Li + CH_3CON(CH_3)_2 \longrightarrow CH_3(CH_2)_9COCH_3 \quad (88\,\%)$$

$$CH_3(CH_2)_2Li + Ph_2C=NPh \longrightarrow \underset{\underset{(CH_2)_2CH_3}{|}}{Ph_2C}-NHPh \quad (77\,\%)$$

The second of the above examples illustrates the preference for addition at the carbonyl group rather than conjugate addition to the $>C=C-C=O$ system. This preference is more marked than in the case of Grignard reagents, probably because the organolithium compound is less bulky than the Grignard reagent and its reactions are thus less subject to steric hindrance.

4.2.2 Organozinc and organocadmium reagents

Neither of these classes of compound is widely used nowadays, except for a few specific reactions. Both classes are less reactive nucleophiles than the corresponding Grignard reagents, and thus are more selective reagents than the latter.

(i) α-Bromoesters react with aldehydes or ketones and metallic zinc to give β-hydroxyesters. Whether this (the **Reformatsky** reaction) is the zinc analogue of a Grignard reaction, with (2) as the nucleophilic species, or whether it is better regarded as the addition of an enolate ion (3) to the carbonyl group, is an interesting mechanistic point but is not strictly relevant in the synthetic context. In practical terms, it is a 'one-pot' reaction, in which the zinc is normally added directly to a mixture of the other reactants. There is no need for the organometallic intermediate to be preformed, as in the case of Grignard reagents:

$$\underset{|}{\overset{R}{\underset{BrCHCO_2R^1}{}}} + Zn \longrightarrow \left[\underset{(2)}{\overset{R}{\underset{BrZnCHCO_2R^1}{}}} \quad or \quad \underset{(3)}{\overset{R}{\underset{CH=C}{}}} \overset{O^-\overset{+}{Z}nBr}{\underset{OR^1}{}} \right]$$

$$\xrightarrow{R^2COR^3} \quad \underset{\overset{|}{R^3} \ \overset{|}{R}}{\overset{O^-\overset{+}{Z}nBr}{\overset{|}{R^2-\overset{|}{C}-CHCO_2R^1}}} \quad \xrightarrow[\text{(work-up)}]{H^+} \quad \underset{\overset{|}{R^3} \ \overset{|}{R}}{\overset{OH}{R^2-\overset{|}{C}-CHCO_2R^1}} \quad (4.13)$$

e.g. $CH_3CHO + \underset{\overset{|}{}}{(CH_3)_2\overset{Br}{\overset{|}{C}}CO_2C_2H_5} \xrightarrow{Zn}$

$$\underset{}{\overset{OH}{CH_3\overset{|}{C}HC(CH_3)_2CO_2C_2H_5}} \quad (70\%)$$

$$PhCOCH_3 + BrCH_2CO_2C_2H_5 \xrightarrow{Zn} \underset{CH_3}{\overset{OH}{Ph\overset{|}{\underset{|}{C}}CH_2CO_2C_2H_5}} \quad (92\%)$$

α,β-Unsaturated carbonyl compounds usually (although not invariably) undergo reaction at the carbonyl group in preference to conjugate addition.

(ii) Organocadmium reagents are used especially for the conversion of acyl chlorides into ketones:

$$[2RMgX + CdCl_2 \longrightarrow] \ R_2Cd + 2R'COCl \longrightarrow 2RCOR' \quad (4.14)$$

e.g. $[CH_3(CH_2)_3]_2Cd + ClCH_2COCl \longrightarrow CH_3(CH_2)_3COCH_2Cl \quad (51\%)$

$[(CH_3)_2CHCH_2CH_2]_2Cd + ClCO(CH_2)_2CO_2CH_3$

$$\longrightarrow (CH_3)_2CH(CH_2)_2CO(CH_2)_2CO_2CH_3 \quad (73\%)$$

These reactions demonstrate the selectivity of the dialkylcadmium reagents: the acyl chloride reacts in each case in preference to the alkyl halide, ester, or ketone.[2]

4.2.3   Organocopper(I) reagents

The reaction of an organolithium compound with a copper(I) halide gives organocopper species, which, depending on the proportions of reagents used, correspond to the empirical formulae RCu and R_2CuLi (cf. section 3.4.1).[3] (As in the case of the other organometallic species, the exact structures are not relevant to this book.) These reagents are formally nucleophilic, and like the other organometallic derivatives in this chapter may be represented by the synthon R⁻; but their selectivity towards electrophiles is of such a remarkable nature that it must be doubtful if their reactions are simple electrophile-nucleophile interactions at all. In most cases the mechanisms have not yet been established beyond doubt, but complex formation and one-electron redox processes are such common features of the chemistry of copper derivatives that their involvement cannot be ignored.

Whatever the mechanisms, the points of synthetic importance are these:

(i) Displacement of halogens is particularly facile, even when the halogen is in a position normally considered 'unreactive' towards nucleophiles:

$$CH_3Cu + CH_3(CH_2)_9I \longrightarrow CH_3(CH_2)_9CH_3 \quad (68\%)$$

$$(CH_3)_2CuLi + \quad \longrightarrow \quad (81\%)$$

$$(CH_2{=}C-)_2CuLi + PhBr \longrightarrow \quad (85\%)$$

$$+ CH_3O\bigcirc I \longrightarrow CH_3O\bigcirc\bigcirc \quad (55\%)$$

These reactions are most successful with the lithium dialkyl- (or dialkenyl-)cuprates and aryl copper compounds. Displacement of other leaving groups (e.g. toluene-p-sulphonate) and the ring-opening of oxirans also occur with lithium dialkylcuprates, although these reactions have been less thoroughly investigated to date.

(ii) As expected from (i), displacement of halogen from an acyl halide occurs very easily (often at −78°), but acyl halides are the only class of carbonyl compound to show appreciable reactivity towards organocopper reagents. Thus the reaction with acyl halides does not proceed beyond the ketone stage, and other carbonyl groups in the molecule are unaffected, e.g.

$$[CH_3(CH_2)_3]_2CuLi + CH_3(CH_2)_4COCl$$

$$\xrightarrow[15\ min]{-78°} CH_3(CH_2)_4CO(CH_2)_3CH_3 \quad (79\,\%)$$

$$(CH_3)_2CuLi + CH_3(CH_2)_4CO(CH_2)_4COCl$$

$$\xrightarrow[15\ min]{-78°} CH_3(CH_2)_4CO(CH_2)_4COCH_3 \quad (95\,\%)$$

$$(CH_3)_2CuLi + I(CH_2)_{10}COCl \xrightarrow[15\ min]{-78°} I(CH_2)_{10}COCH_3 \quad (91\,\%)$$
(excess)

The unreactivity of ketonic carbonyl groups is also illustrated in the following example:

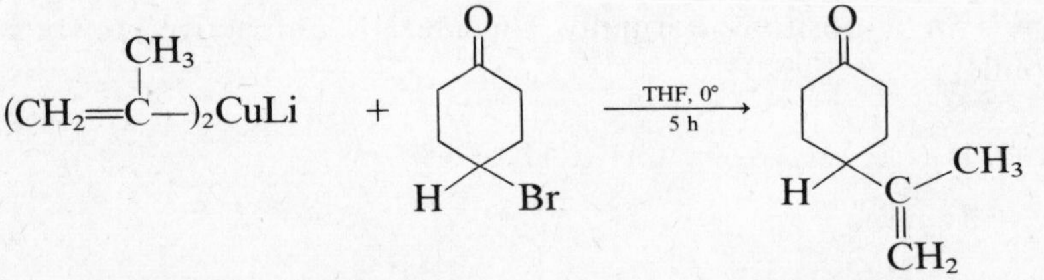

(65 %)

(iii) Although they do not react readily with carbonyl groups (or, perhaps, *because* they do not react readily), organocopper reagents, especially the lithium dialkylcuprates, react with α,β-unsaturated carbonyl compounds to give, almost invariably, the products of conjugate addition, e.g.

$$CH_3Cu + (E)\text{-}CH_3CH{=}CHCOCH_3 \rightarrow (CH_3)_2CHCH_2COCH_3 \quad (85\,\%)$$

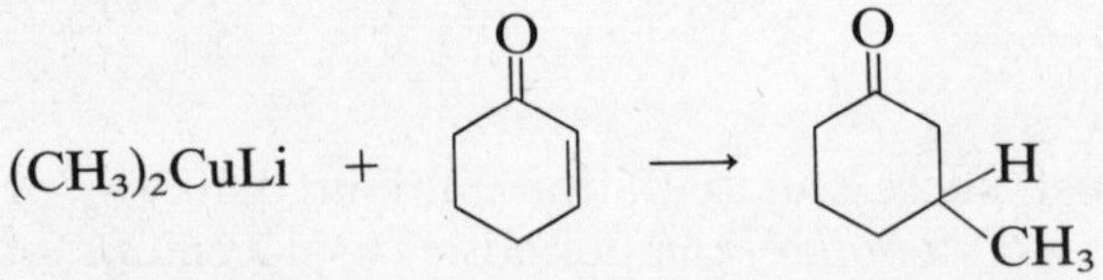

(97 %)

The reaction of Grignard reagents with α,β-unsaturated carbonyl compounds may occur at either of the two electrophilic centres, as already noted (section 4.1.4), but in presence of a small proportion (<10 mole per cent) of a copper(I) salt the addition is almost exclusively at the β-carbon. Organocopper species are the presumed intermediates in such reactions.

(iv) Coupling reactions occur when organocopper(I) reagents are heated (sometimes they even occur at room temperature) and when lithium dialkylcuprates are exposed to oxidising agents (including atmospheric oxygen). These coupling processes are most simply rationalised in terms of a one-electron transfer followed by radical coupling:

$$R\text{—}Cu^{I}[\rightleftharpoons R^{-} + (Cu^{I})^{+}] \rightarrow R^{\cdot} + Cu^{0};\ R^{\cdot} + R^{\cdot} \rightarrow R\text{—}R \qquad (4.15)$$

$$\underset{\displaystyle \quad | \atop \quad R}{R\text{—}(Cu^{I})^{-}} \xrightarrow{\text{oxidation}} R\text{—}(Cu^{II})\text{—}R[\rightleftharpoons R^{-} + (\overset{+}{C}u^{II})R] \rightarrow R^{\cdot} + Cu^{I}\text{—}R;$$

$$R^{\cdot} + R^{\cdot} \rightarrow R\text{—}R \qquad (4.16)$$

[It is, however, by no means certain that 'free' radicals are actually produced in such reactions.]

Examples of this coupling include:

(84 %)

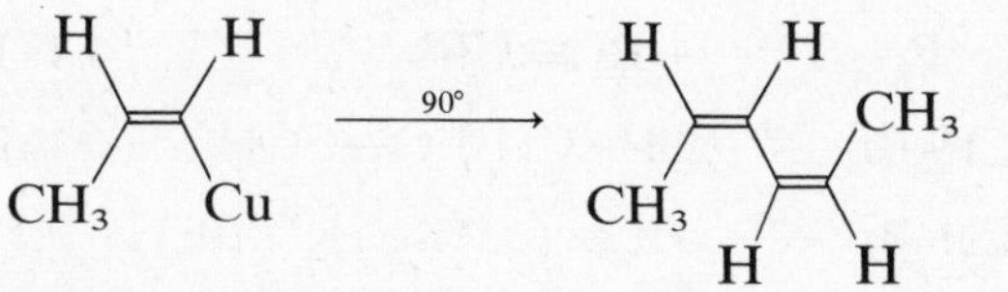

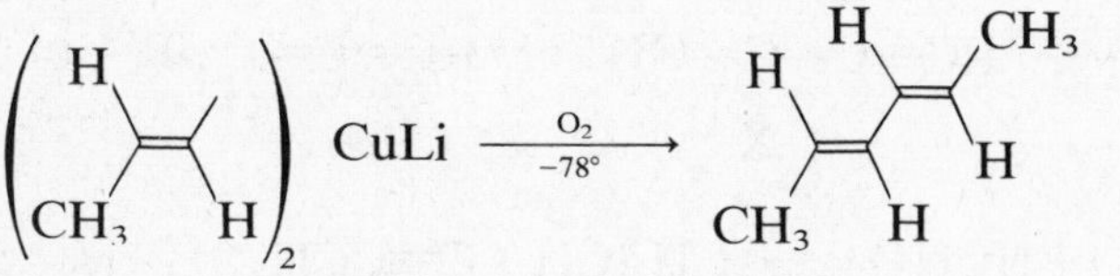

$PhCH_2Cu \xrightarrow{25°} PhCH_2CH_2Ph$ (88 %)

(high yield)

(78 %)

$Ph_2CuLi \xrightarrow[-78°]{O_2} Ph\text{—}Ph$ (75 %)

$\left(C_2H_5\underset{|\atop CH_3}{C}H\text{–}\right)_2 CuLi \xrightarrow[-78°]{O_2} C_2H_5\underset{|\atop CH_3}{C}H\text{—}\underset{|\atop CH_3}{C}HC_2H_5$ (82 %)

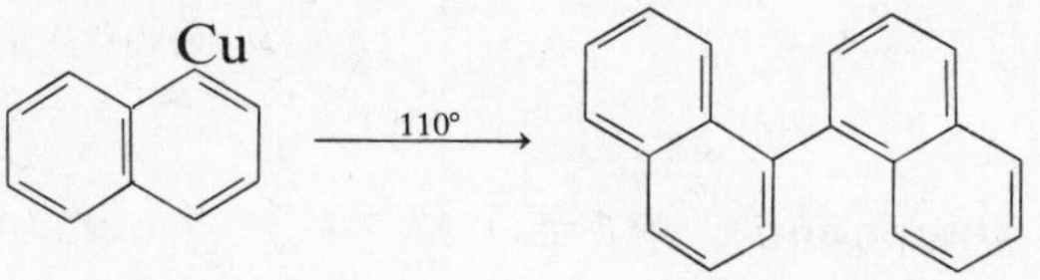

4.3 Reactions of nucleophiles derived from alk-1-ynes

4.3.1 Sodium, lithium and magnesium derivatives

Attention has already been drawn (sections 3.4.2.iii and 4.1.5) to the acidity of alk-1-ynes, and to the consequent formation of carbanionic species from alk-1-ynes and strong bases:

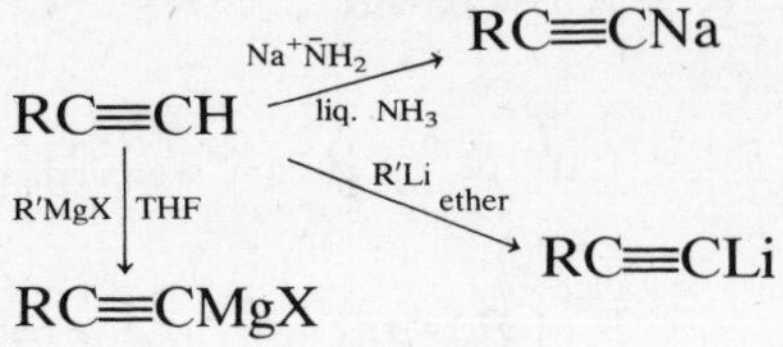

$$
RC\equiv CH \;\xrightarrow[\text{liq. }NH_3]{Na^+\bar{N}H_2}\; RC\equiv CNa
$$

$$
RC\equiv CH \;\xrightarrow[\text{ether}]{R'Li}\; RC\equiv CLi
$$

$$
RC\equiv CH \;\xrightarrow[THF]{R'MgX}\; RC\equiv CMgX
$$

These are less powerful nucleophiles than alkyl- or aryl-lithium compounds, or alkyl or aryl Grignard reagents; nevertheless they undergo the usual range of reactions with electrophiles, as shown below.

(i) **Alkylation:** $RC\equiv C^-M^+ + R'-Y \longrightarrow RC\equiv CR'$ (4.17)

e.g. $CH_3(CH_2)_2C\equiv C^-Na^+ + CH_3Br \xrightarrow{NH_3} CH_3(CH_2)_2C\equiv CCH_3$ (58 %)

$2HC\equiv C^-Na^+ + BrCH_2CH_2Br \xrightarrow{THF} HC\equiv CCH_2CH_2C\equiv CH$ (81 %)

$HC\equiv C^-Na^+ + CH_3(CH_2)_3Br \longrightarrow [CH_3(CH_2)_3C\equiv CH] \xrightarrow{NaNH_2}$
not isolated

$[CH_3(CH_2)_3C\equiv C^-Na^+] \xrightarrow{CH_3CH_2Br} CH_3(CH_2)_3C\equiv C-CH_2CH_3$
(64 % overall)

(ii) **Reactions with carbonyl compounds:** $RC\equiv C^-M^+ + \overset{R'}{\underset{X}{\diagdown}}C=O \longrightarrow$

$$
RC\equiv C-\overset{R'}{\underset{X}{\overset{|}{\underset{|}{C}}}}-O^-M^+ \longrightarrow RC\equiv C-\overset{R'}{\underset{X}{\overset{|}{\underset{|}{C}}}}-OH \text{ or } RC\equiv C-COR' \quad (4.18)
$$

e.g. $CH_3(CH_2)_3C\equiv CLi + HCHO \longrightarrow CH_3(CH_2)_3C\equiv CCH_2OH$ (80 %)

$PhC\equiv CMgBr + CH_2=CH-CHO \longrightarrow PhC\equiv C-\underset{OH}{\overset{|}{C}H}-CH=CH_2$ (52 %)

$ClCH_2C\equiv CLi + CH_3COCH_3 \longrightarrow ClCH_2C\equiv C-\underset{|}{\overset{OH}{\overset{|}{C}}}(CH_3)_2$ (67 %)

Acylation does occur in certain cases, e.g.

$$CH_3C{\equiv}CMgBr + ClCOCHCO_2C_2H_5 \longrightarrow CH_3C{\equiv}CCOCHCO_2C_2H_5$$
(with CH₃ group shown on the carbon bearing CO₂C₂H₅ on both sides)

(65 %)

but such reactions do not often give good yields, and may not stop at the ketone stage, e.g.

$$ClCH_2C{\equiv}CLi + ClCO{\langle}{\bigcirc}{\rangle}NO_2 \longrightarrow ClCH_2C{\equiv}C{-}\underset{\underset{NO_2}{\bigcirc}}{\overset{OH}{C}}{-}C{\equiv}CCH_2Cl$$

(36 %)

Thus acylations are better carried out *via* the copper(I) derivatives (cf. below).

4.3.2   Alkynylcopper(I) compounds

These are much easier to prepare, and with a few exceptions are much more stable, than alkyl- or aryl-copper(I) compounds. They are produced simply by the reaction of the appropriate alk-1-yne with copper(I) chloride, either in aqueous ammonia or in a polar (non-protic) organic solvent such as dimethylformamide or hexamethylphosphoramide, $[(CH_3)_2N]_3PO$.

$$2R{-}C{\equiv}CH + Cu_2Cl_2 \rightarrow 2R{-}C{\equiv}CCu + 2HCl$$

The reactions of these copper derivatives parallel those of the other organocopper(I) reagents discussed earlier (section 4.2.3). The alkynylcopper reagents are thus used in preference to the sodium, lithium, or magnesium analogues for the following types of reactions:

(i) Displacement of halogens from 'unreactive' positions, e.g.

$$PhC{\equiv}CCu + \underset{\underset{I}{}\quad\overset{H}{}}{\overset{H\quad I}{{>}{=}{<}}} \xrightarrow[90{-}100°]{pyridine} PhC{\equiv}C{-}\overset{H}{\underset{H}{C}}{=}\underset{H}{C}{-}C{\equiv}CPh \quad (90\%)$$

$$PhC{\equiv}CCu + I{\langle}{\bigcirc}{\rangle}OCH_3 \xrightarrow[120°,\ 10\,h]{pyridine} PhC{\equiv}C{-}{\langle}{\bigcirc}{\rangle}OCH_3 \quad (98\%)$$

(ii) Conversion of acyl chlorides into ketones, e.g.

$$CH_3(CH_2)_4C\equiv CCu + CH_3COCl \xrightarrow[\text{(catalyst)}]{\text{LiI}} CH_3(CH_2)_4C\equiv CCOCH_3$$

$$(81\%)$$

(iii) Coupling reactions giving conjugated diynes. For symmetrical diynes, oxidative coupling is used: the alkyne is either converted into its copper(I) derivative and oxidised *in situ*, usually by oxygen itself (**Glaser coupling**), or else it is treated with copper(II) acetate in pyridine (**Eglinton–Galbraith coupling**): in both cases the copper(II) ion is apparently the effective oxidant:

$$RC\equiv C^- + Cu^{2+} \longrightarrow RC\equiv C^{\cdot} + Cu^+; \ 2RC\equiv C^{\cdot} \longrightarrow RC\equiv C—C\equiv CR$$

$$[Cu^+ + \text{oxidant}]$$

For example:

$$HO_2CC\equiv CCu \xrightarrow{K_3Fe(CN)_6} HO_2CC\equiv C—C\equiv CCO_2H \quad (60\%)$$

$$PhC\equiv CCu \xrightarrow{O_2} PhC\equiv C—C\equiv CPh \quad (90\%)$$

$$(CH_3)_2\underset{\underset{OH}{|}}{C}—C\equiv CH \xrightarrow[O_2]{Cu_2Cl_2} (CH_3)_2\underset{\underset{OH}{|}}{C}—C\equiv C—C\equiv C—\underset{\underset{OH}{|}}{C}(CH_3)_2 \quad (\text{quantitative})$$

$$HC\equiv CC(CH_3)_2CH_2CO_2CH_3 \xrightarrow[\text{pyridine}]{Cu(OCOCH_3)_2} [CH_3O_2CCH_2C(CH_3)_2C\equiv C—]_2$$

$$(98\%)$$

In the case of unsymmetrical diynes, coupling of an alkynyl copper(I) with a 1-halogenoalk-1-yne is generally used (**Cadiot–Chodkiewicz coupling**); the organocopper derivative is usually generated *in situ*:

$$RC\equiv CCu + X—C\equiv CR' \longrightarrow RC\equiv C—C\equiv CR' \quad (4.20)$$

e.g.

$$HO_2CC\equiv CH + BrC\equiv C \underset{}{\bigcirc} Br \xrightarrow{\underset{C_2H_5NH_2}{Cu_2Cl_2}} HO_2CC\equiv C—C\equiv C \underset{}{\bigcirc} Br$$

$$(92\%)$$

$$PhC\equiv CH + BrC\equiv CCHO \xrightarrow[C_2H_5NH_2]{Cu_2Cl_2} PhC\equiv C—C\equiv CCHO \quad (71\%)$$

4.4 Review

The reactions described in this chapter lead to products of diverse structural types, but all except the oxidative coupling processes [reactions (4.15), (4.16) and (4.19)] conform to a general pattern, *viz.* that they involve the formation of a carbon–carbon single bond, with the organometallic reagent contributing both the electrons of this new bond. In each case, no matter whether the organometallic species is $RMgX$, RLi, R_2Cd, R_2CuLi, etc., the R-group appears in the product, singly bonded to another carbon.

4.4.1   More about disconnections and synthons

In Chapter 3 we introduced the concept of a *disconnection*, as the (imaginary) opposite of a real reaction, and we also introduced the term *synthon* to describe a 'product' (again imaginary) of a disconnection. We now focus attention on the application of these ideas to the reactions described in the present chapter.

As stated above, the general form of these products is $R—C\overset{/}{\underset{\backslash}{—}}$,

and they are formed from a nucleophilic R and an electrophilic C species. So the general disconnection for these products will be

$$R—C\overset{/}{\underset{\backslash}{—}} \implies R^- + {}^+C\overset{/}{\underset{\backslash}{—}} \qquad (4.21)$$

It is worth reiterating that this means, in words,

'The formation of $R—C\overset{/}{\underset{\backslash}{—}}$ implies the reaction of a nucleophilic R-species (which *may be* but *need not be* a carbanion) with an electrophilic $C\overset{/}{\underset{\backslash}{—}}$ -species (which *may be* but *need not be* a carbonium ion).'

We may now write down specific disconnections for the products of some of the numbered reactions in the chapter. These are listed in Table 4.1.

There are three significant omissions from this Table. The disconnections of the oxidative coupling products give radical synthons, e.g.

$$R—R \Rightarrow R^{\cdot} + {}^{\cdot}R; \quad RC{\equiv}C—C{\equiv}CR \Rightarrow RC{\equiv}C^{\cdot} + {}^{\cdot}C{\equiv}CR \qquad (4.22)$$

The disconnection for the product of the Reformatsky reaction

Table 4.1 Disconnections for products of organometallic reactions

Reaction Number	Product		Synthons		Electrophile
4.1	$R-R^1$	$\Longrightarrow$	R^-	R^{1+}	R^1-Y (halide, sulphonate)
4.2	RCH_2CH_2OH	$\Longrightarrow$	R^-	$\overset{+}{C}H_2CH_2OH$	(epoxide)
4.3	$R-\underset{R^2}{\overset{R^1}{\underset{\vert}{\overset{\vert}{C}}}}-OH$	$\Longrightarrow$	R^-	$R^1\overset{+}{\underset{R^2}{\underset{\vert}{C}}}-OH$	$\underset{R^2}{\overset{R^1}{C}}=O$
4.5	RCO_2H	$\Longrightarrow$	R^-	$\overset{+}{C}O_2H$	CO_2
4.6	$RCOR^1$	$\Longrightarrow$	R^-	$\overset{+}{C}OR^1$	$R^1CONR^2_2$
4.7	$RCHO$	$\Longrightarrow$	R^-	$\overset{+}{C}HO$	$HC(OR^1)_3$
4.8	(allylic alcohol, C=C–C(OH)R)	$\Longrightarrow$	R^-	(C=C–$\overset{+}{C}$–OH)	(enone C=C–C=O)
4.9	($R\overset{\vert}{C}-CH$ ... C=O)	$\Longrightarrow$	R^-	($\overset{+}{C}$–CH ... C=O)	(enone C=C–C=O)
4.14	$RCOR^1$	$\Longrightarrow$	R^-	$\overset{+}{C}OR^1$	R^1COCl
4.17	$RC\equiv CR^1$	$\Longrightarrow$	$RC\equiv C^-$	R^{1+}	R^1-Y (halide)
4.18	$RC\equiv C-\underset{X}{\overset{R^1}{\underset{\vert}{\overset{\vert}{C}}}}-OH$	$\Longrightarrow$	$RC\equiv C^-$	$R^1\overset{+}{\underset{X}{\underset{\vert}{C}}}-OH$	R^1COX
	$RC\equiv C-COR^1$	$\Longrightarrow$	$RC\equiv C^-$	$\overset{+}{C}OR^1$	R^1COX
4.20	$RC\equiv C-C\equiv CR^1$	$\Longrightarrow$	$RC\equiv C^-$	$\overset{+}{C}\equiv CR^1$	$R^1C\equiv CBr$

(4.13) is as follows:

$$R^2-\underset{R^3}{\overset{OH}{\underset{\vert}{\overset{\vert}{C}}}}-CH_2CO_2R^1 \Longrightarrow R^2-\underset{R^3}{\overset{OH}{\underset{\vert}{\overset{\vert}{C^+}}}} + \bar{C}H_2CO_2R^1;$$

but this will not be further considered until Chapter 5 (section 5.5).

Reaction (4.4) is somewhat more complicated, since the formation of the product requires two successive attacks by the nucleophile. A series of disconnections is therefore necessary in this case:

$$
\underset{\underset{R}{|}}{\overset{\overset{R'}{|}}{R-C-OH}} \Longrightarrow R^- + \underset{\underset{R}{|}}{\overset{\overset{R'}{|}}{\overset{+}{C}-OH}} \quad \text{[as for product of (4.3)]}
$$

$$
\underset{\underset{R}{|}}{\overset{\overset{R'}{|}}{\overset{+}{C}-OH}} \Longrightarrow \underset{R}{\overset{R'}{>}} C{=}O + H^+
$$

$$
\underset{R}{\overset{R'}{>}} C{=}O \Longrightarrow R^- + \overset{+}{C}OR' \quad \text{[as for products of (4.6) and (4.14)]}
$$

4.4.2 Synthetic equivalents

In Table 4.1, the synthon representing the nucleophile is shown as R^- or $RC{\equiv}C^-$. The actual nucleophile used, however, may vary: for example, in reactions (4.1)–(4.9) it is RMgX, whereas in (4.14) it is R_2Cd. In (4.1)–(4.9), however, it could equally well have been RLi, and in (4.1) and (4.14) it might also have been R_2CuLi. So RMgX, RLi, R_2Cd, and R_2CuLi are all **synthetic equivalents** of the synthon R^-, i.e. they are actual reagents which carry out the function of the synthon R^-. It is important to remember that **there may be more than one synthetic equivalent of any given synthon**.

This is also obvious in the case of the electrophilic synthons. If these are compared with their synthetic equivalents (shown in the last column of Table 4.1), it may be seen that R^{1+} is the synthon corresponding to an alkylating agent (an alkyl halide or a sulphonate ester); R^1CO^+ corresponds to an acylating agent (acyl halide, tertiary amide, and also anhydride or ester or nitrile: sections 4.1.2 and 4.1.3); and so on.

Table 4.2 lists the synthons relating to the reactions in this chapter, and their common synthetic equivalents. (Page references are given for the use of each synthetic equivalent.)

4.5 Problems

We conclude this chapter by considering the application of the disconnection/synthon approach to the solution of a few simple

Table 4.2 Some common synthons and their synthetic equivalents (part 1)

	Synthon	Synthetic equivalent(s)	For example(s) see page(s)
(a) Nucleo-philic synthons	1. R^- (alkyl group)	$RMgX$; RLi; R_2Cd; RCu; R_2CuLi	39–51
	2. R^- (aryl group)	do.	39–51
	3. $RC{\equiv}C^-$	$RC{\equiv}C^-Na^+$; $RC{\equiv}CMgX$; $RC{\equiv}CLi$; $RC{\equiv}CCu$	52–4
(b) Electro-philic synthons	1. R^+ (alkyl group)	RCl, RBr, RI, $ROSO_2R^1$	39, 40, 49, 52
	2. R^+ (aryl group)	RBr, RI, $RN^+_2X^-$	49, 53, 54
	3. $RCH{=}CH^+$	$RCH{=}CHBr$	49–53
	4. $RC{\equiv}C^+$	$RC{\equiv}CBr$	54
	5. $R\overset{+}{C}{=}O$	$RCOCl$, $(RCO)_2O$, RCO_2R^1, $RCONR^1_2$, RCN, (RCO_2H)	41–5, 48–50, 53–4
	6. $H\overset{+}{C}{=}O$	HCO_2R, $HCONR_2$, $CH(OR)_3$	42–3, 47
	7. $\overset{+}{C}\overset{\displaystyle O}{\underset{\displaystyle OH}{}}$	CO_2	42
	8. $\overset{+}{C}H_2OH$	$HCHO$	40, 52
	9. $\overset{+}{C}HOH{\cdot}R$	$RCHO$	40, 52
	10. $^+CR_2OH$	R_2CO	40, 52
	11. $^+CH_2CH_2OH$		40
	12. $^+CH_2CH_2COR$ (CO_2R, CN)	$CH_2{=}CHCOR$ (CO_2R, CN)	45, 50

synthetic problems. Suppose one were asked to devise a synthesis of the following:

$PhCH_2CHOHCH_2CH_3$; $CH_3(CH_2)_{10}CO(CH_2)_2CH_3$; $Ph\overset{\displaystyle CH_3}{\underset{\displaystyle |}{C}}HCH_2CO_2CH_3$;
$\quad\quad$ (4) $\quad\quad\quad\quad\quad\quad\quad\quad$ (5) $\quad\quad\quad\quad\quad\quad\quad$ (6)

$CH_3C{\equiv}C{-}CH_2CH_2OH.$
$\quad$ (7)

One good way of tackling such problems is to write down possible disconnections for the required end-product, and to try to relate the resulting synthons to recognisable synthetic equivalents. With several possible disconnections for each target molecule, are there any general guidelines to assist in choosing the best one?

Table 4.1 indicates that the bond which is disconnected is frequently one joining two carbons of which one or both carries a functional group. So **it is worth considering a disconnection adjacent to a functional group**. The first two examples illustrate this point.

1-Phenylbutan-2-ol (compound 4). Disconnection of the 1,2-bond or of the 2,3-bond (i.e. those flanking the functional group) gives *four* pairs of synthons, as follows:

$$PhCH_2 \dashv CHOHCH_2CH_3 \xrightarrow{(a)} PhCH_2^+ + {}^-\overset{\displaystyle OH}{\underset{\displaystyle |}{C}}HCH_2CH_3$$

$$\text{(4)} \qquad \searrow^{(b)} \qquad PhCH_2^- + {}^+\overset{\displaystyle OH}{\underset{\displaystyle |}{C}}HCH_2CH_3$$

$$PhCH_2CHOH \dashv CH_2CH_3 \xrightarrow{(c)} PhCH_2\overset{\displaystyle OH}{\underset{\displaystyle |}{\overset{+}{C}}}H + {}^-CH_2CH_3$$

$$\text{(4)} \qquad \searrow^{(d)} \qquad PhCH_2^- \overset{\displaystyle OH}{\underset{\displaystyle |}{C}}H + {}^+CH_2CH_3$$

Each of these disconnections gives one synthon which is immediately recognisable, *vis.* $PhCH_2^+$, $PhCH_2^-$, $^-CH_2CH_3$, and $^+CH_2CH_3$, and synthetic equivalents are readily available for all of these (e.g. $PhCH_2Br$, $PhCH_2MgBr$, CH_3CH_2MgBr, and CH_3CH_2Br respectively). However, only two of the four disconnections, *viz.* (b) and (c), give a *pair* of recognisable synthons: $^+\overset{\displaystyle OH}{\underset{\displaystyle |}{C}}HCH_2CH_3$ is, very simply, propanal $+ H^+$, and propanal is the synthetic equivalent (cf. entry 9 Table 4.2b). Similarly $PhCH_2{}^+\overset{\displaystyle OH}{\underset{\displaystyle |}{C}}H$ has phenylethanal, $PhCH_2CHO$, as its synthetic equivalent. There are no obvious synthetic equivalents for the synthons $^-\overset{\displaystyle OH}{\underset{\displaystyle |}{C}}HCH_2CH_3$ and $PhCH_2{}^+\overset{\displaystyle OH}{\underset{\displaystyle |}{C}}H$.

This means that 1-phenylbutan-2-ol should be preparable by the reverse of disconnections (b) and (c):

$$PhCH_2^- + {}^+\overset{\displaystyle OH}{\underset{\displaystyle |}{C}}HCH_2CH_3 \longrightarrow PhCH_2\overset{\displaystyle OH}{\underset{\displaystyle |}{C}}HCH_2CH_3$$

$$\text{(4)}$$

$$PhCH_2\overset{\displaystyle OH}{\underset{\displaystyle |}{\underset{+}{C}}}H + {}^-CH_2CH_3 \nearrow$$

All that remains is to insert the synthetic equivalent in place of each synthon. If there is a choice of synthetic equivalent, some care may be necessary in making that choice, since the synthon approach takes no account of the selectivity of reagents. If there are several suitable equivalents, however, the final choice may well depend on relative cost, and ease of handling. In the present case, the nucleophilic synthons may represent a variety of species (entry 1 in Table 4.2a), but since the Grignard reagent is the easiest to make and it reacts satisfactorily with an aldehyde, it would normally be preferred to, say, the lithium derivative.

So two synthetic routes have emerged to compound (4), *viz.*

$$PhCH_2MgBr + \overset{\overset{\displaystyle O}{\displaystyle \|}}{C}HCH_2CH_3 \quad \text{and} \quad PhCH_2\overset{\overset{\displaystyle O}{\displaystyle \|}}{C}H + BrMgCH_2CH_3.$$

Pentadecan-4-one (compound 5). Again, disconnection on either side of the functional group gives four pairs of synthons:

$$CH_3(CH_2)_{10}CO(CH_2)_2CH_3 \overset{(a)}{\Longrightarrow} CH_3(CH_2)_{10}{}^+ + {}^-\overset{\overset{\displaystyle O}{\displaystyle \|}}{C}(CH_2)_2CH_3$$

(5)

$$\overset{(b)}{\Longrightarrow} CH_3(CH_2)_{10}{}^- + {}^+\overset{\overset{\displaystyle O}{\displaystyle \|}}{C}(CH_2)_2CH_3$$

$$\overset{(c)}{\Longrightarrow} CH_3(CH_2)_{10}\overset{\overset{\displaystyle O}{\displaystyle \|}}{C}{}^+ + {}^-(CH_2)_2CH_3$$

$$\overset{(d)}{\Longrightarrow} CH_3(CH_2)_{10}\overset{\overset{\displaystyle O}{\displaystyle \|}}{C}{}^- + {}^+(CH_2)_2CH_3$$

As in the last example, only two of these disconnections, (b) and (c), give a pair of synthons with recognisable synthetic equivalents, since we have as yet no synthetic equivalent for a synthon of the type

$R\overset{\overset{\displaystyle O}{\displaystyle \|}}{C}{}^-$. So the synthesis of compound (5) should be based on the reverse of either of these disconnections:

$$CH_3(CH_2)_{10}{}^- + {}^+\overset{\overset{\displaystyle O}{\displaystyle \|}}{C}(CH_2)_2CH_3 \longrightarrow CH_3(CH_2)_{10}\overset{\overset{\displaystyle O}{\displaystyle \|}}{C}(CH_2)_2CH_3$$

(5)

$$CH_3(CH_2)_{10}\overset{\overset{\displaystyle O}{\displaystyle \|}}{C}{}^+ + {}^-(CH_2)_2CH_3 \nearrow$$

In this example there is a choice of synthetic equivalent for all four synthons, and so there is a wide choice of possible routes to the product. There are, however, some restrictions: if the acylating agent is an acyl halide, the nucleophilic species must be a dialkyl-cadmium or an alkyl-copper reagent: if the nucleophile is an alkyl-lithium compound, the acylating agent should be a nitrile, a tertiary amide, or a lithium carboxylate, and so on. The following are two possible routes:

$$CH_3(CH_2)_{10}COCl + (CH_3CH_2CH_2)_2CuLi \longrightarrow \text{compound (5)}$$

$$CH_3(CH_2)_{10}CN + CH_3CH_2CH_2MgBr$$

[Both of these involve a C_{12} acylating agent and a C_3 nucleophile; in practice these are preferable to reactions of C_4 acylating agents with C_{11} nucleophiles, because C_{11} halides are relatively expensive while the C_{12} acid (lauric acid) is not.]

Methyl 3-phenylbutanoate (compound 6). Disconnection of this molecule adjacent to the functional group gives only one reasonable pair of synthons, *viz.*

$$\underset{(6)}{PhCHCH_2CO_2CH_3} \Longrightarrow PhCHCH_2^- + {}^+C-OCH_3$$

(with CH_3 substituents shown and O above the carbonyl)

which in turn might suggest the following synthetic route:

$$(PhCHCH_2)_2CuLi + ClC-OCH_3 \longrightarrow \text{compound (6)}$$

(highly selective)   methyl chloroformate

However, esters are often more easily obtained *via* the corresponding acids; even although an extra step is involved, this is counterbalanced by the simplicity of the operation. So another route to compound (6) might be:

$$PhCHCH_2MgBr + CO_2 \longrightarrow \underset{(6a)}{PhCHCH_2CO_2H} \xrightarrow[H^+]{CH_3OH} \text{compound (6)}$$

Both of the above routes suffer from the disadvantage that the nucleophile is prepared from $PhCH(CH_3)CH_2Br$, which would itself require to be prepared. So it is perhaps worth trying other disconnections of the end-product (6) [or the acid (6a)]. We shall see in the next chapter that $\bar{C}H_2CO_2R$ is a synthon which has a number of synthetic equivalents, and so the disconnection

$$(6) \implies \overset{\displaystyle CH_3}{\underset{\displaystyle |}{Ph—\overset{+}{CH}}} + \bar{C}H_2CO_2CH_3$$

appears promising. Two other disconnections, still further removed from the functional group, are worthy of consideration: these involve the carbon bearing the substituent on the chain:

$$\overset{\displaystyle CH_3}{\underset{\displaystyle Ph}{\diagdown}}CHCH_2CO_2CH_3 \implies Ph\overset{+}{C}HCH_2CO_2CH_3 + \bar{C}H_3$$

$$(6) \qquad\qquad \searrow CH_3\overset{+}{C}HCH_2CO_2CH_3 + Ph^-$$

The synthetic equivalents of the electrophilic synthons are α,β-unsaturated esters (cf. entry 12 in Table 4.2b). So the formation of compound (6) is possible by conjugative addition to such an ester, i.e. $CH_3^- + PhCH{=}CHCO_2CH_3$ (methyl cinnamate) or $Ph^- + CH_3CH{=}CHCO_2CH_3$ (methyl crotonate).

In order to ensure that the nucleophile adds conjugatively and not directly to the carbonyl group, the nucleophile of choice is a Grignard reagent containing a small (non-stoicheiometric) amount of a copper(I) salt (cf. section 4.2.3.iii).

Pent-3-yn-1-ol (compound 7). The reader is invited to try this example for him(her)self. Try disconnections of the 1,2- and the 2,3-bond (next to the functional groups). Are the synthetic equivalents of the synthons easily recognisable? Which of the possible routes appears to be the simpler?

We shall present some more problems at the end of Chapter 5.

Notes

1. If RBr is (even moderately) reactive towards nucleophiles, it may react with R′Li to give R–R′ and LiBr [cf. reaction (4.1) for Grignard reagents]. Even if RBr is relatively unreactive towards nucleophiles, the product R′Br may react faster than RBr with R′Li; in this case an excess of R′Li is required:

    $$RBr + 2R'Li \rightarrow RLi + R'{-}R' + LiBr \qquad\qquad (4.12a)$$

    This also helps to minimise interaction of RLi and R′Br to give R–R′.

2. It is commonly supposed (and stated in text-books) that dialkylcadmium reagents do not react with ketones. While it is true that a *purified* (i.e. redistilled or resublimed) dialkylcadmium shows negligible reaction with

ketones, its reaction with acyl chlorides is also very slow and poor yields of ketones result. In presence of a magnesium halide (which is the case when it is generated from $RMgX + CdX_2$ and used *in situ*) a dialkylcadmium will react quite readily, in the manner of a Grignard reagent, with ketones and most other carbonyl groups. It is only its *selectivity* which is the key to reaction (4.4).

3.   Other species such as R_3Cu_2Li, R_3CuLi_2, and $R_5Cu_3Li_2$ have also been recognised; the second of these is claimed to be superior to R_2CuLi in reactions with alkyl, alkenyl, and aryl halides.

5 Formation of carbon–carbon bonds: the use of stabilised carbanions and related nucleophiles

The factors which contribute to carbanion stabilisation, and the different types of stabilised carbanion, have already been described in Chapter 3 (section 3.4.2), and the uses of alkynyl ions in synthesis have been included in Chapter 4 (section 4.3), because of the obvious similarities between the reactions of these ions and those of other organometallic reagents. Now we must consider in detail the reactions of the other groups of stabilised carbanions, in particular those in which the stabilisation is provided by electron-accepting (-M) substituents. Since carbanions stabilised by a carbonyl group may also be written as enolate ions, and indeed are frequently referred to as **enolates** [cf. formula (16) in section 3.4.3], it is convenient to consider the reactions of enols (and enamines) in the same chapter as the reactions of the anions.

We shall divide the reactions into several classes, according to the nature and number of the groups stabilising the carbanion. For each category of carbanion we shall consider reactions under three headings, *viz.* alkylation, acylation, and condensation. (These terms have already been defined in sections 3.3.1 and 3.3.2.) Finally, we shall discuss the reactions of enols and enamines, and their aromatic counterparts.

5.1 Carbanions stabilised by two -M groups

When a —CH_2— or —CHR— group in a molecule is flanked by two -M groups, such a molecule is readily deprotonated by the action of a base, and the resulting anion is stabilised by delocalisation (cf. section 3.4.2.i). The compounds in question are thus relatively strong acids: thus, for example, dinitromethane (1:p$K_a \simeq 4$) is slightly more acidic than acetic acid, and pentan-2,4-dione (2:p$K_a \simeq 9$) is a slightly stronger acid than phenol. The members of this class of compound which are most useful in synthesis, such as diethyl malonate (3) and ethyl acetoacetate (4), have pK_a values of ca. 13 or less.

$CH_2(NO_2)_2$ $CH_3COCH_2COCH_3$ $CH_2(CO_2C_2H_5)_2$
(1)   (2)   (3)

$CH_3COCH_2CO_2C_2H_5$
(4)

From a synthetic point of view there are two important conse-quences of this.

(i) These compounds are deprotonated, essentially completely, by bases such as sodium ethoxide (the pK_a of ethanol being ca. 18). Or, to express this in another way, the equilibrium (5.1) lies far over to the right:

$$X—CH_2—Y + Na^+\bar{O}R \rightleftharpoons X—\bar{C}H—Y\ Na^+ + ROH \qquad (5.1)$$

(X and Y = COR′, CO$_2$R′, CN, NO$_2$, etc.)

[It is of course, important to remember that the electron-accepting substituents (X and Y in reaction 5.1) may themselves be attacked by the base if the latter is also a good nucleophile. It is thus inadvisable to use, for example, hydroxide ion to deprotonate diethyl malonate, because hydrolysis of the ester would almost certainly ensue.]

(ii) These compounds are also deprotonated, if not completely then at least to a significant extent, by organic bases such as piperidine ($pK_a \simeq 11$):

$$X—CH_2—Y + \text{(piperidine, N-H)} \rightleftharpoons X—\bar{C}H—Y \ \text{(piperidinium, N-H}_2^+) \qquad (5.2)$$

5.1.1 Alkylation

This is effected, relatively simply, by reaction of the (pre-formed) anions with the usual range of alkylating agents, halides being the most commonly used. Thus, for example,

$$[CH_2(CO_2C_2H_5)_2 \xrightarrow[C_2H_5OH]{NaOC_2H_5}] Na^+\bar{C}H(CO_2C_2H_5)_2 + CH_3(CH_2)_3Br \rightarrow$$

$$CH_3(CH_2)_3CH(CO_2C_2H_5)_2 \quad (88\%)$$
(5)

$$\left[CH_3COCH_2CO_2C_2H_5 \xrightarrow[C_2H_5OH]{NaOC_2H_5} \right] Na^+\bar{C}H \begin{array}{l} {}^{COCH_3} \\ {}_{CO_2C_2H_5} \end{array}$$

$$+ \ CH_3(CH_2)_2Br \longrightarrow CH_3(CH_2)_2CH \begin{array}{l} {}^{COCH_3} \\ {}_{CO_2C_2H_5} \end{array} \quad (71\%)$$
(6)

$$\left[CH_3COCH_2COCH_3 \xrightarrow[CH_3COCH_3]{K_2CO_3} \right] K^+\bar{C}H(COCH_3)_2 + CH_3I$$

$$\xrightarrow{} \overset{\overset{\displaystyle CH_3}{|}}{CH_3COCHCOCH_3} \quad (75\,\%)$$

$$(7)$$

(Note the use of the weaker base, in view of the greater acidity of the diketone.)

The products of these reactions, (5)–(7), still contain an acidic hydrogen, and the alkylation process may thus be repeated, giving a dialkyl derivative. The second alkylation may be more difficult than the first, because the first alkyl substituent introduced, being electron-repelling, will diminish the acidity of the adjacent hydrogen; and the monoalkylated carbanion will in any case be more sterically hindered than its non-alkylated analogue.

If the two alkyl groups to be introduced are identical, the dialkylation may be carried out as a 'one-pot' reaction, e.g.

$$CH_2(CO_2C_2H_5)_2 \xrightarrow[\text{(ii) 2C}_2\text{H}_5\text{I}]{\text{(i) 2NaOC}_2\text{H}_5} (C_2H_5)_2C(CO_2C_2H_5)_2 \quad (83\,\%)$$

$$CH_2(CN)_2 \xrightarrow[\text{(ii) 2PhCH}_2\text{Br}]{\text{(i) 2NaH}} (PhCH_2)_2C(CN)_2 \quad (75\,\%)$$

$$CH_3COCH_2COCH_3 \xrightarrow[\text{(ii) CH}_3\text{I}]{\text{(i) NaH}} \text{compound 7 (not isolated)} \xrightarrow[\text{(ii) CH}_3\text{I}]{\text{(i) NaH}}$$

$$\underset{\underset{\displaystyle CH_3}{|}}{\overset{\overset{\displaystyle CH_3}{|}}{CH_3COCCOCH_3}} \quad (63\,\%).$$

If the two alkyl groups are different, they may be introduced in a stepwise manner, e.g.

$$CH_2(CO_2C_2H_5)_2 \xrightarrow[\text{(ii) C}_2\text{H}_5\text{Br}]{\text{(i) NaOC}_2\text{H}_5} \underset{(90\,\%)}{C_2H_5CH(CO_2C_2H_5)_2} \xrightarrow[\text{(ii) (CH}_3)_2\text{CHI}]{\text{(i) NaOC}_2\text{H}_5}$$

$$\underset{(45\,\%)}{\overset{\overset{\displaystyle C_2H_5}{|}}{(CH_3)_2CHC(CO_2C_2H_5)_2}} \quad (41\,\%\ \text{overall})$$

The above four examples are instructive in themselves, and so it is worth examining them in greater detail.

(a) The first two may be considered together, since the only important respect in which they differ is the nature of the base required. (Sodium hydride in dimethyl sulphoxide[1] is used in the second case because ethoxide ion is liable to attack cyano-groups.) The procedure consists of addition of the carbanion source (diethyl malonate or malononitrile) to *two* molar equivalents of the appropriate base, and subsequent (gradual) addition of the alkylating agent.

It must be emphasised that the addition of, say, the diethyl malonate to two molar equivalents of sodium ethoxide does not result in the formation of a di-anion: it is an equimolar mixture of the mono-carbanion and ethoxide to which the alkylating agent (ethyl iodide) is initially added. The reader may wonder why, in that case, the ethoxide does not react with the halide to any appreciable extent: the answer lies in the fact that although ethoxide is a stronger *base* than the carbanion under these conditions, the carbanion is a much stronger *nucleophile*. As the alkylation proceeds, with the formation of the mono-alkylated malonate, so this compound is deprotonated by the ethoxide and is thus able to undergo the second alkylation step.

(b) In the third example (dimethylation of pentan-2,4-dione), the choice of base is again noteworthy. Potassium carbonate is sufficiently basic to deprotonate pentan-2,4-dione, as was shown in an earlier example, but not to deprotonate the 3-methyl derivative (7). Sodium ethoxide is basic enough for the latter operation, but is liable to react as a nucleophile at the carbonyl group (cf. section 5.1.2). Hence sodium hydride is again the reagent of choice.

In this reaction, unlike the first two, the two alkyl groups are introduced in separate operations, even although the intermediate mono-alkyl compound is not isolated. It is interesting to speculate on the possible significance of this difference in procedure. It may be, in this particular case, that the stepwise method simply gives better yields, or that the alternative procedure has not been tried. But it may also be (and this is the important point) that diketones like pentan-2,4-dione, and β-keto-esters like ethyl acetoacetate, react with 2 molar equivalents of base, provided that the base is sufficiently strong, to give dianions of the type (8) or (9). These are then alkylated preferentially at the 'wrong' carbon:

$$R^1R^2CHCOCH_2COR \xrightarrow[NH_3(l.)]{2NaNH_2} R^1R^2\bar{C}CO\bar{C}HCOR \xrightarrow{R^3X}$$

(8)

$$\overset{\displaystyle R^3}{\underset{\displaystyle |}{R^1R^2\bar{C}CO\bar{C}HCOR}} \quad (5.3)$$

$$R^1CH_2COCH_2CO_2R \xrightarrow[\text{(ii) } n\text{-}C_4H_9Li]{\text{(i) NaH}} R^1\bar{C}HCO\bar{C}HCO_2R \xrightarrow{R^2X}$$

$$(9)$$

$$\overset{\displaystyle R^2}{\underset{\displaystyle |}{R^1CH\!-\!CO\bar{C}HCO_2R}} \quad (5.4)$$

e.g. $CH_3COCH_2COCH_3 \xrightarrow[\text{(ii) } PhCH_2Cl]{\text{(i) } 2NaNH_2} CH_3COCH_2COCH_2CH_2Ph$

$$(69\%)$$

$$CH_3COCH_2CO_2C_2H_5 \xrightarrow[\text{(ii) } C_2H_5Br]{\text{(i) } NaH+C_4H_9Li} C_2H_5CH_2COCH_2CO_2C_2H_5 \quad (84\%)$$

(c) The final example on p. 66 illustrates the introduction of two different alkyl groups into diethyl malonate, and raises the question of the order in which the groups are inserted. In some cases the order is unimportant. However if the two alkyl groups are very different in bulk, it is advisable to introduce the smaller group first; if the bulky group is put in first, steric hindrance may then inhibit the second alkylation. Also, if the two alkyl groups are very different in their electron-repelling ($+I$) effect, it is advisable to introduce first the group which has the lesser effect, since deprotonation of the alkyl-malonic ester for the second alkylation is made more difficult if the alkyl group is strongly electron-repelling.

Stabilised carbanions are **ambident nucleophiles** (cf. Sykes, p. 96), as implied in the canonical forms (10) and (11) or the delocalised ion (12):

$$\underset{(10)}{\!>\!\bar{C}\!-\!X\!=\!Y} \longleftrightarrow \underset{(11)}{\!>\!C\!=\!X\!-\!\bar{Y}} \qquad\qquad \underset{(12)}{[\ C\!-\!-\!X\!-\!-\!Y]^-}$$

Such ions might therefore have been expected to undergo O-alkylation (since oxygen is the more electronegative 'end' of the delocalised system) rather than the exclusive C-alkylation indicated above. The fact is that O-alkylation of enolates can (and frequently does) occur along with C-alkylation. The proportion of O-alkylated product depends, apparently, on a considerable number of factors: the nature of the alkylating agent, the choice of cation, the solvent, whether the reaction is homogeneous or heterogeneous, whether either of the possible sites for alkylation is sterically hindered, etc. It is not always easy, therefore, to predict the course of a particular alkylation; but in general the use of an alkyl *halide* as the alkylating agent, a *sodium* salt as the nucleophile, and an *alcohol* as solvent (i.e., a solvent which may solvate, and hence deactivate, the oxygen of the enolate) is likely to give mainly C-alkylation.

5.1.2 Hydrolysis of the alkylated products: a route to carboxylic acids and ketones

The chemistry of malonic acid, $CH_2(CO_2H)_2$, and β-keto-acids such as acetoacetic acid, $CH_3COCH_2CO_2H$, is dominated by the ease with which these acids undergo decarboxylation (i.e. lose carbon dioxide) on being heated (cf. Sykes, pp. 280–1):

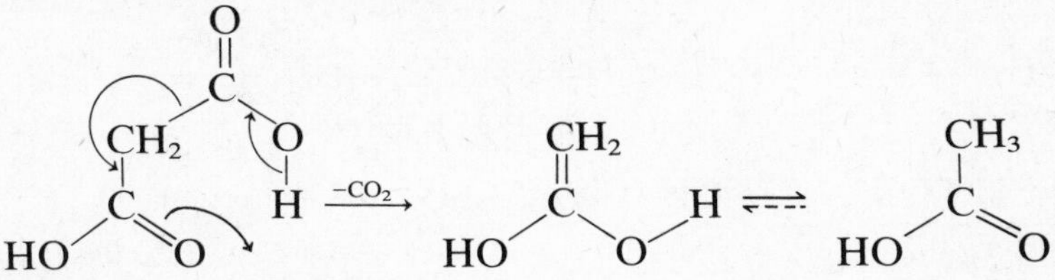

$$\xrightarrow{-CO_2}$$

$$\rightleftharpoons$$

Similarly, $CH_3COCH_2CO_2H \xrightarrow{-CO_2} CH_3COCH_3$.

The mono- and di-alkylated analogues are similarly decarboxylated, e.g.

$$R_2C(CO_2H)_2 \rightarrow R_2CHCO_2H;$$

$$CH_3COCHRCO_2H \rightarrow CH_3COCH_2R.$$

These mono- and di-alkylated acids are of course obtained by hydrolysis of the corresponding esters, the preparation of which has been described in the previous section. Since the alkyl groups are introduced by means of alkyl halides, reaction with the appropriate carbanions followed by hydrolysis and decarboxylation constitutes a method for the conversion of halides into carboxylic acids or ketones:

$$RX \xrightarrow{Na^+\bar{C}H(CO_2C_2H_5)_2} RCH(CO_2C_2H_5)_2 \xrightarrow{hydrolysis} RCH(CO_2H)_2$$

$$\xrightarrow{-CO_2} RCH_2CO_2H \quad (5.4)$$

$$R^1X \xrightarrow{Na^+R\bar{C}(CO_2C_2H_5)_2} RR^1C(CO_2C_2H_5)_2 \xrightarrow{hydrolysis} RR^1C(CO_2H)_2$$

$$\xrightarrow{-CO_2} RR^1CHCO_2H \quad (5.4a)$$

$$R^1X \xrightarrow{Na^+\bar{C}HCO_2C_2H_5 \ (COR)} R^1CH(COR)(CO_2C_2H_5) \xrightarrow{hydrolysis} R^1CH(COR)(CO_2H)$$

$$\xrightarrow{-CO_2} R^1CH_2COR \quad (5.5)$$

$$R^2X \xrightarrow{Na^+R^1\overset{\underset{|}{COR}}{C}CO_2C_2H_5} R^1R^2C\overset{\diagup COR}{\diagdown CO_2C_2H_5} \xrightarrow{\text{hydrolysis}} R^1R^2C\overset{\diagup COR}{\diagdown CO_2H}$$

$$\xrightarrow{-CO_2} R^1R^2CHCOR \quad (5.5a)$$

Thus, for example,

$$CH_3(CH_2)_3Br \xrightarrow{Na^+\bar{C}H(CO_2C_2H_5)_2} CH_3(CH_2)_3CH(CO_2C_2H_5)_2 \xrightarrow[\text{then } H_2SO_4]{KOH,H_2O}$$
$$(5)$$

$$CH_3(CH_2)_3CH(CO_2H)_2 \xrightarrow[-CO_2]{\text{heat}} CH_3(CH_2)_4CO_2H \quad (66 \% \text{ overall})$$

$$CH_3(CH_2)_2Br \xrightarrow{Na^+\,\overset{\underset{|}{COCH_3}}{\bar{C}HCO_2C_2H_5}} CH_3(CH_2)_2CH\overset{\diagup COCH_3}{\diagdown CO_2C_2H_5} \xrightarrow[\text{then } H_2SO_4]{NaOH,H_2O}$$

$$CH_3(CH_2)_2CH\overset{\diagup COCH_3}{\diagdown CO_2H} \xrightarrow[-CO_2]{\text{heat}} CH_3(CH_2)_3COCH_3$$

$$(48 \% \text{ overall})$$

So we now have a method for replacing halogen in a molecule by CH_2CO_2H or CH_2COR. To express this in another way, we have another possible disconnection for carboxylic acids and ketones:

$$RCH_2CO_2H \Rightarrow R^+ \bar{C}H_2CO_2H \tag{5.6}$$
$$RCH_2COR^1 \Rightarrow R^+ \bar{C}H_2COR^1 \tag{5.7}$$

The synthetic equivalent of R^+ is, of course, the alkylating agent RX, as we have seen before. So **the synthetic equivalent of the synthon $\bar{C}H_2CO_2H$ is diethyl malonate,** and **the synthetic equivalent of $\bar{C}H_2COCH_3$** (for example) **is ethyl acetoacetate.** Similarly the synthetic equivalents of $\bar{C}HRCO_2H$ and $\bar{C}HRCOCH_3$ are the appropriately alkylated derivatives of malonic and acetoacetic esters, respectively.

There is one remaining practical point. Hydrolysis of an ester may be carried out under both basic and acidic conditions (Sykes, pp. 234–7), and although the illustrations we have used in this section both involve basic hydrolysis there is no reason why acidic hydrolysis should not be equally effective. Indeed, in some cases basic hydrolysis of β-keto-esters is unsatisfactory, since hydroxide ion may attack the ketonic carbonyl group as well as (or instead of) the ester carbonyl; in such cases hydrolysis under acidic conditions is preferable:

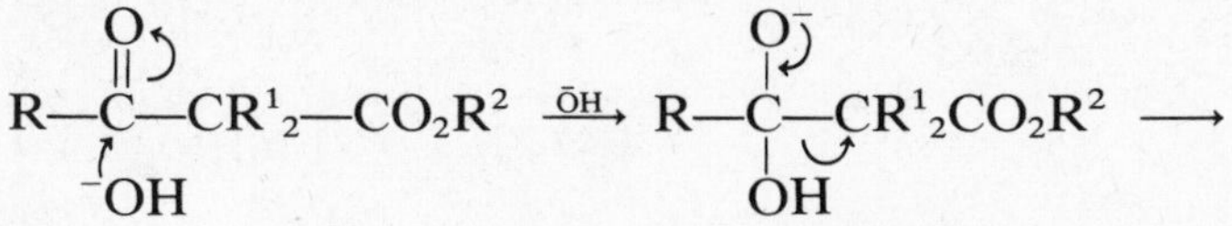

$$R-\overset{\overset{\displaystyle O}{\|}}{\underset{\underset{\displaystyle OH}{\uparrow}}{C}}-CR^1{}_2-CO_2R^2 \xrightarrow{\bar{O}H} R-\overset{\overset{\displaystyle O^-}{|}}{\underset{\underset{\displaystyle OH}{|}}{C}}-CR^1{}_2CO_2R^2 \longrightarrow$$

$$RCO_2H + \bar{C}R^1{}_2CO_2R^2 \rightleftharpoons RCO_2{}^- + R^1{}_2CHCO_2R^2$$

Although this cleavage (deacylation of the β-keto-ester) may occasionally be used to advantage (for example, to bring about a ring-opening: section 7.4.1), it is more often an unwanted side-reaction. It is not usually a serious problem in hydrolysis of mono-alkylated β-keto-esters, since these still contain an acidic hydrogen and are converted by base into delocalised enolate ions; the ketonic carbonyl group is thereby deactivated towards nucleophilic attack. Dialkylated β-keto-esters, on the other hand, contain no such acidic hydrogens, and the keto-group is then particularly prone to attack by hydroxide ion, since it contains the most electrophilic carbon in the molecule.

5.1.3 Acylation

Much of what has been said of alkylation in the two previous sections is equally applicable to acylation. The reactions occur readily, and usually in good yield; and hydrolysis or base-induced cleavage of the product frequently constitutes a useful synthetic procedure.

There are several important differences, however, between alkylation and acylation. Some of these are merely procedural: for example, alcoholic solvents cannot be used for acylations, since alcohols are themselves easily acylated. Also, acylation products of the type $RCOCH\overset{\diagup COR^1}{\diagdown COR^2}$ are strongly acidic, the anion being stabilised by three -*M* groups: this means that such a product may be deprotonated by carbanions of the type $\bar{C}H\overset{\diagup COR^1}{\diagdown COR^2}$, and thus in the acylation of these carbanions the following reactions may be in competition:

$$RCOCl + Na^+\bar{C}H\overset{\diagup COR^1}{\diagdown COR^2} \longrightarrow RCO-CH\overset{\diagup COR^1}{\diagdown COR^2} \qquad (5.8)$$
$$\qquad\qquad (13) \qquad\qquad\qquad\qquad (14)$$

$$RCOCH\overset{\diagup COR^1}{\diagdown COR^2} + Na^+\bar{C}H\overset{\diagup COR^1}{\diagdown COR^2} \rightleftharpoons RCO-\bar{C}\overset{\diagup COR^1}{\diagdown COR^2} Na^+ + CH_2\overset{\diagup COR^1}{\diagdown COR^2}$$
$$\qquad (14) \qquad\qquad (13) \qquad\qquad\qquad (15) \qquad\qquad\qquad (5.8a)$$

$$RCOCl + RCO—\overset{COR^1}{\underset{COR^2}{C}}\ Na^+ \longrightarrow RCO—\overset{COR^1}{\underset{COR^2}{C}}—COR \quad (5.8b)$$

(15)   (16)

Both the main reaction, (5.8), and the first side-reaction, (5.8a), consume the carbanion (13), and so for a good yield of the monoacylated compound (14) either (5.8) must be considerably faster than (5.8a), or else (5.8a) must be suppressed in some way. Fortunately suppression of (5.8a) is straightforward: addition of a second molar equivalent of a strong base [stronger than (13)] replaces (5.8a) by (5.8c):

$$(14) + Na^+(B)^- \rightleftharpoons (15) + (B)H \quad (5.8c)$$

Admittedly (5.8b) is still in principle a possible side-reaction, but (15) is much less nucleophilic than (13), and so the diacylated compound (16) is seldom an important by-product.

One neat method, which both overcomes the solvent problem and also provides the additional mole of base, is shown below. The use of magnesium ethoxide to form the initial carbanion gives a species (17) which, unlike the sodium salt, is ether-soluble, and also is capable of releasing an equivalent of ethoxide at a subsequent stage (corresponding to 5.8c):

$$CH_2(CO_2C_2H_5)_2 \xrightarrow{Mg(OC_2H_5)_2} C_2H_5OMg\overset{+}{\bar{C}}H(CO_2C_2H_5)_2$$

(17)

$$\xrightarrow[\text{dry ether}]{ClCO_2C_2H_5} CH(CO_2C_2H_5)_3 + C_2H_5OMgCl \quad (88\%)$$

$$[\rightleftharpoons Cl\overset{+}{Mg}\bar{C}(CO_2C_2H_5)_3 + C_2H_5OH].$$

Basic hydrolysis of diethyl acylmalonates is of no value as a synthetic method, since it is accompanied by cleavage [deacylation: (5.9)]. Hydrolysis in aqueous acid gives acylmalonic acids and hence, by decarboxylation, methyl ketones (5.10):

$$RCOCH(CO_2C_2H_5)_2 \xrightarrow[H_2O]{NaOH} RCO_2^-Na^+ + CH_2(CO_2^-Na^+)_2 \quad (5.9)$$

$$RCOCH(CO_2C_2H_5)_2 \xrightarrow[H_2O]{H^+} [RCOCH(CO_2H)_2 \xrightarrow{-CO_2}$$

$$RCOCH_2CO_2H \xrightarrow{-CO_2}] RCOCH_3 \quad (5.10)$$

This last reaction, in conjunction with (5.8), provides another method for the conversion of RCOCl into $RCOCH_3$. Although on paper it is more complicated than the reaction with dimethylcadmium (section 4.2.2) or with lithium dimethylcuprate (section 4.2.3.ii), in practice it presents no difficulties, and the overall yields are high, e.g.

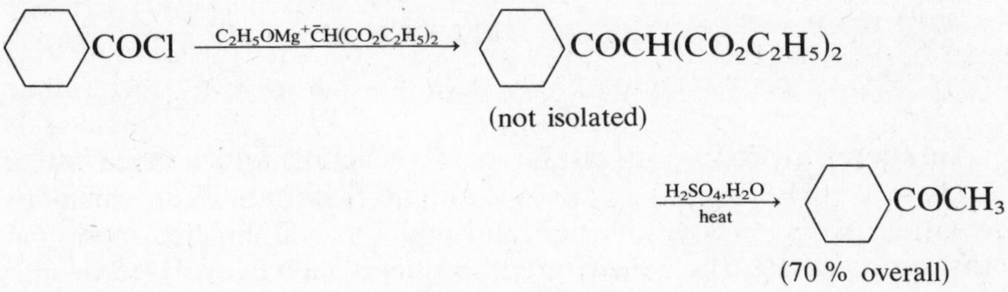

(not isolated)

(70 % overall)

Similarly

(85 % overall)

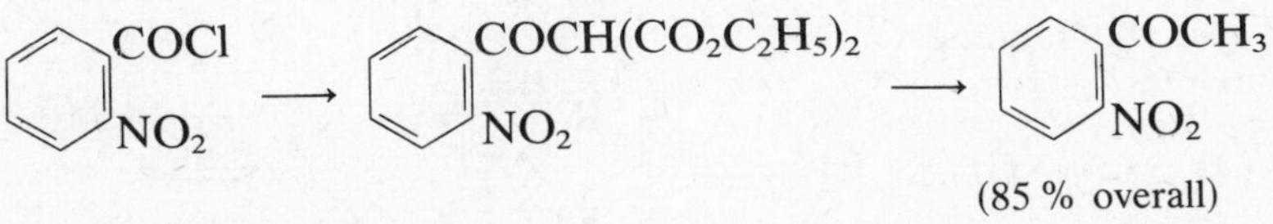

It is sometimes possible to arrest the hydrolysis of the acylmalonic ester at the half-way stage (5.11), and decarboxylation then yields a β-keto-ester; but yields in this reaction are seldom better than 50 %:

(5.11)

Thus,

$$CH_3(CH_2)_{10}COCl \rightarrow CH_3(CH_2)_{10}COCH(CO_2C_2H_5)_2 \xrightarrow[\text{(ii) heat}]{\text{(i) } H_2SO_4,CH_3CO_2H}$$

$$CH_3(CH_2)_{10}COCH_2CO_2C_2H_5 \quad (48\%)$$

$$+ CH_3(CH_2)_{10}COCH_3 \quad (44\%)$$

Acylation of β-keto-esters followed by base-catalysed cleavage is sometimes a useful synthetic procedure. The acylation product, a diketo-ester, undergoes nucleophilic attack at the most electrophilic carbonyl group (one of the keto-functions), and the product, like the starting material, is a β-keto-ester, e.g.

$$CH_3COCH_2CO_2C_2H_5 \xrightarrow[\text{(ii) PhCOCl}]{\text{(i) Na,benzene}} \begin{array}{c} CH_3CO \\ PhCO \end{array}{>}CHCO_2C_2H_5 \xrightarrow[NH_4^+Cl^-]{NH_3,H_2O}$$

$$\text{(70 \%)}$$

$$PhCOCH_2CO_2C_2H_5 \quad \text{(54 \% overall)}$$
$$\text{(77 \%)}$$

An interesting 'one-pot' variant of this reaction, which gives better yields (ca. 70 % overall) involves a simple Schotten–Baumann benzoylation (with benzoyl chloride and aqueous sodium hydroxide) of ethyl acetoacetate; the benzoylated product is then hydrolysed *in situ*.

5.1.4 Condensation reactions[2]

In general terms, carbanions participate in condensation reactions according to the scheme:

$$XCH_2Y + B^- \rightleftharpoons X\bar{C}HY + BH \xrightarrow{RCOR^1} \begin{array}{c} R \\ R^1 \end{array}{>}C{<}\begin{array}{c} O^- \\ CHXY \end{array} \xrightleftharpoons[]{\substack{BH \\ (\text{or } BH^+)}}$$

$$\text{(or } \ddot{B}) \qquad\qquad \text{(or } BH^+)$$

$$\begin{array}{c} R \\ R^1 \end{array}{>}C{<}\begin{array}{c} OH \\ CHXY \end{array} + B^- \xrightleftharpoons[]{-H_2O} \begin{array}{c} R \\ R^1 \end{array}{>}C{=}CXY \qquad\qquad (5.12)$$

$$\text{(or } \ddot{B})$$

Several important points emerge from this general reaction.

(i) The overall stoicheiometry of the reaction is simply

$$XCH_2Y + RCOR^1 \rightarrow RR^1C{=}CXY + H_2O.$$

Thus, even although the base may be concerned in the rate-determining step (i.e. although its concentration may determine the **rate** of reaction), it is **not consumed** in the reaction, but is regenerated in a subsequent step. It is therefore unnecessary to use a stoicheiometric quantity of the base, and a catalytic amount may indeed be sufficient.

(ii) Since the compund XCH_2Y does not require to be converted completely into the carbanion prior to the introduction of the carbonyl compound, it is possible to use a weaker base for a condensation than that required for alkylation or acylation.

(iii) Since all the steps are, in theory, reversible, it may be advantageous to force the reaction to completion by removing the water formed in the last step.

(iv) If the system contains more than one carbanion source, and/or more than one carbonyl group, the condensation occurs preferentially *via* attack of the most stabilised carbanion on the most electrophilic carbonyl carbon atom.

It follows from the above that, if X and Y in reaction (5.12) are both -*M* groups, condensation reactions with aldehydes and ketones should occur in presence of relatively weak bases. We have already pointed out [reaction (5.2)] that amines such as piperidine can deprotonate diethyl malonate, ethyl acetoacetate, etc., to an appreciable extent, and as a result piperidine and other amines successfully bring about condensations involving these highly stabilised carbanions and aldehydes or ketones (generally known as **Knoevenagel condensations**). Thus, for example,

$$PhCHO + CH_2(CO_2C_2H_5)_2 \xrightarrow[0°]{piperidine(0.05 \text{ mol})} PhCH{=}C(CO_2C_2H_5)_2 \quad (75\%)$$

$$\underset{O}{\text{[furan]}}CHO + CH_2(CN)_2 \xrightarrow{PhCH_2NH_2} \underset{O}{\text{[furan]}}CH{=}C(CN)_2 \quad (97\%)$$

$$\overset{\displaystyle CH_3}{\underset{\displaystyle |}{CH_3CH_2CHCHO}} + CH_3COCH_2CO_2C_2H_5 \xrightarrow[0°]{piperidine}$$

$$\overset{\displaystyle CH_3}{\underset{\displaystyle |}{CH_3CH_2CHCH}}{=}C\overset{\displaystyle COCH_3}{\underset{\displaystyle CO_2C_2H_5}{<}} \quad (83\%)$$

In these three cases there is no ambiguity regarding the most acidic hydrogen, and the aldehyde group provides a highly electrophilic carbon: the yields are therefore high. Even when the aldehyde is of the type RCH_2CHO, and can in theory undergo self-condensation (section 5.2.4.1), amines are not sufficiently basic to produce a significant equilibrium concentration of the carbanion $R\bar{C}HCHO$, and so self-condensation is rarely an important side-reaction.

Whereas aldehydes undergo Knoevenagel condensations with a wide variety of carbanion sources (or **active methylene compounds**, as they are often called), the same is not true of ketones. Simple ketones undergo Knoevenagel reactions with malononitrile $[CH_2(CN)_2]$ and ethyl cyanoacetate, but rarely with diethyl malonate (except in presence of titanium tetrachloride) or ethyl acetoacetate.

Whether this selectivity is due to decreased electrophilicity or increased steric hindrance in the ketone, or to increased nucleophilicity or smaller steric demand in the cyano-stabilised nucleophile (or to any combination of these) is not clear.

A remarkable feature of Knoevenagel condensations is the increase in yield which often results from the addition of a catalytic amount of an organic *acid* to the reaction mixture, or alternatively when an ammonium salt (usually the acetate) is used as catalyst in place of the free amine. The exact function of the acid is not fully understood. It may serve to catalyse the formation of a (highly electrophilic) **iminium salt** from the carbonyl compound and the amine [reaction (5.13)]. Alternatively (or additionally) it may serve to promote the dehydration which is the final step in the condensation process (5.12). Its function may well be different in different cases. But its effectiveness is not in doubt, as the following examples show:

$$\diagup C{=}O \underset{}{\overset{R_2NH,H^+}{\rightleftharpoons}} \diagup C{=}\overset{+}{N}R_2 + H_2O \qquad (5.13)$$

$$CH_3(CH_2)_2CHO + CH_2(CO_2C_2H_5)_2 \xrightarrow[\text{boiling benzene}^{[3]}]{\text{piperidinium acetate}}$$

$$CH_3(CH_2)_2CH{=}C(CO_2C_2H_5)_2 \quad (59\,\%)$$

$$CH_3COCH(CH_3)_2 + NCCH_2CO_2C_2H_5 \xrightarrow[\text{boiling benzene}]{CH_3CO_2^-NH_4^+}$$

$$\begin{array}{c} (CH_3)_2CH \\ \hspace{1cm} \diagdown \\ \hspace{1.5cm} C{=}C(CN)CO_2C_2H_5 \quad (64\,\%\text{: possibly a mixture of } E\text{- and } Z\text{-}) \\ CH_3 \diagup \end{array}$$

The other most important variant of the Knoevenagel condensation is that in which one or both of the -*M* groups stabilising the carbanion is carboxyl (CO$_2$H). In this process (usually known as the **Doebner condensation**) malonic or cyanoacetic acid generally furnishes the carbanion, and pyridine or quinoline is used as solvent. A small quantity of a stronger base, e.g. piperidine, may be added, although this is not always necessary. In this reaction (5.14), the condensation is accompanied by decarboxylation:

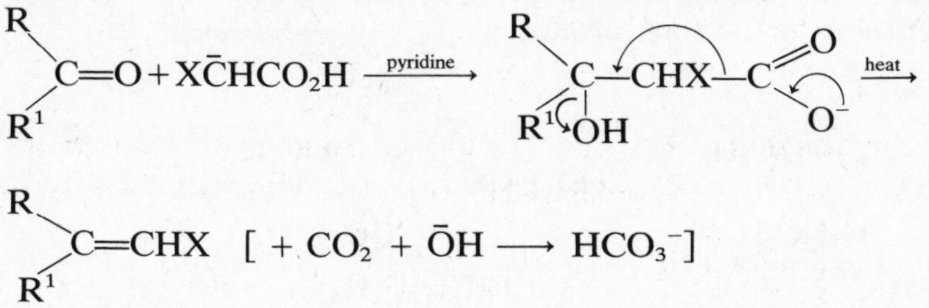

$$\underset{R^1}{\overset{R}{C}}{=}O + X\bar{C}HCO_2H \xrightarrow{\text{pyridine}} \underset{R^1}{\overset{R}{C}}\overset{\displaystyle}{\underset{OH}{C}}{-}CHX{-}C\overset{O}{\underset{O^-}{\diagdown}} \xrightarrow{\text{heat}}$$

$$\underset{R^1}{\overset{R}{C}}{=}CHX \quad [\, + CO_2 + \bar{O}H \longrightarrow HCO_3^-\,] \qquad (5.14)$$

For example,

$$CH_3(CH_2)_5CHO + CH_2(CO_2H)_2 \xrightarrow{\text{pyridine}} CH_3(CH_2)_5CH{=}CHCO_2H$$

(79 %)

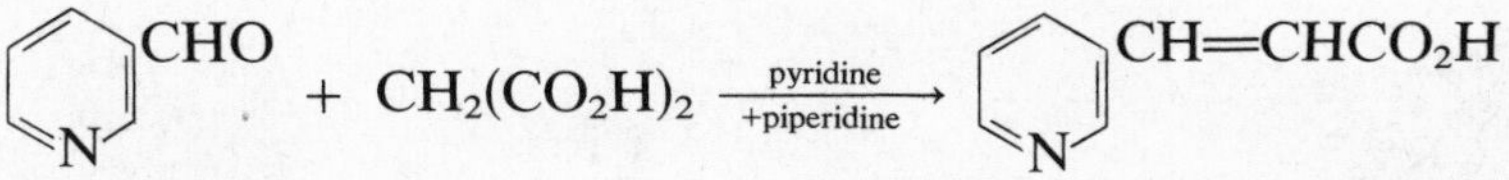

(90 %)

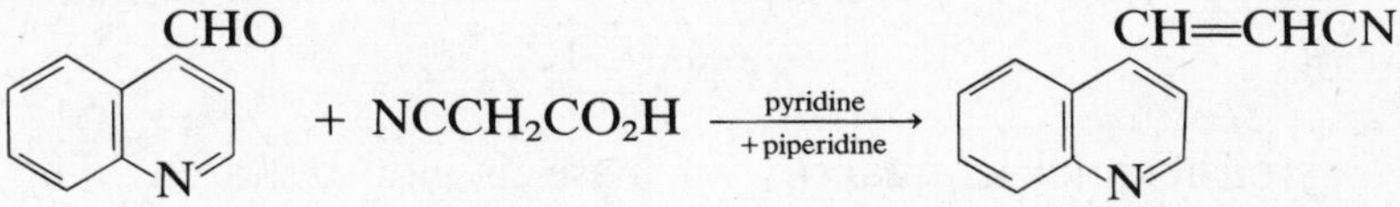

(80 %)

In these reactions the *E*- (i.e. *trans*)-isomer is usually formed.

Alkylmalonic acids may also be used in the Doebner condensation, e.g.

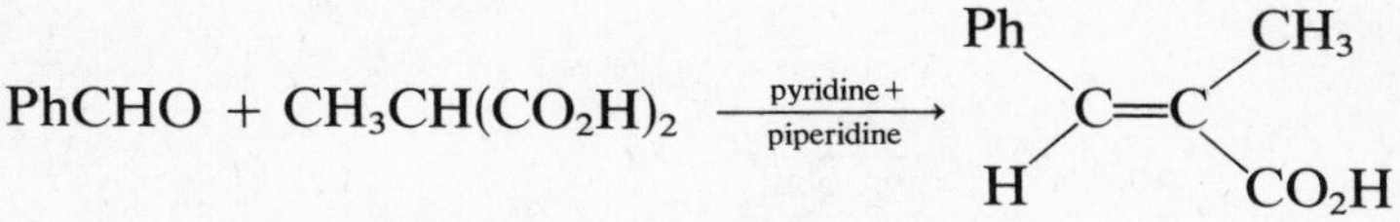

$$PhCHO + CH_3CH(CO_2H)_2 \xrightarrow[\text{piperidine}]{\text{pyridine} +} \underset{H}{\overset{Ph}{C}}{=}\underset{CO_2H}{\overset{CH_3}{C}} \qquad (96 \%)$$

5.1.5 Reactions with $\overset{|}{C}{=}\overset{|}{C}{-}\overset{|}{C}{=}O$ and related systems: the Michael reaction

It sometimes happens that attempts to effect a Knoevenagel condensation between a simple aldehyde and, say, diethyl malonate lead not to the simple condensation product but to one in which one mole of the aldehyde has reacted with two moles of the malonate, e.g.

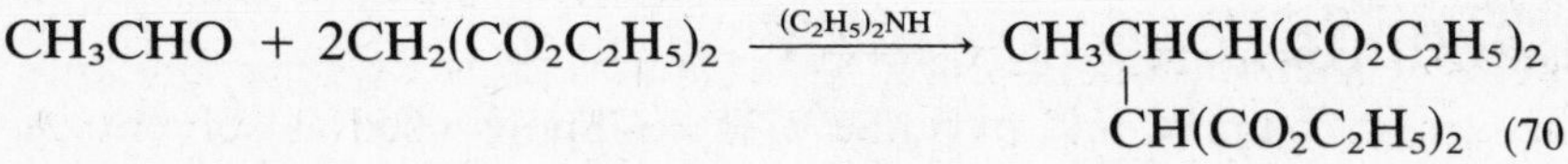

$$CH_3CHO + 2CH_2(CO_2C_2H_5)_2 \xrightarrow{(C_2H_5)_2NH} \underset{\overset{|}{CH}(CO_2C_2H_5)_2}{CH_3\overset{|}{C}HCH(CO_2C_2H_5)_2} \quad (70 \%)$$

Almost certainly, Knoevenagel condensation is the first step, and the condensation product (an α,β-unsaturated ester) then undergoes **conjugate addition** (cf. section 4.1.4) of the second mole of the malonate-derived carbanion. This gives an enolate ion (18) which is protonated to yield the final product:

$$
\begin{array}{ccc}
\underset{\overset{|}{\text{CH}(\text{CO}_2\text{C}_2\text{H}_5)_2}}{\overset{\overset{\displaystyle O}{\parallel}}{\underset{\text{CH}_3\text{CH}=\text{C}}{\text{C}-\text{OC}_2\text{H}_5}}} & \longrightarrow & \underset{\overset{|}{\text{CH}(\text{CO}_2\text{C}_2\text{H}_5)_2}}{\text{CH}_3\text{CH}-\text{C}} \xrightarrow{\text{H}-\text{X}} \text{product}
\end{array}
$$

CH$_3$CH=C with C—OC$_2$H$_5$ (top, O double bond), CO$_2$C$_2$H$_5$, CH(CO$_2$C$_2$H$_5$)$_2$ → CH$_3$CH—C with C—OC$_2$H$_5$ (top, O$^-$), CO$_2$C$_2$H$_5$, CH(CO$_2$C$_2$H$_5$)$_2$ $\xrightarrow{\text{H}-\text{X}}$ product

$$(18)$$

This conjugate addition of a stabilised carbanion to an α,β-unsaturated carbonyl (or cyano- or nitro-) compound has wide applicability, and is generally referred to as the **Michael reaction** (or Michael addition: reaction 5.15). The overall reaction is

$$
\underset{\text{R}^1}{\text{R}_2\text{C}=\text{C}}\overset{\overset{\displaystyle O}{\parallel}}{\text{CR}^2} + \text{XCH}_2\text{Y} \xrightarrow{\text{base}} \underset{\overset{|}{\text{XCHY}}}{\text{R}_2\text{C}}-\underset{\text{R}^1}{\text{CH}}\overset{\overset{\displaystyle O}{\parallel}}{\text{CR}^2} \qquad (5.15)
$$

and so a stoicheiometric quantity of base is not required. The following examples illustrate the generality of the procedure:

$$
\text{PhCH}=\text{CHCOPh} + \text{PhCOCH}_2\text{CO}_2\text{C}_2\text{H}_5 \xrightarrow[\text{or piperidine (0.25 mol)}]{\text{NaOC}_2\text{H}_5(0.33\text{ mol})} \underset{\overset{|}{\text{PhCO}-\text{CHCO}_2\text{C}_2\text{H}_5}}{\text{PhCHCH}_2\text{COPh}}
$$

$$(93\%)$$

$$
\text{CH}_2=\text{CHCN} + \text{CH}_2(\text{CO}_2\text{C}_2\text{H}_5)_2 \xrightarrow[\text{(0.1 mol)}]{\text{NaOC}_2\text{H}_5}
$$

$$
(\text{C}_2\text{H}_5\text{O}_2\text{C})_2\text{CH}-\text{CH}_2\text{CH}_2\text{CN} \quad (55\%)
$$

$$
\text{PhCH}=\text{CHNO}_2 + \text{CH}_2(\text{COCH}_3)_2 \xrightarrow[\text{(3 drops)}]{(\text{C}_2\text{H}_5)_2\text{NH}} \underset{\overset{|}{\text{CH}(\text{COCH}_3)_2}}{\text{PhCHCH}_2\text{NO}_2} \quad (78\%)
$$

Under such conditions, α,β-unsaturated aldehydes may undergo a Knoevenagel-type condensation, or a Michael addition, or (in some cases) both, e.g.

$$
\text{CH}_2=\text{CHCHO} + \text{CH}_2(\text{CO}_2\text{H})_2 \xrightarrow{\text{pyridine}}
$$

$$
\text{CH}_2=\text{CHCH}=\text{CHCO}_2\text{H} \quad (50\%)
$$

$$CH_2\!\!=\!\!CHCHO + CH_2(CO_2C_2H_5)_2 \xrightarrow{\ NaOC_2H_5\ }$$

$$(C_2H_5O_2C)_2CHCH_2CH_2CHO \quad (50\ \%)$$

$$CH_3CH\!\!=\!\!CHCHO + 2CH_2(CO_2C_2H_5)_2 \xrightarrow{\ (C_2H_5)_2NH\ }$$

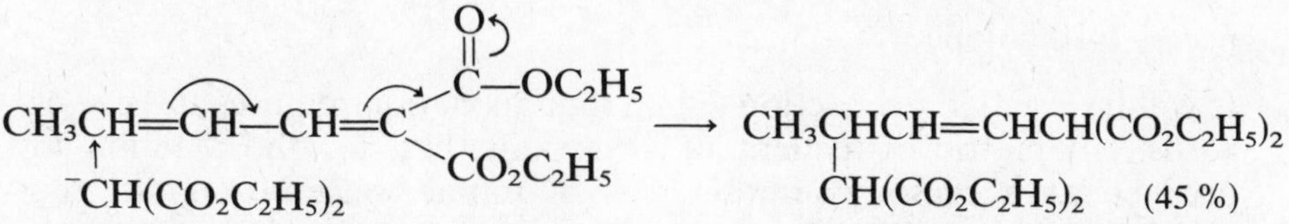

This last example illustrates a further, apparently general, feature of the Michael reaction, namely that when the conjugated system is extended by one or more double bonds, addition occurs preferentially (although not exclusively) at the *end* of the conjugated system. Methyl octa-2,4,6-trienoate similarly undergoes Michael addition mainly at the 7-position (but also at the 3-position):

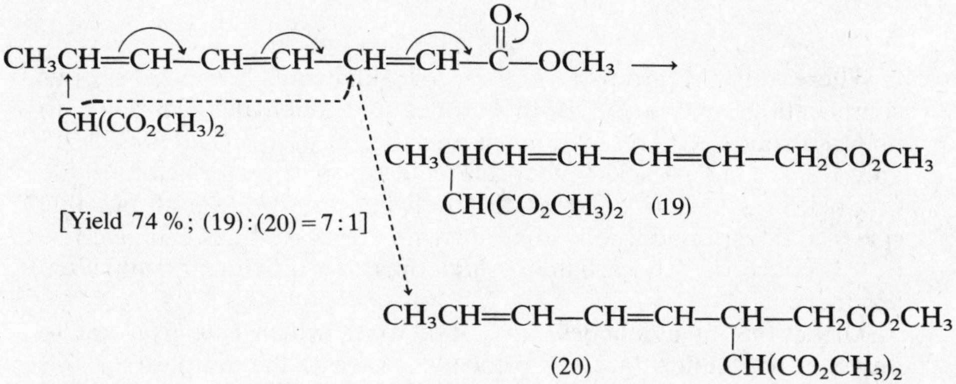

The Michael reaction is also an essential part of many procedures for ring closure, and so will be considered again in Chapter 7 (section 7.1.3).

5.2 Carbanions stabilised by one -*M* group

The reactions in section 5.1 all involve carbanions which are formed comparatively easily by deprotonation of relatively strong carbon acids ($pK_a \le 13$). Compounds of the type RCH_2X or R_2CHX, however (X being a -*M* group) are in general much weaker acids: with the exception of nitroalkanes ($pK_a \approx 9$–10), the compounds in question have pK_a values of approximately 19–27, and thus even moderately strong bases like sodium alkoxides can do no more than produce an equilibrium concentration of the carbanion. If complete conversion into the carbanion is required, an even stronger base must be used, e.g. an alkali metal amide. In such cases, alcohols cannot of course be used as solvents, since they are more acidic than the carbon acids, and so are deprotonated in preference to the latter.

5.2.1 Alkylation

It will be recalled (cf. section 5.1.1) that alkylation requires the use of a stoicheiometric equivalent of base, and that, for the reaction to proceed at a reasonable rate, a high initial concentration of the carbanion is desirable. Thus very strong bases are generally required. Where the stabilising $-M$ group is a cyano- or an ester group, the reactions are straightforward, e.g.

$$CH_3CN + 3CH_3(CH_2)_3Br \xrightarrow[\text{toluene,heat}]{3NaNH_2} [CH_3(CH_2)_3]_3CCN \quad (84\,\%)$$

$$CH_3CH_2CH_2CO_2CH_3 \xrightarrow[\text{(ii) CH}_3CH_2I]{\text{(i) LDA}^{[4]},\text{THF}} (CH_3CH_2)_2CHCO_2CH_3 \quad (96\,\%)$$

Where the $-M$ group is ketonic or aldehydic, however, serious complications may arise. Both ketones and aldehydes are liable to undergo condensation reactions in presence of base (cf. sections 3.3.2 and 5.2.4), and so self-condensation of starting material or of the product, or a 'mixed' condensation between the two, are all possible side-reactions. In addition, some unsymmetrical ketones can give rise to a mixture of two carbanions and thus to a mixture of alkylated products.

Alkylations of aldehydes, and of ketones which can give rise to only one carbanion, are (in principle at least) the simplest of this group, since the only problem is the avoidance of condensation processes. These, of course, involve attack of the carbanion on an unionised carbonyl compound, and occur especially readily in the case of aldehydes; it is therefore important to choose experimental conditions which minimise the concentration of free carbonyl compound throughout the process. The carbanions must be formed quantitatively, in an aprotic solvent, by slow addition of the ketone or aldehyde to a solution of the base (i.e. the base is always in excess), and then an excess (up to ten-fold) of the alkylating agent must be added rapidly (i.e. so that alkylation is kinetically the most favoured process). Examples of such reactions include the following:

$$(CH_3)_2CHCHO \xrightarrow[\text{(ii) BrCH}_2CH=C(CH_3)_2]{\text{(i) KH,THF}} \underset{\underset{CH_2CH=C(CH_3)_2}{|}}{(CH_3)_2C-CHO} \quad (88\,\%)$$

$$PhCOCH_2C_2H_5 \xrightarrow[\text{(ii) C}_2\text{H}_5\text{Br}]{\text{(i) Ph}_3\text{CNa,ether}} PhCOCH(C_2H_5)_2 \quad (62\ \%)$$

Ketones possessing α-hydrogens on both sides of the carbonyl group can (and generally do) give rise to a mixture of two carbanions, and alkylation of such ketones thus gives a mixture of products. Which isomer will predominate is not always easy to predict. If the deprotonation is carried out by slow addition of the ketone to a molar equivalent of the base in an aprotic solvent, the deprotonation is essentially irreversible, and the ratio of carbanions is determined by the relative rates at which the two α-protons are abstracted. This process is *kinetically controlled,* and the less hindered α-proton is generally removed more rapidly, as the following examples show:

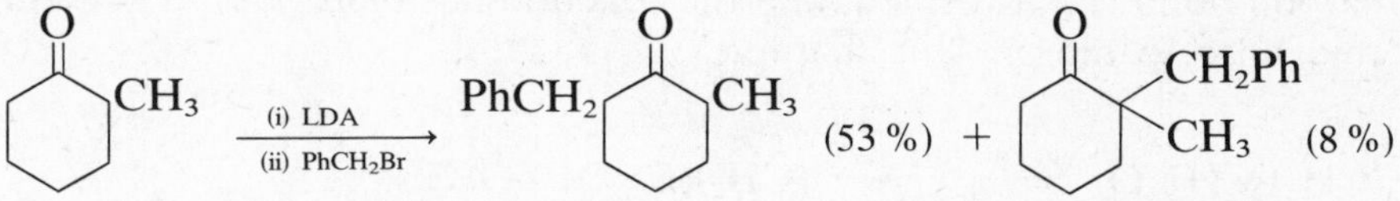

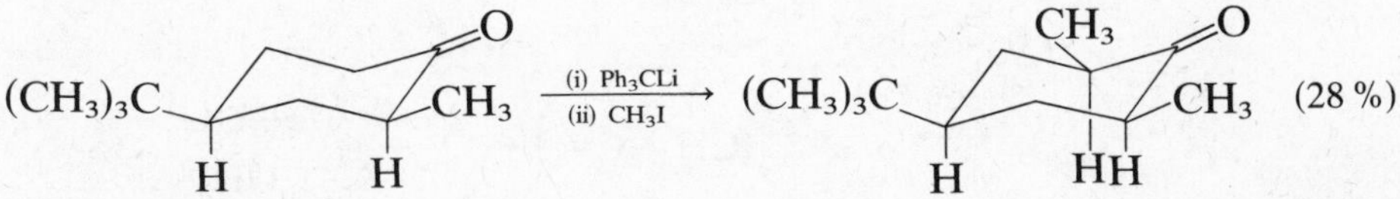

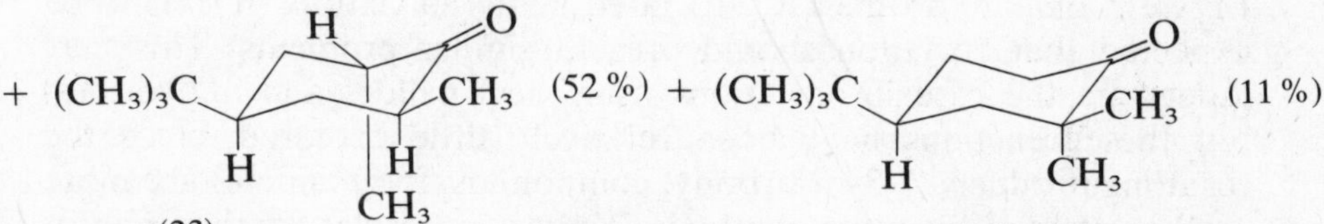

If, on the other hand, the carbanions are generated in the presence of a proton source (even if this is only a small excess of the unionised ketone) the deprotonation step is reversible, and the process is then *thermodynamically controlled.* The product ratios obtained under such conditions may differ substantially from those obtained by the kinetically controlled processes, e.g.

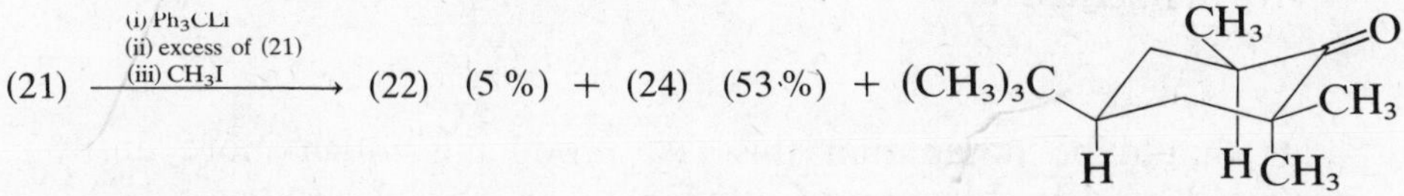

The above, however, is still an over-simplified picture. The carbanion ratio (and hence the product ratio) depends on other factors: for example, the structure of the ketone, the steric demands of the base, the nature of the cation, and the solvent. A full discussion of all these factors is beyond the scope of this book: it is sufficient for our purposes to recognise the difficulties in such reactions, and their limited synthetic utility. In the next section but one (5.2.3) we shall explore some indirect methods for the synthesis of α-alkylated ketones which offer a means of avoiding these difficulties.

Finally, it should be noted that nitroalkanes, although alkylated under much milder conditions than the other types of compound considered above, usually (with very few exceptions) react at *oxygen* rather than at carbon, as in the following example: a new carbon–carbon bond is not formed, and the reaction has therefore no general relevance to the present chapter:

$$(CH_3)_2CHNO_2 \xrightarrow[\text{(ii)}]{\text{(i) NaOC}_2\text{H}_5}$$

5.2.2 Acylation

In view of the problems associated with alkylation, it might be expected that acylation should present similar problems. This may indeed be the case in acylations using acyl halides and anhydrides, but these reactions have been relatively little explored, since the resulting products (1,3-dicarbonyl compounds, for example) are more easily obtained by other methods. Known examples of this type of acylation conform to the expected pattern, however, and require a molar equivalent of a very strong base per mole of acylating agent, e.g.

$$PhCOCH_3 \xrightarrow[\text{(ii) PhCH=CHCOCl}]{\text{(i) NaNH}_2} PhCOCH(COCH{=}CHPh)_2 \quad (66\ \%)$$

$$(CH_3)_2CHCO_2C_2H_5 \xrightarrow[\text{(ii) CH}_3\text{COCl}]{\text{(i) Ph}_3\text{CNa}} (CH_3)_2\underset{\underset{\displaystyle COCH_3}{|}}{C}{-}CO_2C_2H_5 \quad (51\ \%)$$

[It should be noted that this last product is much more easily prepared by dimethylation of ethyl acetoacetate (cf. section 5.1.1).]

The simplest acylation method, in practice at least, for simple ketones and esters is that which uses an ester as the acylating agent. The most familiar form of this acylation (the so-called 'Claisen ester condensation'[5]) involves the formation of ethyl acetoacetate from two molecules of ethyl acetate in presence of sodium ethoxide:

$$CH_3CO_2C_2H_5 + CH_3CO_2C_2H_5 \xrightarrow[-C_2H_5OH]{NaOC_2H_5} CH_3COCH_2CO_2C_2H_5 \quad (75\ \%)$$

It may, however, also be applied to ketones and cyano-compounds, e.g.

$$PhCO_2C_2H_5 + CH_3COPh \xrightarrow{NaOC_2H_5} PhCOCH_2COPh \quad (62\ \%)$$

$$PhCO_2CH_3 + CH_3CN \xrightarrow{NaOCH_3} PhCOCH_2CN \quad (70\ \%)$$

These reactions appear to defy the general rules we have established in earlier sections. The sodium alkoxides are not sufficiently basic to produce more than a small equilibrium concentration of carbanion from any of these very weak acids. However, alkoxides are certainly basic enough to deprotonate the *products* (cf. section 5.1, introduction) and since all the steps in this type of acylation are reversible [reaction (5.16): cf. Sykes, pp. 224–7] the quantitative conversion of the products into their anions provides the driving force for the reactions:

$$RCH_2CO_2R^1 \xrightarrow{\bar{O}R^1} R\bar{C}HCO_2R^1;$$

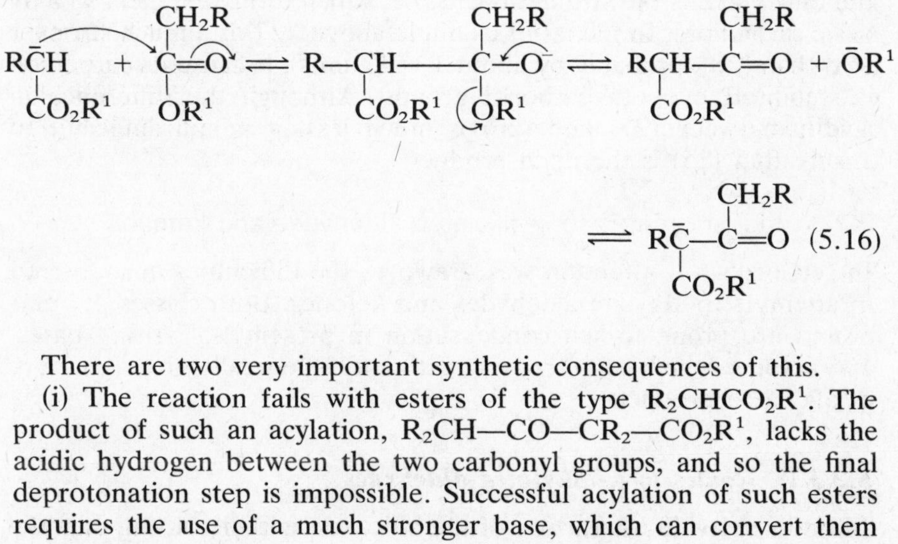

There are two very important synthetic consequences of this.

(i) The reaction fails with esters of the type $R_2CHCO_2R^1$. The product of such an acylation, $R_2CH-CO-CR_2-CO_2R^1$, lacks the acidic hydrogen between the two carbonyl groups, and so the final deprotonation step is impossible. Successful acylation of such esters requires the use of a much stronger base, which can convert them quantitatively into their carbanions:

$$2(CH_3)_2CHCO_2C_2H_5 \xrightarrow{Ph_3CNa} (CH_3)_2\underset{\underset{CO_2C_2H_5}{|}}{C}—COCH(CH_3)_2 \quad (35\%)$$

(ii) Unsymmetrical ketones with α-hydrogens on both sides of the carbonyl group are acylated, almost exclusively, at the less substituted carbon, e.g.

$$CH_3CO_2C_2H_5 + CH_3CO(CH_2)_4CH_3 \xrightarrow{NaNH_2}$$

$$\underset{(25)}{CH_3COCH_2CO(CH_2)_4CH_3} \quad (61\%) + \underset{\underset{(26)}{\underset{COCH_3}{|}}}{CH_3COCH(CH_2)_3CH_3} \quad (0.4\%)$$

(27)   (28)

This is understandable in terms of a mechanism analogous to that of reaction (5.16). Since all the steps are reversible (the reaction is subject to thermodynamic control), the product which accumulates is the one which is the strongest acid (i.e. which forms the most weakly basic carbanion). In the latter example above, (27) is a much stronger acid than (28) (probably by about 10 pK units), because its carbanion is stabilised by both carbonyl groups. Although the difference in acidity between (25) and (26) is much less, it is still sufficient to ensure that (25) is the main product.

5.2.3   Indirect routes to α-alkylated aldehydes and ketones

In section 5.2.1, attention was drawn to the difficulties encountered in attempts to alkylate aldehydes and ketones. Both classes of compound are prone to self-condensation in presence of strong bases, and ketones which can form two carbanions generally give a mixture of alkylated products.

5.2.3.1   Routes to α-alkylated aldehydes

In order to avoid interaction of the aldehyde (either starting material or product) with strong base, the aldehyde may first be converted into a derivative which is less liable to undergo self-condensation. Imines such as (29) may be used in this way (5.17):

$$RCH_2CHO \xrightarrow{R^1NH_2} RCH_2CH{=}NR^1 \xrightarrow[\text{or LDA}]{C_2H_5MgBr^{[6]}} R\bar{C}H{-}CH{=}NR^1$$
(29)

$$\xrightarrow{R^2X} \underset{\underset{R^2}{|}}{RCH}{-}CH{=}NR^1 \xrightarrow{H^+,H_2O} \underset{\underset{R^2}{|}}{RCHCHO} \quad (5.17)$$

$[R^1 = (CH_3)_3C, (CH_3)_2N,$ cyclohexyl]

e.g.

$$(CH_3)_2CHCHO \xrightarrow{\text{(cyclohexyl)}NH_2} (CH_3)_2CHCH{=}N{-}\langle\text{cyclohexyl}\rangle \xrightarrow[\text{(ii) } n\text{-}C_4H_9I]{\text{(i) } C_2H_5MgBr}$$

$$\underset{\underset{(CH_2)_3CH_3}{|}}{(CH_3)_2C}{-}CH{=}N{-}\langle\text{cyclohexyl}\rangle \xrightarrow[\text{H}_2\text{O}]{H^+} \underset{\underset{(CH_2)_3CH_3}{|}}{(CH_3)_2CCHO}$$
(70 % overall)

The other general, indirect route to α-alkylated aldehydes involves the alkylation of a heterocyclic compound of the general type $RCH_2{-}\langle\overset{N}{\underset{X}{}}\rangle$. Several such heterocyclic systems have been exploited in this way, but the best known (and possibly the most versatile) are the dihydro-1,3-oxazines (30: R = H, alkyl, aryl, $CO_2C_2H_5$, etc.). The simple preparation of these compounds, and their use in aldehyde synthesis, are outlined below [reactions (5.18)–(5.20), and the examples which follow].

$$RCH_2CN + \underset{\underset{HO}{}}{HO}\text{—}C(CH_3)_2\text{—}CH_2\text{—}C(CH_3)_2OH \xrightarrow{c. H_2SO_4} (30) \quad (R{=}H{:}65\text{ \% yield;}\ R{=}Ph{:}50\text{ \%})$$
(30)

(5.18)

$$(30) \xrightarrow[\text{THF, }-78°]{n\text{-}C_4H_9Li} \underset{\underset{Li^+}{R\bar{C}H}}{} \xrightarrow{R^1X} \underset{\underset{R^1}{RCH}}{} (31) \xrightarrow[\text{(cf. Section 8.4.6)}]{NaBH_4}$$
(31)

$$\underset{\underset{R^1}{RCH}}{H} \xrightarrow[\text{H}_2\text{O}]{H^+} \underset{\underset{R^1}{|}}{RCHCHO} + \underset{\underset{H_2N}{}}{HO}\text{—}C(CH_3)_2\text{—}CH_2\text{—}C(CH_3)_2NH_2 \quad (5.19)$$

(31) 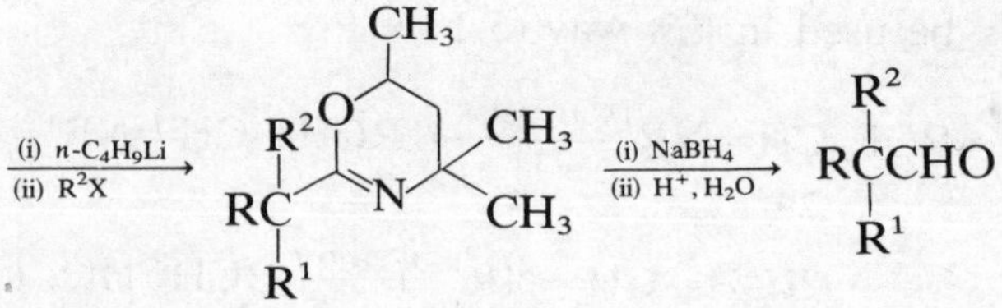 $\xrightarrow[\text{(ii) R}^2\text{X}]{\text{(i) } n\text{-C}_4\text{H}_9\text{Li}}$ [structure] $\xrightarrow[\text{(ii) H}^+, \text{H}_2\text{O}]{\text{(i) NaBH}_4}$ $\underset{R^1}{\overset{R^2}{RCCHO}}$ (5.20)

[structure] $\xrightarrow[\text{(ii) (CH}_3)_2\text{CHI}]{\text{(i) } n\text{-C}_4\text{H}_9\text{Li} \quad \text{(iii) NaBH}_4 \quad \text{(iv) H}^+, \text{H}_2\text{O}}$ $(CH_3)_2CHCH_2CHO$ (49 %)

[structure] $\xrightarrow[\text{(ii) ClCH}_2\text{CH}_2\text{Br}]{\text{(i) } n\text{-C}_4\text{H}_9\text{Li(1 mol)}}$ [bracketed structure]

(i) n-C$_4$H$_9$Li
(ii) [epoxide]

n-C$_4$H$_9$Li

(iii) usual work-up

[bracketed structure]

CHO
|
$PhCHCH_2CH_2OH$ (69 %)

$\xrightarrow[\text{(ii) H}^+, \text{H}_2\text{O}]{\text{(i) NaBH}_4}$ [structure] (57 %)

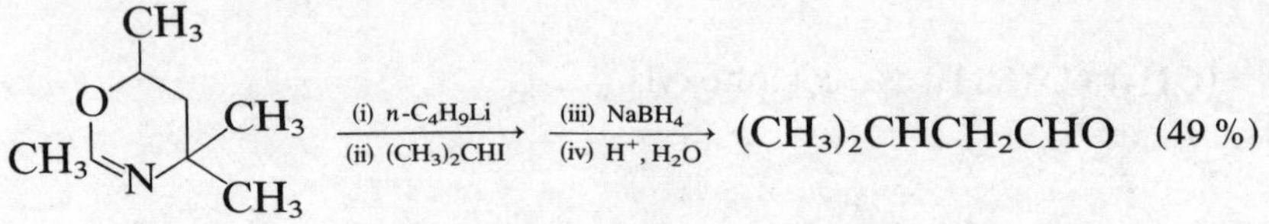

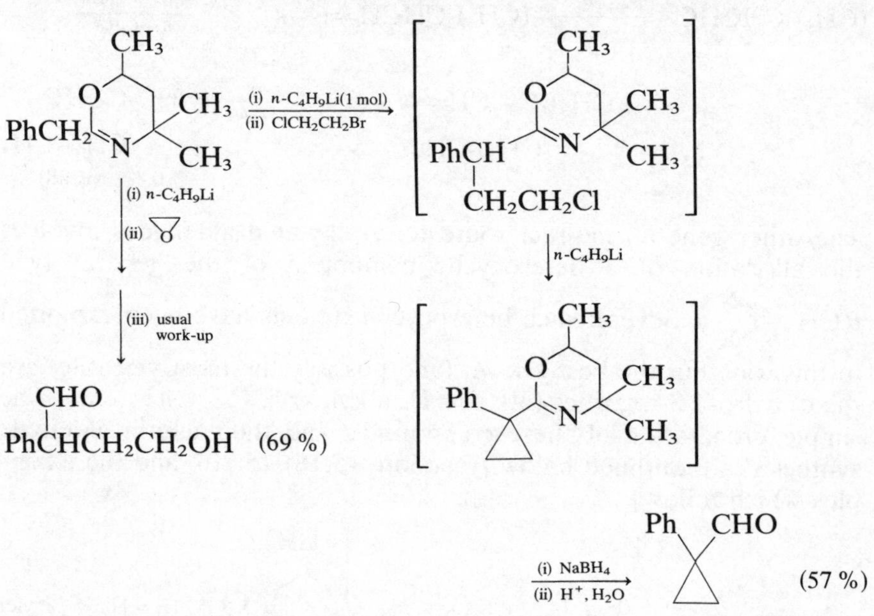

α-Alkylation of aldehydes using enamine intermediates is considered in a later section (5.4.1).

5.2.3.2 Routes to α-alkylated ketones: 'specific enolates'

The self-condensation problem encountered in the alkylation of aldehydes may also arise in connection with alkylation of ketones, although in the latter case it is considerably less important. If contact of the ketone with strong base is the only problem, it may be overcome by protection of the carbonyl group, either as an imine [cf. reaction (5.17)] or as an enamine (section 5.4.1). However, the main problem associated with ketone alkylation is that of **regiospecificity**. We have already shown (section 5.2.1) that unsymmetrical ketones which can be deprotonated at either α-carbon frequently give a mixture of alkyl derivatives, and unless the mixture of isomers is (fortuitously) easy to separate, the synthetic value of the procedure is, to say the least, doubtful. The formation of only one of the carbanions (a '**specific enolate**', as it is often called) from such ketones is thus of considerable importance, since *C*-alkylation of such a carbanion yields a single product.

If direct deprotonation of the ketone, under kinetically or thermodynamically controlled conditions, does not yield the required specific enolate, several other approaches are possible. For example, the ketone may be converted into a β-keto-aldehyde (a Claisen acylation: section 5.2.2) and the latter may then be alkylated at the γ-position [cf. reaction (5.3)] by the action of two moles of strong base and one of alkylating agent. For example,

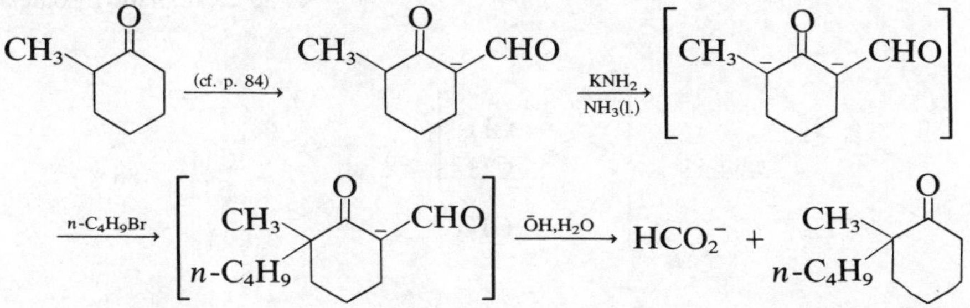

(ca. 50 % overall)

Otherwise, a β-keto-ester may be used as starting material in place of the ketone; this may then be alkylated at either the α- or γ-carbon, according to the conditions [reactions (5.4), (5.5), and (5.5a)]. Hydrolysis and decarboxylation then yield the desired alkylated ketone. Yet another route employs an α,β-unsaturated ketone as starting material; this, when subjected to dissolving metal reduction (section 8.7) yields the specific enolate (32) which may then undergo alkylation [reaction (5.21)]. Alternatively (32) may be generated by conjugate addition (e.g. of a cuprate: cf. section 4.3.2.iii) to the appropriate enone. For example,

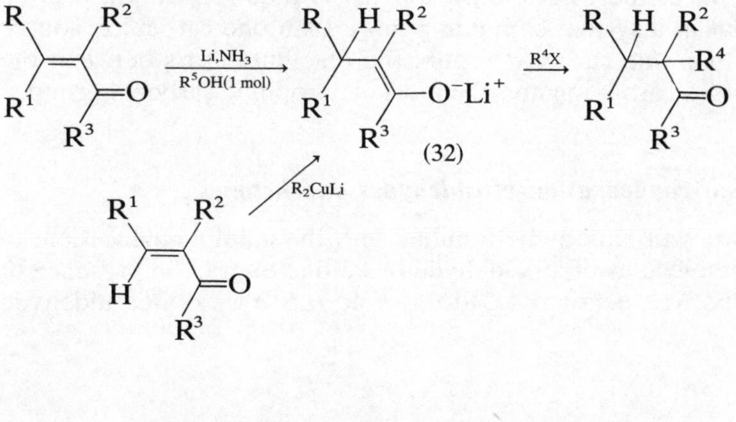

(5.21)

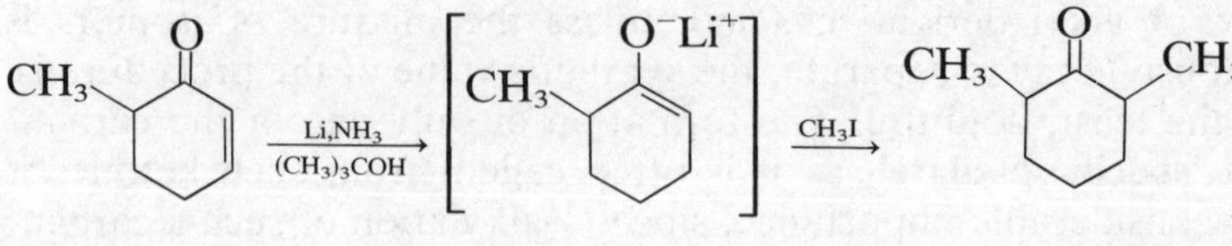

(37 %; no 2,2-dimethyl isomer)

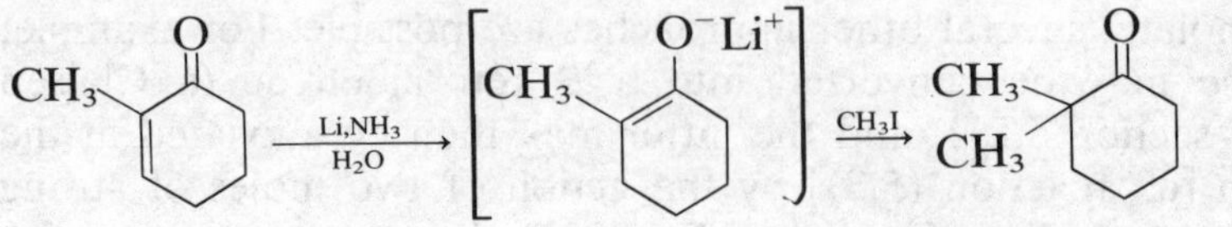

(60 %; no 2,6-dimethyl isomer)

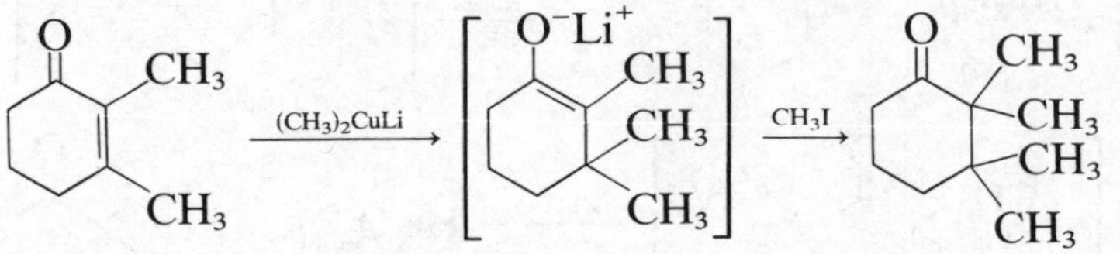

(86 %)

The regiospecificity of alkylations *via* enamines and enol trimethyl-silyl ethers will be considered in later sections (5.4.1 and 13.3.4 respectively).

5.2.4 Condensation reactions[2]

The essential features of condensation reactions have already been set out in sections 3.3.2 and 5.1.4. It will be recalled that it is not necessary (although it may be desirable) to use a stoicheiometric quantity of base, or to use a very strong base; an equilibrium concentration of the carbanion is all that is required. It will also be recalled that in a system containing more than one carbanion source and more than one carbonyl group, the reaction occurs between the most stabilised carbanion and the most electrophilic carbonyl group.

5.2.4.1 Self-condensation of aldehydes and ketones

Most readers will already be familiar with the **aldol condensation**, in which two molecules of an aldehyde or ketone interact in presence of base (or acid: cf. section 5.4.3) to give an α,β-unsaturated aldehyde or ketone [reaction (5.22)]:

$$2RCH_2COR^1 \xrightarrow{\text{base}} RCH_2\underset{\underset{R^1}{|}}{\overset{\overset{OH}{|}}{C}}\text{--}\underset{\underset{H}{|}}{\overset{\overset{R}{|}}{C}}\text{--}COR^1 \xrightarrow{-H_2O} RCH_2\underset{\underset{R^1}{|}}{C}\text{=}\overset{\overset{R}{|}}{C}\text{--}COR^1$$

(33)   (5.22)

Sometimes it is possible to isolate the intermediate addition product (33) in these reactions, and indeed it was the formation of
$CH_3CH(OH)CH_2CHO$ (which is both *ald*ehyde and alco*hol*) from
acetaldehyde which led to the first use of the term 'aldol condensation'. By our definition, however, the formation of aldols must be
described as an addition, and the term **condensation** is reserved for
addition *followed by loss of water*.

Some examples of self-condensation are given below:

$$2CH_3(CH_2)_2CHO \xrightarrow{\text{NaOH,H}_2\text{O}} CH_3(CH_2)_2CH\text{=}\overset{\overset{C_2H_5}{|}}{C}CHO \quad (86\%)$$

$$2PhCH_2CHO \xrightarrow[\text{CH}_3\text{CO}_2\text{H}]{\text{(C}_2\text{H}_5)_2\text{NH,}} PhCH_2CH\text{=}C\overset{\nearrow Ph}{\underset{\searrow CHO}{}} \quad (35\%)$$

$$2CH_3COCH_3 \xrightarrow{\text{Ba(OH)}_2} (CH_3)_2\overset{\overset{OH}{|}}{C}\text{--}CH_2COCH_3 \xrightarrow[\text{I}_2]{\text{heat}} (CH_3)_2C\text{=}CHCOCH_3$$

(52 % overall)

$$2 \quad \text{(cyclopentanone)} \xrightarrow{\text{NaOC}_2\text{H}_5} \text{(bicyclopentylidenone)} \quad (38\%)$$

Although a few of these reactions proceed in good yield, the
majority do not. The products are prone to further reaction with the
carbanion, or may themselves be deprotonated to form other carbanions, which in turn undergo further reactions. So in general, these
processes lead to complex mixtures of products, and they are thus of
little value in laboratory synthesis.

5.2.4.2 Mixed condensations

It follows from the previous section that attempts to condense two
different aldehydes or ketones together may result in even more
complex product mixtures. If both of the compounds can furnish
carbanions equally readily, and if both contain carbonyl groups of

comparable reactivity, four condensation products result: two from self-condensations, e.g. (34) and (35), and two from mixed condensations, e.g. (36) and (37):

$$RCH_2CHO + R^1CH_2CHO \xrightarrow{\text{base}} RCH_2CH{=}\underset{\underset{R}{|}}{C}CHO$$

(34)

$$+ R^1CH_2CH{=}\underset{\underset{R^1}{|}}{C}CHO + RCH_2CH{=}\underset{\underset{R^1}{|}}{C}CHO + R^1CH_2CH{=}\underset{\underset{R}{|}}{C}CHO$$

(35)   (36)   (37)

Mixed condensations are of synthetic value only if they lead to a single product (or, at least, to a mixture containing a preponderance of one product). This is most simply achieved when **one of the reactants contains the most acidic hydrogen and the other contains the most electrophilic carbonyl group**. It should be borne in mind that the order of electrophilicity is aldehyde > ketone > ester, and alkyl-CO- > aryl-CO-; also that acidity of α-hydrogens decreases from aldehyde to ketone to ester.

The problem of mixed condensations is also simplified, of course, if one of the reactants contains no acidic hydrogen. Aromatic (and heteroaromatic) aldehydes, which combine lack of an α-hydrogen with a highly electrophilic carbonyl group, are thus particularly useful as components of mixed condensations,[7] especially when the carbanion source is not also an aldehyde; for example,

$$PhCHO + CH_3COC(CH_3)_3 \xrightarrow[C_2H_5OH, H_2O]{NaOH,} PhCH{=}CHCOC(CH_3)_3 \quad (88\%)$$

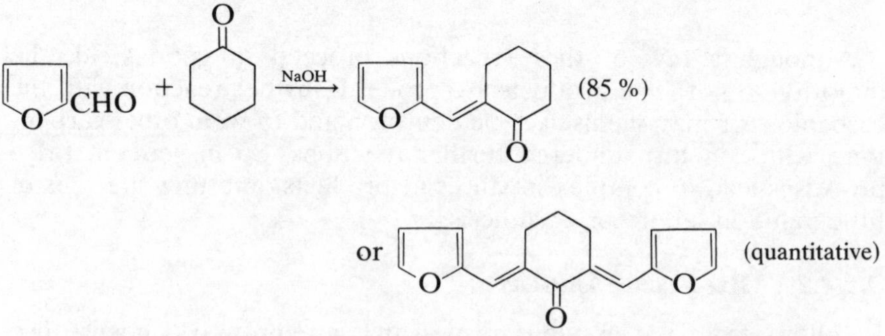

(85 %)

or (quantitative)

(depending on proportions of reactants)

$$PhCHO + CH_3CO_2C_2H_5 \xrightarrow{NaOC_2H_5} PhCH{=}CHCO_2C_2H_5 \quad (68\%)$$

$$PhCHO + CH_3NO_2 \xrightarrow{NaOH} PhCH{=}CHNO_2 \quad (80\%)$$

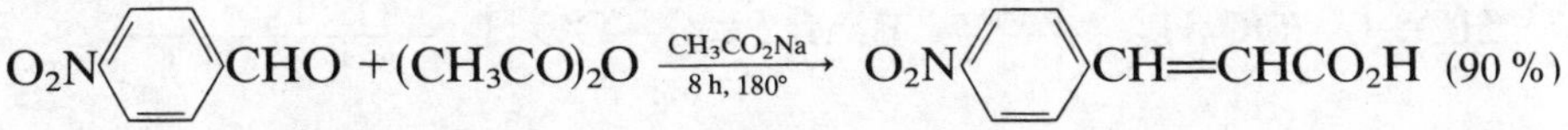

$$O_2N\langle\bigcirc\rangle CHO + (CH_3CO)_2O \xrightarrow[8\,h,\,180°]{CH_3CO_2Na} O_2N\langle\bigcirc\rangle CH{=}CHCO_2H \quad (90\%)$$

Even with other aldehydes as carbanion sources, moderate to good yields of the mixed condensation products may be obtained in some cases (cf. below). In other cases, however, self-condensation of the other aldehyde is the principal reaction:

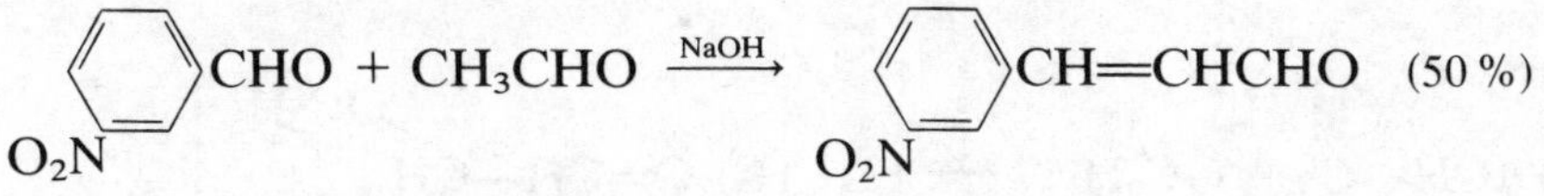

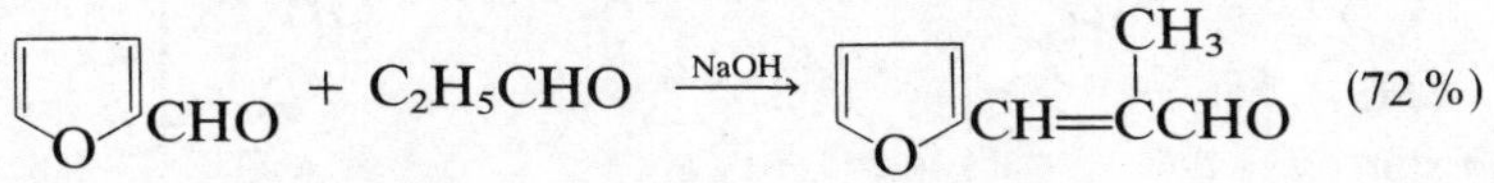

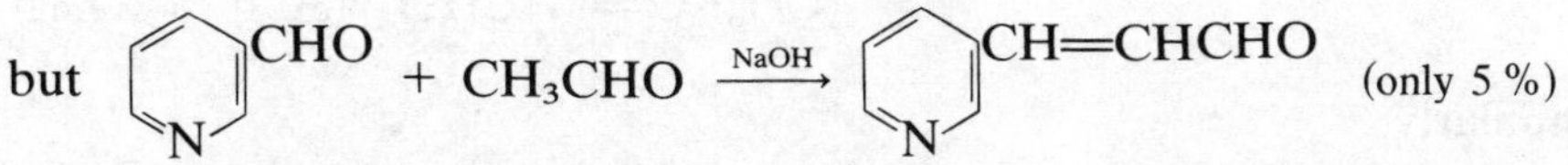

$$\underset{O_2N}{\bigcirc}CHO + CH_3CHO \xrightarrow{NaOH} \underset{O_2N}{\bigcirc}CH{=}CHCHO \quad (50\%)$$

$$\underset{O}{\bigcirc}CHO + C_2H_5CHO \xrightarrow{NaOH} \underset{O}{\bigcirc}CH{=}\overset{\overset{\displaystyle CH_3}{|}}{C}CHO \quad (72\%)$$

but $\underset{N}{\bigcirc}CHO + CH_3CHO \xrightarrow{NaOH} \underset{N}{\bigcirc}CH{=}CHCHO$ (only 5 %)

As a rule, however, aldehyde-derived carbanions do not react with ketonic carbonyl groups, but rather with unionised aldehyde to give self-condensation products. Products of the type $R_2C{=}CHCHO$ or $R_2C{=}C(R^1)CHO$ must therefore be obtained by indirect methods. Some of the methods already described (section 5.2.3.1) for the alkylation of aldehydes may be adapted for this purpose, as the following examples show:

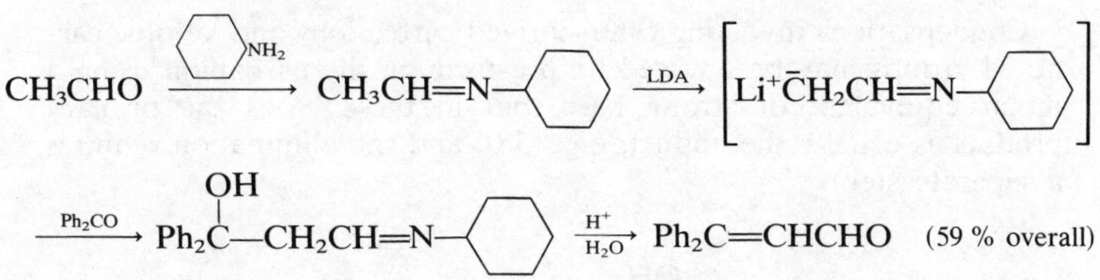

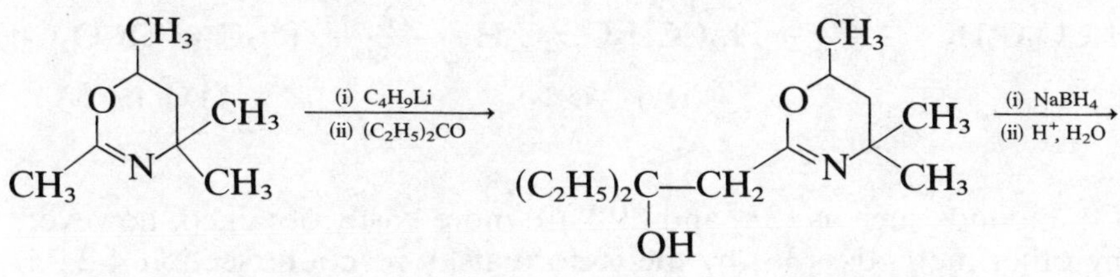

$$CH_3CHO \xrightarrow{\langle\bigcirc\rangle NH_2} CH_3CH{=}N{-}\langle\bigcirc\rangle \xrightarrow{LDA} \left[Li^+\bar{C}H_2CH{=}N{-}\langle\bigcirc\rangle\right]$$

$$\xrightarrow{Ph_2CO} Ph_2\overset{\overset{\displaystyle OH}{|}}{C}{-}CH_2CH{=}N{-}\langle\bigcirc\rangle \xrightarrow[H_2O]{H^+} Ph_2C{=}CHCHO \quad (59\%\ overall)$$

$$\xrightarrow[(ii)\ (C_2H_5)_2CO]{(i)\ C_4H_9Li} \xrightarrow[(ii)\ H^+,\ H_2O]{(i)\ NaBH_4}$$

$$(C_2H_5)_2C{=}CHCHO \quad (62\%\ overall)$$

Another older method makes use of ethoxyethyne, e.g.

$$HC \equiv C-OC_2H_5 \xrightarrow{C_2H_5MgBr} BrMgC \equiv C-OC_2H_5 \xrightarrow[\text{(cf. section 4.3.1. ii)}]{(CH_3)_2CO}$$

$$(CH_3)_2\overset{\displaystyle OH}{\underset{\displaystyle |}{C}}-C \equiv C-OC_2H_5 \xrightarrow[\text{(cf. section 8.4.2)}]{H_2,\ Pd}$$

$$\left[(CH_3)_2\overset{\displaystyle OH}{\underset{\displaystyle |}{C}} \underset{\displaystyle H \quad H}{\overset{\displaystyle OC_2H_5}{\diagdown C = C \diagup}} \right] \xrightarrow{H^+,H_2O}$$

$$\left[(CH_3)_2\overset{\displaystyle \overset{+}{C}OH_2}{C} \underset{\displaystyle H \quad H}{\overset{\displaystyle OC_2H_5}{\diagdown C = C \diagup}} \overset{\displaystyle}{\underset{\displaystyle \overset{\cdot\cdot}{O}H_2}{}} \longrightarrow (CH_3)_2C=CH-CH \underset{\displaystyle \diagdown OH}{\overset{\displaystyle \diagup OC_2H_5}{}} \right] \longrightarrow$$

$$(CH_3)_2C=CHCHO \quad \text{(ca. 10 \% overall)}$$

Similarly,

$$BrMgC \equiv COC_2H_5 \ + \ \text{[structure]} \longrightarrow \longrightarrow \longrightarrow \text{[structure]}$$

citral

(68 % overall)

Condensations involving ester-derived carbanions and ketonic carbonyl groups may be effected by pre-forming the carbanion using a molar equivalent of strong base, but in these cases the primary product is usually the adduct, e.g. (38), and the elimination requires a separate step:

$$CH_3CO_2C_2H_5 \xrightarrow[\text{(ii) Ph}_2\text{CO}]{\text{(i) LiNH}_2} Ph_2\overset{\displaystyle OH}{\underset{\displaystyle |}{C}}CH_2CO_2C_2H_5 \xrightarrow[(-H_2O)]{HCO_2H} Ph_2C=CHCO_2C_2H_5$$

$$\text{(38)} \quad \text{(75 \%)} \qquad\qquad\qquad \text{(39)} \quad \text{(50 \%)}$$

Compounds such as (38) and (39) are more easily obtained, however, by other methods: (38) by the Reformatsky reaction (section 4.2.2.i) and (39) by a Wittig or related reaction (sections 5.3.1.2 and 12.2).

5.2.5 Michael reactions

In principle, conjugate addition to $>\!\!C\!\!=\!\!C\!-\!\!\overset{|}{C}\!\!=\!\!O$ and related systems should be a characteristic of carbanions with one stabilising $-M$ group, just as it is of their doubly stabilised analogues (section 5.1.5). In practice, however, relatively few examples are recorded of this type of Michael reaction. The only exceptions are those in which the carbanion is derived from a nitroalkane (and is therefore produced under relatively mild conditions), and those in which the electrophile is acrylonitrile, $CH_2\!\!=\!\!CHCN$ (which is not only highly reactive and sterically unhindered, but which affords little opportunity for side-reactions). The following examples are typical:

$$CH_3CH_2NO_2 + CH_2\!\!=\!\!CHCOCH_3 \xrightarrow{\text{NaOCH}_3} CH_3\overset{\overset{\displaystyle NO_2}{|}}{C}H\!-\!CH_2CH_2COCH_3$$

(51 %)

$$(C_2H_5)_2CHCHO + CH_2\!\!=\!\!CHCN \xrightarrow{\text{KOH}} (C_2H_5)_2\overset{\overset{\displaystyle CHO}{|}}{C}\!-\!CH_2CH_2CN$$

(75 %)

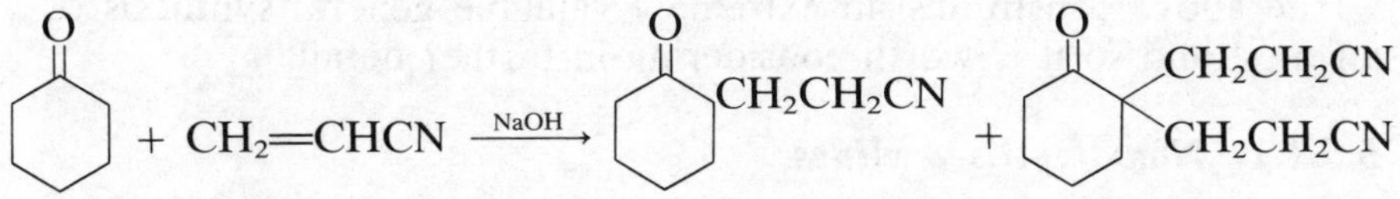

(20 %)   (40 %)

5.3 Carbanions stabilised by neighbouring phosphorus or sulphur

5.3.1 The Wittig reaction

Although the uses of organophosphorus compounds are not considered in detail until Chapter 12, there is one aspect of their chemistry which belongs to the present chapter. This concerns the formation of carbanions by deprotonation of alkyltriphenylphosphonium salts [reaction (5.23a)].

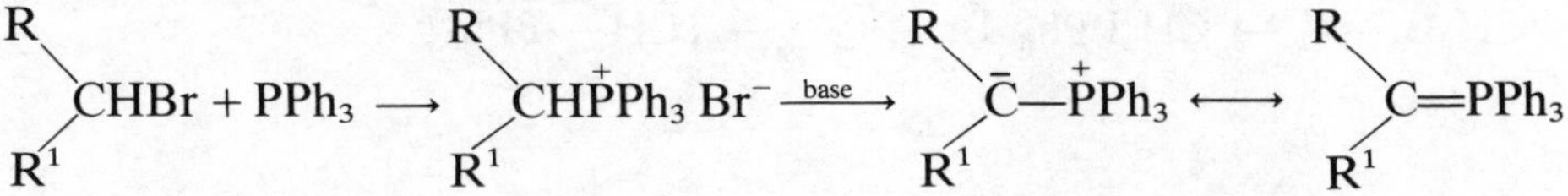

(40)   (41)

(5.23a)

In the products, the negative charge on carbon is balanced by the positive charge on the adjacent phosphorus, and such zwitterions (40) are usually known as **ylides** (or ylids). The alternative **phosphorane** structure (41) implies mesomeric stabilisation of such an ylide by the phosphorus, but as we have already noted (section 3.4.2.ii) the extent of such stabilisation need not concern us here.

The ylides are strong nucleophiles, and as such undergo *C*-alkylation and *C*-acylation with the normal range of reagents. By far their most important reaction from the synthetic viewpoint, however, is their reaction with aldehydes and ketones (the **Wittig reaction**) to give alkenes and triphenylphosphine oxide [reaction (5.23b)]:

$$RR^1C{=}CR^2R^3 + Ph_3P{=}O \quad (5.23b)$$

The above constitutes an extremely valuable general synthesis of alkenes, and so it is worth considering in further detail.

5.3.1.1 Non-stabilised ylides

If R and R^1 in the original alkyl halide are hydrogen or simple alkyl groups, the α-hydrogen of the phosphonium salt is very weakly acidic, and a very strong base (usually butyllithium or phenyllithium) is required to produce the ylide. The ylide, once formed, is a highly reactive compound and is not generally isolable; it is not only strongly basic (deprotonating acids as weak as water), it is strongly nucleophilic in the manner of a Grignard reagent, and reacts rapidly, under mild conditions, to give the adduct (42) effectively irreversibly. This then decomposes spontaneously to give the alkene. If stereoisomerism in the product is possible, a mixture of *E*- and *Z*-isomers is generally obtained. For example,

$$CH_3Br \xrightarrow{PPh_3} CH_3\overset{+}{P}Ph_3 \ \bar{B}r \xrightarrow[\text{(CH}_3)_2\text{SO}]{\text{NaH}} [\bar{C}H_2{-}\overset{+}{P}Ph_3]$$

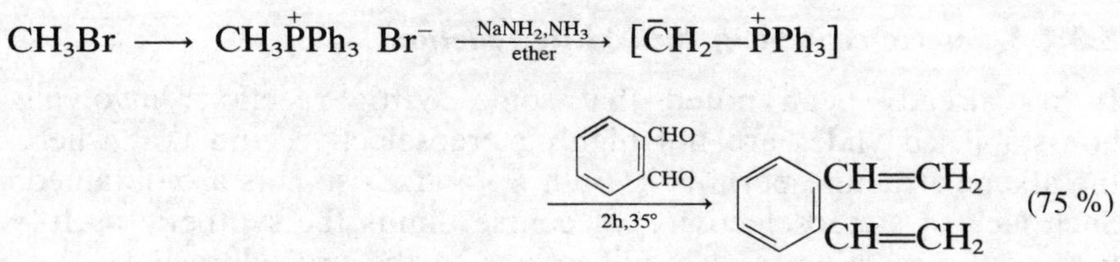

$$CH_3Br \longrightarrow CH_3\overset{+}{P}Ph_3 \ Br^- \xrightarrow[\text{ether}]{NaNH_2,NH_3} [\bar{C}H_2—\overset{+}{P}Ph_3]$$

$$C_2H_5X \rightarrow C_2H_5\overset{+}{P}Ph_3 \ X^- \xrightarrow[\text{or NaNH}_2]{n\text{-}C_4H_9Li} [CH_3\bar{C}H\text{–}\overset{+}{P}Ph_3] \xrightarrow{PhCHO}$$

$$PhCH{=}CHCH_3 \quad (68\text{–}98\,\%)$$
$$(Z\text{- and } E\text{-})$$

[The isomer ratio obtained in this last reaction depends on the nature of X, and of the base used (cf. section 5.3.1.3).]

5.3.1.2 Stabilised ylides

If R or R^1 in reaction (5.23) is a -M group (e.g. an ester), deprotonation of the phosphonium salt is achieved under much less strongly basic conditions, and the resulting ylide (40) [or phosphorane, (41)] is often sufficiently stable to be isolated. It is also sufficiently stable for its addition to carbonyl groups to be reversible [cf. reaction (5.23b)], although it is a relatively weak nucleophile and may not react at all readily with feebly electrophilic carbonyl groups. Where E- and Z-isomers of the final product can exist, it is the E-isomer which usually predominates. For example,

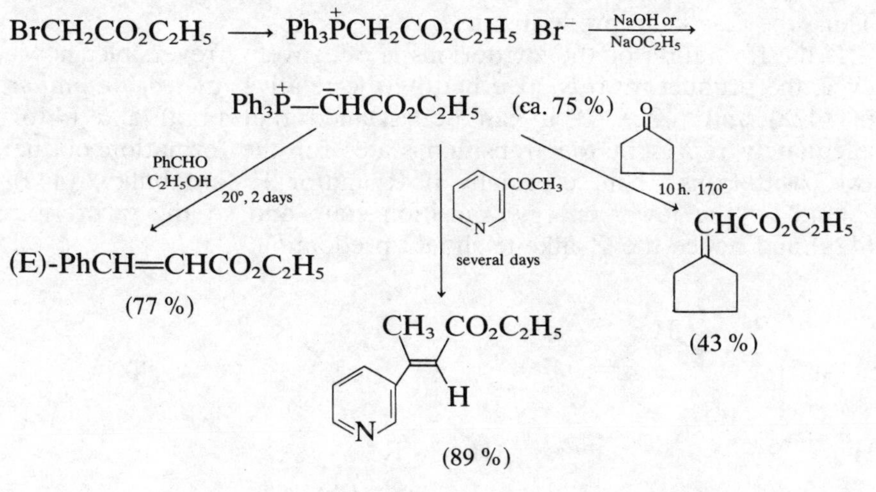

$$BrCH_2CO_2C_2H_5 \longrightarrow Ph_3\overset{+}{P}CH_2CO_2C_2H_5 \ Br^- \xrightarrow[NaOC_2H_5]{NaOH \text{ or}}$$

$$Ph_3\overset{+}{P}—\bar{C}HCO_2C_2H_5 \quad (ca.\,75\,\%)$$

(E)-PhCH=CHCO$_2$C$_2$H$_5$

(77 %)

(89 %)

(43 %)

5.3.1.3   *Steric control in the Wittig reaction*

It has already been noted that some Wittig reactions involving non-stabilised ylides are not highly stereoselective, and that where the alkene structure permits it, both E- and Z-isomers are obtained. Such lack of stereoselectivity, of course, limits the synthetic usefulness of the reaction, and modifications to the procedure have thus been sought in order to improve the steric control.

Since the alkene is formed by decomposition of the zwitterion (42) *via* a cyclic intermediate [cf. reaction (5.23b)], it follows that the stereochemistry of the zwitterion determines the configuration of the alkene. The *erythro*-zwitterion (42a) decomposes to the Z-alkene, and the *threo*-zwitterion (42b) gives the E-alkene:

(42a)   (43a)   (42b)   (43b)

If the formation of the zwitterions (42a) and (42b) is genuinely reversible, as is the case with a stabilised ylide, the product ratio, (43a)/(43b), will be governed by the relative *ease of decomposition* of (42a) and (42b). The eclipsed conformation required to initiate the decomposition is more easily attained in the case of (42b), since there is less steric interaction between the groups R and R^1 than in (42a). Thus, the E-alkene predominates.

If the formation of the zwitterions is effectively irreversible, however, the product ratio is governed by the relative *rates of formation* of (42a) and (42b). If it can be assumed that (44a) and (44b) adequately represent the transition states for the formation of the two zwitterions, then, especially if R and/or R^1 are bulky, (44a) should be the lower-energy transition state and so the zwitterion (42a) and hence the Z-alkene should predominate.

(44a)   (44b)

If a non-stabilised ylide can be obtained [reaction (5.23)] in a solution *free from inorganic salts* (e.g. by using sodamide as the base, and filtering off the sodium halide), its reaction with an aldehyde does indeed give the Z-alkene as the major product, e.g.

$$C_2H_5\overset{+}{P}Ph_3 \ Br^- \xrightarrow{\text{NaNH}_2} [CH_3\overset{-}{C}H\!-\!\overset{+}{P}Ph_3] \xrightarrow{\text{PhCHO}} PhCH\!=\!CHCH_3$$

(98 %; $E:Z = 13:87$)

Similarly,

$$n\text{-}C_3H_7\overset{+}{P}Ph_3 \ Br^- \xrightarrow[\text{(ii) } n\text{-}C_3H_7\text{CHO}]{\text{(i) NaNH}_2} n\text{-}C_3H_7CH\!=\!CHC_2H_5 \quad (49\%; E:Z = 5:95)$$

On the other hand, if the ylide is generated in presence of a lithium halide (e.g. by using an alkyl- or aryl-lithium as the base), the Wittig reaction is much less stereoselective; in addition, the isomer ratio may vary as the halide is changed from bromide to chloride or iodide, and it may also be solvent-dependent. The mechanistic subtleties of these reactions are discussed fully elsewhere[8]; for the purposes of synthesis it is preferable to focus attention on the reactions giving *high* stereoselectivity.

Wittig reactions of non-stabilised ylides may also be modified to yield predominantly E-alkenes. In this modification, the ylide is prepared using phenyllithium, and the addition to the aldehyde is carried out at $-78°$ so that the zwitterions (42a) and (42b) do not undergo the elimination step. Then a second molar equivalent of phenyllithium is added, to form the new ylide (45), and the latter is reprotonated to give (almost exclusively) the more stable *threo*-zwitterion (42b) [reaction (5.24)]. Decomposition of this zwitterion then gives the E-alkene (43b):

(5.24)

For example,

$$C_2H_5\overset{+}{P}Ph_3 \ Br^- \xrightarrow{\text{PhLi}} CH_3\overset{-}{C}H\overset{+}{P}Ph_3 \xrightarrow[\text{(ii) PhLi}]{\text{(i) PhCHO,} -78°} \xrightarrow{\text{(iii) (CH}_3)_3\text{COH}} PhCH\!=\!CHCH_3$$

(73 %; $E:Z = 97:3$)

Similarly,

$$n\text{-}C_6H_{13}\overset{+}{P}Ph_3 \ Br^- + CH_3CHO \rightarrow n\text{-}C_5H_{11}CH=CHCH_3$$

$$(60\,\%;\ E:Z = 96:4)$$

Other phosphorus-containing carbanions and their reactions are discussed in Chapter 12.

5.3.2   Carbanions stabilised by two sulphur atoms

The best-known and most widely used of the carbanions stabilised by divalent sulphur are those derived from 1,3-dithian (46) and its 2-alkyl derivatives. 1,3-Dithian has a pK_a value of 31, and so, although by no means a strong acid, it is converted quantitatively into its carbanion by very strong bases such as butyllithium.

Among the reactions of these carbanions, it is alkylation which to date has proved most useful as a synthetic procedure. Mono- and di-alkylation may be carried out in a stepwise fashion [reaction (5.25)]:

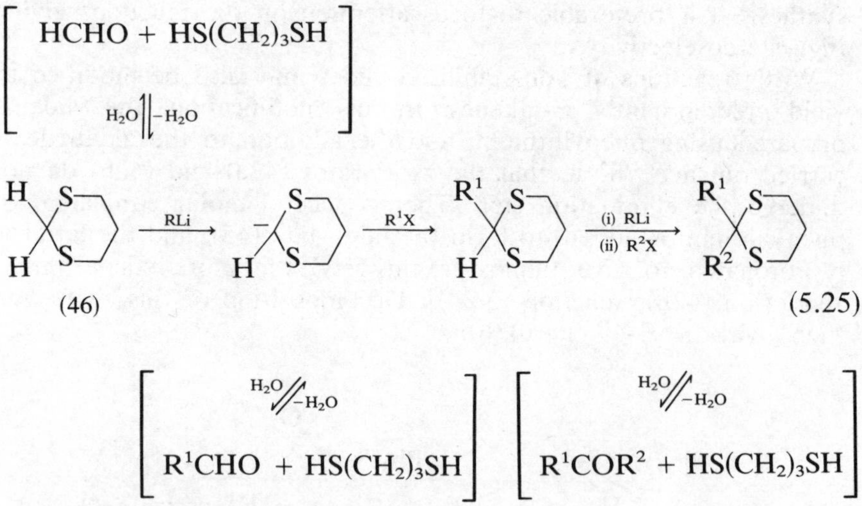

(5.25)

The value of this reaction lies in the fact that 1,3-dithians are dithioacetals or dithioketals, and as such are preparable from, and hydrolysable to, aldehydes or ketones. The method is therefore used to effect the **alkylation of an aldehyde on the carbonyl carbon** using an **electrophilic alkylating agent**, a process not otherwise easily accomplished. The transformation of an electrophilic carbon ($>C=O$) into a nucleophilic one (the deprotonated dithian), and the reverse process, are said to involve **Umpolung**.[9]

Examples of the use of 1,3-dithians include the following:

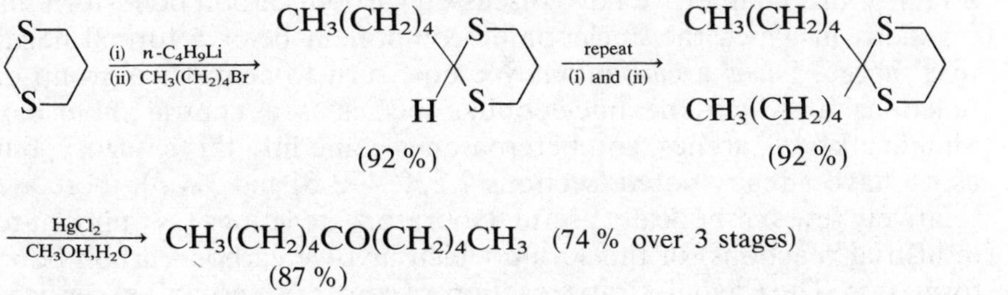

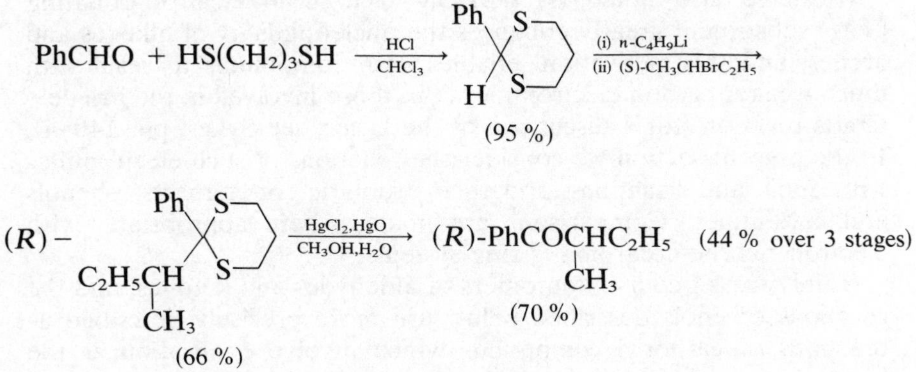

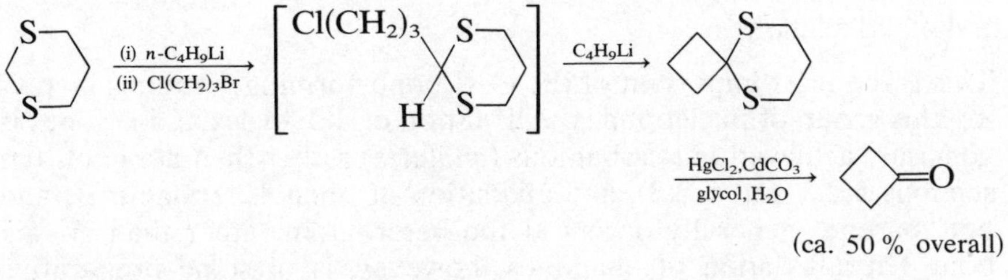

[reaction scheme: 1,3-dithiane → (i) n-C_4H_9Li, (ii) $CH_3(CH_2)_4Br$ → 2-pentyl-1,3-dithiane (92%) → repeat (i) and (ii) → 2,2-dipentyl-1,3-dithiane (92%)]

$$\xrightarrow[CH_3OH,H_2O]{HgCl_2} CH_3(CH_2)_4CO(CH_2)_4CH_3 \quad (74\% \text{ over 3 stages})$$
$$(87\%)$$

$$PhCHO + HS(CH_2)_3SH \xrightarrow[CHCl_3]{HCl}$$ [2-phenyl-1,3-dithiane] (95%) $\xrightarrow[\text{(ii) (S)-}CH_3CHBrC_2H_5]{\text{(i) } n\text{-}C_4H_9Li}$

$$(R)- \text{[2-phenyl-2-(1-methylpropyl)-1,3-dithiane]} \xrightarrow[CH_3OH,H_2O]{HgCl_2,HgO} (R)\text{-}PhCOCHC_2H_5 \quad (44\% \text{ over 3 stages})$$
with C_2H_5CH and CH_3 substituent (66%); product CH_3 (70%)

[reaction scheme: 1,3-dithiane → (i) n-C_4H_9Li, (ii) $Cl(CH_2)_3Br$ → [2-(3-chloropropyl)-1,3-dithiane] → C_4H_9Li → spiro dithiane → $\xrightarrow[\text{glycol, }H_2O]{HgCl_2,CdCO_3}$ cyclobutanone (ca. 50% overall)]

For the synthesis of aldehydes, 1,3,5-trithian (47) provides a convenient (and cheaper!) alternative to 1,3-dithian, e.g.

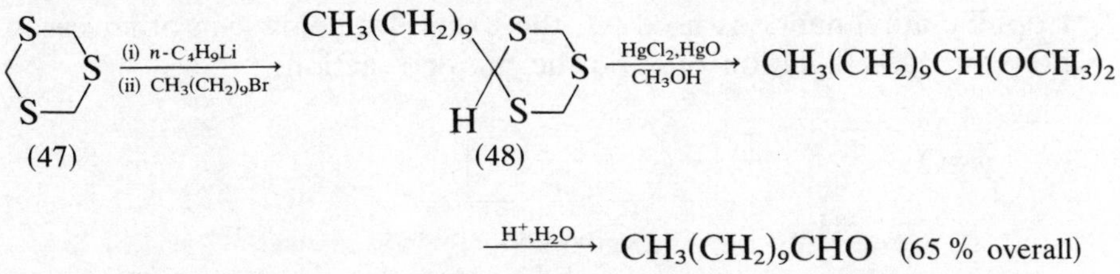

[reaction scheme: (47) 1,3,5-trithiane → (i) n-C_4H_9Li, (ii) $CH_3(CH_2)_9Br$ → (48) 2-decyl-1,3,5-trithiane → $\xrightarrow[CH_3OH]{HgCl_2,HgO} CH_3(CH_2)_9CH(OCH_3)_2$]

$$\xrightarrow{H^+,H_2O} CH_3(CH_2)_9CHO \quad (65\% \text{ overall})$$

The trithian method cannot, however, be used to prepare ketones, since further alkylation of 2-alkyl-1,3,5-trithians [e.g. (48)] occurs at the 4-position.

5.4   Alkene, arene, and heteroarene nucleophiles

So far in this chapter we have discussed carbon–carbon bond-forming reactions in which the nucleophilic component bears a formal negative charge, i.e. is a carbanion. We now turn to consider a group of reactions in which the nucleophilic species is a neutral molecule. Simple alkenes, arenes, and heteroarenes come into this category, but as we have already noted (sections 2.2, 2.4–2.6, and 3.4.3), there are relatively few synthetically useful 'laboratory' reactions (as opposed to industrial reactions) of this group which involve carbon–carbon bond formation. The Friedel–Crafts reaction, of course, is a notable exception.

We have also noted (section 3.4.3) that an electron-donating ($+M$) substituent greatly enhances the nucleophilicity of alkenes and arenes, and this 'activation' enables such compounds to react with much weaker carbon electrophiles than those involved in the Friedel–Crafts reaction (for a discussion of the latter, see Sykes, pp. 140–4). In the present section we consider the reactions of such electrophiles with enols and enamines, and their aromatic counterparts, phenols and arylamines. Comparisons are made, where appropriate, with electron-rich heteroaromatic ring systems.

(Enols are, of course, tautomers of aldehydes and ketones, and the reactions of enols described below are more precisely described as reactions of carbonyl compounds which involve enolisation as the first step.)

5.4.1   Alkylation

This is the least important of the C–C bond-forming reactions involving this group of nucleophiles. Alkylation of aldehydes and ketones is generally achieved *via* carbanions (*enolates*) rather than *via* enols (cf. sections 5.2.1 and 5.2.3), and alkylation of phenols, arylamines, and heteroarenes generally occurs at the heteroatom rather than at carbon. The alkylation of enamines, however, is of some preparative importance, since it provides a method for the indirect α-alkylation of aldehydes and ketones in the absence of strong bases (cf. section 5.2.3). The most useful C-alkylations occur when a highly electrophilic alkyl halide is used (cf. the examples below); in other cases, however, N-alkylation may be the major reaction:

(90 %)

(48 % overall)

(53 %)

Similarly:

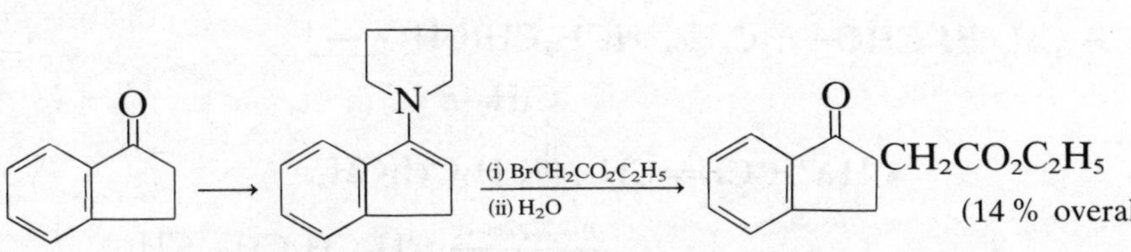

(i) BrCH$_2$CO$_2$C$_2$H$_5$
(ii) H$_2$O

CH$_2$CO$_2$C$_2$H$_5$

(14 % overall)

(35 %)   (41 %)

When an unsymmetrical ketone may give rise to two enamines, it is the more stable of the two which generally predominates. In the case of β-tetralone, the conjugated enamine (49a) is formed, apparently exclusively, at the expense of the non-conjugated isomer (49b); and in the case of 2-methylcyclohexanone the major product formed with pyrrolidine is (50a), since there is a destabilising steric repulsion in the minor isomer (50b) involving the methyl group and the α-hydrogens of the heterocyclic ring:

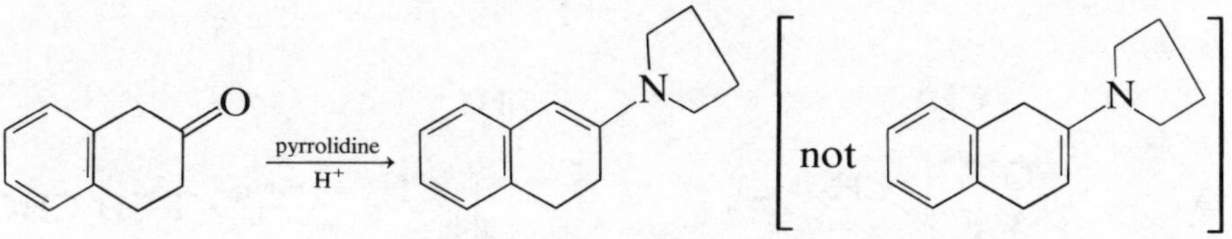

pyrrolidine
H$^+$

not

(49a)   (93 %)   (49b)

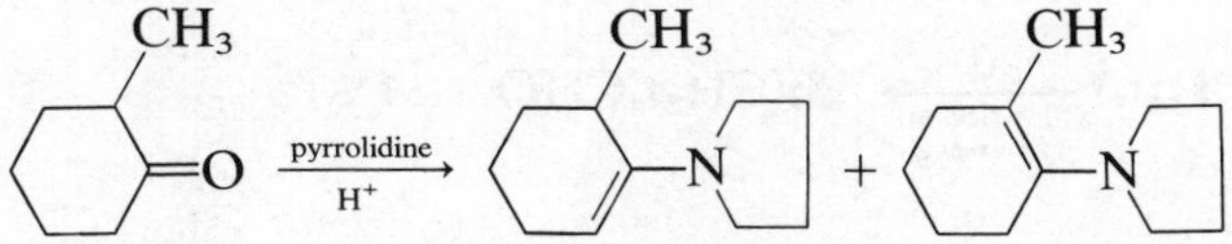

CH$_3$ CH$_3$ CH$_3$

pyrrolidine
H$^+$

+   (50a : 50b = ca. 6 : 1)

(50a)   (50b)

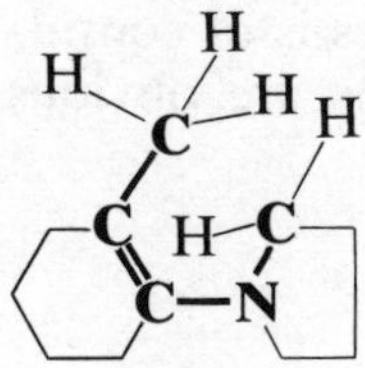

(50b)   (the five boldface atoms are coplanar)

α-Alkylation of aldehydes *via* enamines is of limited usefulness because of the intervention of side-reactions (N-alkylation of the enamine and self-condensation of the aldehyde). In some instances, however, alkylations have been successful, e.g.

$$[CH_3CH_2CH_2CHO + n\text{-}C_4H_9NHCH_2CH(CH_3)_2 \rightarrow]$$

$$CH_3CH_2CH=CH\overset{\displaystyle C_4H_9\text{-}n}{\underset{\displaystyle |}{\text{—N}}}CH_2CH(CH_3)_2$$

$$\xrightarrow[\text{(ii) } H^+/H_2O]{\text{(i) } CH_3CH_2I} CH_3CH_2\underset{\displaystyle |}{\overset{\displaystyle }{CH}}\text{—CHO}$$

$$CH_2CH_3 \quad (41\%)$$

A convenient variant of the dihydro-1,3-oxazine route to aldehydes (section 5.2.3.1) also involves an enamine intermediate (51):

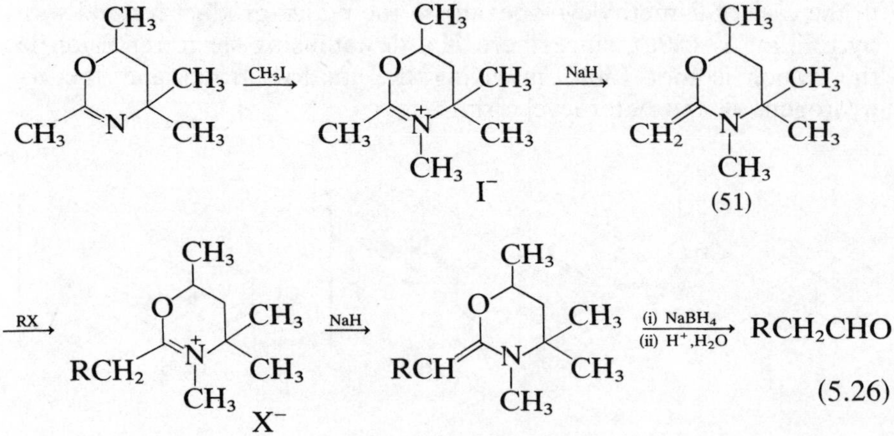

(51)

(5.26)

For example, $Ph(CH_2)_3I \xrightarrow[\substack{\text{(ii) work-up} \\ \text{as above}}]{\text{(i) (51)}} Ph(CH_2)_4CHO$ (51%)

5.4.2 Acylation

In contrast to alkylation, where the synthetic usefulness is limited, acylation of activated alkenes, arenes, and heteroarenes is of considerable synthetic importance. Many of these acylations are obvious variants of the Friedel–Crafts reaction, e.g.

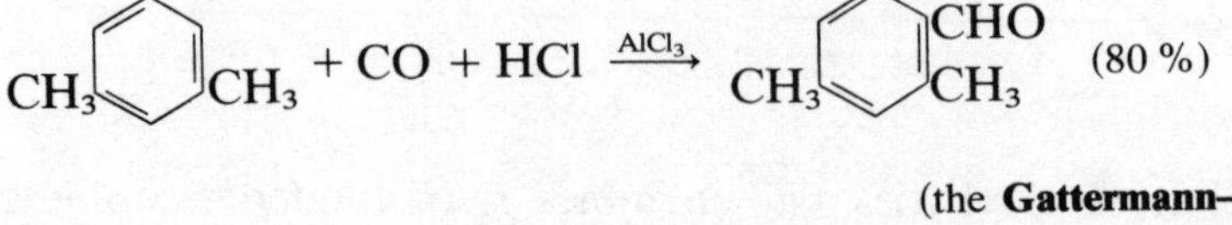

(the **Gattermann–Koch** reaction)

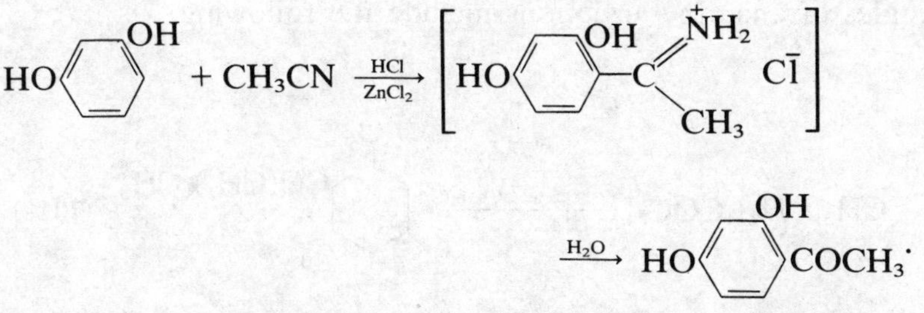

(70 %)

(the **Hoesch** reaction)

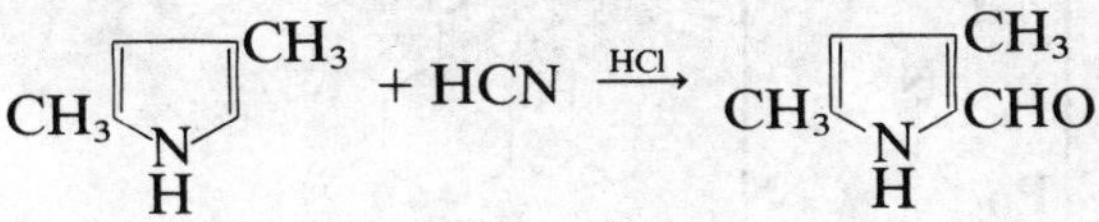

(92 %)

(the **Gattermann** reaction)

(74 %)

$$CH_3COCH_3 \left[\rightleftharpoons CH_3\overset{OH}{\overset{|}{C}}{=}CH_2\right] + (CH_3CO)_2O \xrightarrow{BF_3} CH_3COCH_2COCH_3$$

(70 %)

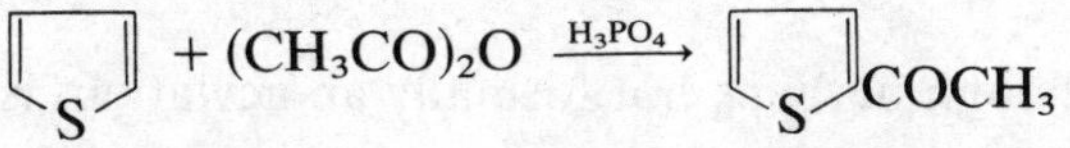

Acylation of enamines, like alkylation, may occur either at carbon or at nitrogen, but since the latter process is easily reversible and gives an *N*-acylammonium salt which can itself act as an acylating agent, it is usually possible to obtain high yields of *C*-acylated products. These acylations are often carried out in presence of an added base, such as triethylamine, since the initial acylation product (52) is moderately acidic and can protonate unreacted enamine in the absence of any stronger base:

(52)

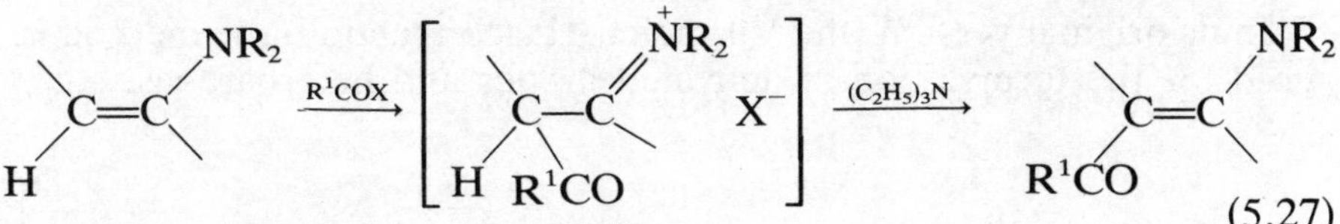

(5.27)

[On the other hand, if the acylating agent contains an α-hydrogen, it may react with the added base giving a keten, which in turn undergoes cycloaddition to the enamine. In such cases, little, if any, acylation product may be obtained.]

Examples of enamine acylations include the following:

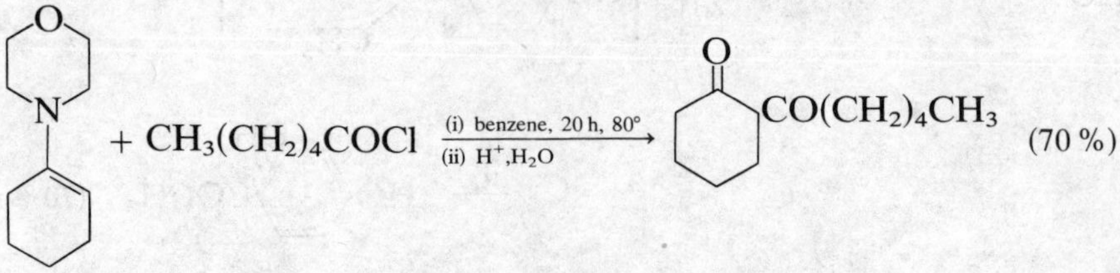

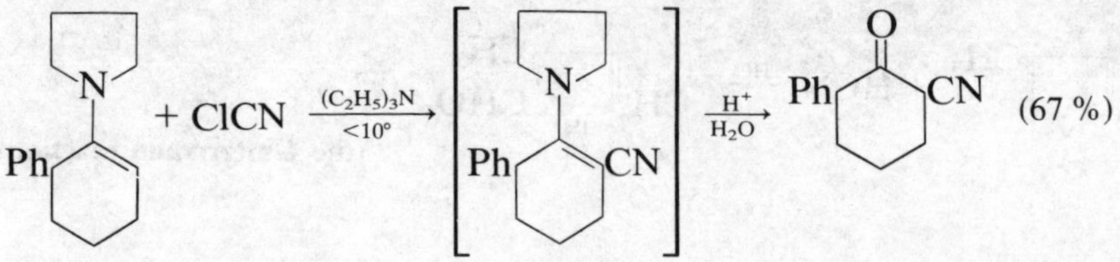

$$+ CH_3(CH_2)_4COCl \xrightarrow[\text{(ii) } H^+,H_2O]{\text{(i) benzene, 20 h, 80°}}$$

$$CO(CH_2)_4CH_3 \quad (70\%)$$

$$+ ClCN \xrightarrow[<10°]{(C_2H_5)_3N}$$

$$\xrightarrow[H_2O]{H^+}$$

$$(67\%)$$

[This latter reaction, although perhaps not formally an acylation, is mechanistically similar.]

Among methods for formylation, the **Vilsmeier–Haack–Arnold** method, using an *N,N*-disubstituted formamide and phosphoryl chloride, is among the most useful. The effective electrophile in these reactions is the chloromethyleneiminium ion (53) [reaction (5.28)]:

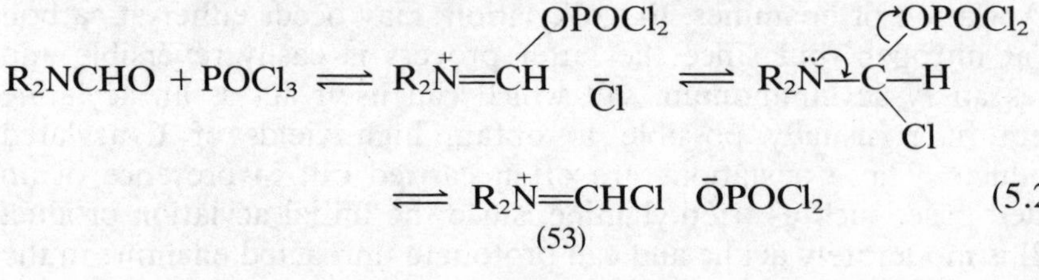

$$R_2NCHO + POCl_3 \rightleftharpoons R_2\overset{+}{N}{=}CH \underset{\bar{Cl}}{\overset{OPOCl_2}{\diagup}} \rightleftharpoons R_2\ddot{N}{-}\overset{OPOCl_2}{\underset{Cl}{\overset{\frown}{C}}}{-}H$$

$$\rightleftharpoons R_2\overset{+}{N}{=}CHCl \quad \bar{O}POCl_2 \qquad (5.28)$$
$$(53)$$

Similarly, $R_2NCHO + COCl_2 \rightarrow R_2\overset{+}{N}{=}CHCl \; Cl^- + CO_2$.

In its original version (the Vilsmeier–Haack reaction) the method is used for the formylation of activated arenes and heteroarenes, e.g.

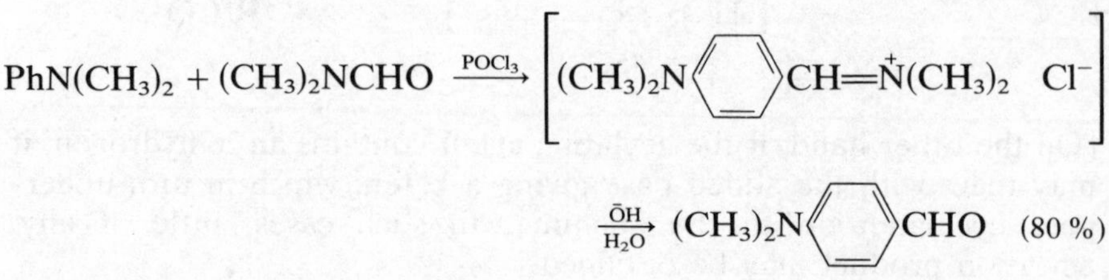

$$PhN(CH_3)_2 + (CH_3)_2NCHO \xrightarrow{POCl_3} \left[(CH_3)_2N\!\!\!\bigcirc\!\!\!CH{=}\overset{+}{N}(CH_3)_2 \;\; Cl^- \right]$$

$$\xrightarrow[H_2O]{\bar{O}H} (CH_3)_2N\!\!\!\bigcirc\!\!\!CHO \quad (80\%)$$

Similarly

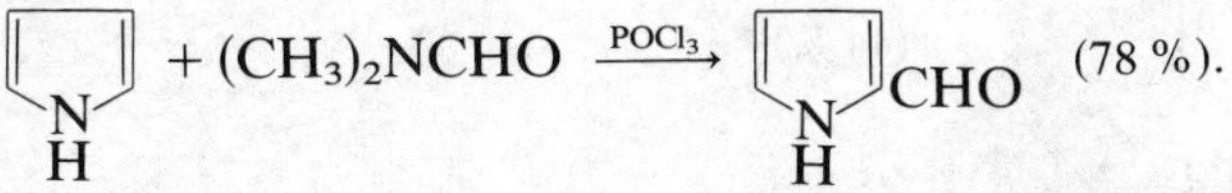

+ $(CH_3)_2NCHO$ $\xrightarrow{POCl_3}$ (78 %).

Activated alkenes also react with these iminium salts very readily, e.g.

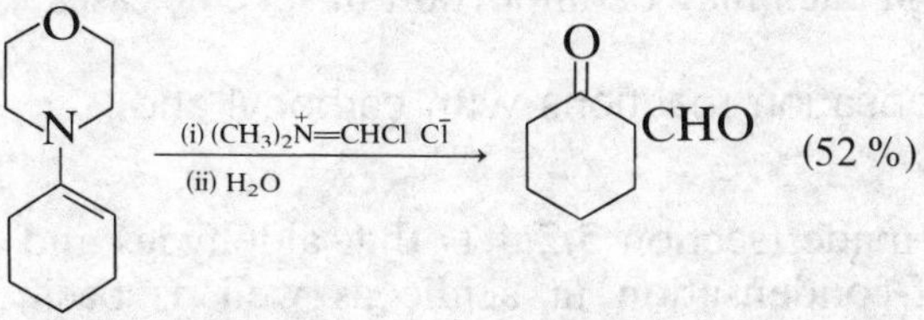

$\xrightarrow[\text{(ii) } H_2O]{\text{(i) } (CH_3)_2\overset{+}{N}=CHCl\ \bar{Cl}}$ (52 %)

However, it appears that in the majority of cases simple formylation products are not obtained.

Two other reactions which are formally acylations are worthy of brief mention. Both are associated principally with acylation of phenols: the **Reimer–Tiemann** reaction, in which the electrophile is dichlorocarbene, and the **Kolbe–Schmitt** reaction, in which the electrophile is carbon dioxide (cf. Sykes, pp. 284–5):

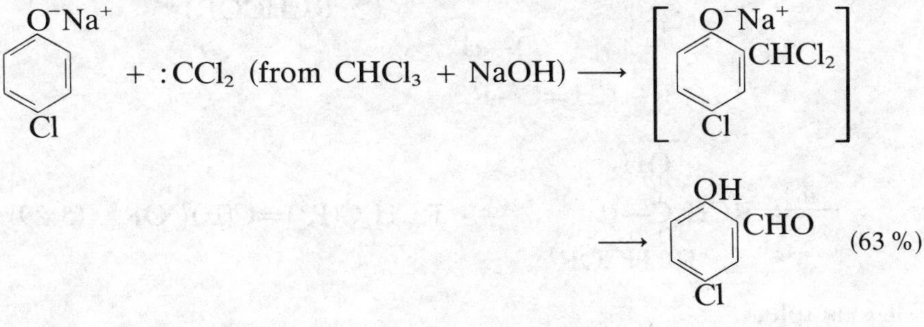

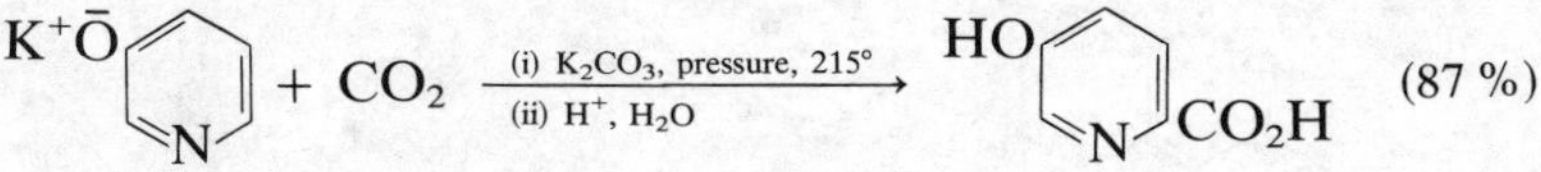

+ $:CCl_2$ (from $CHCl_3$ + NaOH) $\longrightarrow$

$\longrightarrow$ (63 %)

+ CO_2 $\xrightarrow[\text{(ii) } H^+, H_2O]{\text{(i) } K_2CO_3, \text{ pressure, } 215°}$ (87 %)

Both types of reaction have been applied to nucleophiles other than phenoxide ions, e.g.

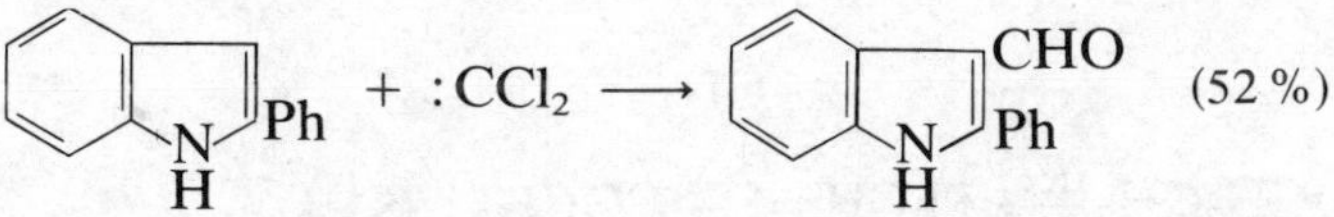

+ $:CCl_2$ $\longrightarrow$ (52 %)

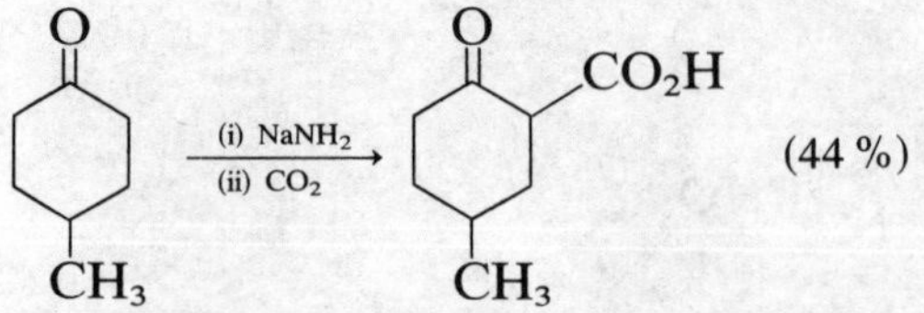

(44 %)

Yields, however, are generally low, and by products are common in the Reimer–Tiemann process. Neither process is therefore a *generally* useful method, although each may be important in specific cases.

5.4.3 Addition and condensation reactions with carbonyl and related compounds

Mention has already been made (section 5.2.4.1) that aldehydes and ketones may undergo self-condensation in acidic as well as basic media. The mechanism of the condensation under acidic conditions clearly cannot involve a carbanionic nucleophile, and such reactions are envisaged as involving an enol as the nucleophile and a protonated carbonyl species as the electrophile [reaction (5.29); cf. Sykes, p. 221]:

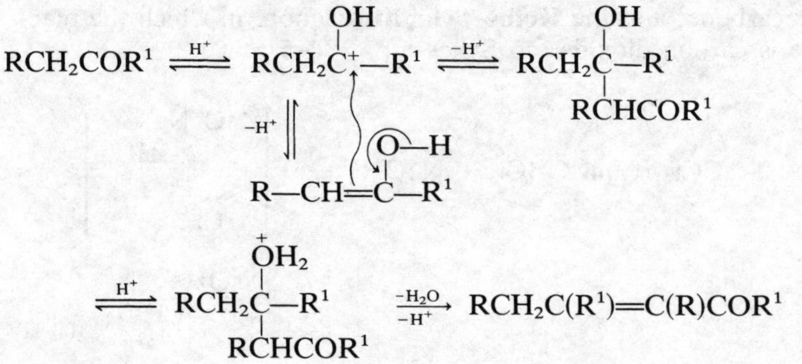

$$RCH_2COR^1 \underset{}{\overset{H^+}{\rightleftharpoons}} RCH_2\overset{+}{C}\text{—}R^1 \underset{}{\overset{-H^+}{\rightleftharpoons}} RCH_2\overset{OH}{\underset{RCHCOR^1}{C}}\text{—}R^1$$

$$\overset{H^+}{\rightleftharpoons} RCH_2\overset{\overset{+}{O}H_2}{\underset{RCHCOR^1}{C}}\text{—}R^1 \xrightarrow[-H^+]{-H_2O} RCH_2C(R^1)\text{=}C(R)COR^1 \quad (5.29)$$

For example,

(83 %)

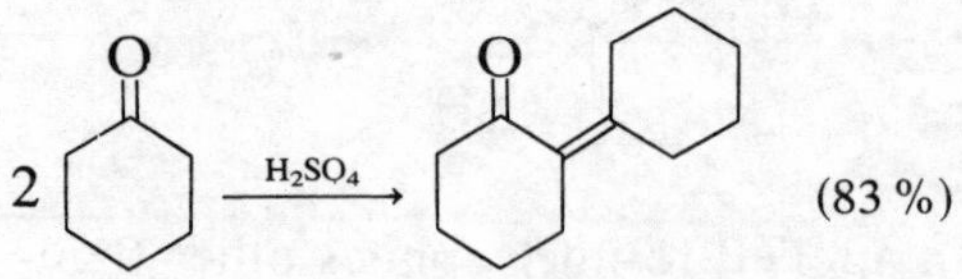

Mixed condensations may also occur, e.g.

$$PhCHO + CH_3COCH_2CH_3 \xrightarrow{HCl} PhCH\text{=}\overset{CH_3}{\underset{}{C}}\text{—}COCH_3 \quad (85 \%)$$

A mechanistically related reaction, of much greater importance in synthesis, is the **Mannich reaction**, in which the electrophilic component is not a protonated carbonyl group but a methyleneiminium ion [produced *in situ* from formaldehyde and a secondary amine in presence of acid: reaction [5.30]]. Unlike the acid-catalysed condensation reactions, Mannich reactions consist of a simple addition step without a final elimination (this presumably reflects the fact that $-\overset{+}{N}HR_2$ is a poorer leaving group than $-\overset{+}{O}H_2$):

$$CH_2=O \overset{H^+}{\rightleftharpoons} \overset{+}{C}H_2OH \overset{R_2NH}{\rightleftharpoons} R_2\overset{+}{\underset{H}{N}}-CH_2OH \rightleftharpoons R_2\overset{\cdot\cdot}{N}-CH_2 \overset{+}{O}H_2 \overset{-H_2O}{\rightleftharpoons}$$

$$R_2\overset{+}{N}=CH_2 \quad (5.30)$$

The nucleophile in the Mannich reaction may be an enol, or an activated arene, or a π-excessive heteroarene, e.g.

$$PhCOCH_3 \left[\rightleftharpoons Ph\overset{OH}{\underset{}{C}}=CH_2 \right] + CH_2=\overset{+}{N}(CH_3)_2 \; \bar{C}l \quad (\text{from } CH_2O + (CH_3)_2NH + HCl)$$

$$\rightarrow PhCOCH_2CH_2N(CH_3)_2 \; (60\%)$$
$$(54)$$

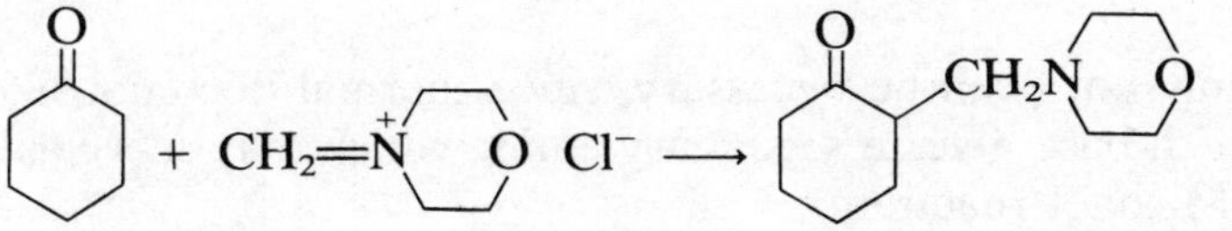

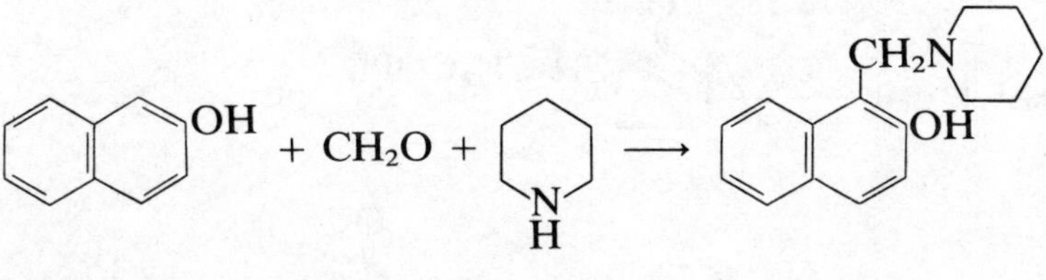

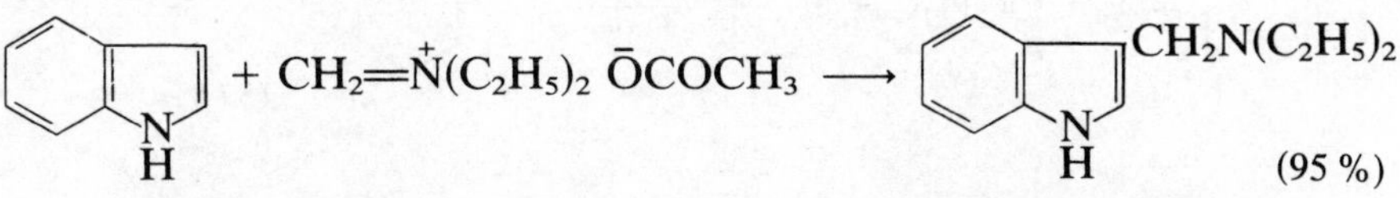

The products of many Mannich reactions (**Mannich bases**, as they are called) are themselves useful synthetic intermediates. β-(Dialkylamino)ketones such as (54) are readily convertible, by an elimination of the Hofmann type (cf. Sykes, p. 250), into vinyl ketones [reactions (5.31) and (5.32)]:

$$RCOCH_2CH_2NR^1_2 \xrightarrow[\text{or distillation}]{\text{base (e.g. }\bar{O}H)} RCOCH{=}CH_2 + R^1_2NH \qquad (5.31)$$

$$RCOCH_2CH_2NR^1_2 \xrightarrow{CH_3I} RCOCH_2CH_2{-}^+\overset{\displaystyle CH_3}{\underset{\displaystyle |}{N}}R^1_2\,I^- \xrightarrow{\text{base}}$$

$$RCOCH{=}CH_2 + R^1_2NCH_3 \quad (5.32)$$

Since vinyl ketones are valuable electrophiles in the Michael reaction (sections 5.1.5 and 5.2.5) but are rather unstable (being liable to polymerisation), their generation *in situ* from Mannich bases provides a useful variant of the Michael reaction, e.g.

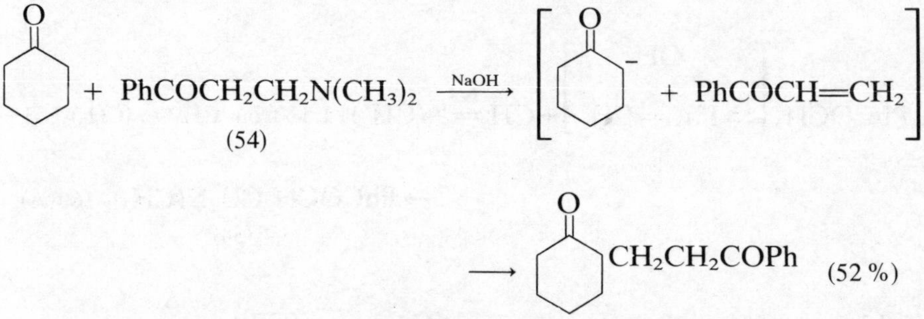

(54)

(52 %)

Added base may not even be necessary, since thermal decomposition of the Mannich base gives a secondary amine which may serve as catalyst for the Michael reaction:

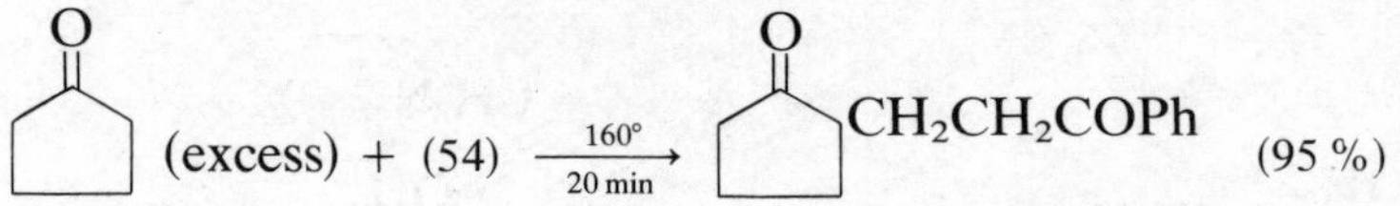

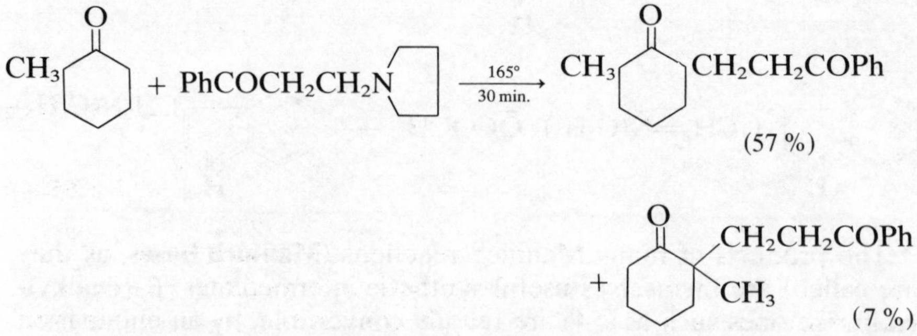

(95 %)

(57 %)

(7 %)

These so-called 'thermal Michael reactions' are of particular mechanistic interest: thermal decomposition of (54) at 160°, for example, produces dimethylamine (b.p. 7°) which should be lost as vapour and thus be unable to catalyse the addition. It is supposed that transamination occurs between the Mannich base and the other ketone, giving the enone and an enamine [reaction (5.33)]. The latter then functions as the nucleophile in the Michael addition:

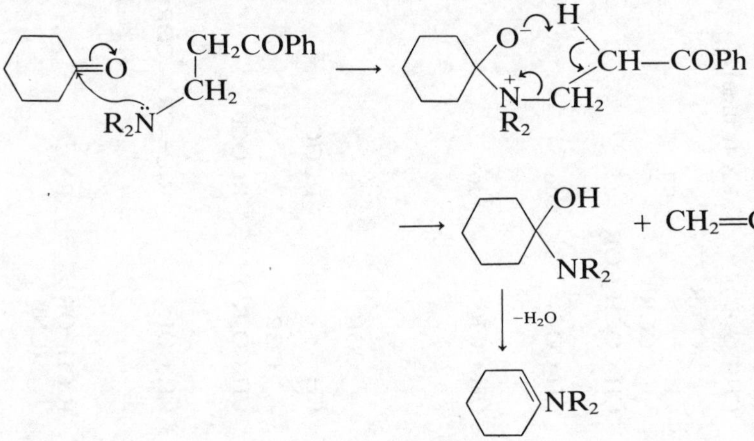

$$\text{[reaction scheme (5.33)]}$$

CH$_2$COPh   CH$_2$ R$_2$N⋯ ⟶ O⁻ H ⟶ N⁺—CH$_2$ R$_2$ ⟶CH—COPh

⟶ OH NR$_2$ + CH$_2$=CHCOPh

$\xrightarrow{-H_2O}$

NR$_2$ (5.33)

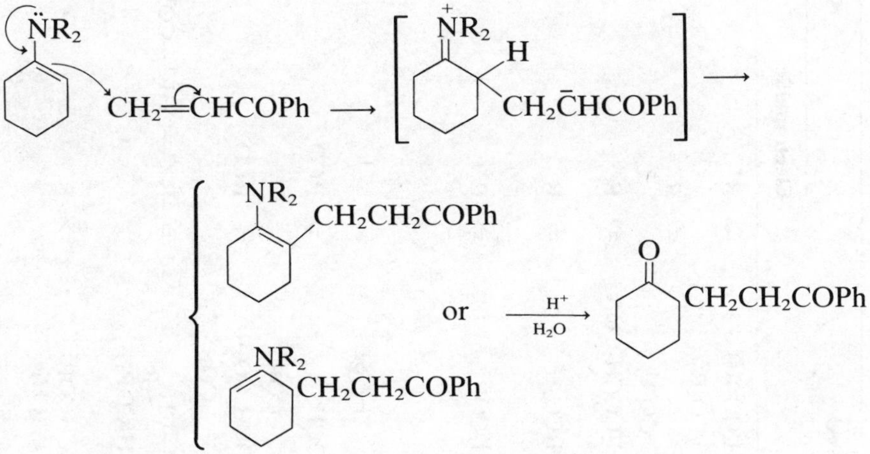

N̈R$_2$ CH$_2$=CHCOPh ⟶ [NR$_2$ H   CH$_2$C̄HCOPh] ⟶

{ NR$_2$ CH$_2$CH$_2$COPh

NR$_2$ CH$_2$CH$_2$COPh }   or $\xrightarrow[H_2O]{H^+}$ O CH$_2$CH$_2$COPh

5.5 Review

The majority of the reactions contained in Chapter 4 lead to monofunctional products, and involve nucleophilic synthons which are devoid of functionality. In the present chapter, however, the majority of the products contain two or more functional groups, and (since the nucleophiles in each case are stabilised by adjacent atoms

Table 5.1 Disconnections for some products formed from stabilised carbanions and related species

Section number	Product	Synthons — Electrophilic	Synthons — Nucleophilic	Synthetic equivalents — Electrophilic	Synthetic equivalents — Nucleophilic
5.1.1 (pp. 65–8)	$RCH(CO_2R^1)_2$	$\Uparrow\ R^+$	$\overline{C}H(CO_2R^1)_2$	RX (halide)	$CH_2(CO_2R^1)_2$
do.	$RCH(CO_2R^2)(COR^1)$	$\Uparrow\ R^+$	$\overline{C}H(CO_2R^2)(COR^1)$	do.	$CH_2(CO_2R^2)(COR^1)$
do (pp. 67–8)	$RCH_2COCH_2COR^1$	$\Uparrow\ R^+$	$\overline{C}H_2COCH_2COR^1$	do.	$CH_3COCH_2COR^1$
5.1.2, reaction (5.4)	RCH_2CO_2H	$\Uparrow\ R^+$	$\overline{C}H_2CO_2H$	do.	$CH_2(CO_2R^1)_2$
do. (reaction (5.5))	RCH_2COR^1	$\Uparrow\ R^+$	$\overline{C}H_2COR^1$	do.	$CH_2(CO_2R^2)(COR^1)$
5.1.3, reaction (5.8)	$RCOCH(COR^1)(COR^2)$	$\Uparrow\ RCO^+$	$\overline{C}H(COR^1)(COR^2)$	RCOCl	$CH_2(COR^1)(COR^2)$
reaction (5.11)	$RCOCH(COR^2)CO_2R^1$	$\Uparrow\ RCO^+$	$\overline{C}H_2CO_2R^1$	RCOCl	$CH_2(CO_2R^1)_2$
5.1.5, reaction (5.15)	$R_2C{-}CH(R^1){-}CH(COR^3)_2$	$\Uparrow\ R_2\overset{+}{C}{-}CHR^1{-}COR^2$	$\overline{C}H(COR^3)_2$	$R_2C{=}CR^1{-}COR^2$	$CH_2(COR^3)_2$
5.2.1 (pp. 80–1 also 5.2.3.2)	RR^1CHCOR^2 (CN)	$\Uparrow\ R^+$	$R^1\overline{C}HCOR^2$ (CN)	RX	$R^1CH_2COR^2$ (CN)
5.2.2 (pp. 82–4)	$RCOCHR^1CO_2R^2$	$\Uparrow\ RCO^+$	$R^1\overline{C}HCO_2R^2$	RCO_2R^2	$R^1CH_2CO_2R^2$

Table 5.1. (cont.)

Reference	Target		Synthon	Reagent	Reagent equivalent
5.2.3.1, reaction (5.17)	RR²CHCHO	⇒ (R²)⁺	RC̄HCHO	R²X	RCH₂CH=NR¹
do. reaction (5.19)	RR¹CHCHO	⇒ (R¹)⁺	RC̄HCHO	R¹X	[oxazine ring: C(CH₃)₂, N, CH₃, O, RCH₂, RCH₂]
5.3.2, reaction (5.25)	R¹COR²	⇒ (R²)⁺	R¹C̄O	R²X	[1,3-dithiane: R¹···H]
5.4.1, (pp. 100–2)	RR¹CHCOR²	⇒ R⁺	R¹C̄HCOR²	RX	R²CH=C—NR³₂, R¹CH=
5.4.1, reaction (5.26)	RCH₂CHO	⇒ R⁺	C̄H₂CHO	RX	[N-CH₃ oxazine: C(CH₃)₂, N, O, CH₂, CH₃]
5.4.2, (pp. 103–4)	RCH(COR¹)(COR²), R¹	⇒ R¹CO⁺	RC̄HCOR²	RCOCl	R²CH=C—NR³₂, RCH=
5.4.3, (pp. 108–9)	RCOCH₂CHCOR², R¹	⇒ RCOCH₂CH₂⁺	R¹C̄HCOR²	RCOCH₂CH₂NR³₂	R¹CH₂COR²

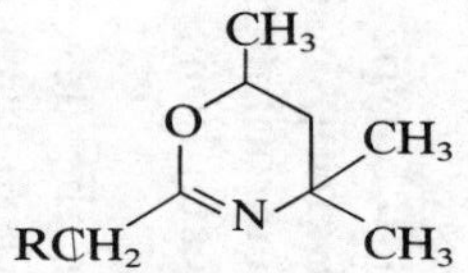

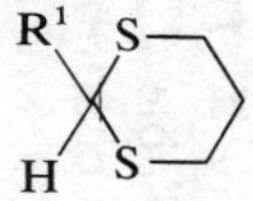

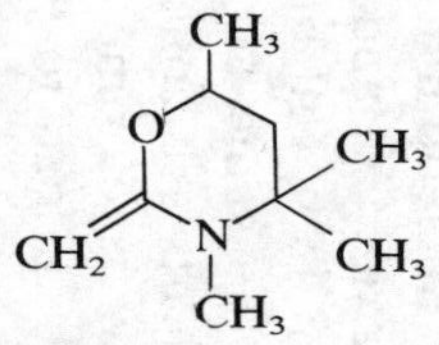

or groups) these nucleophilic synthons are all of the type $\bar{C}H_2X$ or $\bar{C}HXY$ or $\bar{C}XYZ$ (X, Y, and Z being functional groups). These are shown in Table 5.1; it will be noted that the electrophilic synthons are, in general, the same as in Chapter 4. The synthons and their synthetic equivalents are collected in Table 5.2.

The other new development in this chapter is the description of two processes which lead to the formation of carbon–carbon double bonds, *viz.* condensations and Wittig reactions. So now we have to devise a system for the disconnection of double bonds.

Neither of these processes is, of course, a one-step reaction. Each consists of an addition step, which forms a single carbon–carbon bond, followed by an elimination step in which the second carbon–carbon bond is formed. So disconnection of a double bond is also a two-stage process: (i) addition to the double bond, i.e. the opposite of elimination, and (ii) disconnection of the resulting C–C single bond. Thus,

$$
\overset{OH}{\underset{}{>}}C{=}\overset{}{C}{-}\overset{}{C}{=}O \implies \overset{OH}{>}C{+}CH{-}\overset{}{C}{=}O \implies \overset{OH}{>}C^{+} + {}^{-}CH{-}\overset{}{C}{=}O
$$

$$(5.34)$$

$$
\overset{\overset{+}{P}Ph_3}{>}C{=}C{<} \implies \overset{\overset{+}{P}Ph_3}{>}C{-}\overset{O^-}{C}{<} \implies \overset{\overset{+}{P}Ph_3}{>}C^{-} + \left[\overset{O^-}{{}^{+}C} \longleftrightarrow \overset{O}{\underset{}{C}} \right]
$$

$$(5.35)$$

[There are, of course, other methods of forming double bonds, by functional group interconversion. The partial reduction of a triple bond (sections 8.4.2 and 11.5) is among the most familiar of these, and in certain cases it may be attractive to use an alkyne as the synthetic precursor of an alkene – especially since a wide variety of substituted alkynes may be easily prepared (cf. section 4.3)].

5.6 Problems

5.6.1 Strategy of disconnection

In the 'Problems' section of Chapter 4 we demonstrated that synthetic routes to monofunctional compounds could frequently be revealed by performing a disconnection on the end-product (the 'target molecule') adjacent to the functional group. Similarly we showed that another useful disconnection may be one adjacent to a point of chain branching, i.e. one adjacent to a tertiary (or quaternary) carbon.

Table 5.2 Some common synthons and their synthetic equivalents (part 2)
This table is a continuation of Table 4.2 (p. 58), and the entries are numbered consecutively with those of that previous Table.

	Synthon	Synthetic equivalents	For example(s), see page(s)
(a) Nucleo-philic synthons	4. $R\bar{C}HCHO$	RCH_2CHO; $RCH_2CH{=}NR^1$; 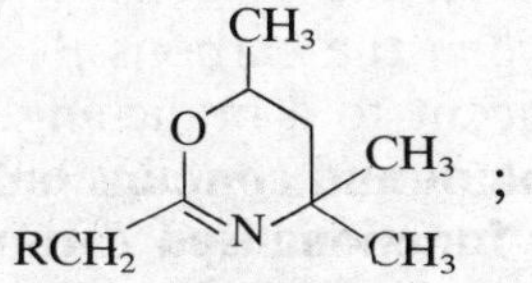 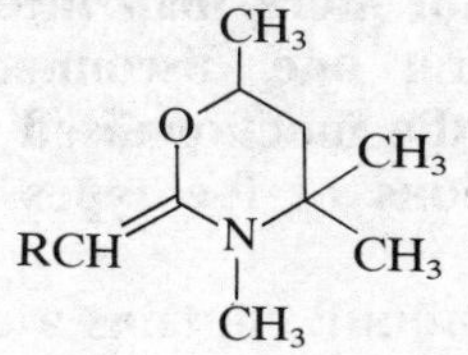 [oxazine structure: RCH_2–, with CH_3, O, N, CH_3, CH_3] ; [structure: RCH–, with CH_3, O, N, CH_3, CH_3, CH_3]	80, 84–6, 88–92, 101–2
	5. $R\bar{C}HCOR^1$	RCH_2COR^1; $RCH{=}\overset{\displaystyle R^1}{\underset{\displaystyle CO_2R^2}{C}}{-}NR^2{}_2$; $RCHCOR^1$	69–74, 80–4, 86–90, 100–1, 104–9
	6. $R\bar{C}HCO_2H$	$RCH_2CO_2R^1$; $RCH(CO_2H)_2$; $RCH(CO_2R^1)_2$	69–72, 76–7
	7. $R\bar{C}HCO_2R^1$	$RCH_2CO_2R^1$; $RCH(CO_2R^1)_2$; $RCH{=}\overset{\displaystyle CO_2R^1}{\underset{\displaystyle OR^1}{C}}$ $\bar{O}\overset{+}{Z}nBr$ (section 4.2.2)	69–72, 80, 82–4 90, 92 47–8
	8. $R\bar{C}HCN$	RCH_2CN; $RCHCN$	77, 80, 83
	9. $R\bar{C}HCO$-CH_2COR^1	$RCH_2COCH_2COR^1$	67–8, 87
	10. $\bar{C}HO$	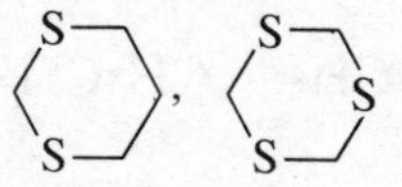 [1,3-dithiane ring], [1,3-dithiolane/dithiane ring]	98–9
	11. $R\bar{C}O$	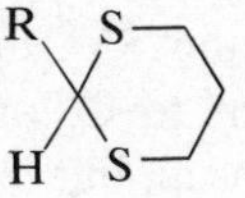 [dithiane ring with R, H]	98–9
	12. $R\bar{C}(COR^1)_2$, $R\bar{C}(CO_2R^1)_2$, etc.	$RCH(COR^1)_2$, $RCH(CO_2R^1)_2$, etc.	64–79
(b) Electro-philic synthons	6. $H\overset{+}{C}O$	HCO_2R; $HCONR_2$; $CH(OR)_3$; $ClCH{=}\overset{+}{N}R_2X^-$	84, 87, 102–5
	12. $\overset{+}{C}H_2CH_2COR$	$CH_2{=}CHCOR$; $R^1{}_2NCH_2CH_2COR$; $R^1{}_3\overset{+}{N}CH_2CH_2COR\ X^-$	77–9, 93, 108–9
	13. $\overset{+}{C}HCl_2$	$CHCl_3$ (via: CCl_2)	105

We are now able to extend the list of potentially useful disconnections, and to consider disconnections of difunctional compounds.

(i) **If the compound contains only carbon–carbon single bonds, try the following disconnections:**
   (a) Adjacent to a functional group
   (b) Between the carbons α- and β- to a functional group
   (c) Between the carbons β- and γ- to a functional group
   (d) Adjacent to a branching point in a carbon chain.

(ii) **If the compound contains only carbon–carbon single bonds, and two functionalised carbon atoms close together (separated by not more than three other carbons), it is usually worth trying one disconnection between the functional groups. If the functionalised carbons are farther apart, two disconnections of the types shown in (i) are likely to be required.**

(iii) **If the compound contains a carbon–carbon double bond, it is worth considering a disconnection of this bond.** If it is an isolated (i.e. non-conjugated) double bond, it may imply a Wittig reaction; if it is conjugated to a $-M$ group, it may imply a condensation reaction or a Wittig reaction with a stabilised ylide. **If each of the doubly bonded carbons is attached to hydrogen, it may be worth considering if the corresponding alkyne is readily accessible.**

5.6.2   Examples

As in Chapter 4, we conclude with some worked examples. Once again we consider four target molecules:

$$CH_3COCH(CH_3)(CH_2)_4CH_3; \quad PhCOCH_2CH_2CN;$$
$$(55)(56)$$

$$(CH_3)_2CHCH{=}CHCO_2CH_3; \quad CH_3CH{=}CH(CH_2)_5CHO.$$
$$(57)(58)$$

3-Methyloctan-2-one (compound 55). Of all the possible disconnections for this molecule, those adjacent to C_3 appear particularly attractive: not only is C_3 α- to the functional group, but it is also the branching point of the chain. Since C_3 is joined to three other carbons, there are six pairs of synthons to be considered, *viz.*

$$CH_3COCH(CH_3)(CH_2)_4CH_3 \xrightarrow{(a)} CH_3CO\overset{+}{C}H(CH_2)_4CH_3 + CH_3{}^-$$
$$(55)$$
$$\xrightarrow{(b)} CH_3CO\overset{-}{C}H(CH_2)_4CH_3 + CH_3{}^+$$

$$CH_3CO \overset{\text{(c)}}{\underset{|}{}} \overset{CH_3}{\underset{|}{CH}}(CH_2)_4CH_3 \overset{\text{(c)}}{\Longrightarrow} CH_3CO^+ + \overset{CH_3}{\underset{|}{}}{}^-CH(CH_2)_4CH_3$$

$$\overset{\text{(d)}}{\Longrightarrow} CH_3CO^- + \overset{CH_3}{\underset{|}{}}{}^+CH(CH_2)_4CH_3$$

$$\overset{CH_3}{\underset{|}{}}CH_3COCH \overset{\text{(e)}}{}(CH_2)_4CH_3 \overset{\text{(e)}}{\Longrightarrow} \overset{CH_3}{\underset{|}{}}CH_3COC^+H + {}^-(CH_2)_4CH_3$$

$$\overset{CH_3}{\underset{|}{}}\overset{\text{(f)}}{\Longrightarrow} CH_3CO^-CH + {}^+(CH_2)_4CH_3$$

The twelve synthons all have recognisable synthetic equivalents, and so the reverse of any one of the six disconnections may form the basis of a successful synthesis.

Disconnections (a) and (e) offer the least attractive (or, in any case, the most difficult) possibilities. In each case the electrophile is an α-halogenoketone ($CH_3COCHXR$) and the nucleophile an organometallic reagent – most probably a cuprate, since Grignard and lithium derivatives would react with the ketone as well as the halide. So the two syntheses would be:

$$\overset{Br}{\underset{|}{}}CH_3COCH(CH_2)_4CH_3 + (CH_3)_2CuLi$$

and

$$\overset{CH_3}{\underset{|}{}}CH_3COCHBr + [CH_3(CH_2)_4]_2CuLi$$

$\Big\}$ compound (55)

Difficulties arise, however, in the preparation of the bromoketones. Bromination of butanone and octan-2-one does indeed give the required 3-bromo-derivatives, but also the isomeric 1-bromo-compounds and di and poly-brominated products: purification of the 3-bromo-compounds can thus be difficult.

Disconnection (c) indicates a synthesis of compound (55) from an acylating agent and an organometallic derivative of 2-bromoheptane. This is similar to the synthesis of pentadecan-4-one already discussed (section 4.5, p. 60) and is not considered further here. Disconnection (d) indicates a synthesis from an acyl anion equivalent (i.e. a dithian) and an alkylating agent:

$$\xrightarrow{\text{hydrolysis}} \text{compound (55)}$$

The remaining two disconnections, (b) and (f), indicate synthesis of (55) by the reactions of alkylating agents with stabilised carbanions or related species. The critical factor in each is the choice of the synthetic equivalent of the nucleophilic synthon. In case (f), for example, the most obvious synthetic equivalent is butanone, and the problem then becomes that of generating only one of the two possible carbanions (i.e. specific enolate formation: section 5.2.3.2). It is not easy to ensure that a mixture of anions, and hence a mixture of isomeric products, will not be obtained. The same difficulty arises with the enamine route, since butanone may give two isomeric enamines by reaction with any given amine.

The simplest way of ensuring reaction only at the required position is to 'activate' that position (i.e. attach another $-M$ group to C_3). So

$$CH_3$$
$$|$$

the synthetic equivalent of choice is $CH_3COCH{-}CO_2C_2H_5$, and the synthesis may be summarised as follows:

$$\overset{\displaystyle CH_3}{\underset{|}{CH_3COCH}}{-}CO_2C_2H_5 \xrightarrow[\text{(ii) } n\text{-}C_5H_{11}Br]{\text{(i) NaOC}_2H_5} \underset{\underset{\displaystyle (CH_2)_4CH_3}{|}}{\overset{\displaystyle CH_3}{\underset{|}{CH_3COCCO_2C_2H_5}}} \xrightarrow[\substack{\text{(ii) heat} \\ -CO_2}]{\text{(i) H}^+\text{,H}_2\text{O}}$$

$$(55a)$$

$$\overset{\displaystyle CH_3}{\underset{|}{CH_3COCH}}(CH_2)_4CH_3$$

A similar argument may be used for case (a): the same intermediate, (55a), is involved.

The syntheses outlined here are not, of course, the only possible routes to 3-methyloctan-2-one. Which is the best in practice has never been determined, and such determination may well be no more than a process of trial and error.

4-oxo-4-phenylbutanonitrile (compound 56). In this molecule the two functional groups are sufficiently close to each other to make it worth considering a single disconnection between the two. There are four such disconnections which are reasonably obvious:

$$PhCOCH_2CH_2{\overset{\text{(a)}}{\dashv}}CN \Longrightarrow PhCOCH_2\overset{+}{C}H_2 + \overset{-}{C}N$$
$$(56)$$

$$PhCOCH_2{\dashv}CH_2CN \underset{\text{(c)}}{\overset{\text{(b)}}{\rightleftarrows}} \begin{array}{l} PhCOCH_2^{\ -} + \overset{+}{C}H_2CN \\[1ex] PhCOCH_2^{\ +} + \overset{-}{C}H_2CN \end{array}$$

$$PhCO{\dashv}CH_2CH_2CN \overset{\text{(d)}}{\Longrightarrow} PhCO^- + \overset{+}{C}H_2CH_2CN$$

These in turn imply the following possible syntheses for compound (56):

$$PhCOCH{=}CH_2 + KCN \longrightarrow$$

$$\underset{\underset{CO_2C_2H_5}{|}}{PhCOCH_2} \xrightarrow[\text{(ii) ClCH}_2\text{CN}]{\text{(i) NaOC}_2\text{H}_5} \underset{\underset{CO_2C_2H_5}{|}}{PhCOCH{-}CH_2CN} \xrightarrow[\text{(ii) heat}]{\text{(i) hydrolysis}}$$

$$PhCOCH_2Br + \bar{C}HCN \rightarrow \underset{\underset{CO_2C_2H_5}{|}}{PhCOCH_2CHCN} \xrightarrow[\text{(ii) heat}]{\text{(i) hydrolysis}}$$
$$\phantom{PhCOCH_2Br + \bar{C}HCN \rightarrow}\underset{CO_2C_2H_5}{|}$$

$$\overset{Ph}{\underset{Li^+}{}}\!\!\left\langle\!\!\begin{array}{c}S{-}\\ S{-}\end{array}\!\!\right\rangle + CH_2{=}CHCN \longrightarrow$$

The first possibility, involving conjugate addition of cyanide ion to an enone, is the simplest of the four. The Mannich base (54), which we have already encountered (p. 107), may be used as a substitute for the enone. The second and third possibilities are also acceptable, since they both involve the alkylation of a doubly stabilised carbanion by means of a highly reactive halide, and since esters are much more easily hydrolysed than nitriles. The final possibility, involving the dithian, is the least attractive of the four: organolithium reagents generally prefer direct attack on functional groups at the expense of conjugate addition (section 4.2.1). On the other hand, other synthetic equivalents of $PhCO^{-}$[10] may be capable of successful conjugate addition and provide a useful synthesis of (56).

Methyl 4-methylpent-2-enoate (compound 57). This molecule contains a carbon–carbon double bond conjugated with an ester carbonyl group, and so is conceivably the product of a condensation or of a Wittig reaction involving a stabilised ylide. Thus,

$$(CH_3)_2CHCH \underset{\overset{|}{OH}}{\overset{\cdot}{-}} CH_2CO_2CH_3 \Longrightarrow$$

$$(CH_3)_2CHCH{=}CHCO_2CH_3$$
$$(57)$$

$$\xrightarrow{\text{(a)}} (CH_3)_2CH\underset{\overset{|}{OH}}{\overset{\cdot}{C}H} + \bar{C}H_2CO_2CH_3$$

$$\xrightarrow{\text{(b)}} (CH_3)_2CHCH\underset{\overset{|}{O^-}}{\overset{\cdot}{-}}\underset{\overset{|}{\overset{+}{P}Ph_3}}{CHCO_2CH_3} \Longrightarrow$$

$$(CH_3)_2CHCHO + {}^-CH\underset{\overset{|}{\overset{+}{P}Ph_3}}{}CO_2CH_3$$

The Wittig synthesis is therefore:

$$BrCH_2CO_2CH_3 \xrightarrow{PPh_3} Ph_3\overset{+}{P}CH_2CO_2CH_3\ Br^- \xrightarrow{NaOCH_3}$$

$$Ph_3\overset{+}{P}-\overset{-}{C}HCO_2CH_3 \xrightarrow{(CH_3)_2CHCHO} \text{compound (57)}$$

The condensation method is less straightforward. The most obvious synthetic equivalents for the two synthons suggest the following:

$$(CH_3)_2CHCHO + CH_3CO_2CH_3 \xrightarrow{base}$$

This, however, is unsatisfactory, since the *aldehyde,* not the ester, contain the most acidic hydrogen; so the ester must be 'activated'. A satisfactory procedure uses dimethyl malonate, and another makes use of the Reformatsky reaction:

$$(CH_3)_2CHCHO + CH_2(CO_2CH_3)_2 \rightarrow (CH_3)_2CHCH{=}C(CO_2CH_3)_2, \text{ etc.}$$

$$\longrightarrow \longrightarrow \text{compound (57)}$$

$$(CH_3)_2CHCHO + BrCH_2CO_2CH_3 \xrightarrow{Zn} (CH_3)_2CH\overset{\overset{\displaystyle OH}{|}}{C}HCH_2CO_2CH_3 \xrightarrow[-H_2O]{H^+}$$

$$\text{compound (57)}$$

The reader may be able to devise other methods.

Non-7-enal (compound 58). As in the last chapter, the final problem is left unsolved as a challenge to the reader. There are several possible syntheses, and the method of choice may depend, in practice, on availability of starting materials. The principal difficulty in the synthesis of this compound is the sensitivity of the aldehyde group towards oxidation, reduction, and condensation, and so it may be advantageous to introduce the aldehyde group at a late stage in the synthesis.

Compound (58) is similar in its functionality to the insect pheromone which we introduced at the beginning of Chapter 3 (compound A, p. 24). The reader is now invited to devise a route to Compound A, and to compare this route with those already published: these are summarised in Chapter 14 (section 14.2).

Notes

1.   Sodium hydride, although a strong base, is a surprisingly poor nucleophile; if hydride is required to act as a nucleophile (for example, to react with a carbonyl group), a complex hydride, such as $\overline{A}lH_4$ or $\overline{B}H_4$, should be used (cf. section 8.2). In any case the reactive species pro-

duced by solution of sodium hydride in dimethyl sulphoxide is $CH_3SOCH_2^-Na^+$.

2. The reader is reminded that we have chosen in this book to adopt a restricted definition of the term 'condensation reaction', *viz.* a reaction in which addition of a carbanion (or other nucleophile) to a carbonyl compound (or other electrophile) is followed by elimination of water (cf. section 3.3.2, footnote 3).

3. The use of boiling benzene as solvent facilitates removal of water (by azeotropic distillation).

4. LDA = lithium di-isopropylamide, $Li^+\bar{N}[CH(CH_3)_2]_2$.

5. This does not conform to our restricted definition of a 'condensation reaction' (cf. footnote 2, and section 3.3.2, footnote 3).

6. Addition of the Grignard reagent to the double bond of (29) is not, apparently, an important side-reaction.

7. The condensation of an aromatic aldehyde with another aldehyde or ketone is generally known as the **Claisen–Schmidt condensation**, and that with an anhydride as the **Perkin condensation**.

8. For example, by B. J. Walker, in *Organophosphorus Chemistry* (Penguin Books, 1972), pp. 144–50.

9. German: reversal of polarity.

10. So far we have made no mention of other synthetic equivalents for the synthon $R\bar{C}O$, but several others are in fact known. One such is involved in the formation of benzoin from benzaldehyde and cyanide ion:

$$\overset{\displaystyle OH}{\underset{\displaystyle |}{}}$$
$$2PhCHO + \bar{C}N \rightarrow PhCHCOPh$$

The disconnection corresponding to this reaction is

$$\overset{OH}{\underset{|}{PhCH}} {\Large\shortmid} COPh \Rightarrow \overset{OH}{\underset{|}{PhCH}} + \bar{C}OPh$$

The actual nucleophile, formed from benzaldehyde and $\bar{C}N$, is $Ph\overset{\displaystyle OH}{\underset{\displaystyle CN}{\overset{|}{\underset{|}{C}}}}-$

(cf. Sykes, p. 227), and this nucleophile (a stabilised carbanion) readily participates in the Michael reaction (section 5.2.5). Thus, compound (56) may be obtained as follows:

$$PhCHO + \bar{C}N \longrightarrow \left[\overset{OH}{\underset{CN}{Ph\overset{|}{\underset{|}{C}}-}}\right] \xrightarrow{CH_2=CHCN} \left[Ph-\overset{O}{\underset{\bar{C}N}{C}}\ \ \overset{H}{\underset{CH_2}{\bar{C}H-CN}}\right]$$

$$\xrightarrow{-CN^-} PhCOCH_2CH_2CN \quad (80\%)$$
$$(57)$$

6 Formation of carbon–heteroatom bonds: the principles

In the last three chapters we have been concerned with the formation of carbon–carbon bonds, with a view to constructing the molecular framework of some particular target compound. This is all very well if the target compound has a skeleton composed entirely of carbon atoms. But there are, of course, very many organic compounds for which this is not true: in this connection one thinks particularly of *heterocyclic* compounds, which by definition have molecular skeletons containing *heteroatoms*, i.e. atoms other than carbon. Before we deal with methods of forming cyclic compounds, therefore, it is appropriate to consider a few general points in relation to carbon–heteroatom bond formation.

6.1 Carbon–halogen bonds

One does not normally think of a halogen atom in an organic molecule as constituting part of the molecular framework, but rather as a substituent attached to that framework. One may also consider that the principal methods for forming a carbon–halogen bond are simply matters of functionalisation or functional group interconversion and as such have been covered already in Chapter 2. So why return to the subject here?

The answer is simple – to recall an important mechanistic point. If one were asked to express in general terms how carbon–halogen bonds are usually formed, one would tend to think first of the reaction of an electrophilic carbon with a halide ion:

$$R{-}X \Rightarrow R^+ + X^- \tag{6.1}$$

It is all too easy to ignore the reaction of a *nucleophilic* carbon with an *electrophilic* halogen species:

$$R{-}X \Rightarrow R^- + X^+ \tag{6.2}$$

although there are possibly just as many useful syntheses of this latter type as of the former [the halogenation of benzene derivatives (scheme 2.3) being one of the most familiar examples]. In addition it will be recalled that, whereas halide ions are rather weak nucleophiles and thus require strongly electrophilic carbon species for reaction, some of the positive halogen species are highly potent electrophiles.

Finally it must be remembered that there is a third possible disconnection for the R–X bond:

$$R \dashv X \Rightarrow R^{\cdot} + {}^{\cdot}X \qquad (6.3)$$

and radical reactions constitute another important method for carbon–halogen bond formation (section 2.1; cf. Sykes, pp. 314–18).

6.2 Carbon–oxygen and carbon–sulphur bonds

Unlike the halogens, oxygen and sulphur atoms are able to form *two* covalent bonds to carbon in an uncharged molecule, and can therefore be incorporated into the skeleton of organic compounds as well as contributing towards functional groups attached to the skeleton.

In the case of carbon–oxygen bond formation, the vast majority of the reactions are those of electrophilic carbon with nucleophilic oxygen:

$$R \dashv O{-}R' \Rightarrow R^{+} + {}^{-}O{-}R' \qquad (6.4)$$

$$RCO \dashv O{-}R' \Rightarrow R\overset{+}{C}O + \bar{O}{-}R' \qquad (6.5)$$

[It must be remembered, of course, that although the synthon $\bar{O}$–R′ appears in the above, it is not necessary for the synthetic equivalent to bear a negative charge: alcohols (or water, if R′ = H) may be sufficiently nucleophilic to react with the electrophile.]

Those bond-forming reactions between a nucleophilic carbon and an electrophilic oxygen are much less common: from the synthetic viewpoint the most useful of this type are oxidative procedures such as the formation of oxirans from alkenes (epoxidation: cf. scheme 2.1 and section 9.2.5.1; also Sykes, p. 186) and the Baeyer–Villiger reaction (section 9.5.3; cf. Sykes, pp. 126–8). Similarly C—O bond formation *via* radical reactions is of relatively limited synthetic value (but see section 9.4).

With regard to carbon–sulphur bonds, the position is complicated by the different oxidation states in which sulphur is commonly encountered. The formation of a $\geq$C—S—C$\leq$ (or $\geq$C—S—H) grouping is almost invariably one requiring electrophilic carbon and nucleophilic sulphur. However, the oxides of sulphur are electrophiles [cf. the sulphonation of benzene (section 2.1)], and so the formation of a $\geq$C—SO$_2$X or $\geq$C—SO$_2$—C$\leq$ grouping may well involve a carbon nucleophile and a sulphur electrophile, e.g.

$$CH_3(CH_2)_{11}MgBr + SO_2 \rightarrow CH_3(CH_2)_{11}SO_2H \quad (80\%)$$

$$PhSO_2Cl + PhH \xrightarrow{FeCl_3} PhSO_2Ph \quad (80\%)$$

6.3 Carbon–nitrogen bonds

Carbon–nitrogen bond formation is more complicated still. An uncharged nitrogen atom in an organic molecule forms three covalent bonds, and so nitrogen forming part of the molecular framework may be singly bonded to three different atoms (as in amines); or it may be doubly bonded to one atom and singly bonded to another (e.g. $>$C=N—C$\leq$); or it may replace –CH in an aromatic compound, so that its bonding may be represented as $>$C⋯N⋯C$\leq$. [Nitrogen triply bonded to a single carbon, of course, constitutes a cyano-group, and the formation of cyano-compounds is already covered in Chapter 2 (schemes 2.9, 2.10, 2.12)]. The fact that positively charged nitrogen is tetra-covalent might be regarded as a further complication, but in fact this complication is much more apparent than real.

The bond-forming reactions may conveniently be grouped under several headings, according to the reaction mechanism.

6.3.1 Nucleophilic nitrogen and electrophilic carbon

This is by far the most important process for carbon–nitrogen bond formation. Ammonia and amines are good nucleophiles, by virtue of possessing a lone pair of electrons, and they react with electrophiles in similar fashion to carbon nucleophiles:

(i) Alkylation: $\quad$ —N:$\curvearrowright$R—X $\longrightarrow$ $\geq\overset{+}{N}$—R $\; X^-$ $\hfill$ (6.6)

$$H—N:\curvearrowright R—X \longrightarrow —\overset{\overset{H}{|}}{N}—R \; X^- \rightleftharpoons \; \overset{..}{N}—R \; + \; HX \quad (6.6a)$$

(ii) Acylation:

$$\underset{}{\overset{R}{\underset{X}{\diagdown}}}\text{N:} \quad \overset{R}{\underset{X}{\diagup}}\text{C=O} \longrightarrow \underset{}{\overset{R}{\underset{X}{-\overset{+}{\text{N}}-\overset{|}{\text{C}}-\text{O}^-}}} \longrightarrow \overset{R}{\underset{-\overset{+}{\text{N}}}{\diagup}}\text{C=O} \quad X^- \tag{6.7}$$

$$\overset{H}{\underset{}{-\ddot{\text{N}}}} + \overset{R}{\underset{X}{\diagup}}\text{C=O} \longrightarrow \overset{R}{\underset{\overset{+}{\text{H-N}} \quad X^-}{\diagup}}\text{C=O} \longrightarrow \overset{R}{\underset{}{\diagdown}}\ddot{\text{N}}-\overset{|}{\text{C}}=\text{O} + \text{HX} \tag{6.7a}$$

(iii) Condensation:

$$\overset{H}{\underset{H}{\diagdown}}\text{N:} \quad \overset{R}{\underset{R'}{\diagup}}\text{C=O} \longrightarrow \overset{H}{\underset{H}{\diagdown}}\overset{R}{\underset{R'}{-\overset{+}{\text{N}}-\overset{|}{\text{C}}-\text{O}^-}}$$

$$\rightleftharpoons \overset{H}{\underset{}{\diagdown}}\ddot{\text{N}}-\overset{R}{\underset{R'}{\overset{|}{\text{C}}}}-\text{OH} \xrightarrow{-\text{H}_2\text{O}} \text{N=C}\overset{R}{\underset{R'}{\diagdown}} \tag{6.8}$$

Thus it follows that when a molecular skeleton contains amino-nitrogen (i.e. —N< singly bonded to three different atoms) the correct disconnection is almost always:

$$-\overset{|}{\underset{|}{\text{C}}}-\text{N}\diagup \Longrightarrow -\overset{|}{\underset{|}{\text{C}}}\diagdown\overset{+}{\text{N}}\overset{H}{\diagup} \Longrightarrow -\overset{|}{\underset{|}{\text{C}}^+} + :\text{N}\overset{H}{\diagdown} \tag{6.9}$$

Similarly for amides,

$$-\overset{O}{\overset{\|}{\text{C}}}\diagdown\text{N}\diagup \Longrightarrow -\overset{O}{\overset{\|}{\text{C}}}-\overset{+H}{\text{N}}\diagup \Longrightarrow -\overset{O}{\overset{\|}{\text{C}}^+} + :\text{N}\overset{H}{\diagdown} \tag{6.10}$$

and for carbon–nitrogen double bonds, by far the most common disconnection is:

$$\text{C=N}- \Longrightarrow \overset{OH}{\underset{}{\text{C}}}-\text{N}\overset{H}{\diagup} \Longrightarrow \overset{OH}{\underset{}{\text{C}}}\diagdown\overset{+}{\text{N}}-\text{H}$$

$$\Longrightarrow \overset{OH}{\underset{}{\text{C}^+}} + :\text{N}-\text{H} \tag{6.11}$$

The reader is entitled to ask why we have chosen to write (6.9) in an extended form rather than the simpler form (6.9a) shown below, since in this latter form there is an obvious analogy with (6.1) and (6.4):

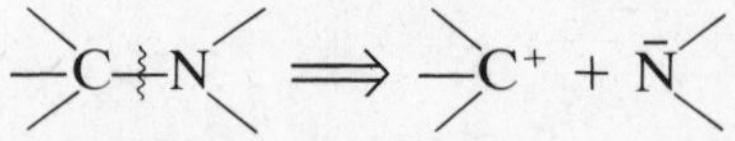

$$\mathrm{\overset{|}{\underset{|}{C}}{+}N\diagup} \implies \mathrm{\overset{|}{\underset{|}{C}^{+}} + \bar{N}\diagup} \tag{6.9a}$$

In fact, there is no reason why one should not use the simplified form (6.9a), provided that one remembers that $\bar{N}\diagdown$ **is only a synthon** (representing nucleophilic nitrogen) and that **it does not represent amide ions.** Admittedly, alkali metal amides are occasionally used to form C–N single bonds [for example, in the Tschitschibabin amination of pyridine (scheme 2.5)], but they are much too strongly basic to be generally useful as nucleophiles, since they are liable to cause eliminations, rearrangements, and other unwanted side-reactions.

6.3.2   Electrophilic nitrogen and nucleophilic carbon

As far as amino-nitrogen is concerned, this type of interaction is rarely important. The two notable exceptions (and neither is particularly common) are the formation of aziridines from alkenes and nitrenes (cf. section 7.2.3) and the Beckmann rearrangement (Sykes, pp. 122–5).

On the other hand, nitrogen electrophiles occupy an important place in the chemistry of aromatic compounds, NO_2^+, NO^+, and ArN_2^+ being the most familiar. Of these, the first two are of value only for the introduction of functional groups, and need not concern us further here (but see section 6.3.3). However, arenediazonium ions may be used in skeleton-forming reactions, not only with 'electron-rich' aromatic systens like phenols (scheme 2.9) but also with enolates and other stabilised carbanions, e.g.

$$PhN_2^+Cl^- + PhCOCH_2COPh \xrightarrow[\text{or pyridine}]{CH_3CO_2Na} Ph\overset{H^+}{N}{=}N{-}\overset{H}{C}(COPh)_2 \rightleftharpoons$$

$$PhNH{-}N{=}C(COPh)_2 \quad (>80\%)$$

$$PhN_2^+HSO_4^- + CH_2(SO_2CH_3)_2 \xrightarrow{NaOH} PhNHN{=}C(SO_2CH_3)_2 \quad (56\%)$$

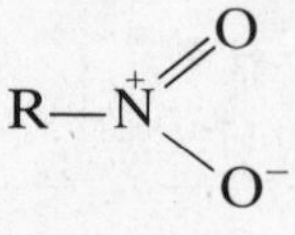

In principle, it should be possible for nitro-compounds, $R{-}\overset{+}{N}\overset{\diagup O}{\diagdown O^-}$, to function as sources of electrophilic nitrogen, and one might expect reaction with, for example, a carbanion as follows (cf. condensation with a $>C=O$ group):

$$H-\overset{|}{\underset{|}{C}}-\overset{O}{\underset{O^-}{\overset{||}{N^+}}}-R \xrightarrow[\text{(solvent)}]{H-B} H-\overset{|}{\underset{|}{C}}-\overset{OH}{\underset{O^-}{\overset{C}{N^+}}}-R \xrightarrow{-H_2O} \overset{\diagdown}{\diagup}C=\overset{+}{\underset{O^-}{N}}-R \quad (6.12)$$

In practice, however, although the process is useful in certain areas of heterocyclic chemistry (section 7.1.4.2) the generality of the reaction is insufficient to merit further consideration here.

6.3.3   Nitroso-compounds, including nitrites

In principle, the nitroso-group may act as a source of either electrophilic or nucleophilic nitrogen; for although the $N=O$ bond is polar, and the nitroso-group may thus be regarded as the nitrogen analogue of an aldehyde (i.e. with *electrophilic* nitrogen), the nitrogen also carries an unshared pair of electrons which may confer *nucleophilic* character upon it. In practice, most of the useful synthetic procedures involve nitroso-compounds as electrophilic species, as the following examples show:

$$PhNO + PhMgBr \rightarrow PhN\underset{|}{-}Ph \quad (48\%)$$
$$OH$$

$$(C_2H_5)_2N\langle \bigcirc \rangle NO + PhCH_2CN \xrightarrow{NaOH} Ph\overset{CN}{\underset{}{C}H}-\overset{OH}{\underset{}{N}}-\langle \bigcirc \rangle N(C_2H_5)_2$$

$$\xrightarrow{-H_2O} \overset{NC}{\underset{Ph}{\diagdown}}C=N\langle \bigcirc \rangle N(C_2H_5)_2 \quad (90\%)$$

This latter reaction indicates another possible disconnection for a carbon–nitrogen double bond:

$$\overset{\diagdown}{\diagup}C=N- \implies H-\overset{|}{\underset{|}{C}}\overset{OH}{\underset{}{\vdots}}N- \implies H-\overset{\diagdown}{\diagup}C^- + \overset{OH}{\underset{}{+}}N- \quad (6.13)$$

It must be emphasised, however, that this is relatively uncommon. Nitroso-compounds are themselves often difficult to prepare, and once prepared they may be highly reactive and difficult to handle. In the vast majority of cases, the correct disconnection for C=N bonds is that of (6.11).

We shall return to this and related reactions in the chapter on Oxidation (section 9.2.3).

Just as carboxylate esters may act as acylating agents (cf. sections 4.1.2 and 5.2.2) so nitrite esters are nitrosating agents:

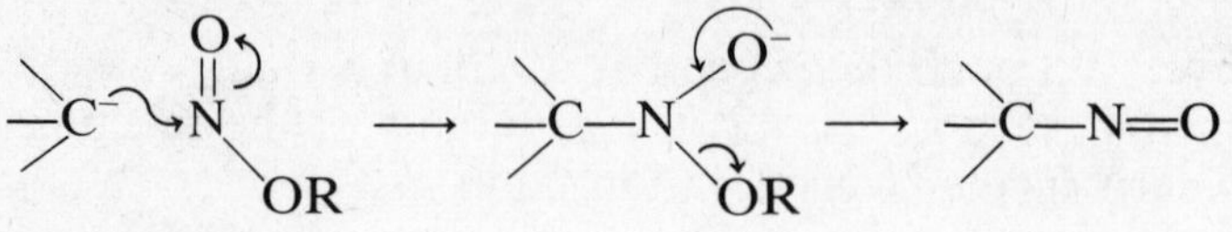

$$
\underset{OR}{\overset{O}{\underset{\text{C}-\text{N}}{}}} \longrightarrow \underset{OR}{\overset{O^-}{\underset{\text{C}-\text{N}}{}}} \longrightarrow \underset{}{\text{C}-\text{N}=\text{O}} \qquad (6.14)
$$

For example,

$$
\text{PhCOCH}_2\text{Ph} + (\text{CH}_3)_2\text{CHCH}_2\text{ONO} \xrightarrow{\text{NaOC}_2\text{H}_5} \text{PhCOCHPh}^{\text{NO}}
$$

$$
\overset{\text{NOH}}{\underset{\rightleftharpoons}{}} \text{PhCOCPh}
$$

(61 %)

We shall also return to this reaction in Chapter 9 (section 9.5.2).

7 Ring closure (and ring opening)

So far in this book, little attention has been paid to those bond-forming reactions which lead to the creation of a cyclic molecule, and so, in this final chapter dealing with the construction of molecular frameworks, we consider reactions resulting in ring closure.

The first (and, undoubtedly, the largest) group of ring forming reactions comprises nothing more than **intramolecular** variants of reactions described elsewhere in the book in *inter*molecular terms. In these processes, an *n*-membered ring is formed by cyclisation of a chain of *n* atoms. The second group of reactions is **intermolecular,** involving the simultaneous formation of *two* bonds between (usually) two different molecules. Such processes are usually called **cycloadditions,** the **Diels–Alder reaction** (Sykes, pp. 193–5) being the best-known example. [A careful distinction should be made between these genuinely concerted, intermolecular ring closures and the large group of apparently intermolecular cyclisations which in reality consist of two separate steps, ring closure being the second.] The third group consists of **electrocyclic** reactions, which are *intra*molecular and related mechanistically to cycloadditions.

By comparison with ring closure, ring opening is a relatively little-used process in synthesis, but it is of considerable value in a few special situations, as will be seen in section 7.4.

7.1 Intramolecular cyclisation by electrophile-nucleophile interaction

7.1.1 Introduction

Many of the bond-forming reactions described in earlier chapters may be adapted to produce cyclic compounds, as the following examples show:

Alkylation (cf. section 5.1.1):

$$Br(CH_2)_3Br + CH_2(CO_2C_2H_5)_2$$

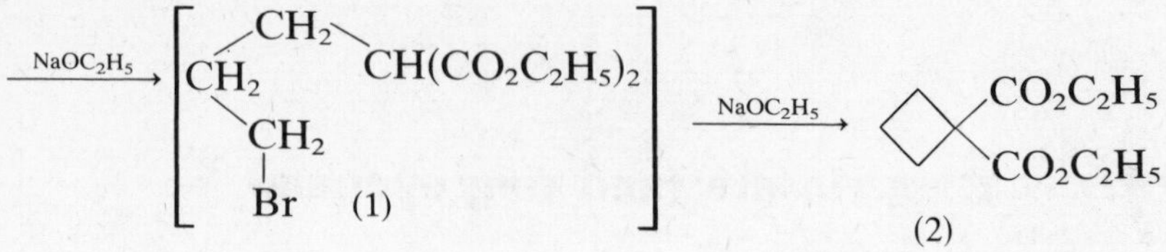

$\xrightarrow{NaOC_2H_5}$ [formula (1)] $\xrightarrow{NaOC_2H_5}$ [formula (2)]

(ca. 40 % overall)

Acylation (cf. section 5.2.2):

$$C_2H_5O_2C(CH_2)_4CO_2C_2H_5 \xrightarrow{NaOC_2H_5}$$ [formula (4)] (81 %)

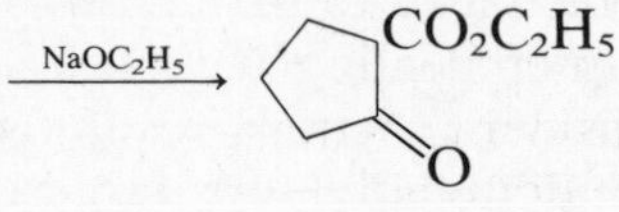

(3)

This intramolecular equivalent of the Claisen acylation is generally known as the **Dieckmann reaction** (see also Sykes, p. 226).

Condensation (cf. section 5.2.4):

$$CH_3CO(CH_2)_4COCH_3 \xrightarrow{KOH}$$ [formula] (83 %)

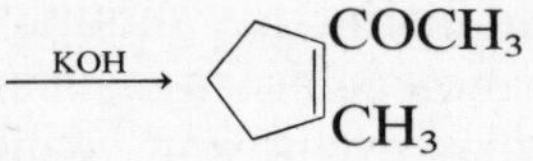

Monocyclic compounds may similarly be converted into bicyclic compounds, e.g.

Electrophilic aromatic substitution (section 2.5):

$$Ph(CH_2)_4OH \xrightarrow{H_3PO_4}$$ [formula] $\longrightarrow$ [formula] (50 %)

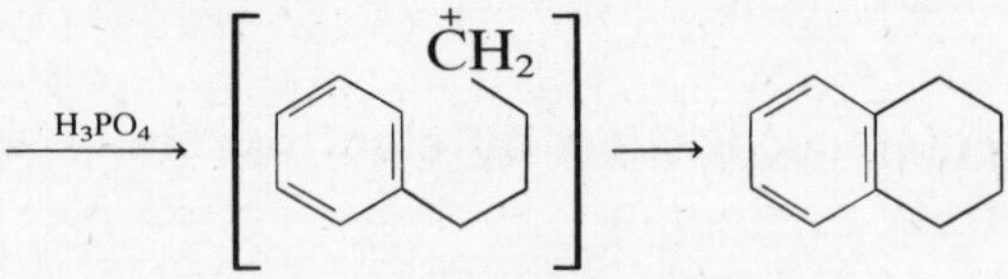

$$Ph(CH_2)_2COCl \xrightarrow{AlCl_3}$$ [formula] (90 %)

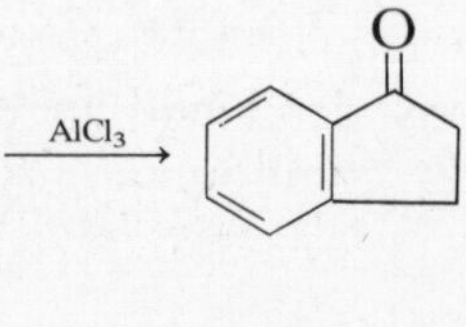

Alkylation:

$$\text{(cyclopentane with H, CH}_2\text{OTs groups)} + CH_2(CO_2C_2H_5)_2 \xrightarrow{2NaOC_2H_5} \text{(bicyclic product with } CO_2C_2H_5 \text{ groups)} \quad (75\%)$$

Acylation:

$$\text{(cyclohexanone with } CH_2CH_2CO_2CH_3 \text{ chain)} \xrightarrow{NaOCH_3} \text{(bicyclic diketone)} \quad (73\%)$$

Condensation:

$$\text{(cyclopentanone with } CH_2COCH_2CH_3 \text{ chain)} \xrightarrow{KOH} \text{(bicyclic product with } CH_3\text{)} \quad (85\%)$$

Heterocyclic compounds may also be prepared by analogous methods: these generally involve carbon–heteroatom bond formation with the heteroatom as the nucleophilic centre.

7.1.2 Facility of intramolecular ring closure: Baldwin's rules

There are several factors which influence the ease with which intramolecular ring closure occurs. First, there is the 'distance factor'. For the formation of an n-membered ring, the new bond must be formed between two atoms which are separated by $(n-2)$ other atoms, and it follows that, as n increases, there is a decreasing probability of the molecule adopting a conformation in which the 'reactive' atoms are sufficiently close for bond formation to occur. Second, there are various kinds of 'strain factor'. Angle strain (i.e. distortion of the normal bond angles) in the cyclised compound may destabilise it relative to its acyclic precursor, and *if the ring closure is reversible* the equilibrium may then lie in favour of the latter. Unfavourable steric interactions in the product (e.g. 1,3-diaxial repulsion between substituents[1]) may have the same effect. Angle strain and/or unfavourable steric interactions in the *transition state* for the ring closure step are of much wider significance. It is a consideration of the geometry of these transition states which has led to the formulation of **Baldwin's rules for ring closure**.[2] If a transition state cannot be attained without a serious distortion of normal bond angles or distances, it follows that the ring closure will occur only with difficulty (or not at all), and such processes are described by Baldwin as *disfavoured*.

Most readers will already be familiar with the geometry of the transition state for nucleophilic substitution (S_N2 reaction) at a tetrahedral carbon atom. If the overall reaction is $X^- + RCH_2Y \rightarrow RCH_2X + Y^-$, for example, the optimum direction of approach of the incoming nucleophile X^- is along the C–Y axis, and the resulting transition state (5) has an X–C–Y angle of 180°.

$$X^- \overset{R}{\underset{H}{\overset{|}{\underset{H}{C}}}}Y \rightleftharpoons X^{\delta-}\text{---}\overset{R}{\underset{H \ H}{\overset{|}{C}}}\text{---}Y^{\delta-} \longrightarrow X\text{---}\overset{R}{\underset{H \ H}{C}} \quad Y^- \qquad (7.1)$$

$$(5)$$

In the corresponding reactions in which the electrophilic carbon is trigonal (as in a carbonyl group) or digonal (e.g. in an alkyne or a cyano-group), the optimum direction of approach of the nucleophile is at an angle of 109° to the C=Y bond, and 60° to the C≡Y bond respectively [structures (6) and (7)].

$$\overset{X^-}{\underset{109°}{\ }}\ \overset{}{C}{=}Y \longrightarrow \overset{X}{\underset{}{C}}\text{--}Y^- \qquad (7.2)$$

$$(6)$$

$$\overset{X^-}{\underset{60°}{\ }}\ 120°\ \overset{}{-}C{\equiv}Y \longrightarrow 120°\ -C \overset{X}{\underset{Y^-}{\ }}120° \qquad (7.3)$$

$$(7)$$

In Baldwin's terminology, ring closures are classified according to three criteria: (i) the size of the ring being formed, (ii) whether the atom or group Y lies outside the ring being formed or else is part of the ring system, and (iii) whether the electrophilic carbon is tetrahedral, trigonal, or digonal. So a reaction of the type (7.4) would be classified as 5-*Exo-Tet* (5-membered ring, Y outside the ring being

$$X \overset{}{\underset{|}{C}}{-}Y \longrightarrow X{-}\overset{}{C} + Y^- \qquad (7.4)$$

formed, tetrahedral carbon undergoing substitution). Similarly, an intramolecular Michael reaction of the type (7.5) would be classified as

$$(7.5)$$

6-*Endo-Trig* [6-membered ring, Y (=carbon in this case) forming part of the ring, trigonal carbon undergoing addition], and reactions (1) → (2) and (3) → (4) (p. 128) are 4-*Exo-Tet* and 5-*Exo-Trig* respectively.

Baldwin's Rules are as follows. They apply to cyclisations in which the nucleophilic atom X is a first-row element (e.g. C, N, or O).

Rule 1 3- to 7-*Exo-Tet* processes are all favoured. 5- and 6-*Endo-Tet* processes are disfavoured.[3]

Rule 2 3- to 7-*Exo-Trig* processes are all favoured. 3- to 5-*Endo-Trig* processes are disfavoured; 6- and 7-*Endo-Trig* processes are favoured.

Rule 3 3- and 4-*Exo-Dig* processes are disfavoured; 5- to 7-*Exo-Dig* processes are favoured. 3- to 7-*Endo-Dig* processes are favoured.

It does not follow, of course, that because a process is 'favoured' it will necessarily occur readily in every case. The other factors mentioned earlier may all exert an influence. In general, however, a 'favoured' process occurs more readily than one which is 'disfavoured', and five- and six-membered ring compounds are formed more easily than their analogues with smaller or larger rings.

The reader is invited to describe the other cyclisations in this section using the Baldwin notation.

7.1.3 Michael addition in ring closure processes

Intramolecular cyclisation of the type described above involves the interaction of an electrophilic and a nucleophilic centre which are already joined by a chain of, say, $(n-2)$ other atoms. The electrophilic and nucleophilic properties of those centres result from the presence of adjacent functional groups, and the construction of the chain of n atoms with the functional groups correctly positioned is often the most difficult part of the synthesis.

In this regard, the Michael reaction in one or other of its various forms (sections 5.1.5, 5.2.5, and 5.4.3) has proved particularly useful, since it leads to a product in which two -*M* groups are separated by (usually) three carbon atoms. The basic conditions necessary for the Michael reaction may also serve to promote a subsequent condensation or similar ring closure step, as the following examples show:

+ PhCH=CHCOCH$_3$
NaNH$_2$
CH$_3$
O
O
Ph
O
(43 %)
Ph

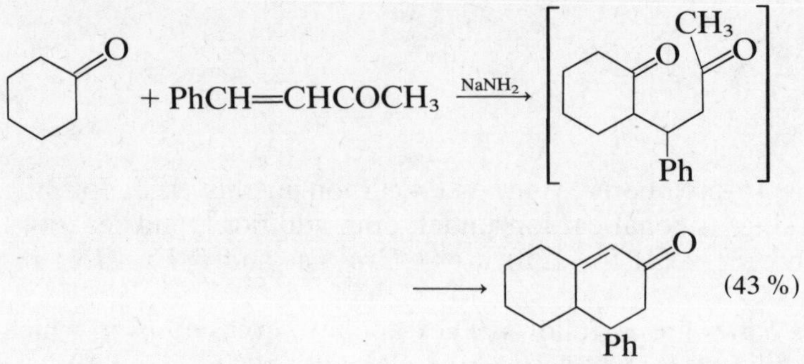

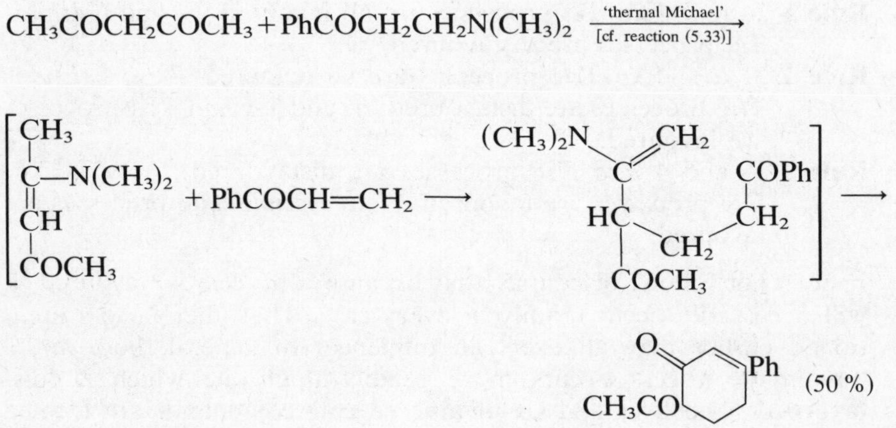

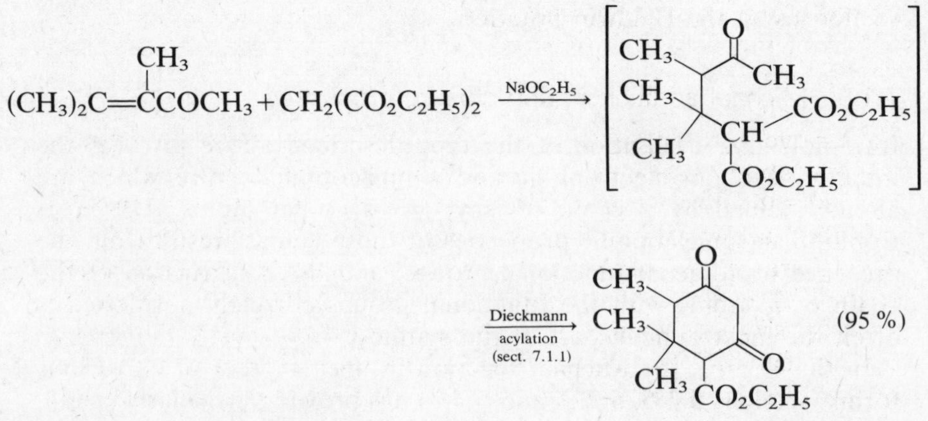

CH$_3$COCH$_2$COCH$_3$ + PhCOCH$_2$CH$_2$N(CH$_3$)$_2$
'thermal Michael'
[cf. reaction (5.33)]
CH$_3$
C—N(CH$_3$)$_2$
CH
COCH$_3$
+ PhCOCH=CH$_2$
(CH$_3$)$_2$N
CH$_2$
C
COPh
HC
CH$_2$
CH$_2$
COCH$_3$
O
Ph
CH$_3$CO
(50 %)

CH$_3$
(CH$_3$)$_2$C=CCOCH$_3$ + CH$_2$(CO$_2$C$_2$H$_5$)$_2$
NaOC$_2$H$_5$
O
CH$_3$
CH$_3$
CH$_3$
CO$_2$C$_2$H$_5$
CH
CH$_3$
CO$_2$C$_2$H$_5$
Dieckmann
acylation
(sect. 7.1.1)
O
CH$_3$
CH$_3$
O
CH$_3$
CO$_2$C$_2$H$_5$
(95 %)

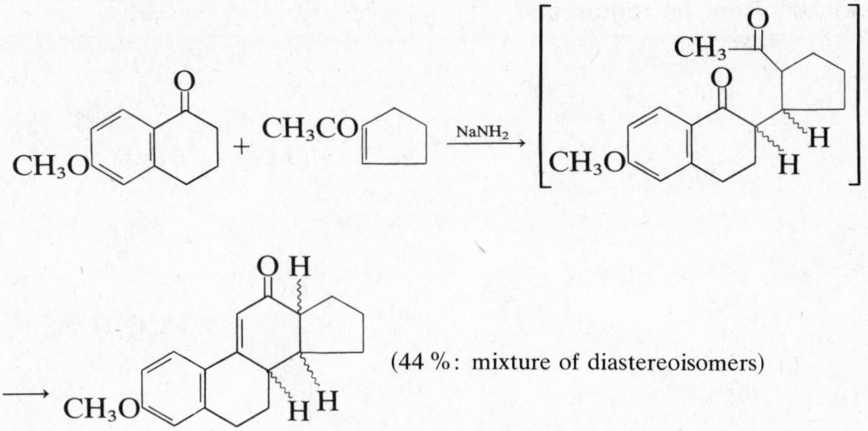

(44 %: mixture of diastereoisomers)

Michael addition followed by intramolecular condensation, as illustrated in the first and last of the above reactions, is sometimes referred to as **Robinson annelation.** The use of a Mannich base in place of an enone in the Michael reaction was also introduced by Robinson, and such reactions (like the second example above) are sometimes called **Michael–Robinson additions.**

7.1.4 Cyclisation leading to aromatic and heteroaromatic rings

7.1.4.1 Carbocyclic rings
Such a large number of benzene derivatives, with a wide variety of functional groups, may be obtained commercially that the preparation of other benzene derivatives usually amounts to nothing more than functionalisation and/or interconversion of functional groups. Methods for preparing benzene derivatives from acyclic precursors are seldom of practical importance, at least on a 'laboratory' scale, and are therefore not considered further in this book.

Naphthalene derivatives are also available in reasonable variety, but relatively few representatives of other polycyclic aromatic systems are obtainable other than by laboratory syntheses. Such syntheses generally involve benzene or naphthalene derivatives as starting materials, and two reactions commonly employed for the actual ring-closure step are a variant of the Friedel–Crafts reaction (cf. section 7.1.1) or an arylation (cf. section 2.5), as the following examples show. It should be noted that the Friedel–Crafts method may not lead initially to the formation of a fully conjugated molecule, and that a subsequent dehydrogenation step (or steps: section 9.2.4) may therefore be required:

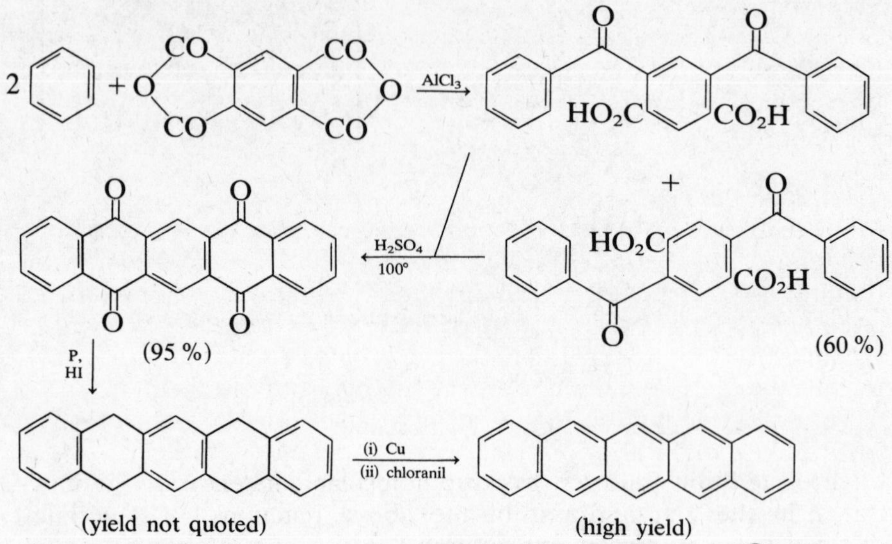

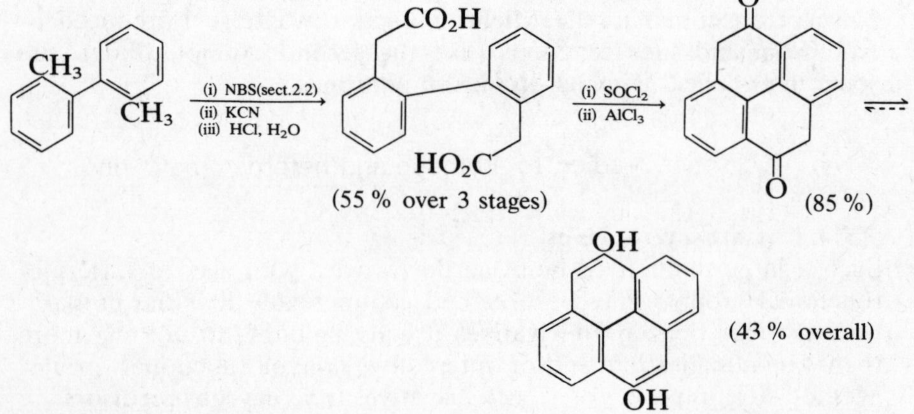

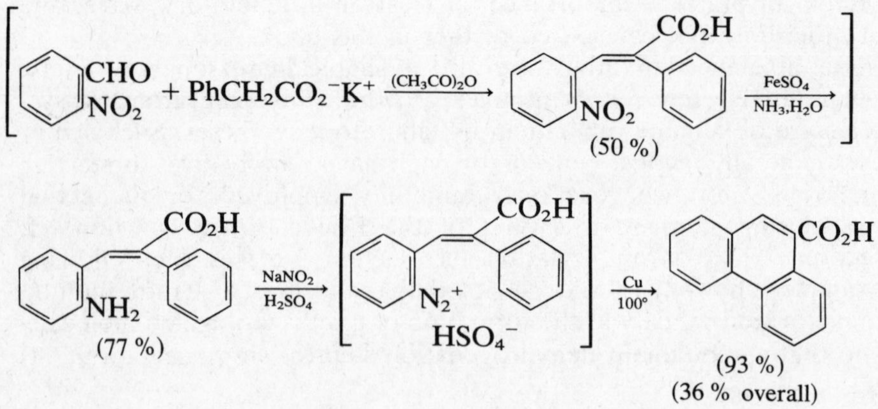

(95 %)

(60 %)

(yield not quoted)

(high yield)

(55 % over 3 stages)

(85 %)

(43 % overall)

(50 %)

(77 %)

(93 %)
(36 % overall)

(Intramolecular arylation of this type is usually referred to as the **Pschorr reaction.**)

The Diels–Alder reaction may also be used in certain cases to prepare polycyclic systems (cf. section 7.2.1), and in other instances (cf. section 7.3) electrocyclic processes have been employed successfully.

7.1.4.2 Heterocyclic rings

Methods for the synthesis of heterocyclic compounds are so numerous, and of such variety, that a separate volume would probably be required to cover the topic adequately. What follows here is an attempt to offer a few general guidelines; the coverage is restricted to the most common ring sizes and heteroatoms, *viz.* five- and six-membered rings containing oxygen, sulphur, and nitrogen.

The following features of these reactions should be noted:

(i) In the synthesis of a monocyclic compound, the ring closure step very often (although by no means always) involves carbon–heteroatom bond formation.

(ii) If the system contains two adjacent heteroatoms, it is unusual for the ring closure step to involve heteroatom–heteroatom bond formation [except when the electrophilic group is nitroso- or nitro- (cf. sections 6.3.2 and 6.3.3), a nitrene (cf. section 12.4.2), or a diazonium group (cf. section 6.3.2)].

(iii) If the target molecule is bicyclic, with the heterocyclic ring fused to a benzene ring, the starting compound is almost invariably a pre-formed benzene derivative.

A. *Monocyclic compounds* For the formation of heterocycles containing **oxygen** or **sulphur,** the majority of ring closure procedures involve an enol or enethiol as the nucleophile, and a carbonyl group as the electrophile, e.g.

Specific examples include the following:

1. $(CH_3)_3CCO \quad COC(CH_3)_3$ $\xrightarrow{\text{TsOH}}$ $(CH_3)_3C \ldots C(CH_3)_3$ (80 %)

2. $CH_3CO \quad COCH_3$ $\xrightarrow[\text{ZnCl}_2]{(CH_3CO)_2O}$ $CH_3 \ldots CH_3$ (62 %)

(8)

3. 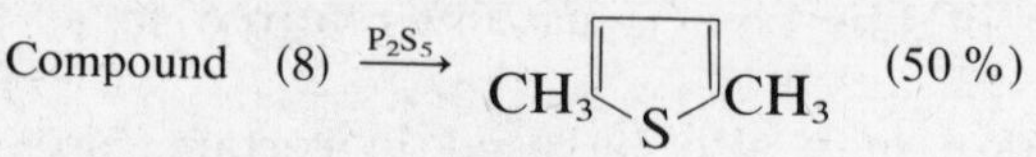 Compound (8) $\xrightarrow{P_2S_5}$ CH$_3$⟨thiophene ring⟩CH$_3$ (with S) (50 %)

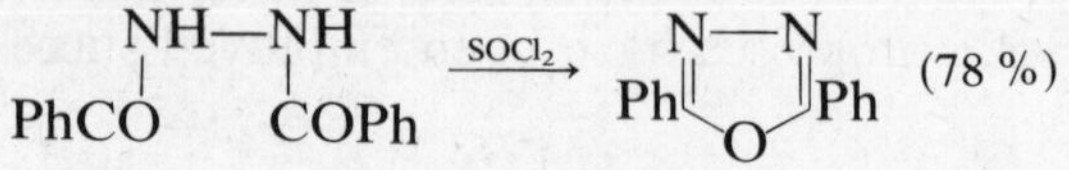

4.
NH—NH
| |
PhCO   COPh $\xrightarrow{SOCl_2}$ N—N / Ph⟨oxadiazole ring⟩Ph (with O) (78 %)

5. $CH_3COCH_2COCH_3 + H_2NOH \xrightarrow{HCl}$

$$\left[\begin{array}{c} CH_3C-CH_2 \\ \ \ \ \ | \quad \quad | \\ \ \ \ \ N \quad \ COCH_3 \\ \ \ \ \ OH \end{array} \right] \longrightarrow CH_3\text{⟨isoxazole ring, N–O⟩}CH_3$$

(62 %)

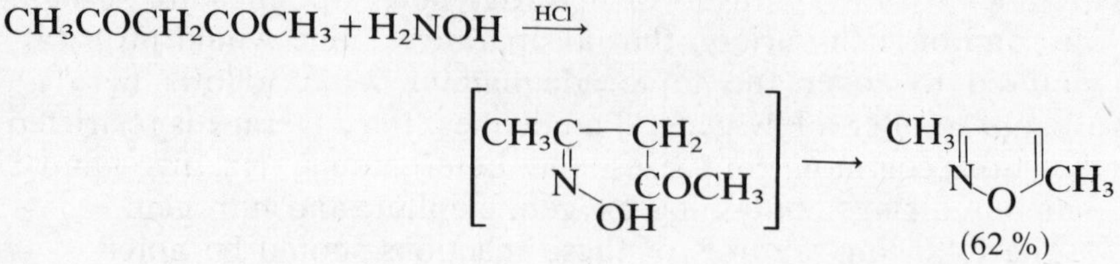

6. 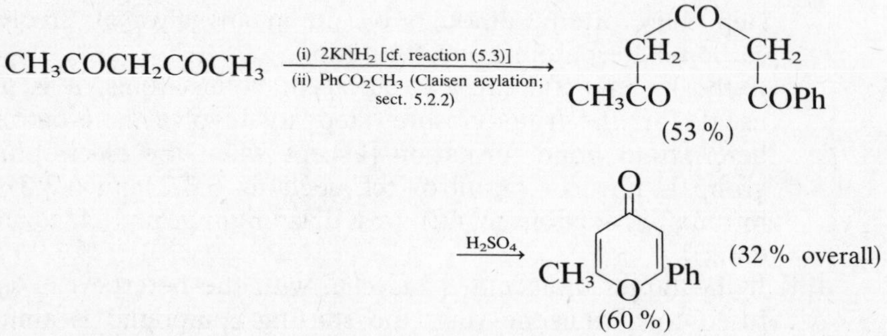 $CH_3COCH_2COCH_3 \xrightarrow[\text{(ii) PhCO}_2\text{CH}_3 \text{ (Claisen acylation; sect. 5.2.2)}]{\text{(i) 2KNH}_2 \text{ [cf. reaction (5.3)]}}$

CO / CH$_2$ CH$_2$ / CH$_3$CO   COPh

(53 %)

$\xrightarrow{H_2SO_4}$ ⟨pyranone ring⟩ CH$_3$ · · Ph (32 % overall)

(60 %)

Exceptions to this general mode of ring closure are more common in the sulphur-containing series; in such cases intramolecular condensation usually serves as the cyclisation step, e.g.

7. 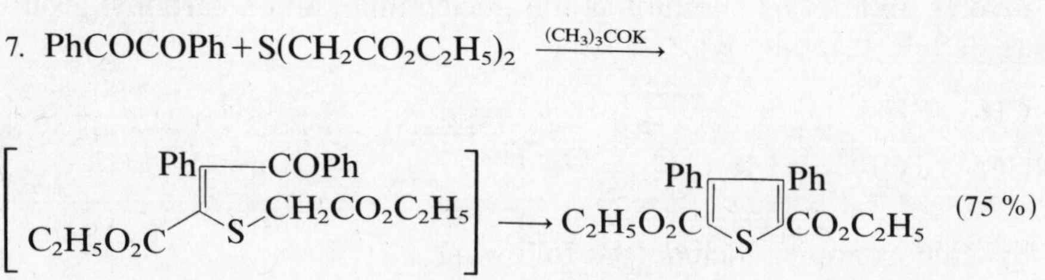 $PhCOCOPh + S(CH_2CO_2C_2H_5)_2 \xrightarrow{(CH_3)_3COK}$

$$\left[\begin{array}{c} Ph—COPh \\ \ \ \ \ \ \ \ \ CH_2CO_2C_2H_5 \\ C_2H_5O_2C \quad S \end{array} \right] \longrightarrow C_2H_5O_2C\text{⟨thiophene ring⟩}CO_2C_2H_5 \ \ (Ph, Ph)$$

(75 %)

8. 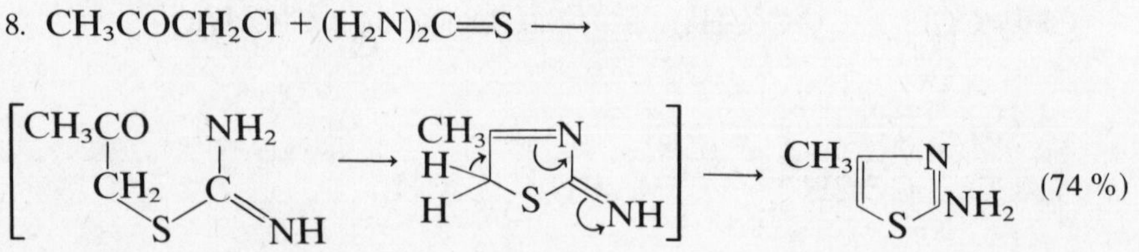 $CH_3COCH_2Cl + (H_2N)_2C{=}S \longrightarrow$

$$\left[\begin{array}{c} CH_3CO \quad NH_2 \\ \ \ \ \ | \quad \quad \ \ | \\ \ \ \ \ CH_2 \quad C \\ \ \ \ \ \ S \quad \ NH \end{array} \right] \longrightarrow \left[\begin{array}{c} CH_3 \ \ N \\ H \ \text{⟨ring⟩} \\ H \ \ S \ \ NH \end{array} \right] \longrightarrow CH_3\text{⟨thiazole ring⟩}NH_2$$

(74 %)

Example 8 involves carbon–nitrogen bond formation as the cyclisation step, and interactions of carbonyl and amino-groups (acylation and condensation) undoubtedly constitute the majority of cyclisations leading to **nitrogen** heterocycles, as the following examples also show:

9.

$$\underset{NH_2}{\overset{CHO}{\diagup}} + \underset{COCH_3}{\overset{CH_2CO_2C_2H_5}{|}} \xrightarrow[CH_3CO_2NH_4]{(C_2H_5)_3N} \left[\underset{NH_2}{\overset{CO_2C_2H_5}{\diagup\diagdown COCH_3}} \right] \longrightarrow$$

$$\underset{N}{\overset{CO_2C_2H_5}{\bigcirc}} \quad (50\%)$$

10.

$$\underset{CH_3CO}{\overset{CH_3}{\underset{|}{CO}}} \overset{CH_2CN}{+ \underset{NH_2}{CO}} \xrightarrow{\text{piperidine}}$$

$$\left[\underset{CH_3CO}{\overset{CH_3}{\diagdown}} \underset{CO}{\overset{CN}{\diagup}} \underset{NH_2}{\diagup} \longrightarrow \underset{CH_3}{\overset{CH_3}{\underset{N}{\diagup}}} \underset{O}{\overset{CN}{\diagup}} \right] \longrightarrow \underset{CH_3}{\overset{CH_3}{\underset{N}{\diagup}}} \underset{O}{\overset{CN}{\bigcirc}} \quad (66\%)$$

11.

$$\underset{CH_3CO}{\overset{CH_2-COCH_3}{|}} + H_2NNH_2 \longrightarrow$$

$$\left[\underset{CH_3}{\overset{-COCH_3}{\diagdown N-NH_2}} \rightleftharpoons \underset{CH_3}{\overset{-COCH_3}{\underset{H}{\diagdown N-NH_2}}} \right] \longrightarrow \underset{CH_3}{\overset{CH_3}{\underset{H}{\diagup N-N}}}$$

$$(73\%)$$

12.

$$\underset{S}{\overset{NH_2}{\underset{C}{\diagup}}} \overset{CH_3}{+ \underset{CO_2C_2H_5}{CO-CH_2}} \xrightarrow[H_2O]{K_2CO_3} \left[\underset{S}{\overset{CH_3}{\underset{C}{\diagup N}}} \underset{NH_2}{\overset{CO_2C_2H_5}{\diagup}} \rightleftharpoons \right.$$

$$\left. \underset{S}{\overset{CH_3}{\underset{HN}{\diagdown}}} \underset{NH_2}{\overset{CO_2C_2H_5}{\diagup}} \right] \longrightarrow \underset{S}{\overset{CH_3}{\underset{H}{\diagdown}}} \underset{O}{\overset{}{\diagup}} \rightleftharpoons \underset{HS}{\overset{CH_3}{\underset{N}{\diagdown}}} \underset{OH}{\overset{}{\diagup}} \quad (95\%)$$

Nitrogen analogues of examples 1–6 involve enamines as nucleophiles in place of enols or enethiols, e.g.

13. 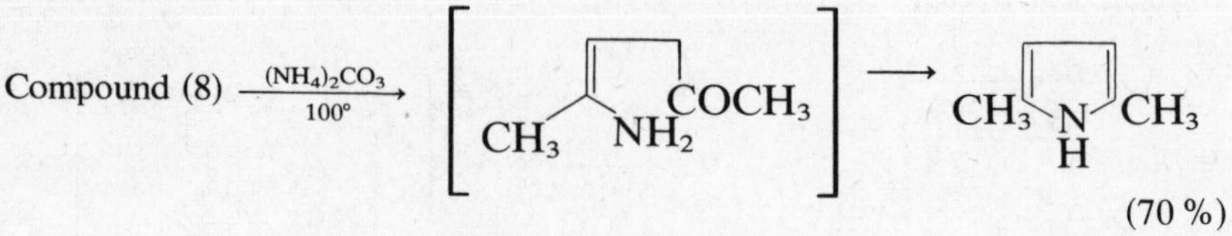
Compound (8)
$(NH_4)_2CO_3$
100°
CH_3
NH_2
$COCH_3$
CH_3
N
H
CH_3
(70 %)

14. 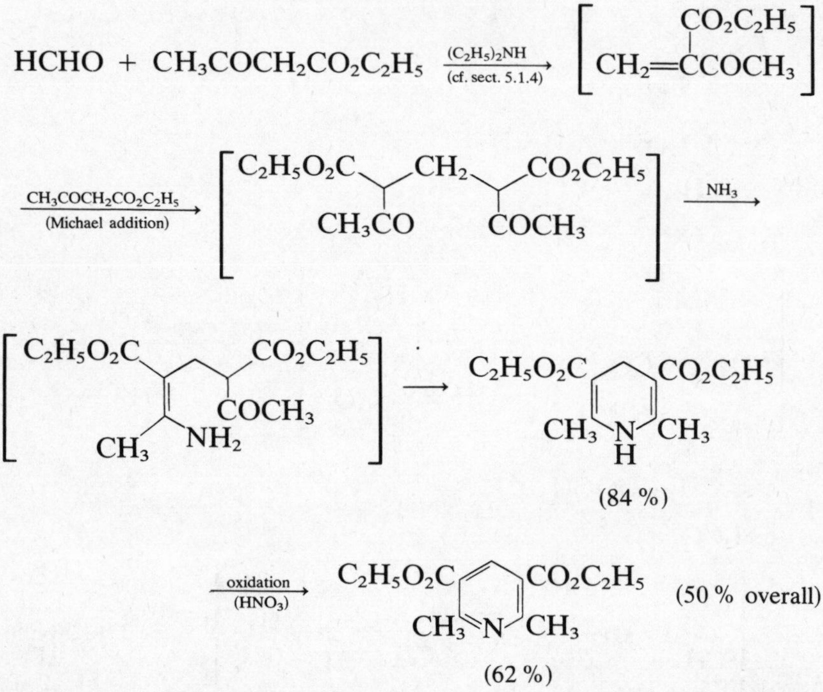
HCHO + $CH_3COCH_2CO_2C_2H_5$
$(C_2H_5)_2NH$
(cf. sect. 5.1.4)
$CO_2C_2H_5$
$CH_2{=}CCOCH_3$
$CH_3COCH_2CO_2C_2H_5$
(Michael addition)
$C_2H_5O_2C$
CH_2
$CO_2C_2H_5$
CH_3CO
$COCH_3$
NH_3
$C_2H_5O_2C$
$CO_2C_2H_5$
$COCH_3$
CH_3
NH_2
$C_2H_5O_2C$
$CO_2C_2H_5$
CH_3
N
H
CH_3
(84 %)
oxidation
(HNO_3)
$C_2H_5O_2C$
$CO_2C_2H_5$
CH_3
N
CH_3
(50 % overall)
(62 %)

B. *Benzo-fused compounds* In these cyclisations, the starting material is generally an *ortho*-disubstituted benzene (examples 15–18), in which case the ring closure is effected by one of the methods outlined in section A above; otherwise only one substituent on the benzene ring is incorporated (examples 19–20), and ring closure is then brought about by electrophilic aromatic substitution, very often of the Friedel–Crafts or related type.
Thus,

15. 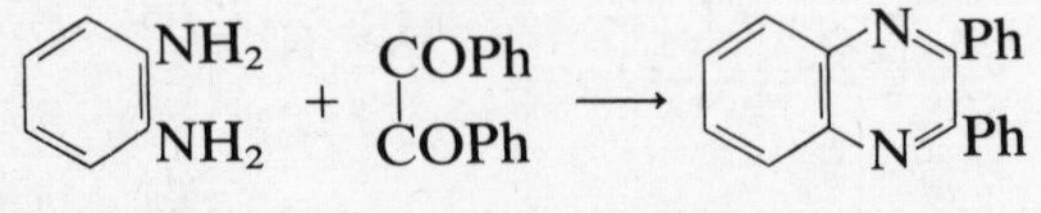
NH_2
NH_2
+
COPh
COPh
N
Ph
N
Ph

(51 %)

16.

17.

18.

19.

20. 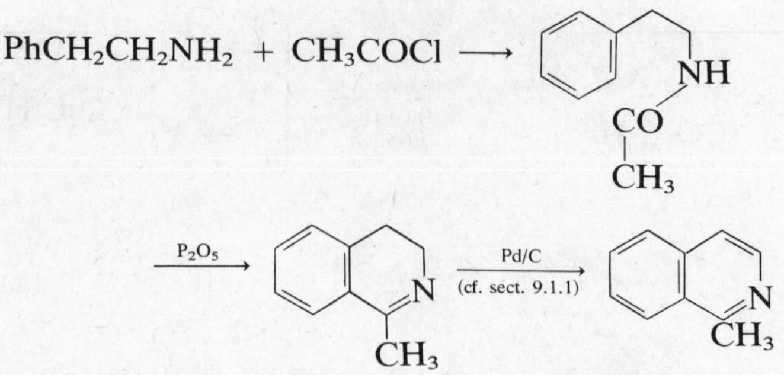 $PhCH_2CH_2NH_2 + CH_3COCl \longrightarrow$

$\xrightarrow{P_2O_5}$

$\xrightarrow[\text{(cf. sect. 9.1.1)}]{Pd/C}$

(73 % overall)

7.1.5 Formation of medium and large rings

It has already been pointed out in section 7.1.2 that one of the factors on which the ease of intramolecular cyclisation depends is the so-called 'distance factor': the larger is the ring to be formed, the less is the probability that the acyclic precursor will adopt a conformation which brings the electrophilic and nucleophilic atoms sufficiently close for cyclisation to be possible. Under such circumstances, *inter*-molecular reaction between two molecules of the precursor becomes much more probable than *intra*molecular cyclisation.

For the formation of medium (8–11-membered) and large rings (12-membered and over), therefore, special methods may be required in order to promote cyclisation at the expense of intermolecular reactions. In the usual procedure, normally referred to as the 'high dilution' technique, the acyclic precursor is introduced very slowly into the reaction medium, so that its concentration is always very low (often 10^{-3} M or less); at this concentration the probability of inter-molecular reaction is greatly reduced. Under such high dilution conditions, Dieckmann and related acylation reactions lead to acceptable yields of medium- and large-ring compounds, e.g.

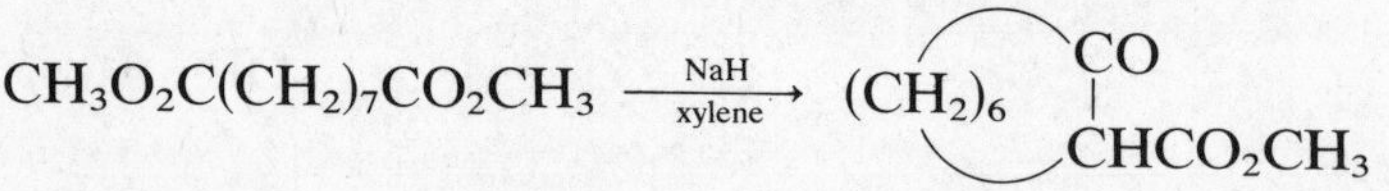

$$CH_3O_2C(CH_2)_7CO_2CH_3 \xrightarrow[\text{xylene}]{NaH}$$

(48 %)

[The ester (1 M solution) is added dropwise over 9 days to a stirred suspension of the hydride (2.5-fold excess; ca. 1 M).]

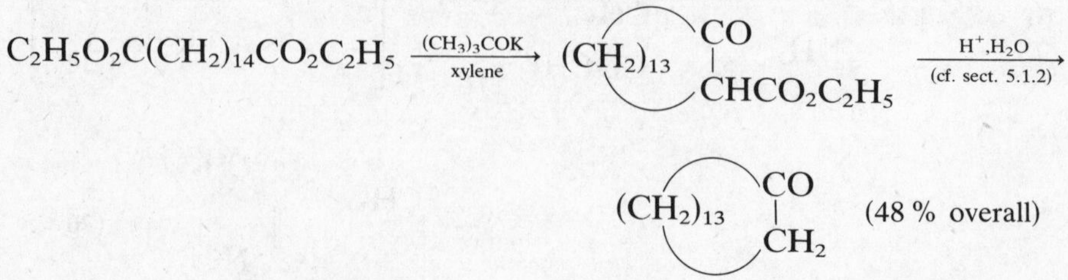

$$C_2H_5O_2C(CH_2)_{14}CO_2C_2H_5 \xrightarrow[\text{xylene}]{(CH_3)_3COK}$$

$$\xrightarrow[\text{(cf. sect. 5.1.2)}]{H^+, H_2O}$$

(48 % overall)

[Ester (4 M solution) added to base (4.8-fold excess; also ca. 4 M) dropwise over 24 h.]

$$NC(CH_2)_{20}CN \xrightarrow[\text{ether}]{PhN(CH_3)Na} (CH_2)_{19}\overset{CHCN}{\underset{C\diagdown NH}{|}} \rightleftharpoons (CH_2)_{19}\overset{\overset{CN}{|}}{\underset{C\diagdown NH_2}{C}}$$

(9)

$$\xrightarrow[H_2O]{H^+} (CH_2)_{19}\overset{CHCN}{\underset{CO}{|}} \longrightarrow (CH_2)_{19}\overset{CH_2}{\underset{CO}{|}} \quad (70\% \text{ overall})$$

(10)

In this version (the **Thorpe–Ziegler reaction**) the intermediate cyano-enamine [such as (9)] or cyano-ketone [such as (10)] may be isolated if desired.

High dilution methods may also be applied to the preparation of macrocyclic esters (lactones), e.g.

$$Br(CH_2)_{10}CO_2H \xrightarrow[CH_3COC_2H_5]{K_2CO_3} (CH_2)_{10}\overset{CO}{\underset{O}{|}} \quad (85\%)$$

[Bromoacid (0.15 M solution) added over 2 days to the base (large excess: ca. 0.25 M).]

One reaction which has been applied with great success to the preparation of medium and large rings is the **acyloin reaction.** This, in its simplest form, is an analogue of the bimolecular reduction of ketones (section 8.3.2) and is also related to the Bouveault–Blanc reduction (section 8.4.4). It involves one-electron reduction of an ester by metallic sodium, and dimerisation of the resulting radical anion. This then loses alkoxide ions, giving a diketone (11), and the latter then undergoes further reduction to the dianion (12) of the acyloin (13).

$$2R-\overset{O}{\overset{\|}{C}}-OR' \xrightarrow{2e^-} 2R-\overset{O^-}{\overset{|}{\underset{\cdot}{C}}}-OR' \longrightarrow \overset{O^- \; O^-}{\underset{OR' OR'}{RC-CR}} \xrightarrow{-2\bar{O}R'} \overset{O \quad O}{\overset{\| \quad \|}{RC-CR}}$$

(11)

$$\xrightarrow{2e^-} \overset{O^- \; O^-}{RC-CR} \longrightarrow \overset{O^- \; O^-}{RC=CR} \xrightarrow[\text{(work-up)}]{2H^+} \overset{O \quad OH}{\overset{\| \quad |}{RC-CHR}}$$

(12)   (13)

Long-chain *diesters* give cyclic acyloins; since the reaction is heterogeneous, taking place on the surface of the metal, there is not the same need to employ high dilution methods, e.g.

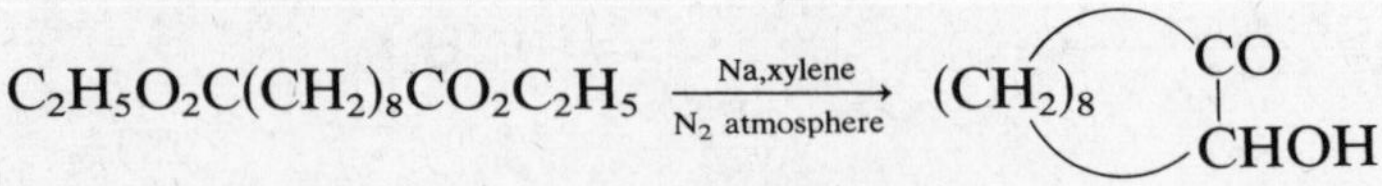

$$C_2H_5O_2C(CH_2)_8CO_2C_2H_5 \xrightarrow[\text{N}_2 \text{ atmosphere}]{\text{Na,xylene}} (CH_2)_8 \begin{matrix} CO \\ | \\ CHOH \end{matrix} \quad (46\%)$$

[Ester (undiluted) added over 3 hours to a suspension of sodium (4-fold excess) in xylene.]
Similarly,

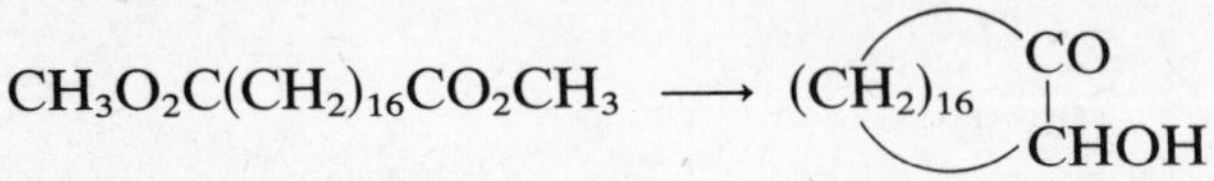

$$CH_3O_2C(CH_2)_{16}CO_2CH_3 \longrightarrow (CH_2)_{16} \begin{matrix} CO \\ | \\ CHOH \end{matrix} \quad (96\%)$$

If a long chain of atoms contains one or more rigid sections, in which free rotation about bonds is not possible, there may be an increased chance of cyclisation to form a medium or large ring, e.g.

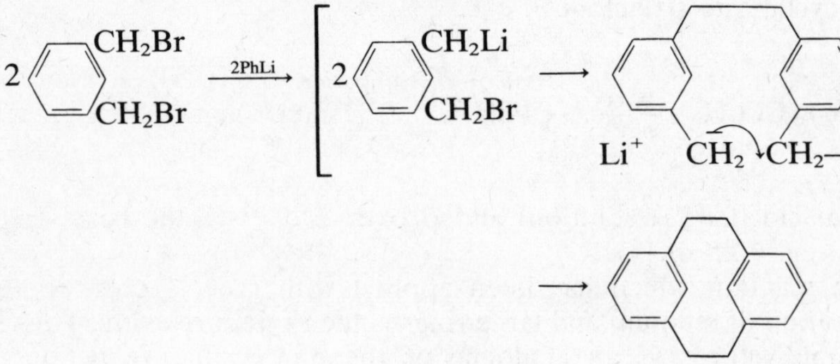

(39 %)

(10-membered ring)

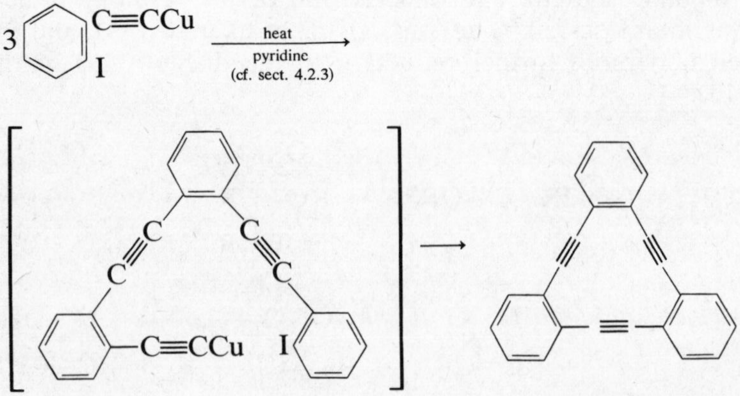

(26 %)

(12-membered ring)

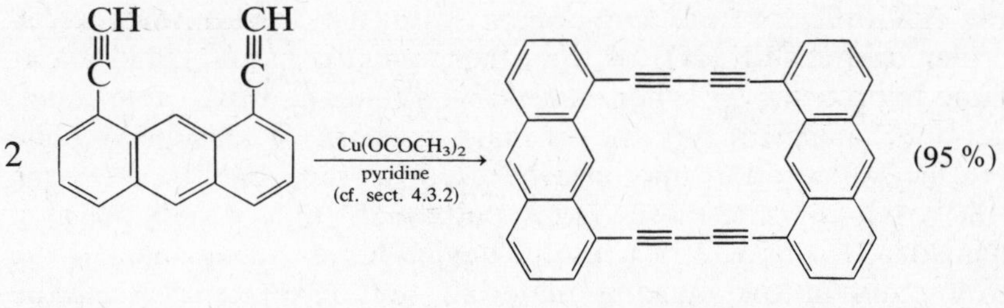

(18-membered ring)

This last process, oxidative coupling of alkynes, has been of particular value in the synthesis of **annulenes** (for an example, see Chapter 14).

Bicyclic compounds (fused or bridged systems) may also serve as precursors of medium- or large-ring monocyclic compounds. Such reactions are formally ring opening procedures, and as such are discussed in section 7.4.

7.2 Cycloaddition

Most readers will already be familiar with the Diels–Alder reaction, which in its simplest form consists of the reaction of a conjugated diene with a monoene (usually conjugated with a -M group) to give a cyclohexene derivative [reaction (7.6)]:

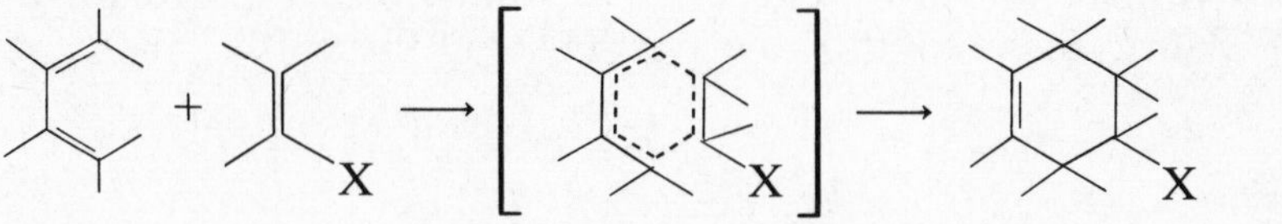

(7.6)

e.g. $CH_2{=}CHCH{=}CH_2 + CH_2{=}CHCHO \xrightarrow{100°}$ 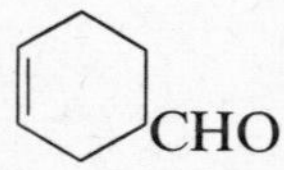 (quantitative)

Most readers will also have appreciated that reactions of this type cannot be described adequately in terms of electrophile-nucleophile interactions, and it is equally clear that they do not involve radical pathways. They are representatives of a large group of reactions which involve the interaction of π-electron systems, in a concerted manner and *via* a cyclic transition state, and such reactions are generally described as **pericyclic** or **symmetry-controlled.**

The mechanisms of these reactions are considered in Sykes, Chapter 12 (pp. 328–44) in terms of **frontier orbitals.**[4] The reactions are considered to arise by interaction of the **highest occupied molecular** orbital **(HOMO)** of one component with the **lowest unoccupied molecular orbital (LUMO)** of the other component. We shall adopt the same approach here when necessary, although, as we have stated before, mechanism is not the primary concern of this book. The 'curved arrow' notation may also be used [as in (7.6a)] to show the overall result of these reactions: although not strictly correct mechanistically, this is still a useful device for ensuring that all the electron pairs in the starting materials are accounted for in the products.

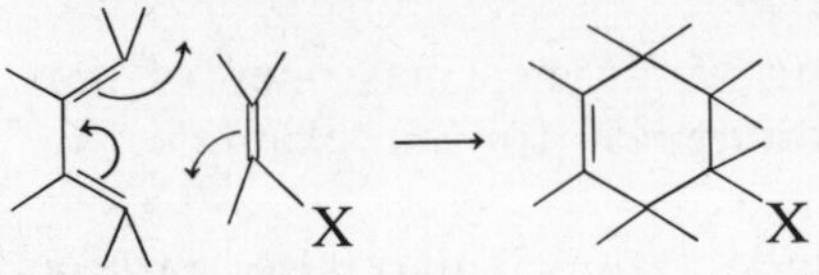

$$(7.6a)$$

In the remainder of this section, and in section 7.3, we shall consider some of the most important pericyclic reactions from the synthetic viewpoint.

7.2.1 The Diels–Alder reaction

The main features of this reaction have already been set out by Sykes (pp. 193–5 and 337–9): the process involves the interaction of a 4π-electron system (the diene) and a 2π-electron system (the monoene or **dienophile,** as it is often called), and so the overall reaction is a $[4+2]$-cycloaddition. The reaction is stereospecifically *syn*- with respect to both diene and dienophile, as expected for a HOMO-LUMO interaction of the type $(14)+(15)$ or $(16)+(17)$.

HOMO (14)   LUMO (16)

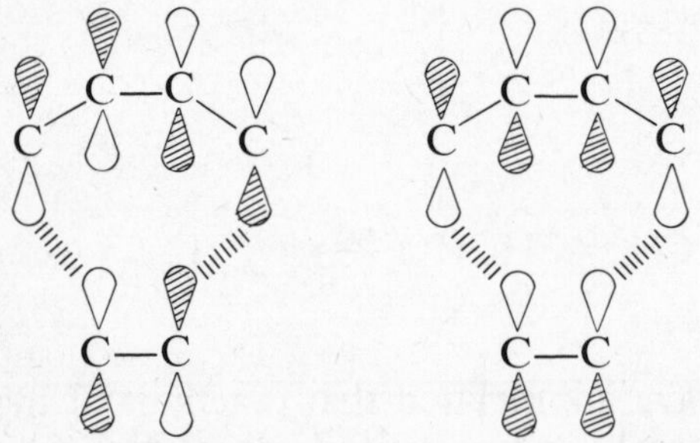

LUMO (15)   HOMO (17)

Thus, the relative configuration of the starting materials is retained in the product, e.g.

1. 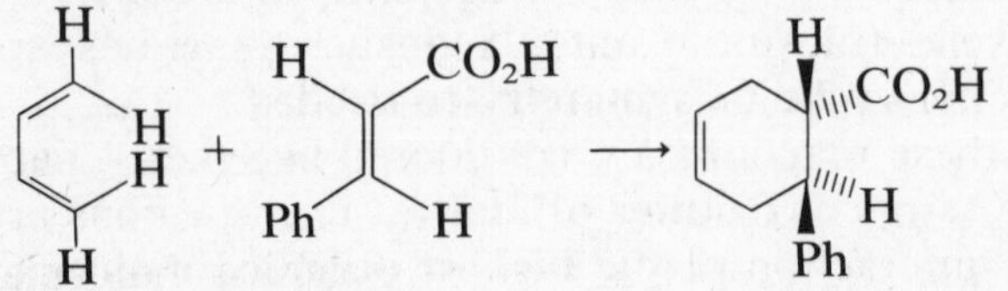 (ca. 35 %; no *cis*-isomer)

2. 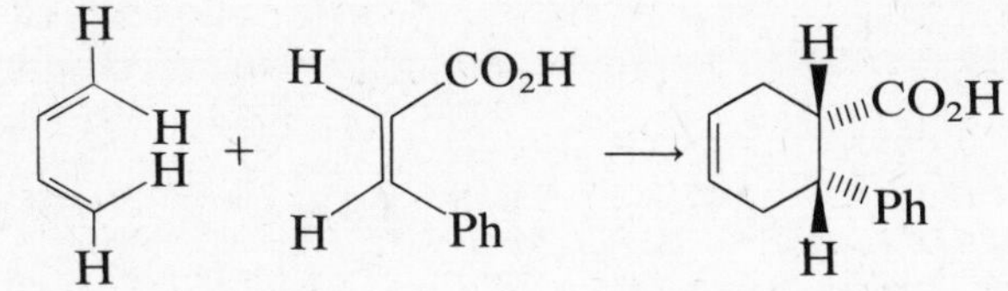 (ca. 10 %; no *trans*-isomer)

The diene must be able to adopt the *cisoid* conformation in order that reaction should occur. Dienes which are fixed in the *transoid* conformation, e.g. (18), cannot undergo the Diels–Alder reaction. If

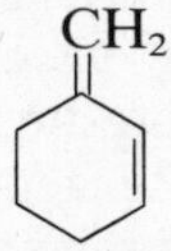

(18)

the adoption of the *cisoid* conformation leads to unfavourable steric interactions (as in example 4, between CH_3 and H) the reaction may be very slow (contrast examples 3 and 4).

3. 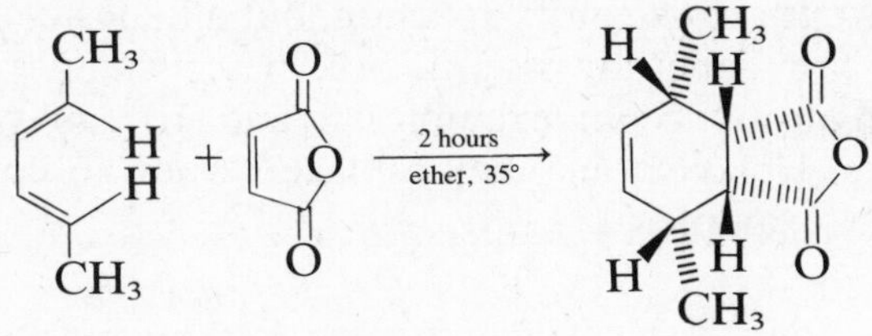 [quantitative; no (20)]

(19)

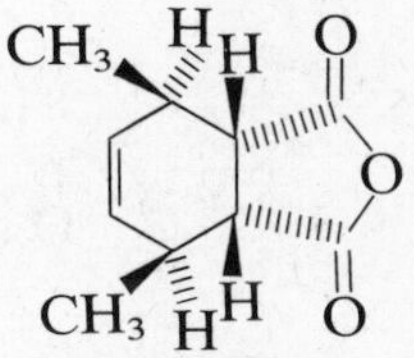

(20)

4. 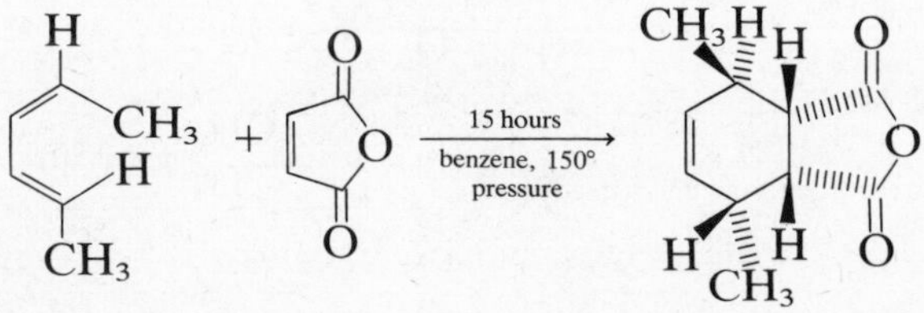 (ca. 32 %)

If the *syn*-addition of the diene and dienophile can lead to two possible adducts, as·in example 3, it is usually the product of *endo*-addition [in this case, (19)] which predominates over the *exo*-addition product [in this case, (20)]. This preference for *endo*-addition is attributed to additional orbital overlap between the components in the transition state [cf. (21)] which cannot occur in the *exo*-transition state [cf. (22)]:

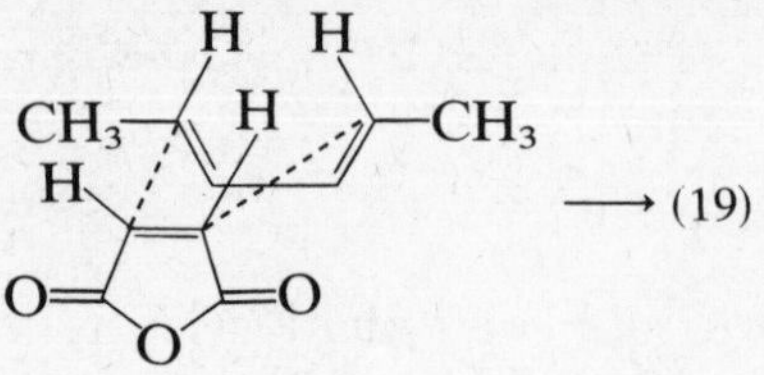

                    (21) 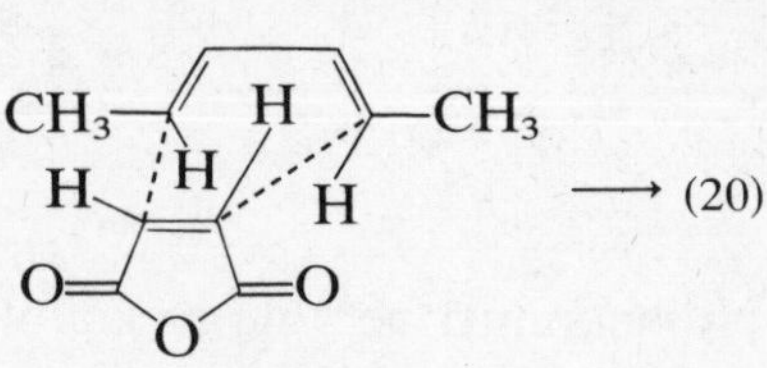   (22)

A wide variety of components may participate in the Diels–Alder reaction, and so the procedure is of considerable synthetic importance. For example, the diene may be carbocyclic (example 5) or heterocyclic (examples 7 and 8). Benzene derivatives do not participate readily in Diels–Alder reactions, since the adducts would be non-aromatic, but polycyclic compounds such as anthracene readily form adducts (examples 9 and 10) since an additional benzenoid ring is thereby formed. The dienophile component may be subject to equally wide variation: simple alkenes like ethylene require high temperature and pressure for satisfactory reaction, but alkenes conjugated to a -M group are generally useful. Alkynes (example 7), including benzyne (dehydrobenzene; examples 8 and 10), may be used in place of alkenes. Heteroatoms may replace carbon in either the diene or dienophile (examples 11–13):

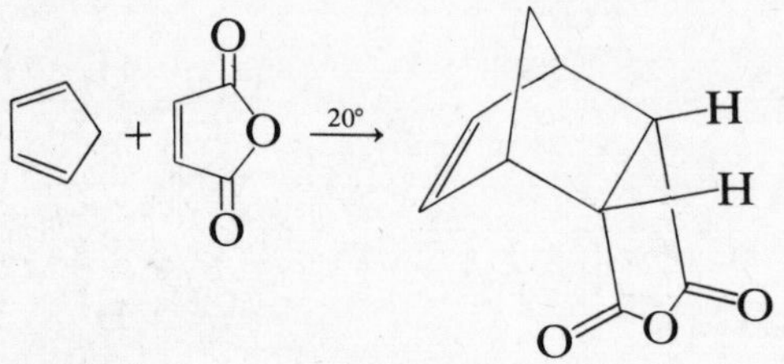

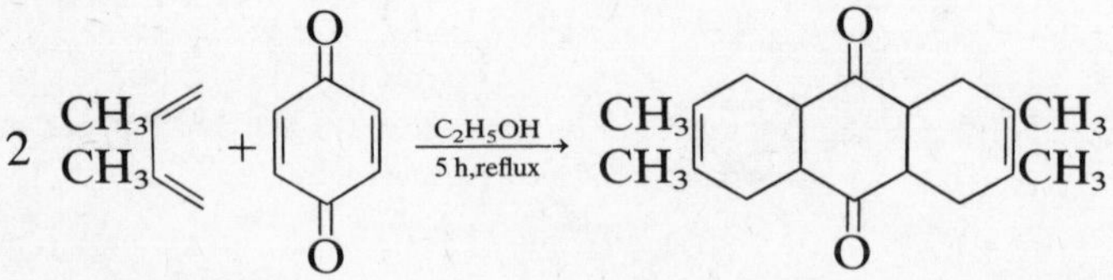

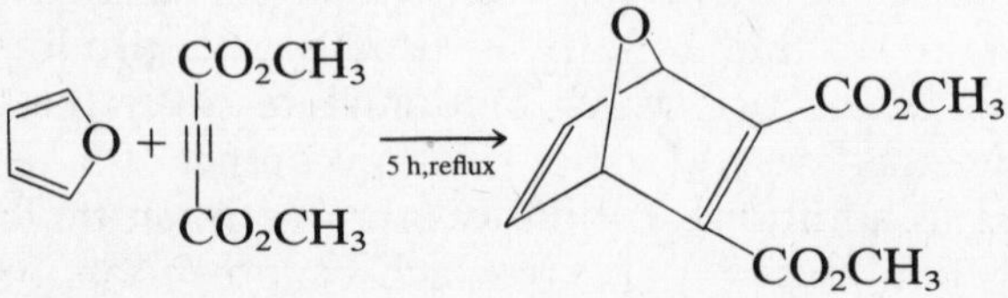

8. 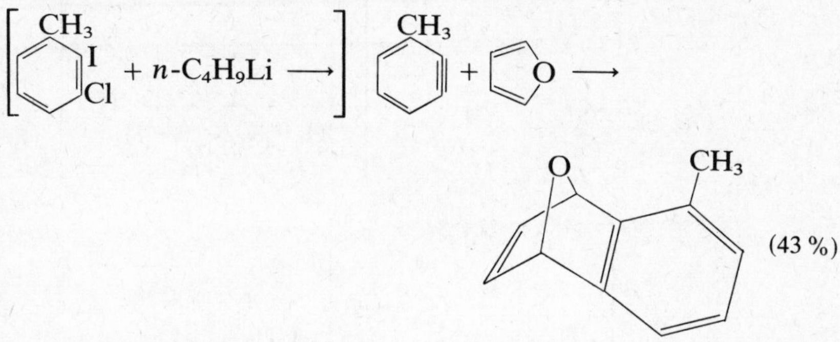 $\left[\begin{array}{c} CH_3 \\ \hline \end{array} I \\ Cl \end{array} + n\text{-}C_4H_9Li \longrightarrow \right]$ $\begin{array}{c} CH_3 \\ \hline \end{array}$ + $\begin{array}{c} O \end{array}$ $\longrightarrow$

(43 %)

9. 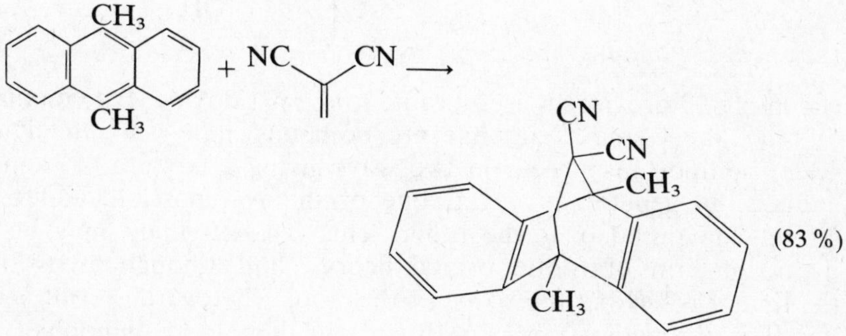 $\begin{array}{c} CH_3 \\ \hline \\ CH_3 \end{array}$ + $NC \quad CN$ $\longrightarrow$

(83 %)

10. 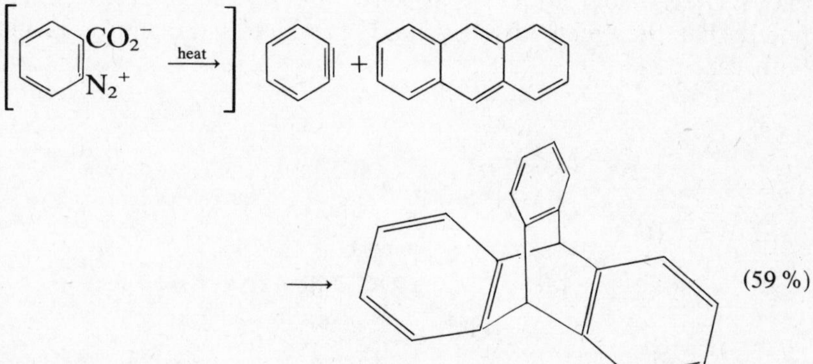 $\left[\begin{array}{c} CO_2^- \\ \hline \\ N_2^+ \end{array} \right] \xrightarrow{heat}$ $\begin{array}{c} \\ \hline \end{array}$ + $\begin{array}{c} \\ \hline \end{array}$

$\longrightarrow$

(59 %)

11. 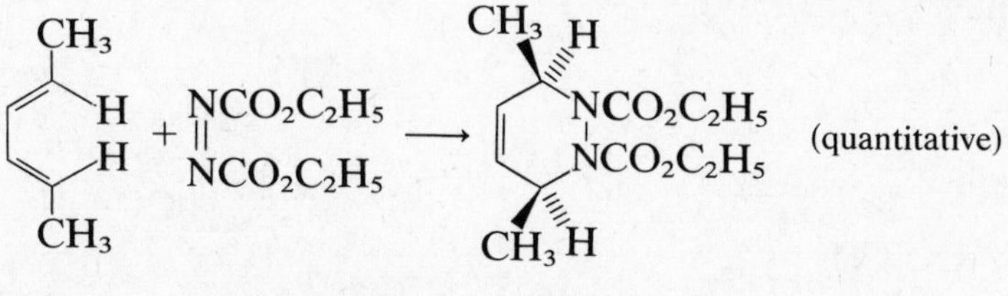 $\begin{array}{c} CH_3 \\ \text{---H} \\ \text{---H} \\ CH_3 \end{array}$ + $\begin{array}{c} NCO_2C_2H_5 \\ \| \\ NCO_2C_2H_5 \end{array}$ $\longrightarrow$ $\begin{array}{c} CH_3 \; H \\ N\text{---}CO_2C_2H_5 \\ N\text{---}CO_2C_2H_5 \\ CH_3 \; H \end{array}$ (quantitative)

12. 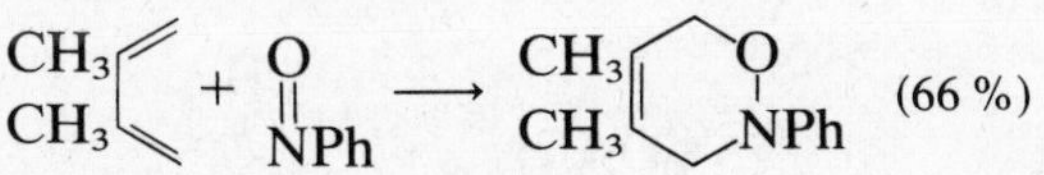 $\begin{array}{c} CH_3 \\ CH_3 \end{array}$ + $\begin{array}{c} O \\ \| \\ NPh \end{array}$ $\longrightarrow$ $\begin{array}{c} CH_3 \quad O \\ CH_3 \quad NPh \end{array}$ (66 %)

13.

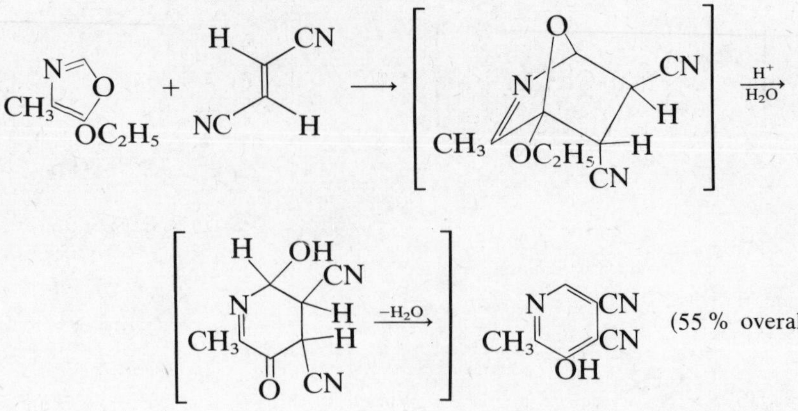

[Example 13 provides a useful route to the pyridoxine (B_6) vitamins.]

If the diene and dienophile are both unsymmetrical, the Diels–Alder addition may occur in two ways giving a mixture of isomeric adducts; in general, however, one of the two possible adducts is strongly favoured over the other. This *regioselectivity* may be explained in terms of frontier orbital theory[5], and although discussion of the theory is beyond the scope of this book, the overall result is not. For the vast majority of substituents in diene and dienophile, the major Diels–Alder adduct is as shown in (7.7) and (7.8). It should be noted that the '*meta*-disubstituted' product is the minor isomer in both cases:

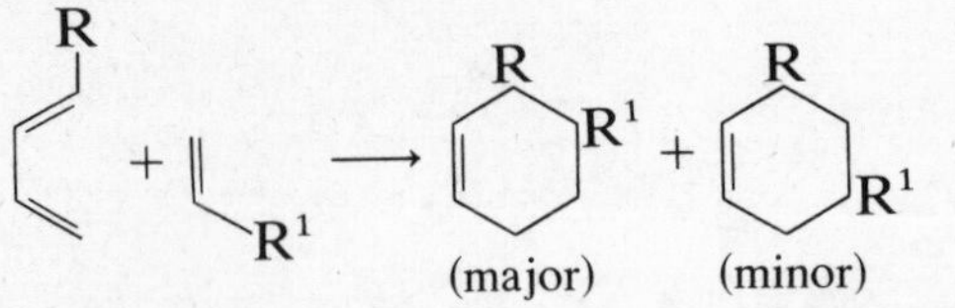

(7.7)

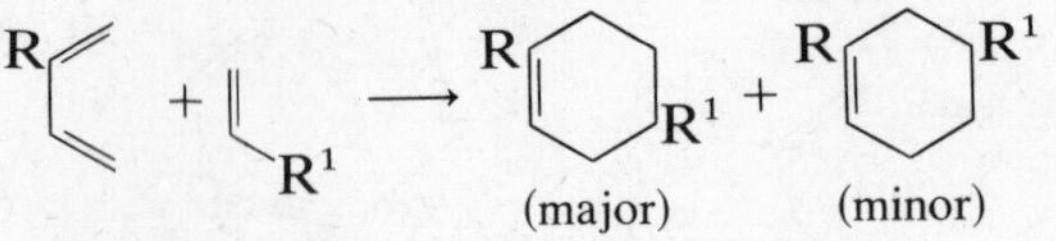

(7.8)

For example,

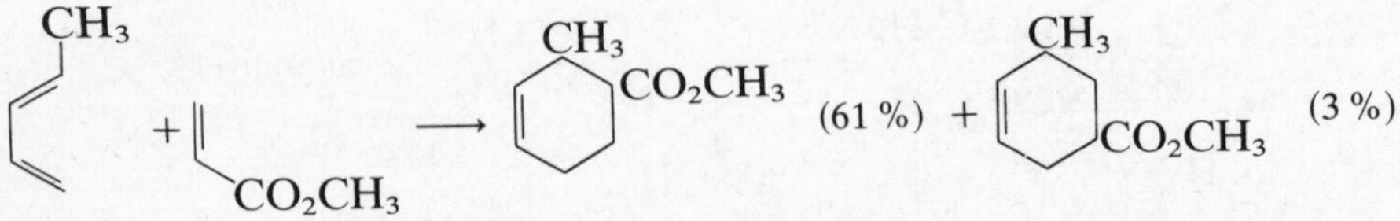

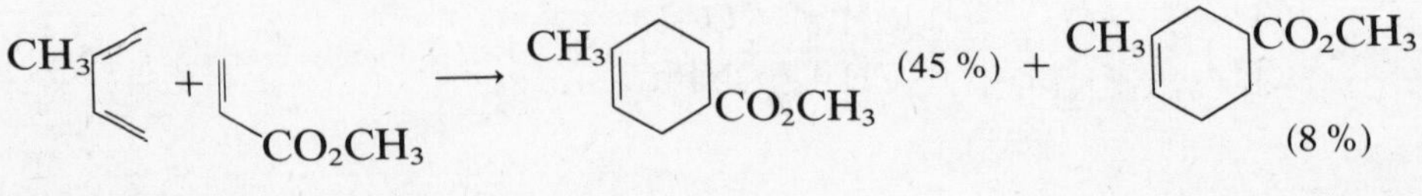

he regioselectivity of additions to unsymmetrical alkenes (or
r **dipolarophiles**) is much more difficult to explain, and is not
idered further here.

he following are representative examples of 1,3-dipolar cycload-
ns:

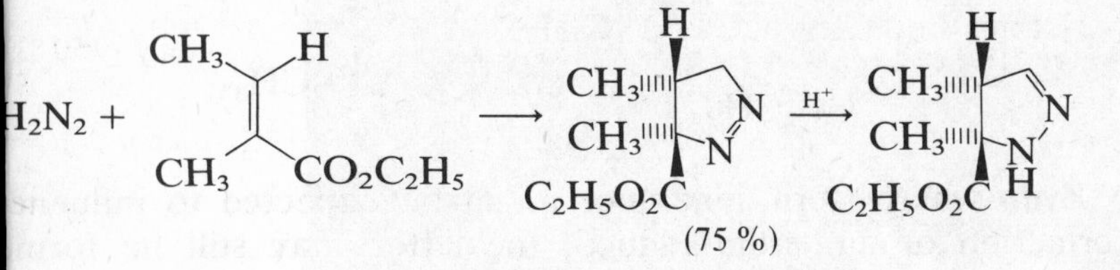

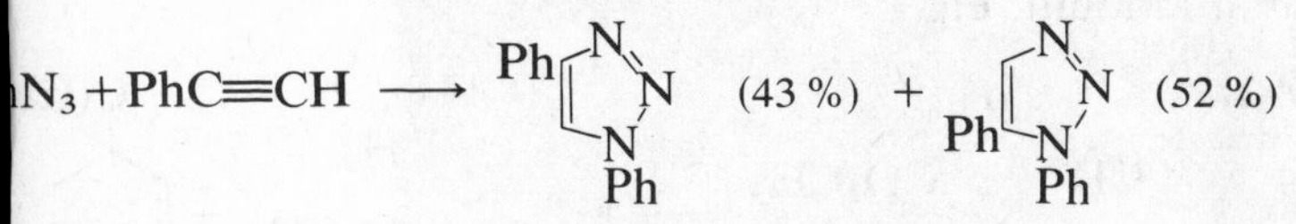

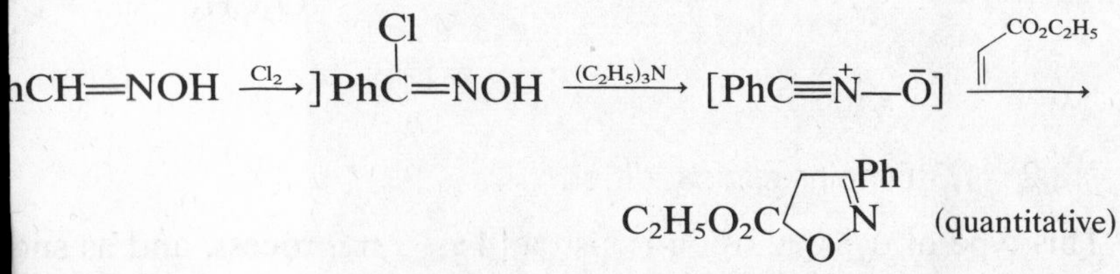

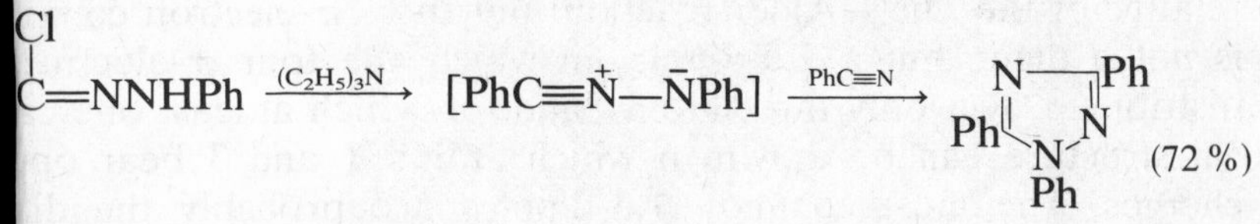

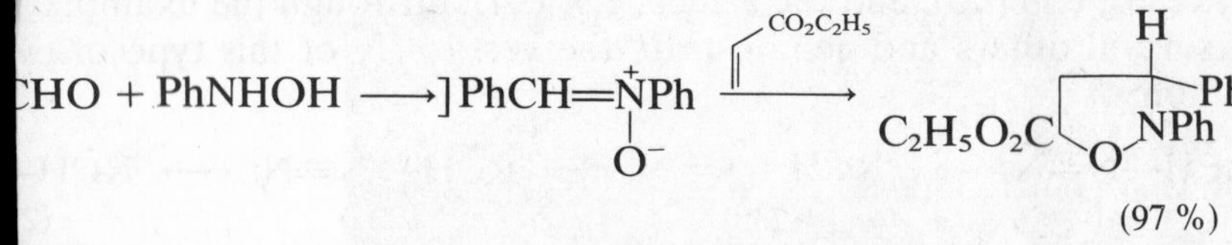

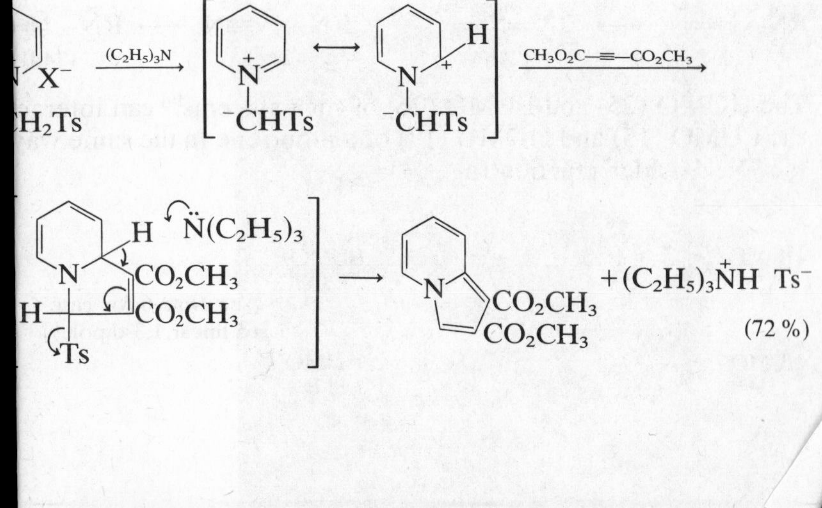

H_2N_2 +

(75 %)

N_3 + PhC≡CH ⟶

(43 %) +   (52 %)

$hCH=NOH \xrightarrow{Cl_2}] PhC=NOH \xrightarrow{(C_2H_5)_3N} [PhC≡\overset{+}{N}—\bar{O}]$

$C_2H_5O_2C$ (quantitative)

$C=NNHPh \xrightarrow{(C_2H_5)_3N} [PhC≡\overset{+}{N}—\bar{N}Ph] \xrightarrow{PhC≡N}$

(72 %)

$CHO + PhNHOH \longrightarrow] PhCH=\overset{+}{N}Ph$
$\overset{|}{\underset{O^-}{}}$

$C_2H_5O_2C$ (97 %)

$N^+ X^- \xrightarrow{(C_2H_5)_3N}$

CH_2Ts

$\xrightarrow{CH_3O_2C—≡—CO_2CH_3}$

⟶ $+ (C_2H_5)_3\overset{+}{N}H \ Ts^-$

(72 %)

Even where steric hindrance might be exp[ected?] formation of an 'ortho-adduct', the latter m[ ] substantial amount, e.g.

(54 %)

7.2.2 1,3-Dipolar cycloaddition

This type of cycloaddition is also a $[4\pi + 2\pi]$p[ ] relative of the Diels–Alder reaction; but the 4[ ] is not a diene but a **1,3-dipole,** in which the[ ] distributed over only three atoms, and for whi[ch] cal structure can be drawn in which atoms [ ] charges. The most common 1,3-dipoles are [ ] kanes, e.g. (23), and the azides, e.g. (24), altho[ugh] several others and demonstrate the versatility[ ] tion:

$$\overset{+}{R\overset{}{C}H}{-}\overset{\cdot\cdot}{N}{=}\overset{\cdot\cdot}{\overset{}{N}}: \longleftrightarrow RCH{=}\overset{+}{N}{=}\overset{\cdot\cdot}{\overset{}{N}}: \longleftrightarrow R\overset{}{\overset{}{C}}H{-}\overset{+}{N}$$

(23a)   (23b)   (23c)

$$\overset{+}{R\overset{\cdot\cdot}{N}}{-}\overset{\cdot\cdot}{N}{=}\overset{\cdot\cdot}{\overset{}{N}}: \longleftrightarrow R\overset{}{N}{=}\overset{+}{N}{=}\overset{\cdot\cdot}{\overset{}{N}}: \longleftrightarrow R\overset{\overline{\overline{\cdot\cdot}}}{N}{-}\overset{+}{N}{\equiv}$$

(24a)   (24b)   (24c)

The HOMO (25) and LUMO (26) of such syst[ ] the LUMO (15) and HOMO (17) of a monoe[ ] the Diels–Alder reaction (p. 144):

7.2.3 Addition of carbenes and nitrenes to alkenes

Carbenes (Sykes, pp. 259–61) are uncharged electron-deficient car-bon species, $R_2C:$. Among their most characteristic, and synthetically useful, reactions is addition to alkenes [reaction (7.9)] to give cyclo-propanes. **Nitrenes,** which are the nitrogen analogues of carbenes, similarly undergo addition to alkenes giving aziridines [reaction (7.10)].

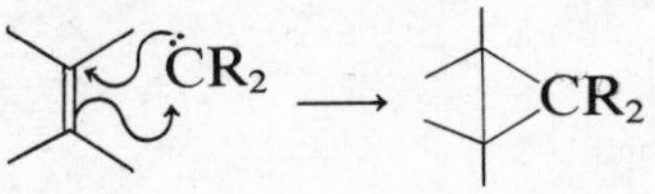

$$(7.9)$$

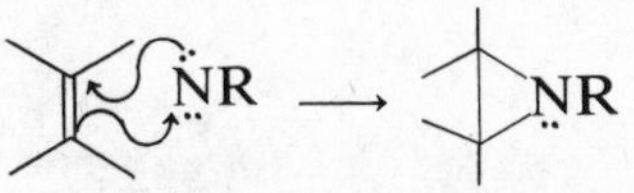

$$(7.10)$$

The precise mechanism of the addition depends on the arrange-ment of the non-bonding electrons in the carbene or nitrene. If both electrons are in one orbital and the other is empty [the so-called *singlet state*: e.g. structure (27)], the addition may be regarded as a [2+2] cycloaddition involving a HOMO-LUMO interaction as shown below [(27)+(15) or (27)+(17)]. If the two electrons are in different orbitals [the *triplet state*: e.g. (28)], the addition follows a radical pathway and is not concerted but stepwise.

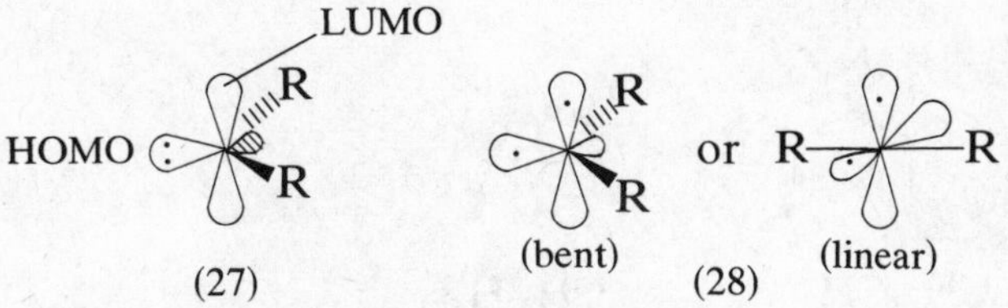

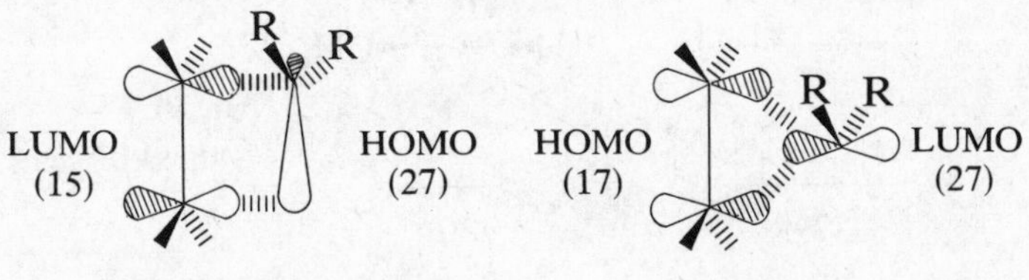

In many cases, the detailed mechanism is unimportant for our purposes; the result is the same, whichever mechanism operates. In other cases, however, the stereochemistry of the product may depend on the mechanism: the concerted addition is stereospecific, the rela-tive configurations in the alkene being retained in the product, but the stepwise radical addition is not stereospecific and may lead to a mixture of diastereoisomeric cyclopropanes [reaction (7.11)]:

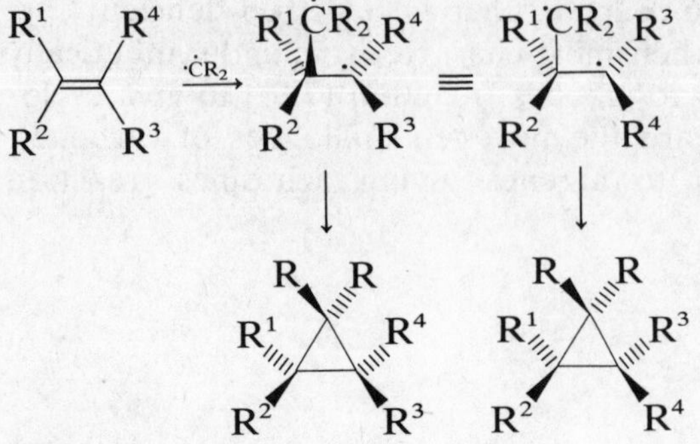

$$\text{(7.11)}$$

The following are representative examples:

1. 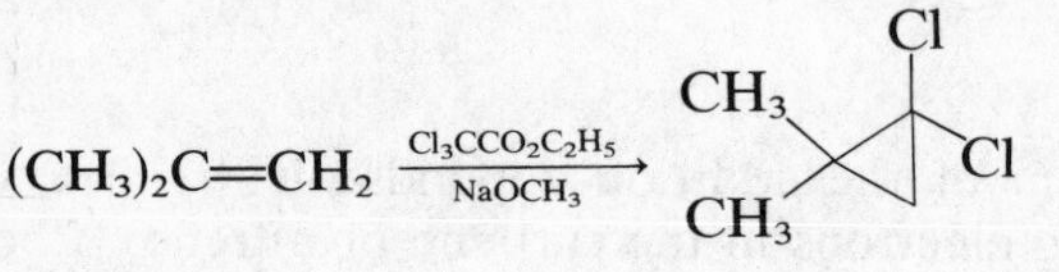 $(CH_3)_2C=CH_2 \xrightarrow[\text{NaOCH}_3]{\text{Cl}_3\text{CCO}_2\text{C}_2\text{H}_5}$ (65 %)

2. 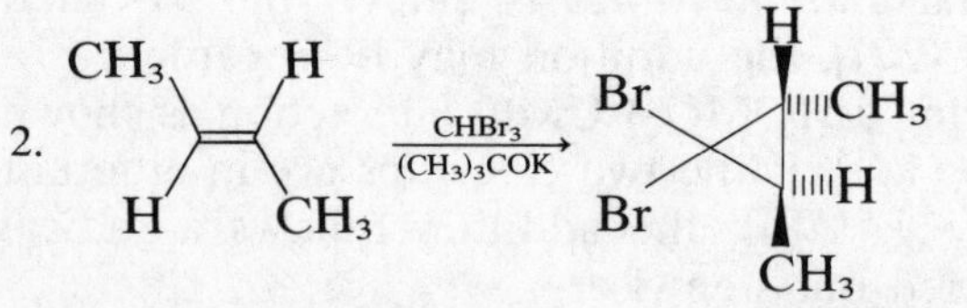 $\xrightarrow[\text{(CH}_3\text{)}_3\text{COK}]{\text{CHBr}_3}$ (68 %)

3. 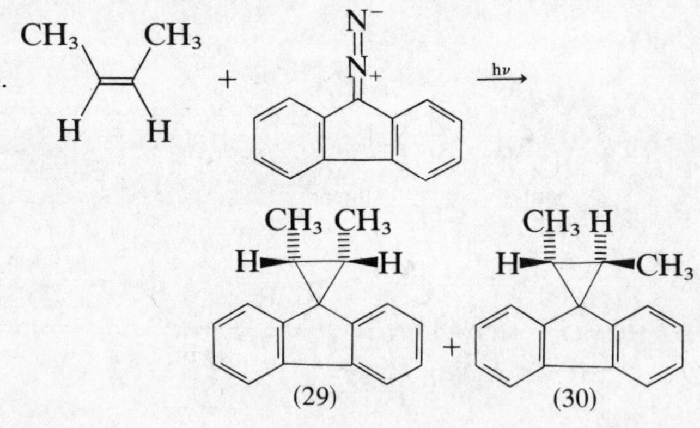 $+$ $\xrightarrow{h\nu}$

$+$ [ca. 50 %; (29):(30) = 66:34]

(29)   (30)

4. 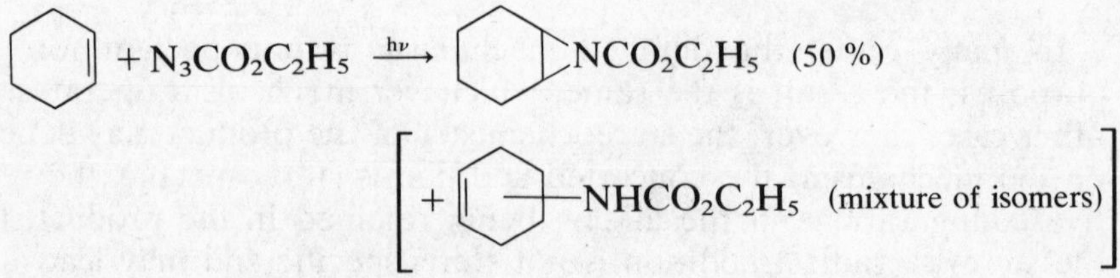 $+ N_3CO_2C_2H_5 \xrightarrow{h\nu}$ $NCO_2C_2H_5$ (50 %)

$\left[+ \right.$ $-NHCO_2C_2H_5$ (mixture of isomers) $\left.\right]$

An alternative method for the formation of cyclopropanes from alkenes is provided by the **Simmons–Smith reaction** [reaction (7.12)], which involves reaction of the alkene with a dihalogenomethane and zinc (usually in presence of copper). This is most simply rationalised in terms of a carbene intermediate, but the available evidence suggests that the more likely intermediate is (31) and that a free carbene is not involved:

$$CH_2I_2 + Zn \longrightarrow ICH_2ZnI$$

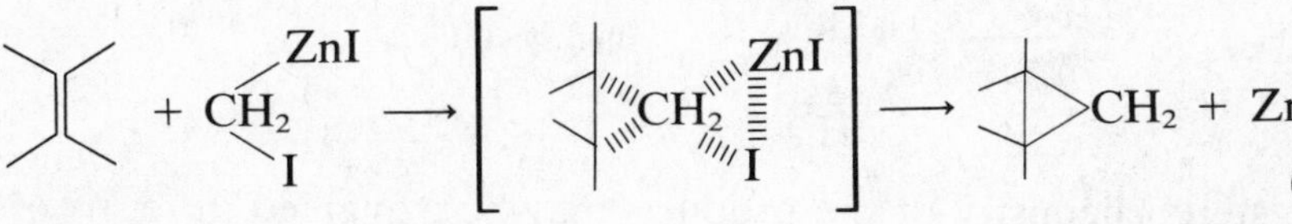

(7.12)

(31)

Thus, for example,

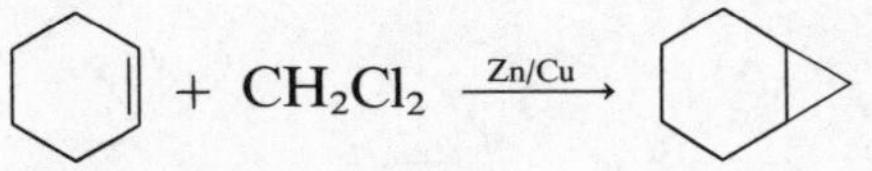

(65 %)

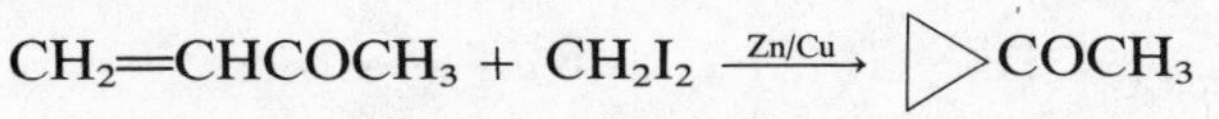

$$CH_2{=}CHCOCH_3 + CH_2I_2 \xrightarrow{Zn/Cu} \,$$ $COCH_3$ (50 %)

7.3 Electrocyclic ring closure

The Diels–Alder reaction and 1,3-dipolar cycloaddition, which are described in the preceding section, each involve the redistribution of six π-electrons *via* a cyclic transition state. If these six π-electrons are contained *within the same molecule*, an analogous redistribution may take place intramolecularly, and such an intramolecular pericyclic process is referred to as an **electrocyclic** reaction:

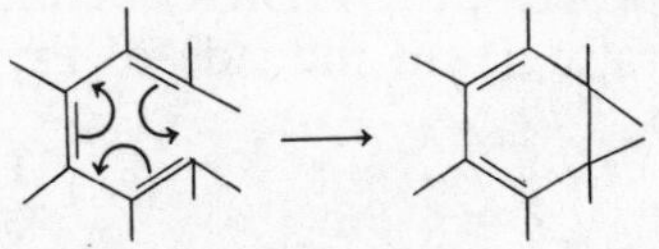

(7.13)

The reaction is stereospecific, like the Diels–Alder and 1,3-dipolar cycloadditions, e.g.

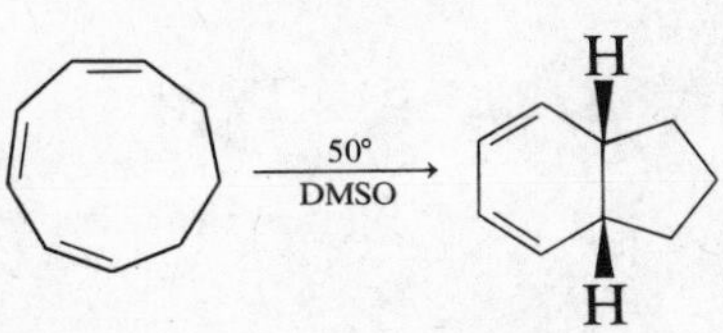

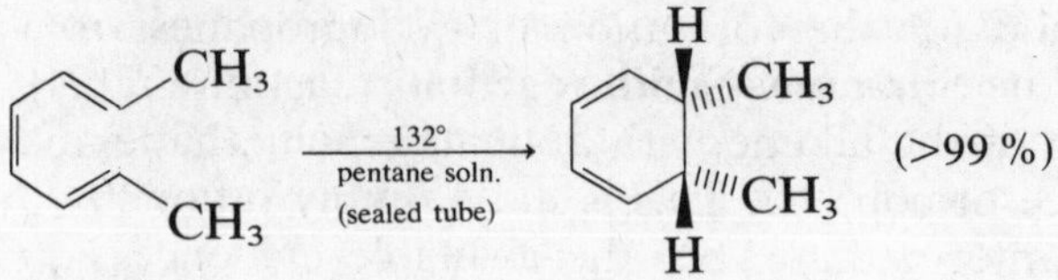

(32)

Similarly,

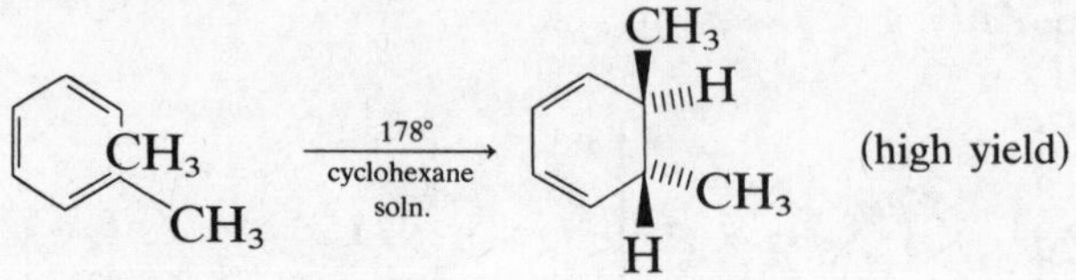

(high yield)

The stereochemistry of the products may be explained in terms of frontier orbital theory (cf. Sykes, pp. 332–4). The HOMO for a conjugated triene is (33), and ring closure is thus a **disrotatory** process:

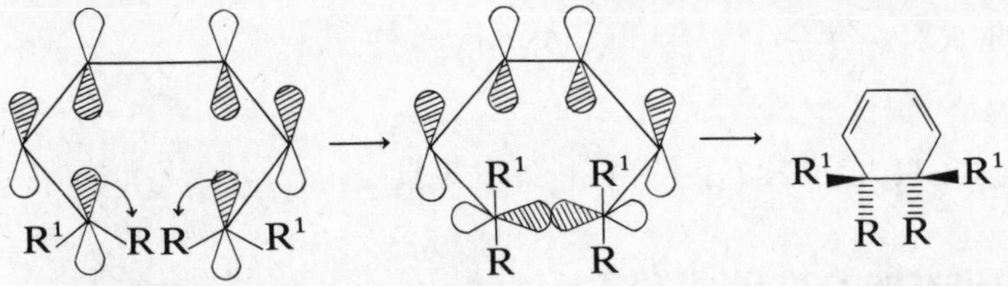

(33)

These reactions are, however, reversible (as, indeed, are the Diels–Alder and dipolar cycloaddition reactions), and the examples above are equilibria which happen to favour the cyclised isomer. In other cases the equilibrium favours the acyclic isomer, and this occasionally provides a useful method of ring *opening* (cf. section 7.4.3). One might, for example, expect a conjugated diene to be capable of cyclisation to a cyclobutene [a **conrotatory** process, the HOMO being (14)], but in such cases the equilibria generally lie on the side of the diene.

(7.14)

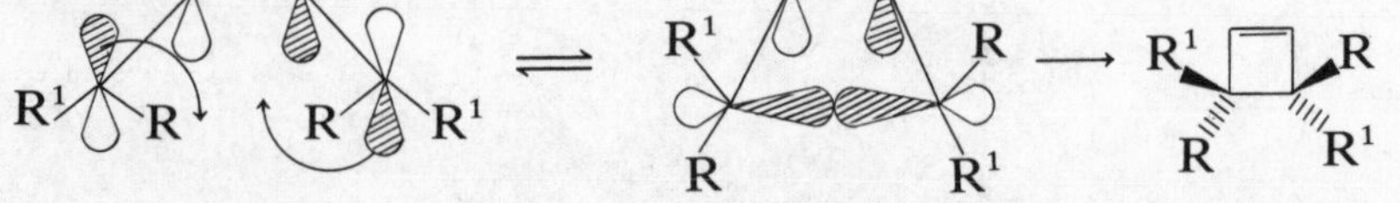

(14)

Electrocyclic ring closure may also be brought about by photo-chemical means. In such reactions the stereochemistry of the product is the opposite of that obtained by thermal cyclisation, e.g.

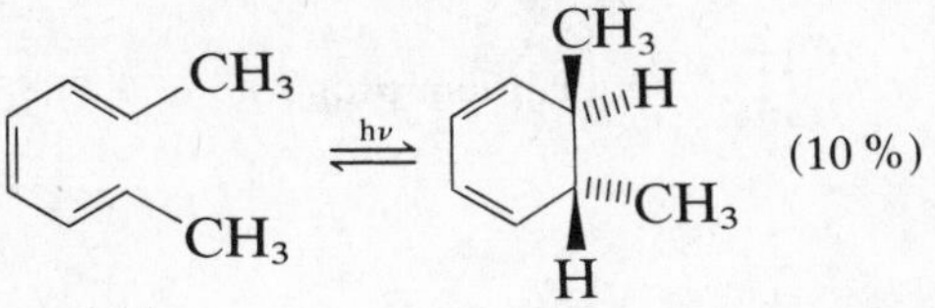

(32)

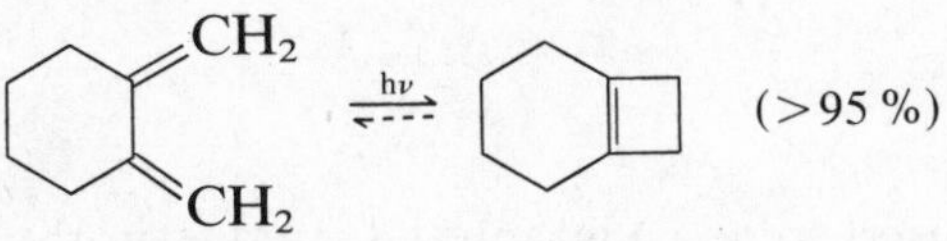

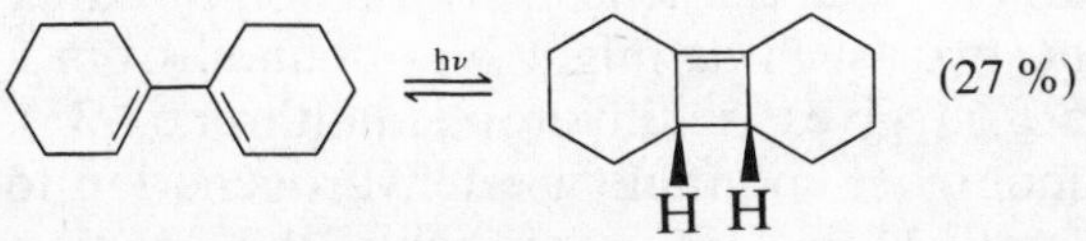

(10 %)

(>95 %)

(27 %)

Irradiation of the substrate results in the promotion of an electron into the orbital of next higher energy level, i.e. the ground-state LUMO. This now becomes the HOMO for the photochemical ring closure [(16) for a diene and (34) for a triene], and the resultant ring closures are disrotatory and conrotatory respectively (cf. Sykes, pp. 334–5).

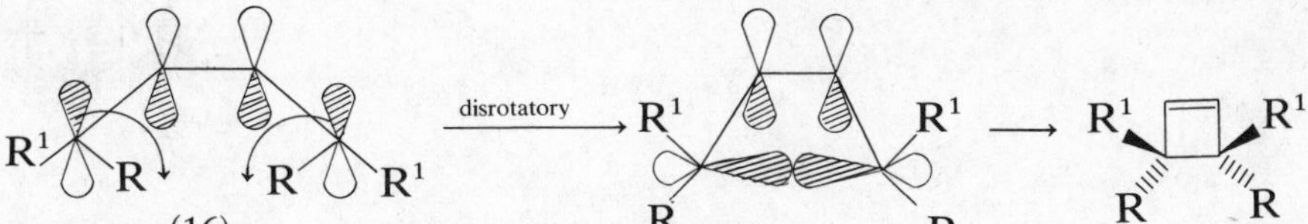

(16)

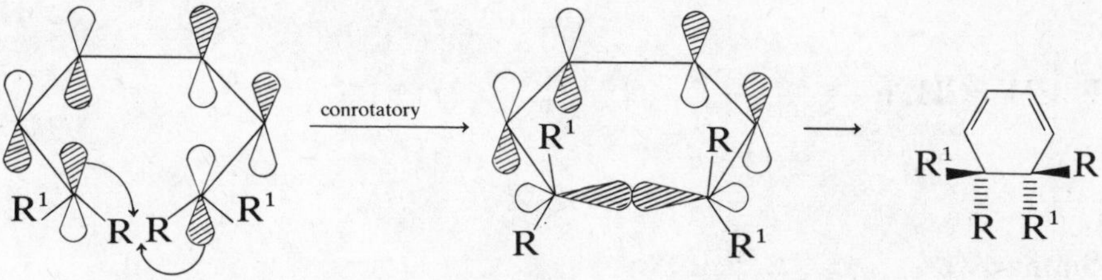

disrotatory

(34)

conrotatory

Irradiation of alkenes, however, also leads to the interconversion of *E*- and *Z*-isomers, and so a diene or triene in which the double bond configurations are not fixed (e.g. in a ring system) may undergo this type of isomerisation as well as ring closure, e.g.

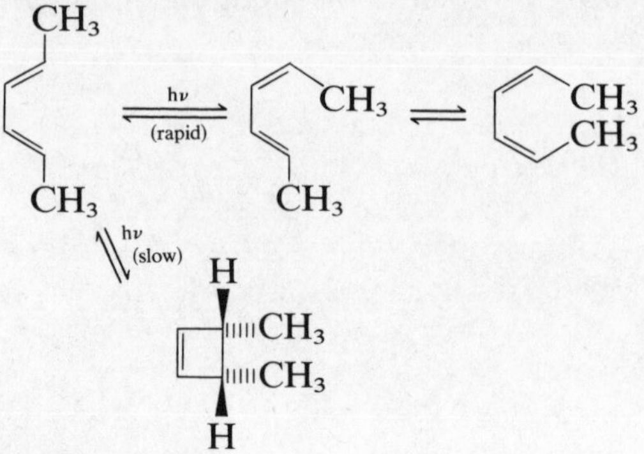

(equilibrium proportion 36:30:34)

The $E \rightarrow Z$ isomerisation is used to considerable advantage in the photocyclisation of stilbenes (1,2-diarylethenes). Irradiation of either isomer, or a mixture of the two (such as might be obtained from a Wittig reaction: section 5.3.1) gives a dihydrophenanthrene (35), which in presence of air undergoes spontaneous dehydrogenation to the phenanthrene [reaction (7.15)]. This is probably the simplest route to phenanthrene derivatives currently available.

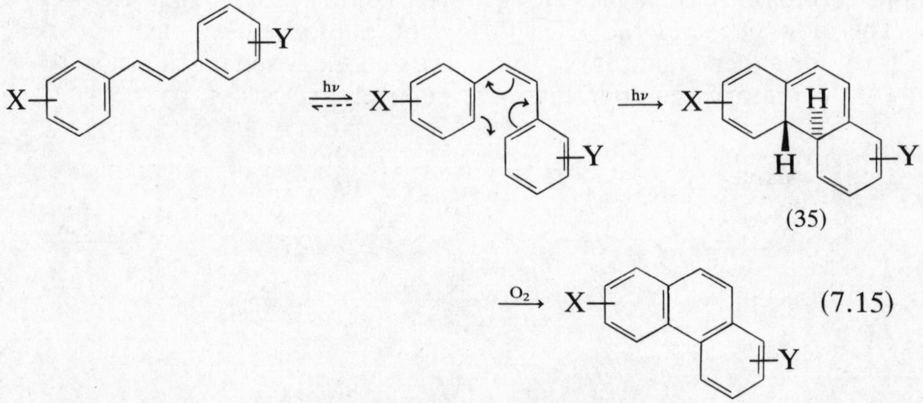

(35)

(7.15)

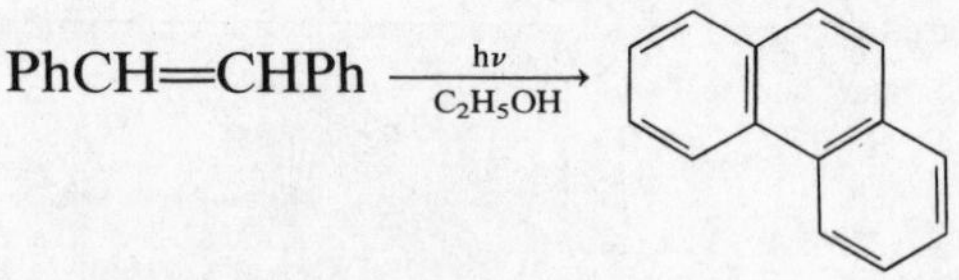

$$PhCH{=}CHPh \xrightarrow[C_2H_5OH]{h\nu}$$

(73 %)

Similarly

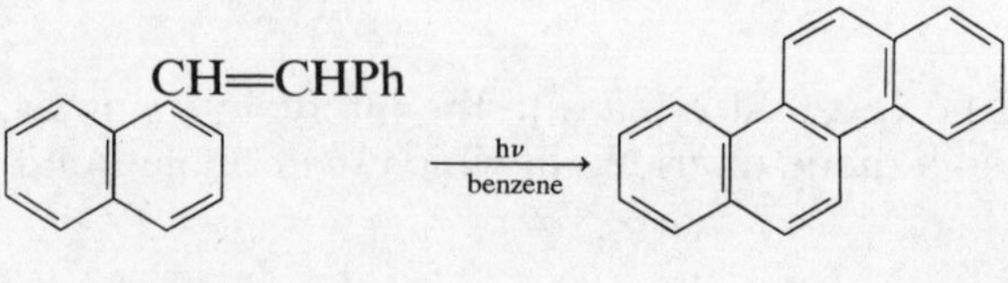

(77 %)

The method has also been adapted with considerable success for the synthesis of **helicenes** (cf. Ch. 14).

7.4 Ring opening

The value of ring opening as a synthetic procedure is not as obvious as that of ring closure: indeed we have discussed synthesis so far only in terms of bond formation, and examples of bond cleavage (e.g. the decarboxylation of malonic or β-keto-acid derivatives, or the release of carbonyl groups from 1,3-dithians or dihydro-1,3-oxazines) have been incidental to the main theme. In Chapter 10 we shall encounter bond cleavage in connection with the removal of protective groups. In the present section, however, we consider bond cleavage in a specific context – ring opening – and as a synthetic method in its own right.

Apart from the above, the two main synthetic uses of ring opening are as follows:

(i) The atoms at either end of the bond which is broken will bear functional groups in the ring-opened product; ring opening may thus provide a route to difunctional molecules in which the functional groups are separated by several other atoms.

(ii) In a bi- or polycyclic molecule, cleavage of a bond which is common to two rings may lead to a medium- or large-ring molecule which is otherwise difficult to prepare.

We shall classify ring opening processes according to their reaction type.

7.4.1 Hydrolysis, solvolysis, and other electrophile-nucleophile interactions

This is a large and diverse group, and the examples are merely illustrative of this diversity:

1.

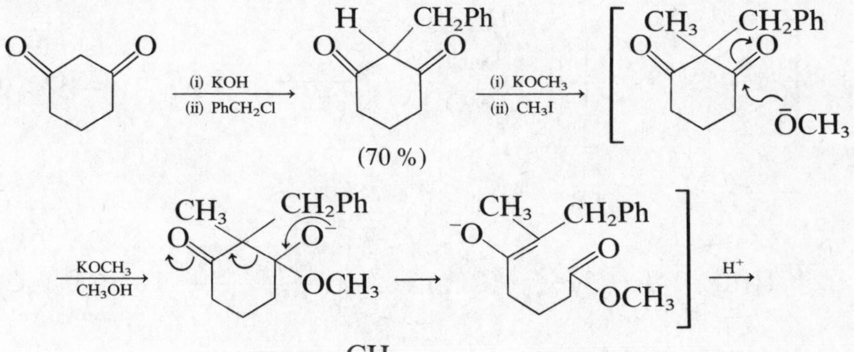

$PhCH_2\overset{\underset{\displaystyle |}{CH_3}}{CH}CO(CH_2)_3CO_2CH_3$ (50 % overall; cf. section 5.1.2)

2. 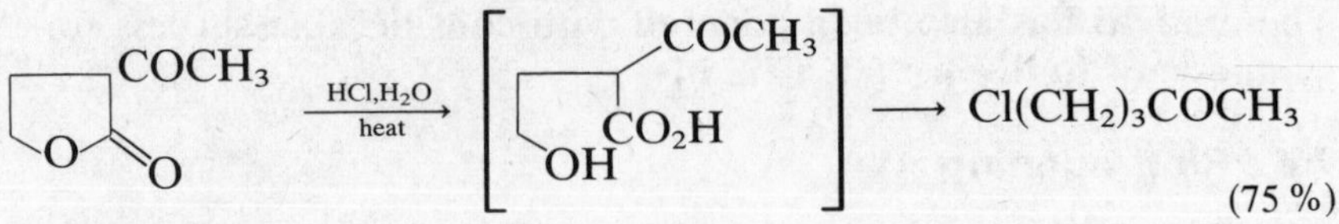 $\xrightarrow[\text{heat}]{\text{HCl,H}_2\text{O}}$ $\longrightarrow$ $\text{Cl(CH}_2)_3\text{COCH}_3$

(75 %)

3. $\xrightarrow{\text{heat}}$ $(\text{CH}_3)_2\text{N(CH}_2)_3\text{CH}=\text{CH}_2$ (80 %)

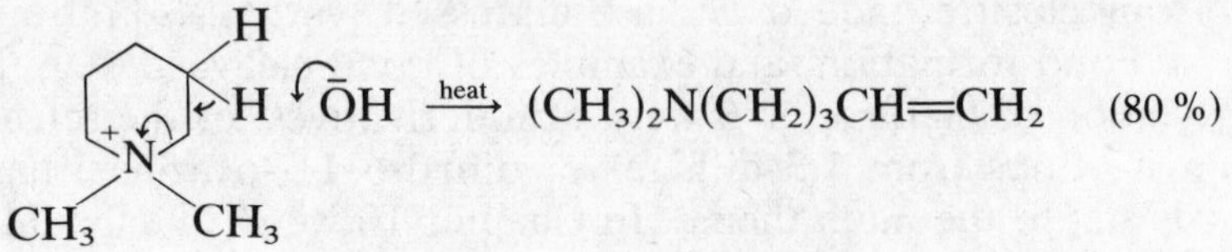

4. 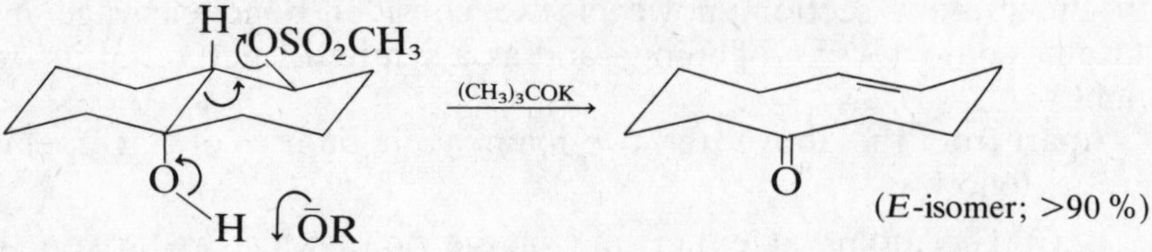 $\xrightarrow{\text{(CH}_3)_3\text{COK}}$ (*E*-isomer; >90 %)

7.4.2   Oxidative and reductive ring opening

Oxidative ring opening of a synthetically useful kind is generally that of a cycloalkene or a cycloalkanone. These reactions are discussed at greater length in sections 9.2.6 and 9.5.3, but the examples below serve to illustrate the potential of the methods:

1. 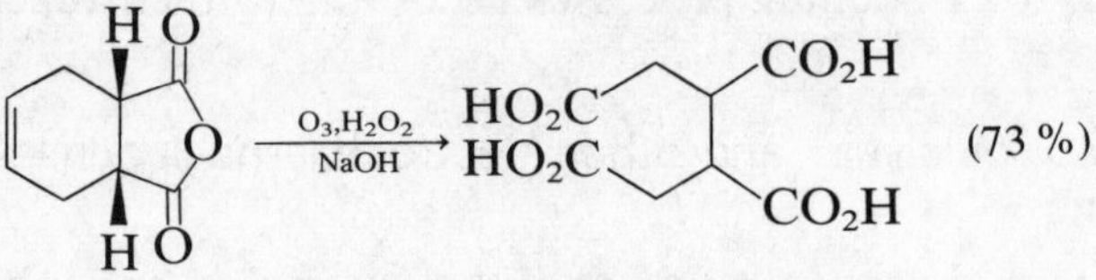 $\xrightarrow[\text{NaOH}]{\text{O}_3,\text{H}_2\text{O}_2}$ (73 %)

2. 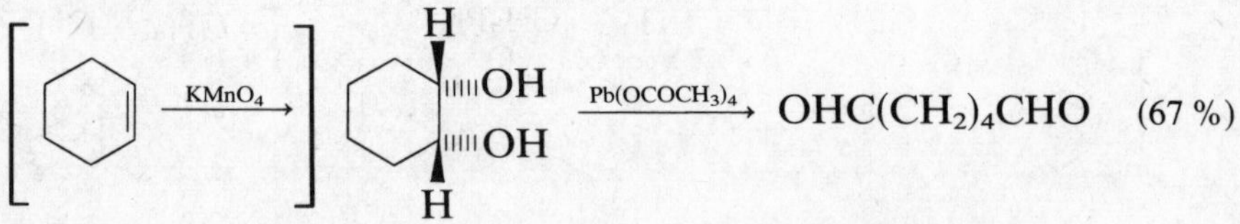 $\xrightarrow{\text{KMnO}_4}$ $\xrightarrow{\text{Pb(OCOCH}_3)_4}$ $\text{OHC(CH}_2)_4\text{CHO}$ (67 %)

3. $(\text{CH}_2)_{14}$ $\overset{\text{CO}}{}$ $\xrightarrow[\text{CH}_3\text{CO}_2\text{H}]{\text{H}_2\text{SO}_5}$ $(\text{CH}_2)_{14}$ $\xrightarrow[\text{(ii) H}^+,\text{H}_2\text{O}]{\text{(i) NaOH,H}_2\text{O}}$ $\text{HO(CH}_2)_{14}\text{CO}_2\text{H}$

(quantitative)

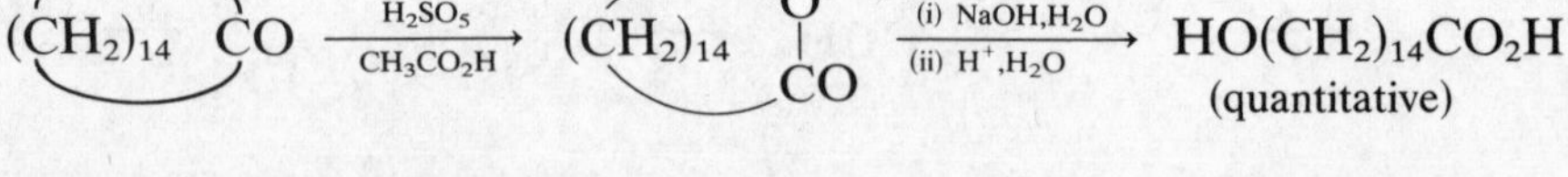

Reductive ring opening is of less general value, although hydrogenolysis of some sulphur-containing compounds (cf. also section

8.4.3.3) provides a notable exception, e.g.

4. 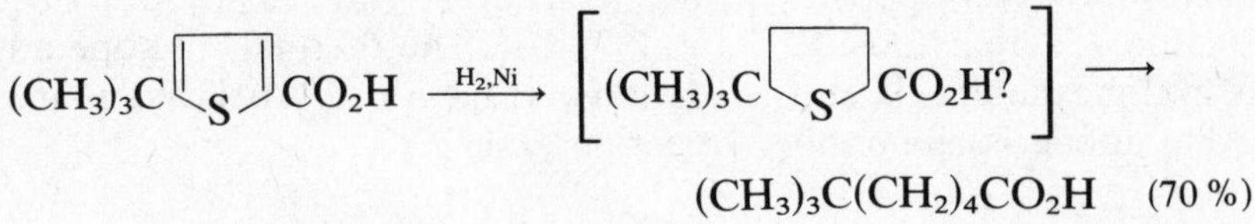 $(CH_3)_3C(CH_2)_4CO_2H$ (70 %)

7.4.3 Pericyclic ring opening

We have already pointed out that Diels–Alder and related cycloadditions are, in principle, reversible, and the **retro-Diels–Alder reaction** has some useful synthetic applications. Many of these involve the cleavage of a bicyclic Diels–Alder adduct which has itself been formed from a cyclic diene and a dienophile. The cleavage reaction becomes effectively irreversible if one of the cleavage products is volatile. For example,

1. 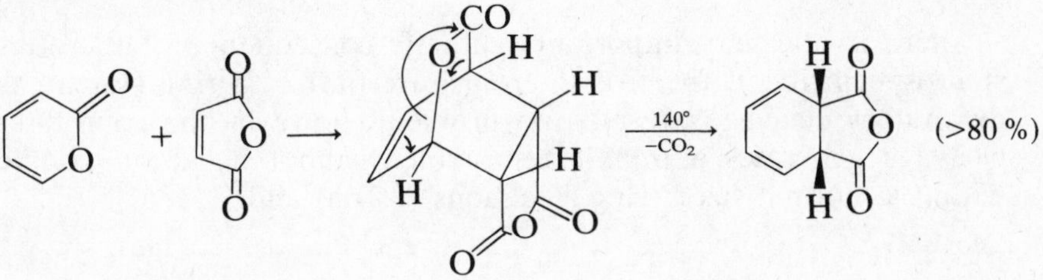 (>80 %)

2.

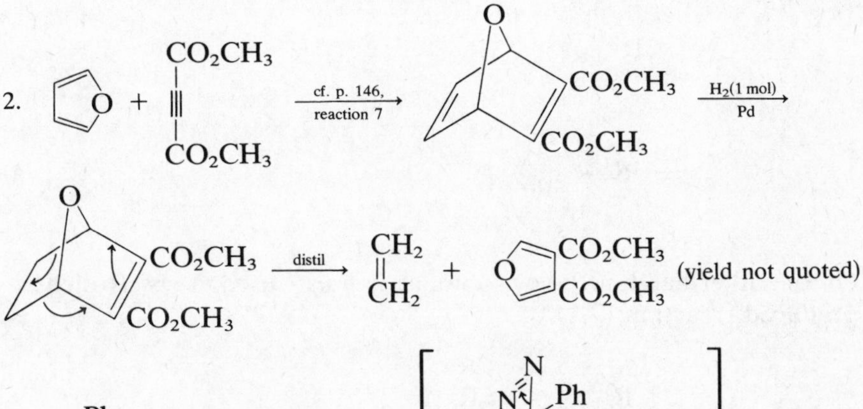

(yield not quoted)

3.

(83 %)

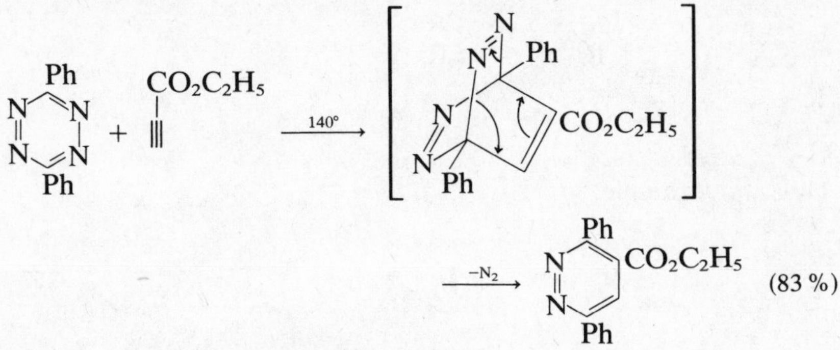

Electrocyclic ring opening is another important pericyclic reaction, and is the exact opposite of the electrocyclic ring closure described in section 7.3 [reactions (7.13) and (7.14)]. The thermal ring opening of a cyclohexadiene is thus *disrotatory* [reaction (7.16)] and that of a cyclobutene is *conrotatory* [reaction (7.17)].

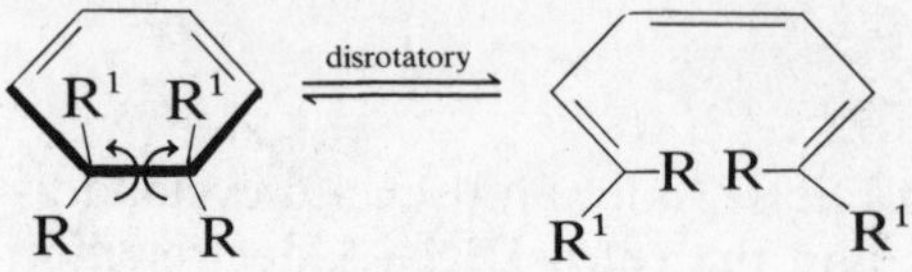

$$\text{(7.16)}$$

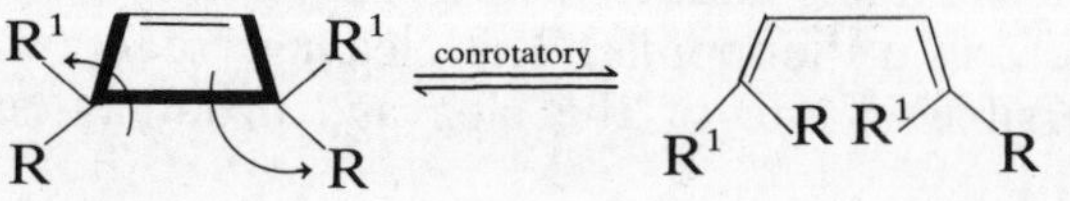

$$\text{(7.17)}$$

There is one very important synthetic consequence of the above stereospecificity. If the two R groups are part of a ring system, the disrotatory cleavage presents no problem; however the conrotatory cleavage generates a *trans*-alkene which cannot be accommodated within a 'normal-sized' ring [reactions (7.16a) and (7.17a)]:

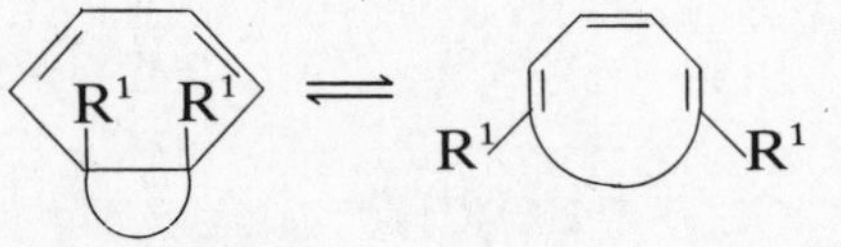

$$\text{(7.16a)}$$

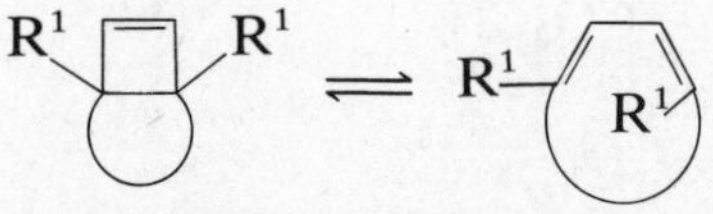

$$\text{(7.17a)}$$

On the other hand, if the cyclobutene is *trans*-fused, a *cis,cis*-diene is produced [reaction (7.18)]:

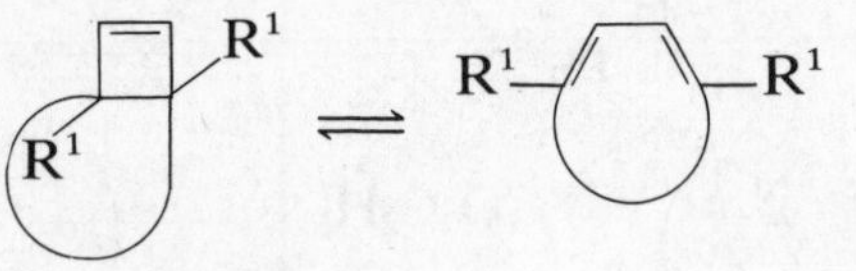

$$\text{(7.18)}$$

Thus, for example,

1. 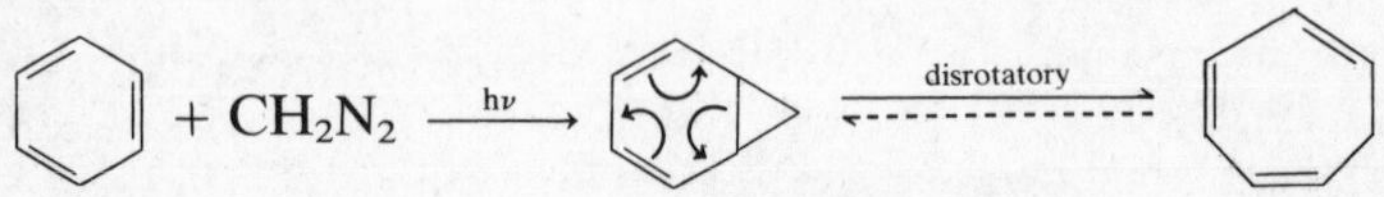 $\bigcirc + CH_2N_2 \xrightarrow{h\nu} \quad \xrightarrow{\text{disrotatory}} \quad$ (32 %)

2. 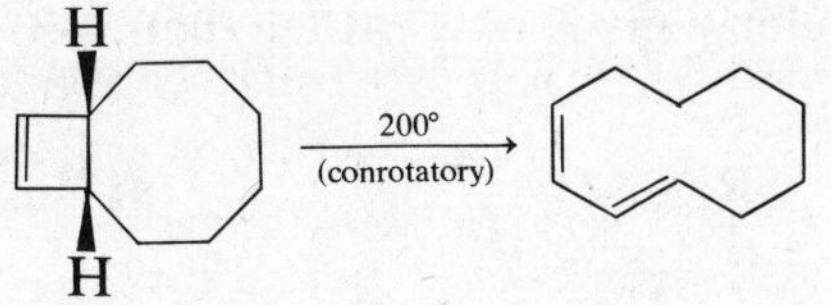 $\xrightarrow[\text{(conrotatory)}]{200°}$ (95 %)

3. 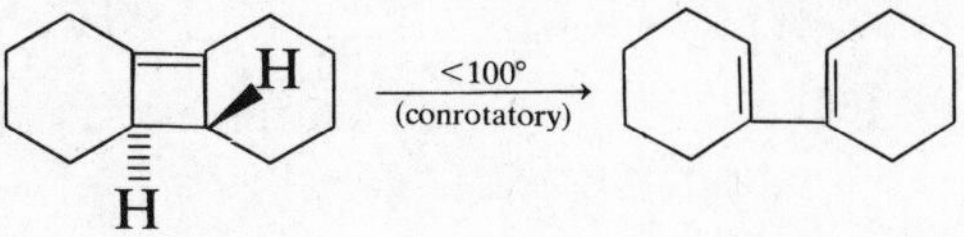 $\xrightarrow[\text{(conrotatory)}]{<100°}$ (high yield)[7]

(36)   (37)

The final type of pericyclic process to be considered in this chapter is one involving a **Cope rearrangement** (cf. Sykes, pp. 342–3). In this reaction, which also involves a 6-membered cyclic transition state [reaction (7.19)], a 1,5-diene is rearranged to another 1,5-diene by concerted formation of a C_1—C_6 single bond, breaking of the C_3—C_4 single bond, and migration of both double bonds:

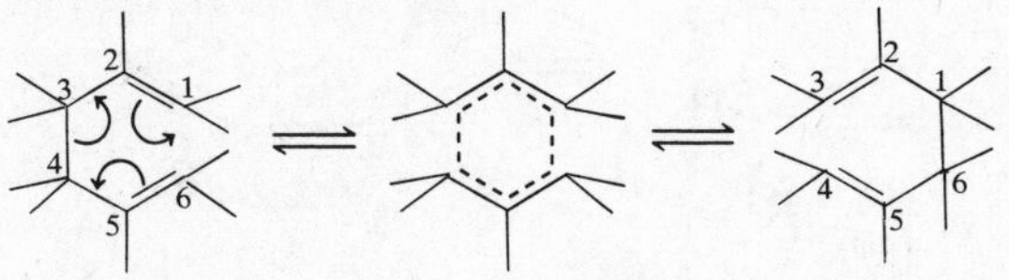

(7.19)

It follows that if the original 3,4-bond is part of a ring system the method may be used for a type of ring opening [reaction (7.19a)].

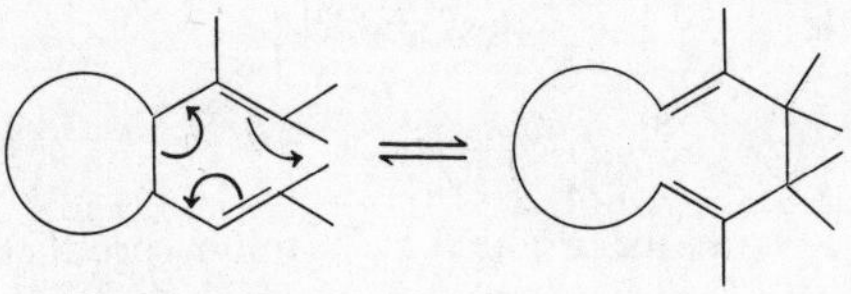

(7.19a)

The rearrangement is generally stereospecific, although the configuration of the product is not always predictable, depending as it does on the conformation in the transition state. A chair conformation (38) is preferred to a boat [(39) or (40)] where both may reasonably be formed: thus a diene of type (41) gives (42) rather than (43) or

(44), and a diene such as (45) similarly gives (44) rather than (43) or (42):

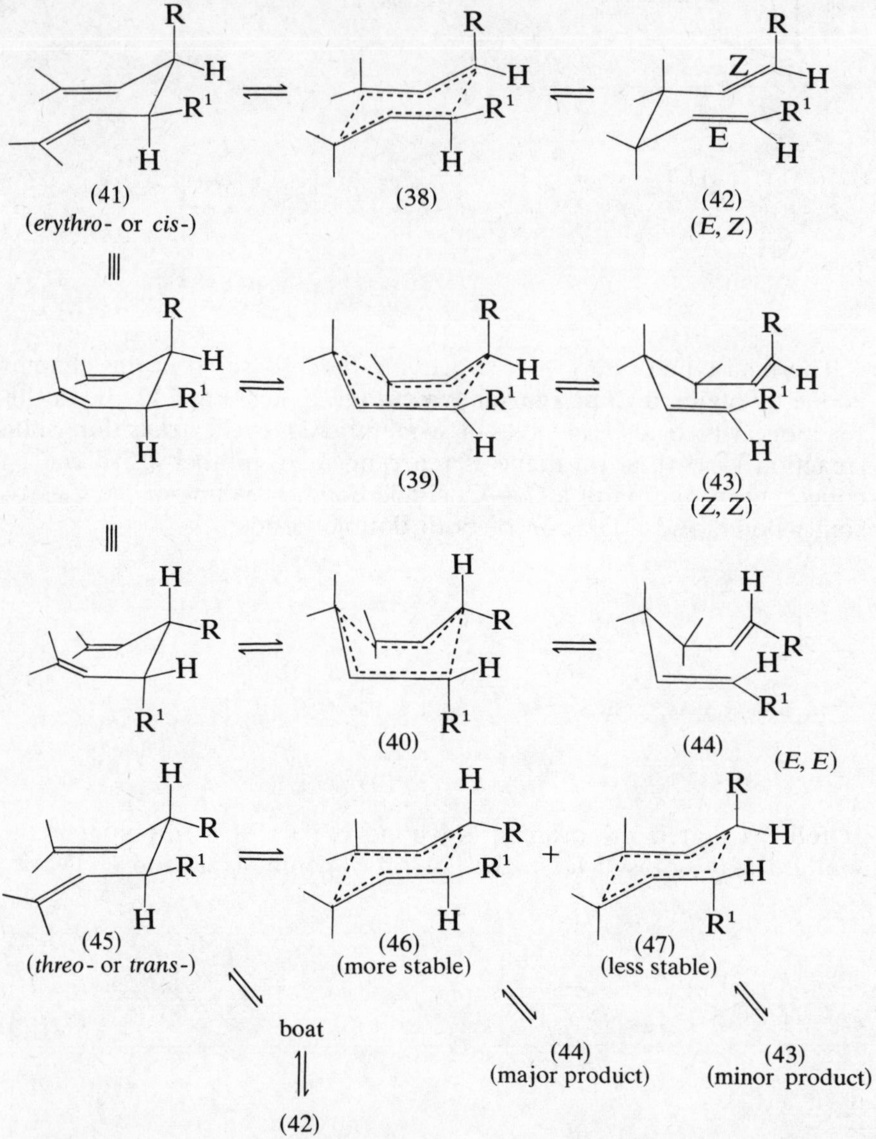
R
H
R1
H
(41)
(erythro- or cis-)
R
H
R1
H
(38)
R
Z
H
R1
E
H
(42)
(E, Z)
R
H
R1
H
R
H
R1
H
(39)
R
H
R1
H
(43)
(Z, Z)
H
R
H
R1
H
R
H
R1
(40)
H
R
H
R1
(44)
(E, E)
H
R
R1
H
(45)
(threo- or trans-)
H
R
R1
H
(46)
(more stable)
+
R
H
H
R1
(47)
(less stable)
boat
(42)
(44)
(major product)
(43)
(minor product)

The preference for the chair-like transition state is explicable in frontier-orbital terms[8]. If the formation of the new single bond is regarded as a HOMO-LUMO interaction, the formation of the *boat-like* transition state (48) requires an unfavourable orbital interaction between C_2 and C_5; such an interaction is absent in the chair-like transition state (49).

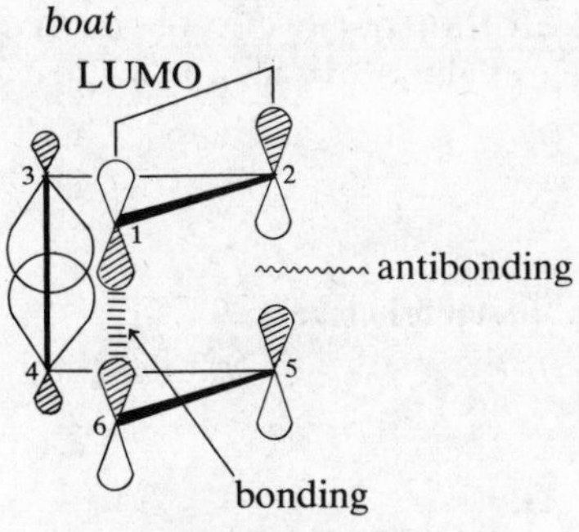

(48)

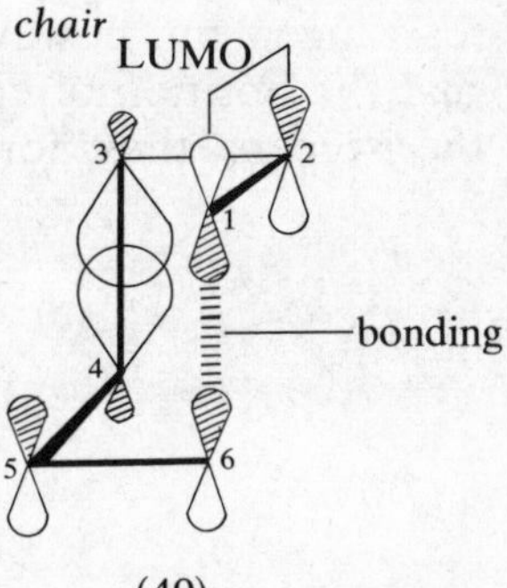

(49)

Examples of the Cope rearrangement include the following:

1. (erythro-) $\xrightarrow[via\ chair]{280°}$ (97 %)

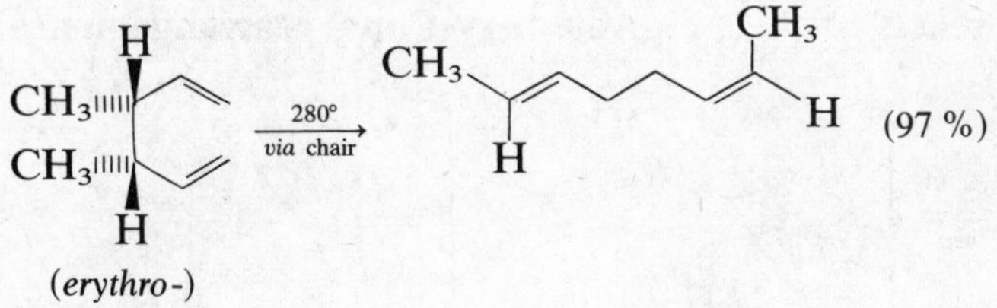

2. (threo-) $\xrightarrow[via\ 2\ chairs]{180°}$ (87 %)

+ (10 %)

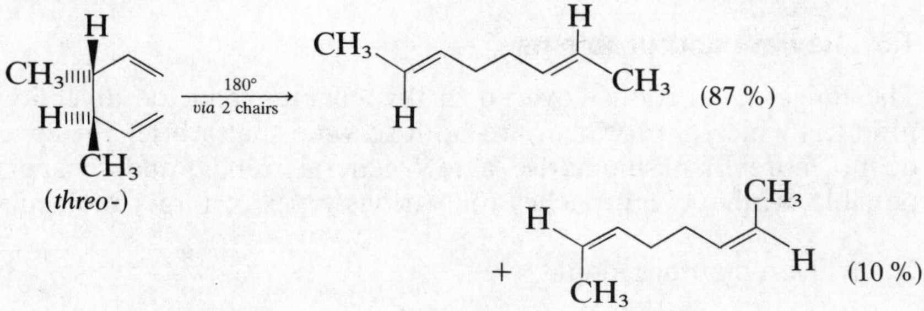

3. $\xrightarrow[\substack{via\ boat\\(chair\ too\ strained)}]{120°}$ (91 %)

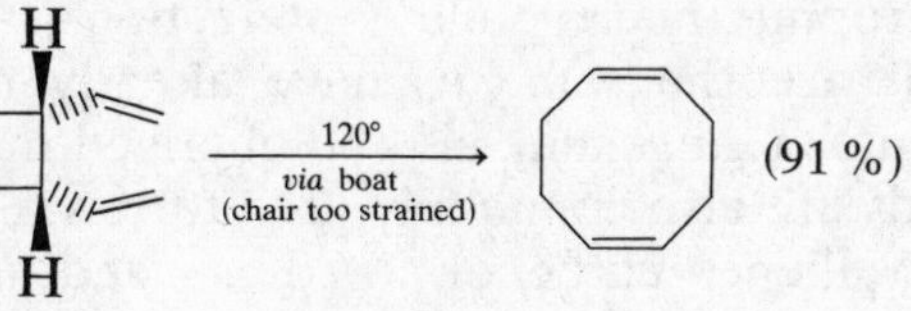

4. $\xrightarrow[via\ boat]{room\ temp.}$ —$(CH_2)_3CH_3$ (quantitative)

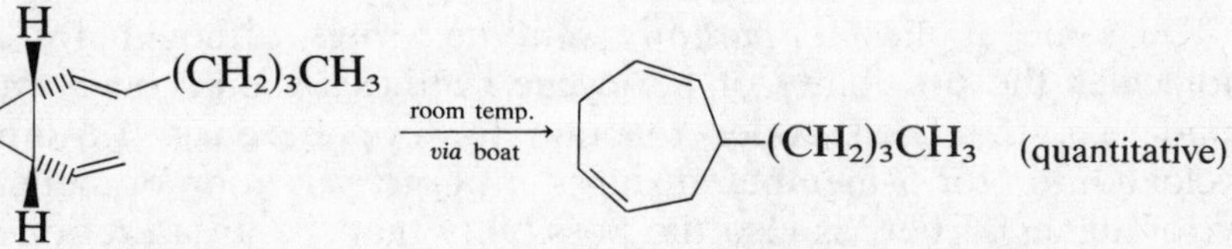

The Cope rearrangement, however, like other pericyclic processes, is reversible, and the position of equilibrium depends on the relative stabilities of the isomers; thus, for example,

5.

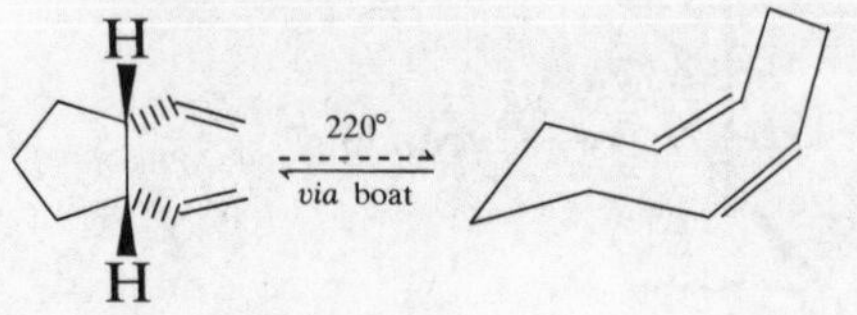

$$\xrightarrow[\text{via boat}]{220°}$$

(Z, Z)   [equilibrium ratio 95:5]

6.

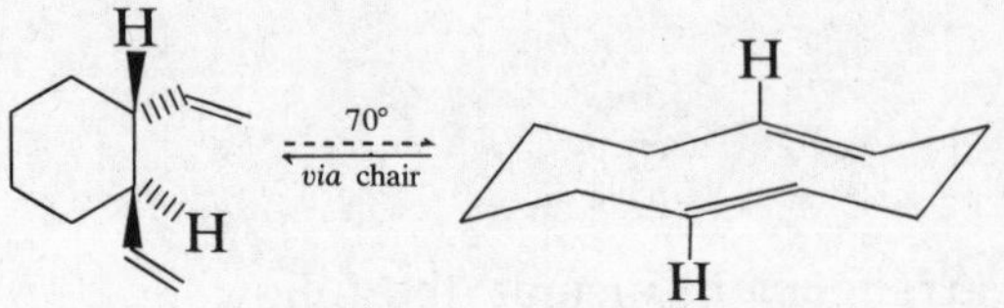

$$\xrightarrow[\text{via chair}]{70°}$$

(E,E)   [equilibrium ratio >99:1]

In such cases the forward reaction can be made to predominate only if the product reacts further, e.g. the **'oxy-Cope' rearrangement:**

7.

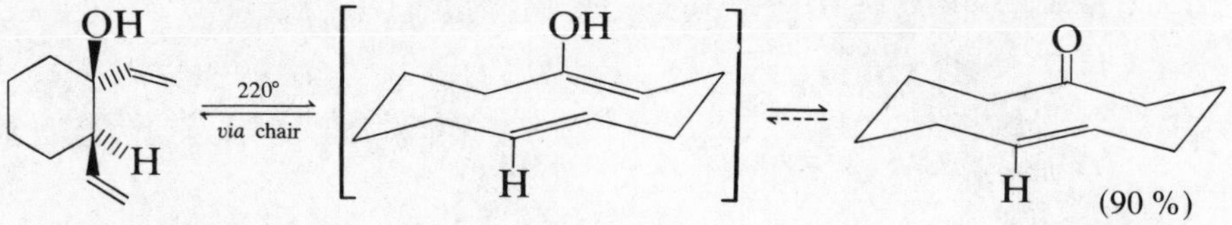

$$\xrightleftharpoons[\text{via chair}]{220°}$$

(90 %)

7.5   Review and problems

The range of reactions covered in this chapter, and the diversity of products which are formed, are both so wide that a brief review can do no more than summarise a few general trends, and so suggest possible synthetic approaches for various types of target molecule.

7.5.1   Non-aromatic rings

Saturated rings of 'normal' size (5- and 6-membered) are generally made by standard electrophile-nucleophile interactions; the guidelines for disconnections are the same as have already been described in section 5.6.1. Smaller rings may also be obtained in this way, although special methods are also commonly used for each ring system (e.g. cyclopropane ⇒ alkene + carbene). Medium and large rings, as we have already seen, require special methods (section 7.1.5).

The same applies to *partially saturated rings*, although for such molecules the possibility of pericyclic synthesis should be borne in mind (e.g. the Diels–Alder reaction for cyclohexenes, 1,3-dipolar cycloaddition for 5-membered rings, or photocyclisation of dienes for cyclobutenes). There is also the possibility that the unsaturation may be introduced *via* an elimination reaction (cf. section 9.2.4), or that the molecule results from partial hydrogenation of an aromatic or other fully conjugated species (cf. section 8.8).

7.5.2 Aromatic rings

We have already indicated that monocyclic benzene derivatives are usually made from simpler benzene derivatives by functionalisation and/or functional group interconversion, and that benzo-fused compounds (whether carbocyclic or heterocyclic) are usually synthesised from a mono- or (*ortho*-di)-substituted benzene; so the correct disconnection for a benzo-fused molecule is either next to a ring junction or one atom removed from a ring junction. For 5-membered heteroaromatic rings the correct disconnection is usually that of a carbon–heteroatom bond.

7.5.3 Problems

We conclude this chapter, like Chapters 4 and 5, with four synthetic exercises. Two of the target molecules are carbocyclic and two are heterocyclic:

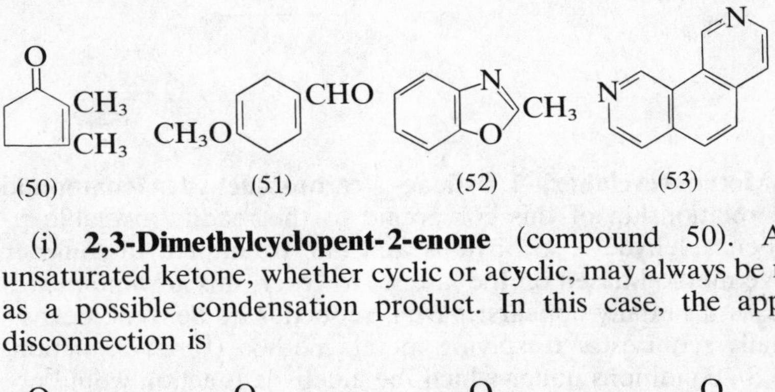

(50)   (51)   (52)   (53)

   (i) **2,3-Dimethylcyclopent-2-enone** (compound 50). An α,β-unsaturated ketone, whether cyclic or acyclic, may always be regarded as a possible condensation product. In this case, the appropriate disconnection is

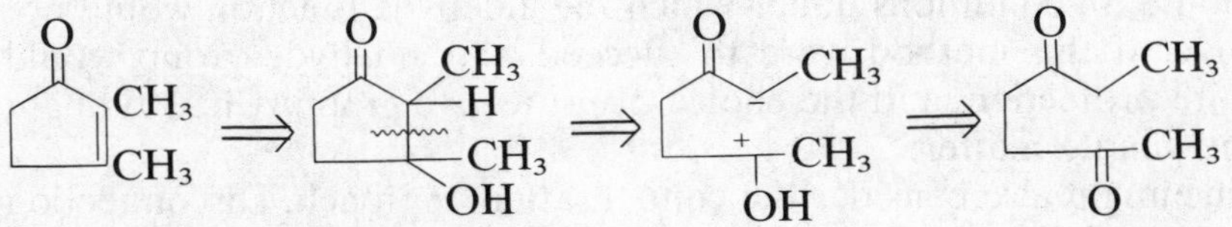

So the synthesis would be:

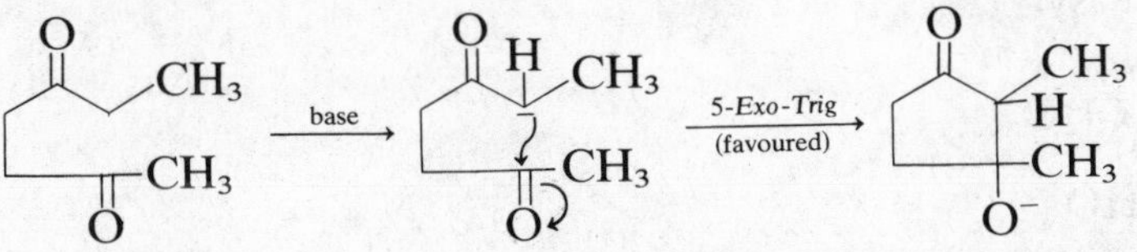

$\longrightarrow$ $\longrightarrow$ compound (50)

Heptane-2,5-dione is not readily available commercially, and so will require to be made: problems of this kind have already been discussed in section 5.6.1 and are not considered further here. It is of importance, however, that deprotonation of heptane-2,5-dione may occur not only at position 6, but also at positions 1, 3, and 4, and thus compound (50) may not be the sole product. Deprotonation at position 1 might lead ultimately to 3-ethylcyclopent-2-enone (54), a process which is a possible competitor with the cyclisation to (50), but in practice (50) is the main product of reaction of heptane-2,5-dione with a variety of bases. (Deprotonation at position 3 or 4 cannot lead to unstrained cyclic products – only to *inter*molecular reaction – and is of no major significance.)

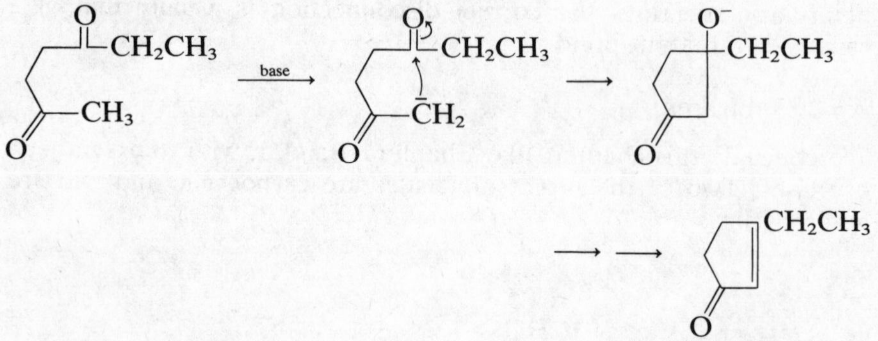

(54)

(ii) **4-Methoxycyclohexa-1,4-diene-1-carboxaldehyde** (compound 51). The relationship of this compound to the readily available *p*-methoxybenzaldehyde is so obvious that one is tempted to consider a reductive method based on the latter. However, this is by no means as simple as it initially appears: partial reduction of benzene derivatives usually requires a 'dissolving metal' method (Birch reduction: section 8.8), conditions under which the aldehyde function would not survive. If this method were to succeed, the aldehyde group would require protection, and the choice of protective group (Ch. 10) might be no simple matter.

One might also consider the condensation approach. Disconnection in the usual manner gives (55) as the required acyclic precursor, and although the synthesis of this dialdehyde is not impossible, it is certainly not easy.

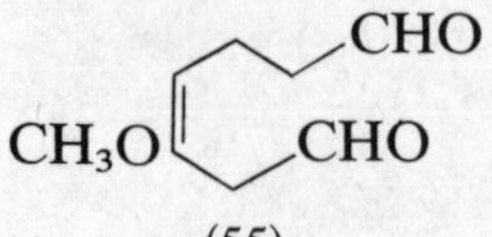

(55)

One is then left with the possibility of a pericyclic process, and the Diels–Alder reaction is worth considering for the formation of a partially saturated 6-membered ring. Disconnection of a Diels-Alder adduct is precisely the same as performing a retro-Diels–Alder reaction (section 7.4.3), and when this is applied to (51) the results are as follows:

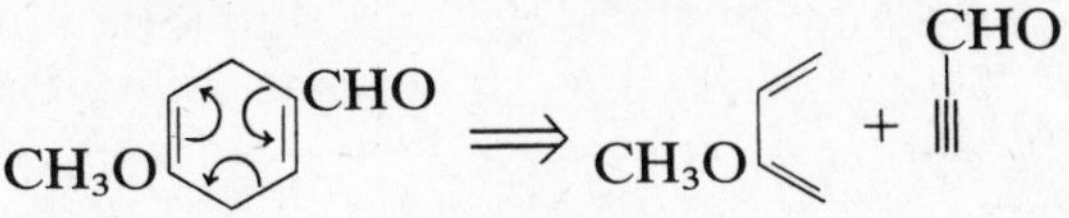

or

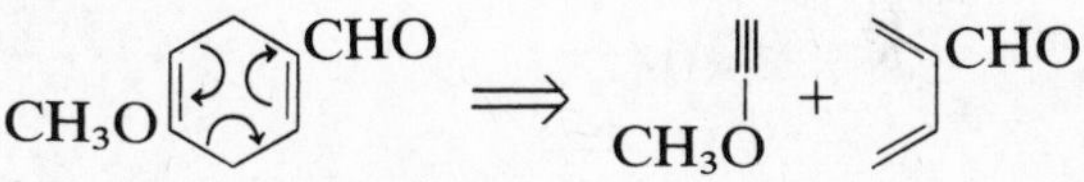

In theory, either of these Diels–Alder reactions should offer a practicable route to (51), especially since the substituents in the product are '*para-*' to each other. In practice the first is preferred because of the availability of the starting materials: 2-methoxybutadiene is obtainable from but-1-en-3-yne, and propynal from prop-2-yn-1-ol, as follows:

$$HC\equiv C-CH=CH_2 \xrightarrow[\substack{HgO \\ BF_3}]{3CH_3OH} CH_3\underset{\underset{OCH_3}{|}}{\overset{\overset{OCH_3}{|}}{C}}CH_2CH_2OCH_3 \xrightarrow[\substack{(i.e. \ mild \ acid) \\ heat}]{KHSO_4}$$

$$CH_2=\underset{\underset{|}{OCH_3}}{C}-CH=CH_2 \quad (39\% \text{ overall})$$

[The intermediate, 1,3,3-trimethoxybutane, is also commercially available.]

$$HC\equiv CCH_2OH \xrightarrow[\text{(cf. sect. 9.3.1.1)}]{CrO_3, \ H^+} HC\equiv CCHO \quad (40\%)$$

This Diels–Alder reaction constitutes the only route to (51) which is recorded in the literature at the time of writing (1980).

(iii) **2-Methylbenzoxazole** (compound 52). The correct disconnection for a benzo-fused heterocycle usually involves a carbon–

heteroatom bond, e.g.

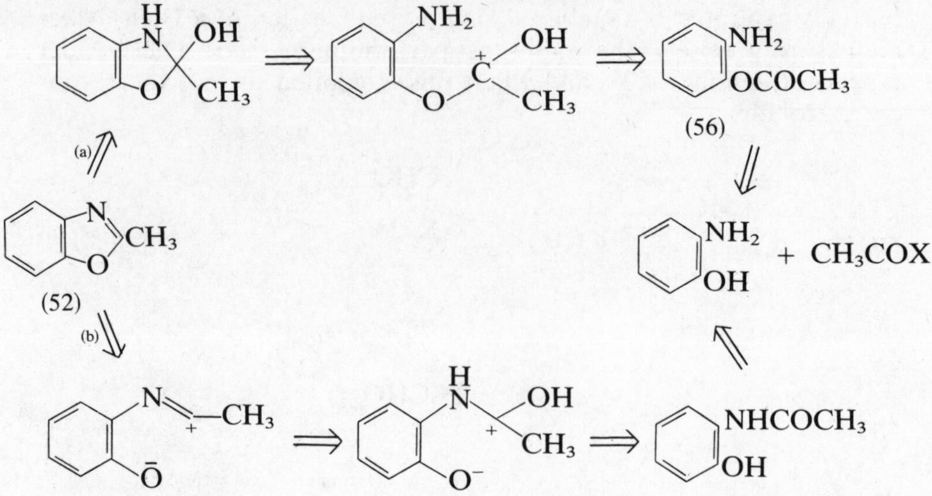

(52)

Possible syntheses of (52) might therefore be:

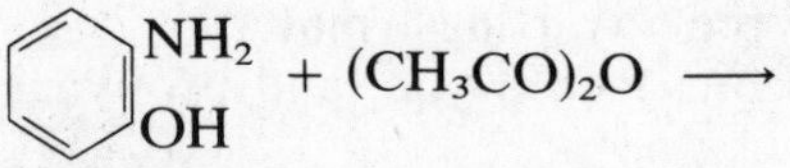

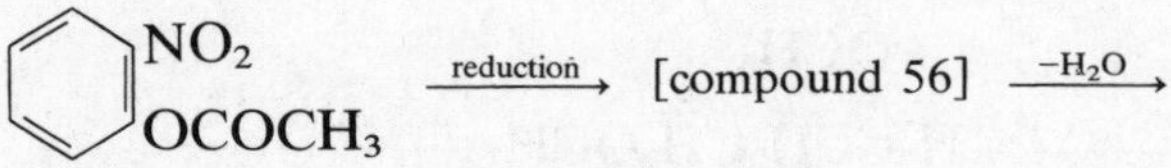

(iv) **2,9-Phenanthroline** (compound 53). As in Chapters 4 and 5, the solution of the final problem is left to the reader. The symmetry of the product is noteworthy, and two synthetic approaches are worthy of consideration:

a. The product may be regarded as an isoquinoline, and one of the standard isoquinoline syntheses applied;

b. The product may be regarded as a phenanthrene derivative and a standard phenanthrene synthesis applied.

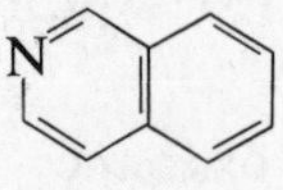

isoquinoline 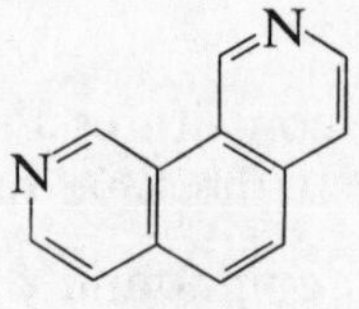　　　　　　(53) 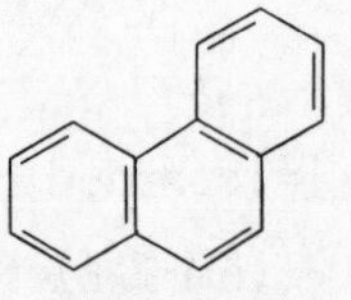　　　　　　phenanthrene

Notes

1.

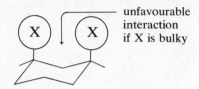

unfavourable
interaction
if X is bulky

2.   Cf. J. E. Baldwin, *J. Chem. Soc. Chem. Comm.*, 1976, 734.

3.   An *Endo-Tet* process is not formally a ring closure at all, but the term could be used to describe a reaction such as

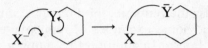

4.   As Sykes has pointed out, of course, the **Woodward–Hoffmann rules** [R. B. Woodward and R. Hoffmann, *Angew. Chem., Int. Edition*, **8,** 781 (1969)] provide a much more rigorous treatment of these reactions, in that the symmetry of all the orbitals of reactants and products are considered. The frontier orbital method is a simplified approach.

5.   Cf. I. Fleming, *Frontier Orbitals and Organic Chemical Reactions*, Wiley, 1976, pp. 132–40.

6.   Cf. Fleming, *op. cit.*, p. 93.

7.   The *cis*-isomer of (36) is decomposed to (37) only at a much higher temperature (>250°): this may be a radical reaction rather than an unfavourable disrotatory cleavage.

8.   Cf. Fleming, *op. cit.*, pp. 102–3, 108.

8 Reduction

In this chapter we shall discuss the reduction of a number of multiply bonded functional groups and some examples of reductive cleavage of single carbon–heteroatom bonds. In addition to a number of fairly specific reactions for reduction of certain functional groups, there are three methods which may be used for the reduction of many functional groups: (a) catalytic hydrogenation, (b) metal hydride reduction, and (c) dissolving metal reactions. These three general methods are considered first.

8.1 Catalytic hydrogenation

Catalytic hydrogenation is a widely applicable technique for the reduction of organic compounds. Usually, the reaction is carried out by stirring or shaking a solution of the compound, in presence of a heterogeneous catalyst, under an atmosphere of hydrogen. It is convenient to discuss the catalyst and the solvent in terms of two types of reaction: (a) low pressure hydrogenation and (b) high pressure hydrogenation. The former involves the use of pressures of hydrogen usually in the range 1 to 4 atm at 0 to 100° and the latter 100 to 300 atm pressure at up to 300°.

Low pressure hydrogenation is carried out in presence of a catalyst such as Raney nickel, platinum (usually produced *in situ* by hydrogenation of PtO_2 – Adams' Catalyst), or palladium or rhodium on a support which can be, in order of decreasing activity, carbon, barium sulphate, or calcium carbonate. Solvents can affect the activity of a catalyst, the activity increasing from neutral non-polar solvents such as cyclohexane to polar acidic solvents such as acetic acid.

Depending on the physical properties of the compound to be reduced, high pressure hydrogenation can be carried out with or without a solvent in presence of a catalyst such as Raney nickel, copper chromite,[1] or palladium on carbon. Table 8.1 lists hydrogenation products of various functional groups in an approximate order of ease of hydrogenation, the acid chloride being the most reactive and the arene the least.

Table 8.1 Products of catalytic hydrogenation

Functional group	Hydrogenation product(s)
RCOCl	RCHO
RNO_2	RNH_2
RC≡CR	RCH=CHR (*Z*, *cis*)
RCHO	RCH_2OH
RCH=CHR	RCH_2CH_2R
RCOR	RCH(OH)R
$ArCH_2X$	$ArCH_3$
RC≡N	RCH_2NH_2
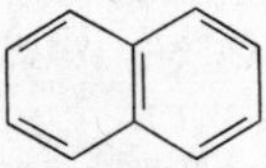	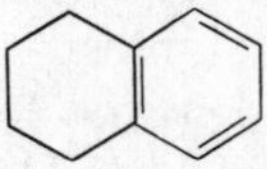
RCO_2R'	$RCH_2OH + R'OH$
RCONHR	RCH_2NHR
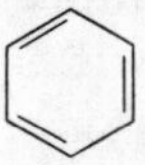	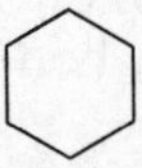

Hydrogenation using a heterogeneous catalyst may sometimes lead to isomerisation of the substrate (cf. section 8.4.1). Isomerisation may be minimised by use of a homogeneous catalyst, e.g. tris(triphenylphosphine)-rhodium chloride (1) since the intermediate complex (2) is less susceptible to rearrangement than its counterpart in the heterogeneous reaction. This can be of considerable importance in, for example, deuteriation. The ease of separation of the catalyst from the reaction mixture is sacrificed when using a homogeneous catalyst, but polymer-bound analogues may combine ease of removal with the formation of products of high purity:

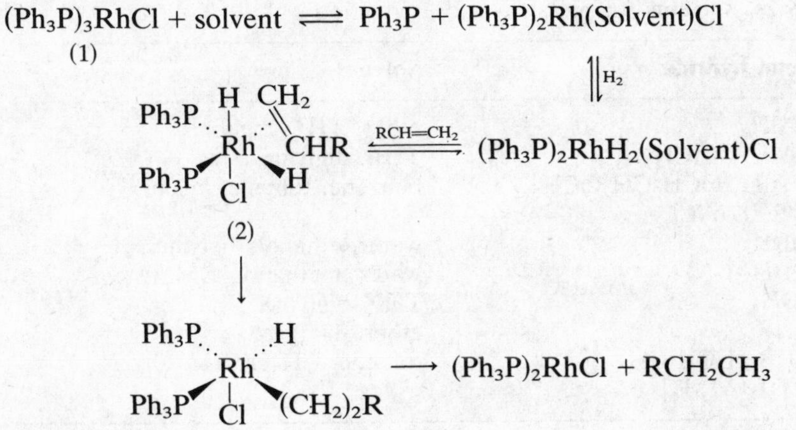
(Ph3P)3RhCl + solvent ⇌ Ph3P + (Ph3P)2Rh(Solvent)Cl
(1)
H2
Ph3P
H
CH2
Rh
CHR
Ph3P
Cl
H
RCH=CH2
(Ph3P)2RhH2(Solvent)Cl
(2)
Ph3P
H
Rh
Ph3P
Cl
(CH2)2R
(Ph3P)2RhCl + RCH2CH3

(8.1)

8.2 Metal hydride reductions

Certain metal hydrides are synthetic equivalents of the hydride ion
(H^-) synthon and as such are powerful reducing agents which react
preferentially at electron-deficient centres. The more strongly basic
hydrides (e.g. NaH and CaH_2), however, are not reducing agents.
Some of the many commercially available hydride reducing agents
(table 8.2) react violently with water and readily with alcohols, and so
reactions must be carried out in anhydrous ethereal or hydrocarbon
solvents. The most commonly used solvents for each reagent are also
given in the table.

Table 8.3 lists some selected reductions which can be achieved
using the reagents listed in table 8.2. Unless otherwise stated, the
product(s) obtained are those indicated in the left-hand column.

As is indicated in table 8.3, $LiAlH_4$, RED-AL, and AlH_3 are
non-selective reagents. Only in the case of an α,β-unsaturated carbonyl compound is AlH_3 preferred to $LiAlH_4$. In fact, the more
selective reagent, DIBAL-H, is probably a better choice.

$$Ph(CH_2)_3OH \xleftarrow{LiAlH_4} PhCH{=}CHCHO \xrightarrow{AlH_3} PhCH{=}CHCH_2OH$$

RED-AL has the following advantages over $LiAlH_4$: (i) it does not
ignite in moist air and is stable in dry air, (ii) it is thermally stable up
to 200°, and (iii) it is very soluble in aromatic hydrocarbon solvents.
Sodium cyanoborohydride is a very selective reagent: for example,
under neutral conditions in hexamethylphosphoramide as solvent, it
will reduce primary alkyl halides in presence of aldehydes.

Choice of solvent may also play a part in selectivity. For example,
sodium borohydride in diglyme solution is a very mild reagent,

Table 8.2 Solvents for metal hydride reductions

no.	Metal hydride	Solvent
1	$LiAlH_4$	ether, THF, diglyme
2	$LiAlH[OC(CH_3)_3]_3$	THF, diglyme
3	$NaAlH_2(OCH_2CH_2OCH_3)_2$ [RED-AL]	benzene, toluene, xylene
4	$NaBH_4$	water, ethanol, diglyme
5	$NaBH_3(CN)$	water, methanol, DMSO
6	$LiBH_4$	THF, diglyme
7	AlH_3	ether, THF
8	$AlH[CH_2CH(CH_3)_2]_2$ [DIBAL-H]	toluene, $CH_3O(CH_2)_2OCH_3$

Table 8.3 Products of metal hydride reductions

Reduction	Reducing agent							
	1	**2**	**3**	**4**	**5**	**6**	**7**	**8**
RCHO → RCH$_2$OH	✓	✓	✓	✓	✓	✓	✓	✓
RCOR → RCH(OH)R	✓	✓	✓	✓	✓	✓	✓	✓
RCOCl → RCH$_2$OH	✓	(a)	✓	✓		✓	✓	
lactone → diol	✓	×	✓	(b)	×	✓	✓	(c)
epoxide → 1,2-diol	✓	×	✓	(b)	×	✓	✓	
RCO$_2$R′ → RCH$_2$OH + R′OH	✓	(d)	✓	(b)	×	✓	✓	(a)
RCO$_2$H → RCH$_2$OH	✓	×	✓	×	×	✓	✓	(a)
RCONR$_2$ → RCH$_2$NR$_2$	(e)	×	✓	×	×	×	✓	×
RC≡N → RCH$_2$NH$_2$	✓	×	×	×	×	×	✓	(a)
RNO$_2$ → RNH$_2$	(f)	×		×	×	×	×	
RX$^{(g)}$ → RH	✓	×	✓	×	✓	×	×	
RC≡CR → RCH=CHR Z, *cis*								✓

(a) reduction proceeds to the aldehyde stage only
(b) very slow reaction
(c) reduction proceeds to lactol stage only
(d) phenyl esters give aldehydes
(e) some amides are reduced to aldehydes
(f) R is aliphatic
(g) X = halogen or OSO$_2$R′

reducing aldehydes but not ketones. This selectivity can also be achieved by use of lithium triacetoxyborohydride as reducing agent.

It is, therefore, important to make a careful choice of reagent, solvent and reaction conditions to obtain the desired selectivity in reduction of molecules having more than one reducible functional group.

8.3 Dissolving metal reductions

This type of reaction, formerly thought to involve 'nascent' hydrogen, involves electron transfer to the substrate from a metal such as lithium, sodium, potassium, magnesium, calcium, zinc, tin or iron. A proton donor (e.g. water or ethanol) may either be present during electron transfer or be added at a later stage. Reduction of the carbonyl group can result in the formation of three types of product depending on the reaction conditions used (8.2). Reduction to the alcohol takes place in presence of a proton donor, when the initially formed radical anion (3), (8.3), is first protonated and then converted into the carbanion (4) by a second electron transfer. In the absence of

$$
\begin{array}{ccc}
 & \underset{\displaystyle \diagup}{\overset{\diagdown}{C}}{=}O & \\
\text{Zn/HCl}\nearrow & \Big\downarrow\text{Na/EtOH} & \text{Mg/Hg}\searrow
\end{array}
$$

$$
\underset{\diagup}{\overset{\diagdown}{C}}H_2 \qquad\qquad \underset{\diagup}{\overset{\diagdown}{C}}H(OH) \qquad\qquad \underset{OH}{\overset{\diagdown}{C}}{-}\underset{OH}{\overset{}{C}}\diagup \qquad\qquad (8.2)
$$

(Clemmensen reduction, section 8.4.3.3)

(section 8.4.3.1)

(Bimolecular reduction, section 8.4.3.2)

a proton donor, (3) dimerises to the pinacolate dianion (5). The Clemmensen procedure involves successive electron transfers to the protonated ketone adsorbed on the surface of the metal. In this case, low concentrations of ketone at the metal surface are desirable to minimise bimolecular reduction.

$$
\underset{\diagup}{\overset{\diagdown}{C}}{=}O + M \longrightarrow \underset{\diagup}{\overset{\diagdown}{C}}{-}O^-\ M^+ \longrightarrow \underset{O^-\ \ O^-}{\overset{\diagdown}{C}{-}\overset{}{C}}\diagup\ \ 2M^+
$$

$$
(3) \qquad\qquad\qquad (5)
$$

$$
\Big\downarrow\text{ROH} \qquad\qquad \Big\downarrow\text{H}^+
$$

$$
\underset{\diagup}{\overset{\diagdown}{\bar{C}}}{-}OH \xleftarrow{\ M\ } \underset{\diagup}{\overset{\diagdown}{\cdot C}}{-}OH \qquad\qquad \underset{OH\ \ OH}{\overset{\diagdown}{C}{-}\overset{}{C}}\diagup \qquad\qquad (8.3)
$$

(4)

$$
\Big\downarrow\text{H}^+
$$

$$
\underset{\diagup}{\overset{\diagdown}{C}}HOH
$$

Under certain circumstances, reductions analogous to those carried out under dissolving metal conditions can be carried out electrochemically. Some examples of dissolving metal and of electrochemical reductions will be found in the later sections of this chapter.

8.4 Reduction of specific functional groups

8.4.1 Reduction of alkenes

Alkenes are rapidly hydrogenated in presence of a catalyst, usually platinum, Raney nickel, or palladium or rhodium on carbon, to the corresponding alkane. Although such reactions are normally regarded as being stereospecifically *cis* additions, rearrangements occurring on the catalyst surface make this statement an oversimplification. For example, 1,2-dimethylcyclohexene (6) isomerises to 2,3-dimethylcyclohexene (7) and the result of catalytic hydrogenation of

(6) is a mixture of *cis*-(8) and *trans*-(9) dimethylcyclohexanes. Platinum catalysts tend to cause less isomerisation than palladium as can be seen in (8.4). Isomerisation of this type often results in the formation of complex product mixtures when catalytic deuteriations are attempted.

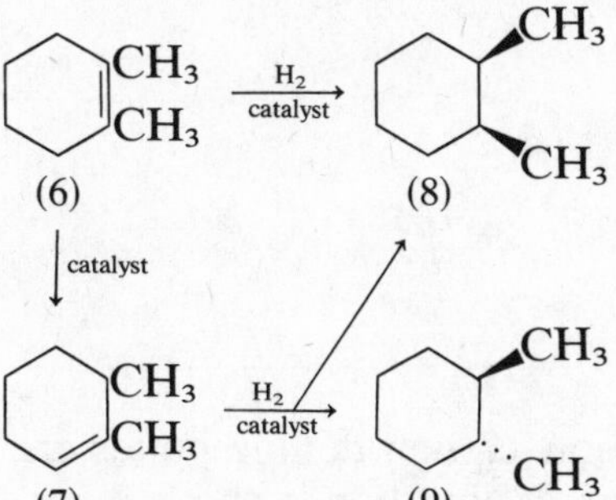

$$(6) \xrightarrow[\text{R.T., acetic acid}]{\text{1 atm } H_2, \text{ PtO}_2} \quad \underset{(82\%)}{(8)} + \underset{(18\%)}{(9)} \quad (100\%)$$

(8.4)

$$(6) \xrightarrow[\text{R.T., acetic acid}]{\text{1 atm } H_2, \text{ Pd/Al}_2\text{O}_3} \quad \underset{(23\%)}{(8)} + \underset{(77\%)}{(9)} \quad (100\%)$$

In hindered alkenes, addition takes place on the less hindered side. For example, in the case of hydrogenation of the bicyclo[2.2.1]hept-2-enecarboxylic acid (10) there is less steric hindrance to adsorption on the catalyst surface on the face of the molecule *cis* to the methylene bridge. The product is then the *endo* isomer (11) rather than the *exo* isomer (12).

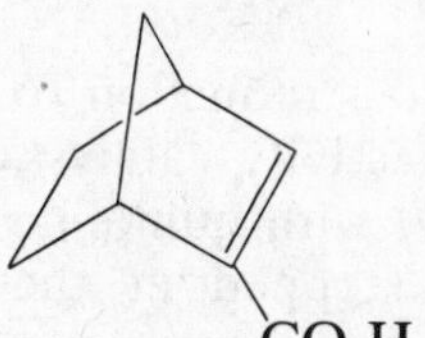

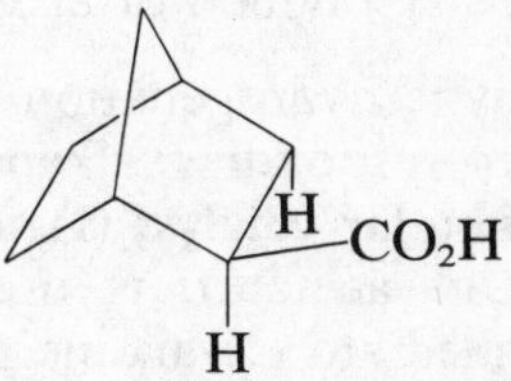

$$(10) \xrightarrow[\text{acetic acid}]{H_2, \text{ PtO}_2} (11) \quad (83\%)$$

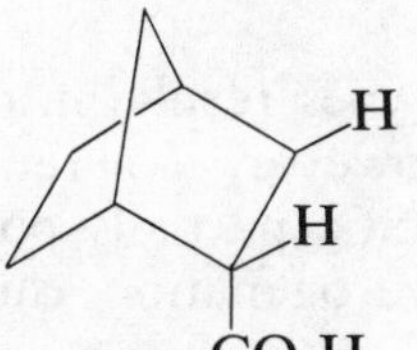

It is possible to reduce double bonds selectively in presence of esters and ketones and even, in some instances, aldehydes provided that the reaction conditions are carefully controlled. However, a greater degree of selectivity is achieved by use of the homogeneous catalyst tris(triphenylphosphine)rhodium chloride:

$$PhCH=CHCOPh \xrightarrow[MeCO_2Et]{H_2,Pt} PhCH_2CH_2COPh \quad (90\%)$$

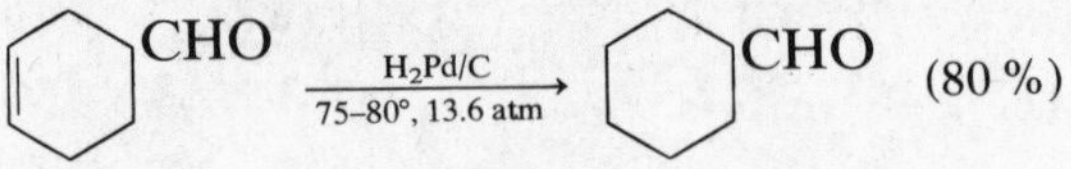

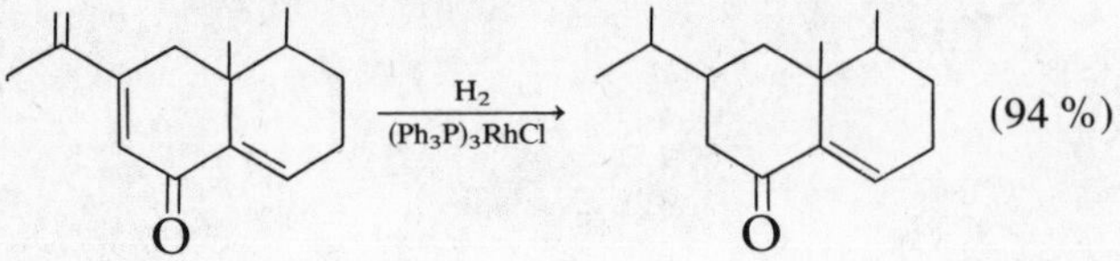

$\xrightarrow[75-80°, 13.6 \text{ atm}]{H_2Pd/C}$ (80%)

$\xrightarrow[(Ph_3P)_3RhCl]{H_2}$ (94%)

Non-polar and moderately polar carbon–carbon double bonds are reduced by di-imide (13) whereas polar double bonds such as carbonyl groups and the carbon–carbon double bonds in α,β-unsaturated ketones are not affected. Although the first reports of reductions involving di-imide used hydrazine in presence of an oxidising agent, a more useful source is the decomposition of azodicarboxylate salts in an acidic medium (8.5):

$$^-O_2CN=NCO_2^- + 2H^+ \rightarrow HN=NH + 2CO_2 \qquad (8.5)$$
$$(13)$$

$$CH_2=CHCH_2OH + 2K^+\bar{O}_2CN=NCO_2^- \xrightarrow[\substack{\text{methanol} \\ \text{room temp.}}]{CH_3CO_2H}$$

$$CH_3CH_2CH_2OH + N_2 + 2CO_2$$
$$(78\%)$$

8.4.2 Reduction of alkynes

Catalytic hydrogenation of alkynes results in complete reduction to the corresponding alkanes. However, if a reduced-activity catalyst, the **Lindlar catalyst** (Pd on $BaSO_4$ partially poisoned with quinoline or lead acetate), is used, hydrogenation can be stopped at the intermediate *cis*-alkene stage.

$$CH_3O_2C(CH_2)_3C\equiv C(CH_2)_3CO_2CH_3 \xrightarrow[\substack{\text{quinoline} \\ \text{methanol}}]{H_2,Pd/BaSO_4}$$

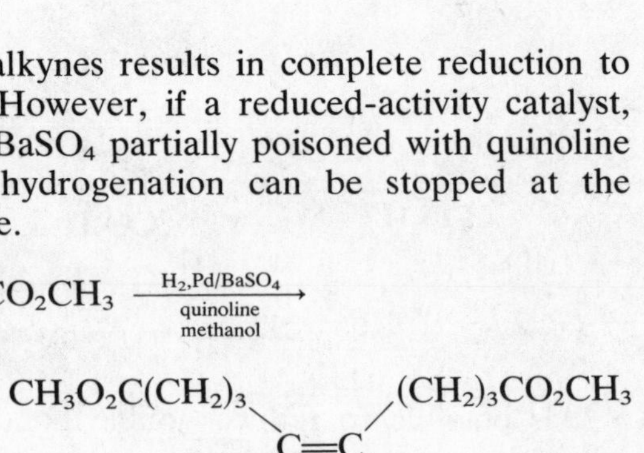

(97%)

Lithium aluminium hydride does not readily reduce aliphatic alkynes but α-hydroxyalkynes and α,β-alkynoic acids are both reduced to *trans*-allylic alcohols.

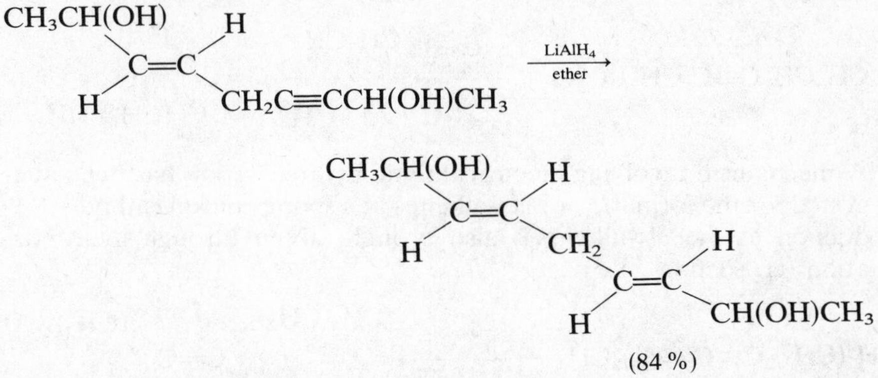

(84 %)

DIBAL-H does, however, react more readily with aliphatic alkynes, the product in this case being the *cis*-alkene:

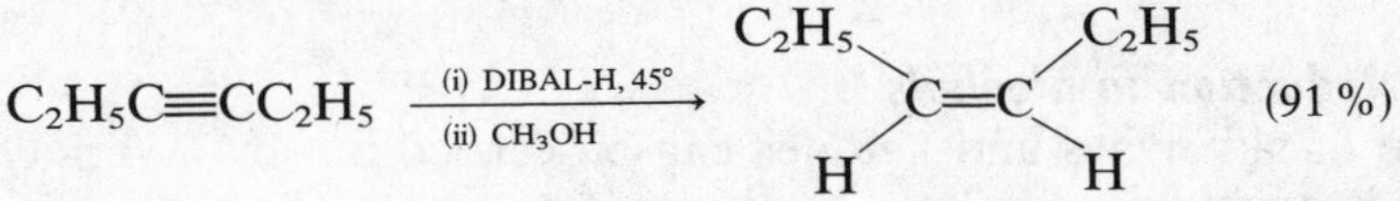

$$C_2H_5C{\equiv}CC_2H_5 \xrightarrow[\text{(ii) } CH_3OH]{\text{(i) DIBAL-H, 45°}}$$ (91 %)

A more general method for reducing non-terminal alkynes to *trans*-alkenes is the **dissolving metal procedure** using lithium or sodium in liquid ammonia followed by protonation (8.6). Reduction of diarylalkynes is more complex; either *cis*- or *trans*-alkenes can be formed depending on the metal used.

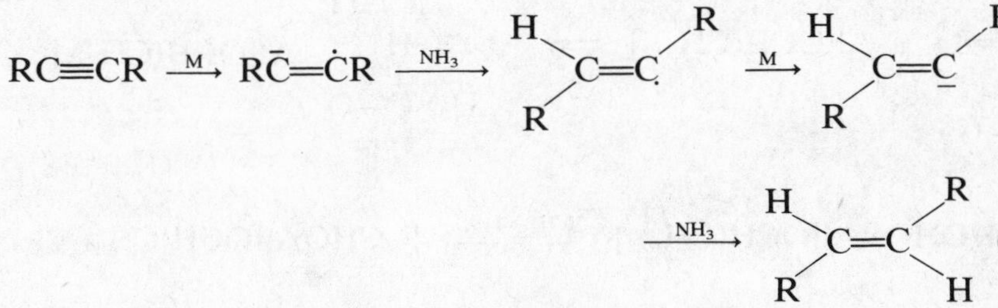

$$RC{\equiv}CR \xrightarrow{M} R\bar{C}{=}\dot{C}R \xrightarrow{NH_3}$$ (8.6)

$$CH_3C{\equiv}C(CH_2)_3C{\equiv}CCH_3 \xrightarrow[NH_3]{Na}$$ (72 %)

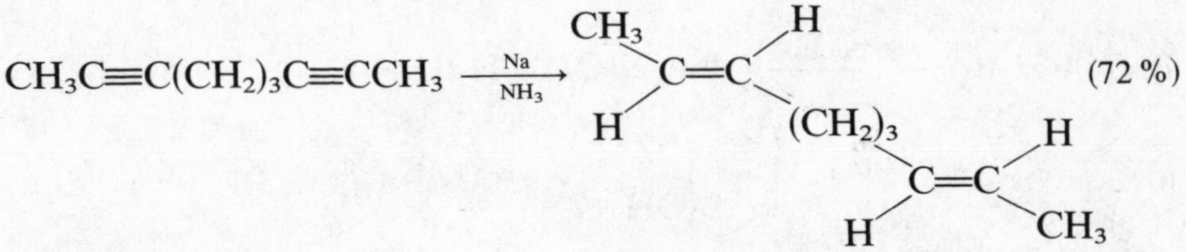

Electrochemical reduction of alkynes with lithium chloride in methylamine may involve formation of lithium atoms which reduce the alkyne by electron transfer. This results in formation of *trans*-alkenes.

$$CH_3CH_2C{\equiv}C(CH_2)_3CH_3 \xrightarrow[\substack{\text{Pt cathode} \\ \text{undivided cell} \\ \text{LiCl/MeNH}_2}]{} \begin{array}{c} CH_3CH_2 H \\ \diagdown C{=}C \diagup \\ \diagup \diagdown \\ H (CH_2)_3CH_3 \end{array} \quad (58\,\%)$$

A mechanism involving electrocatalytic hydrogenation has been suggested for the formation of *cis*-alkenes at a spongy nickel cathode. Reduction to the *cis*-alkene is also brought about through hydroboration (cf. section 11.5).

$$CH_3(CH_2)_2C{\equiv}C(CH_2)_2CH_3 \xrightarrow[\substack{\text{spongy Ni} \\ \text{cathode} \\ \text{ethanol/H}_2\text{SO}_4}]{} \begin{array}{c} CH_3(CH_2)_2 (CH_2)_2CH_3 \\ \diagdown C{=}C \diagup \\ \diagup \diagdown \\ H H \end{array} \quad (80\,\%)$$

8.4.3 Reduction of aldehydes and ketones

8.4.3.1 *Reduction to alcohols*

Reduction of aldehydes and ketones can be carried out by a variety of methods including catalytic hydrogenation, metal hydride, dissolving metal, and aluminium isopropoxide (the **Meerwein–Ponndorf–Verley** reaction). Unless stereochemical considerations, which will be discussed later, are important, all methods result in the same product from acyclic ketones and aldehydes. However, the Meerwein–Ponndorf–Verley reaction (8.7) is useful when the carbonyl group is to be reduced in presence of other reducible groups:

$$R_2C{=}O + Al[OCH(CH_3)_2]_3 \rightleftharpoons \begin{array}{c} R_2C{=}O \\ H) \diagdown \\ | Al[OCH(CH_3)_2]_2 \\ (CH_3)_2C{-}O \end{array}$$

$$\Updownarrow \qquad (8.7)$$

$$R_2CHOH + Al[OCH(CH_3)_2]_3 \xrightleftharpoons{(CH_3)_2CHOH} R_2CHOAl[OCH(CH_3)_2]_2$$

$$+ (CH_3)_2C{=}O$$

$$PhCOCH_2Br \xrightarrow{Al[OCH(CH_3)_2]_3} PhCH(OH)CH_2Br \quad (85\,\%)$$

(70 %)

Recently it has been suggested that the reduction of aldehydes and ketones, dissolved in an inert solvent, by propan-2-ol catalysed by dehydrated neutral alumina may be a suitable general method. The advantages claimed for the method are:

(i) α,β-unsaturated aldehydes are reduced to allylic alcohols,
(ii) aldehydes can be reduced in presence of some ketones,
(iii) many labile functional groups (e.g. nitro, cyano and halogeno) survive the reaction,
(iv) the propanol/alumina reagent can be stored in a sealed vial for long periods,
(v) low cost of reagents and ease of isolation of the products.

$$\underset{NO_2}{\overset{CHO}{\bigcirc}} \xrightarrow[Al_2O_3,\ CCl_4]{(CH_3)_2CHOH} \underset{NO_2}{\overset{CH_2OH}{\bigcirc}} \quad (84\%) \qquad (8.8)$$

$$(CH_3)_2C{=}CH(CH_2)_2CH{=}CHCHO \xrightarrow[Al_2O_3,\ CCl_4]{(CH_3)_2CHOH}$$

$$(CH_3)_2C{=}CH(CH_2)_2CH{=}CHCH_2OH \quad (88\%)$$

When stereochemical factors are involved, a more complex situation is encountered. In the case of hydride reductions which can give two epimeric alcohols, it has been postulated that such reactions are governed by two factors: (a) product development (or stability) control[2] and (b) steric approach control. Thus when the incoming reagent is small, the former has the greater effect and the more stable alcohol predominates. For bulkier reagents, steric factors dominate and attack is from the less hindered side of the molecule. These effects have also been recognised in electrochemical and in dissolving metal reductions leading to a product distribution similar to that found in unhindered hydride reductions. Usually, therefore, the experimentally simpler technique of lithium aluminium hydride reduction is preferred to these other methods.

Catalytic hydrogenation results in *cis* addition from the less hindered side of the molecule. The products formed on reduction of 4-t-butylcyclohexanone (14) and of 3,3,5-trimethylcyclohexanone (17) under a variety of conditions are listed in tables 8.4 and 8.5 respectively and serve to illustrate the importance of choice of reagent when stereochemical factors have to be considered. It should be noted that the relative thermodynamic stabilities of (15) and (16) are approximately 4:1 and of (18) and (19) approximately 16:1.

Table 8.4 Products of reduction of 4-t-butylcyclohexanone

Reducing agent	Ratio of (15) : (16)
LiAlH$_4$	9:1
LiAlH[OC(CH$_3$)$_3$]$_3$/THF	9:1
Li/liq NH$_3$ in ether/butanol	49:1
H$_2$/Raney Ni/ethanol	1:1
Al[OCH(CH$_3$)$_2$]$_3$	3:1

Table 8.5 Products of reduction of 3,3,5-trimethylcyclohexanone

Reducing agent	Ratio of (18) : (19)
LiAlH$_4$/ether	1:1
LiAl[OC(CH$_3$)$_3$]$_3$/THF	1:8
H$_2$/Raney Ni/ethanol	1:9
Pt cathode/LiCl in hexamethylphosphoramide/ ethanol	10:1
Li/liq. NH$_3$/ethanol	99:1

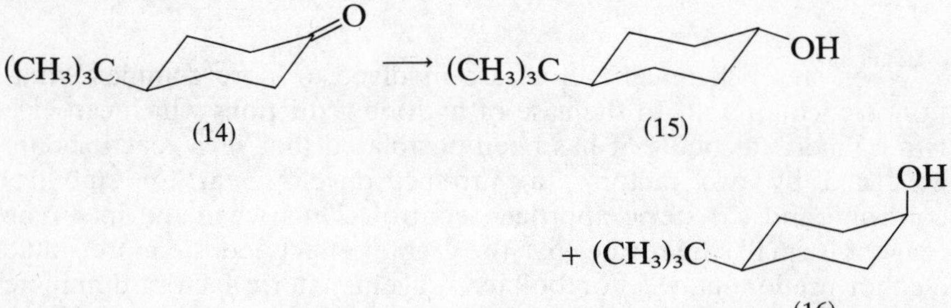

(14) → (15)

+ (16)

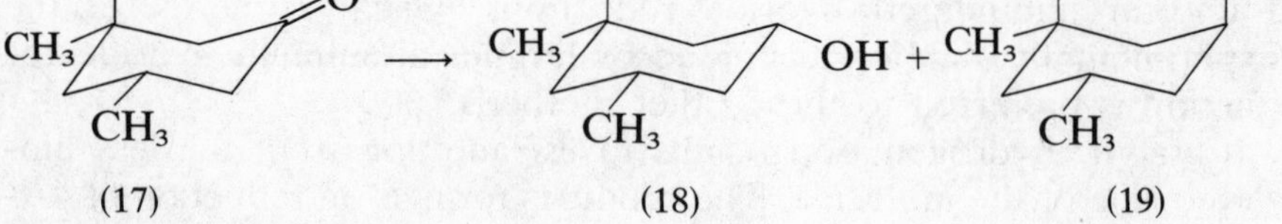

(17) → (18) + (19)

8.4.3.2 Bimolecular reduction

When ketones react with magnesium, zinc, or aluminium (used often as amalgams) in the absence of proton donors, the initially formed radical ions dimerise to the dianion of a 1,2-diol. Bimolecular reduction (8.9) competes with other reductions such as the Clemmensen reaction:

$$CH_3COCH_3 \xrightarrow[\substack{benzene \\ reflux}]{Mg/Hg} ((CH_3)_2\dot{C}-\bar{O})_2Mg^{2+} \longrightarrow \begin{matrix} (CH_3)_2C-O^- \\ | \\ (CH_3)_2C-O^- \end{matrix} Mg^{2+}$$

$$\Big\downarrow H_2O \qquad (8.9)$$

$$(CH_3)_2C(OH)C(OH)(CH_3)_2$$

$$(50\,\%)$$

8.4.3.3 *Reduction of ketones to methylene groups*

Reaction of ketones with zinc amalgam in presence of mineral acid (the **Clemmensen** reaction) reduces the carbonyl group to methylene. Often the reaction is carried out in presence of toluene to produce a three-phase system in which most of the ketone remains in the upper hydrocarbon layer, and the protonated carbonyl compound in the aqueous layer is reduced on the metal surface by the mechanism shown in (8.10):

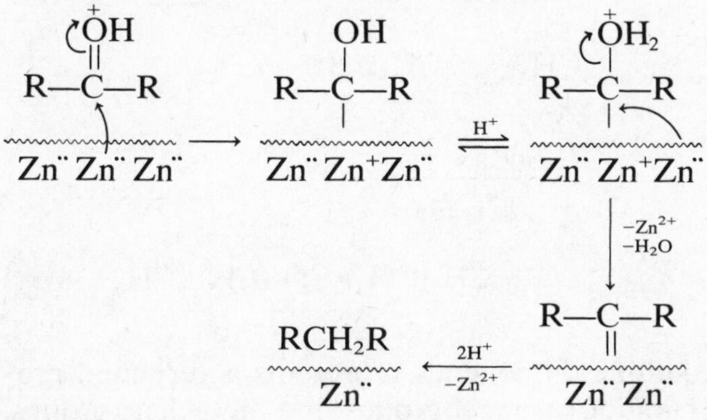

$$(8.10)$$

The purpose of carrying out the reaction in a three-phase system is to minimise bimolecular reduction by maintaining only a low concentration of protonated carbonyl compound at the metal surface. The following examples demonstrate the scope of the reaction. (The method cannot, of course, be used for the reduction of acid-sensitive compounds.)

$$PhCOCH_2CH_2CO_2H \xrightarrow[\substack{HCl \\ toluene}]{Zn/Hg} Ph(CH_2)_3CO_2H \quad (85\,\%)$$

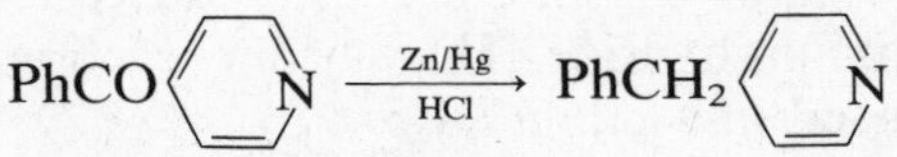

$$PhCO\langle\underline{\quad}\rangle N \xrightarrow[HCl]{Zn/Hg} PhCH_2\langle\underline{\quad}\rangle N \quad (80\,\%)$$

Complementary to the Clemmensen reduction is the **Wolff–Kishner** reaction in which the ketone hydrazone is treated with a strong base. Several modifications of the reaction have been used, one of the more successful being the **Huang–Minlon** procedure. In this case, the carbonyl compound, hydrazine hydrate, and potassium hydroxide are heated together in a high boiling solvent. It has also been shown that use of dimethyl sulphoxide as solvent causes reaction to take place at substantially lower temperatures. For base-sensitive compounds an alternative procedure involves the reaction of a tosylhydrazone of the ketone with sodium cyanoborohydride. Due to the slow reduction of carbonyl compounds under these conditions, it is unnecessary to preform the hydrazone:

$$R_2C{=}O \xrightarrow{N_2H_4} R_2C{=}NNH_2 \xrightarrow{OH^-} R_2C{=}N\bar{N}H \underset{\xleftarrow{\hspace{1cm}}}{\xrightarrow{R'OH}} R_2CHN{=}NH$$

$$\Big\downarrow OH^-$$

$$R_2CH_2 \xleftarrow{R'OH} R_2C\bar{H} + N_2 \longleftarrow R_2CHN{=}\bar{N}$$

$$CH_3\underset{S}{\Big\langle\!\!\!\diagup}COCH_3 \xrightarrow[\substack{KOH, digol\\200°}]{N_2H_4} CH_3\underset{S}{\Big\langle\!\!\!\diagup}CH_2CH_3 \quad (83\%)$$

$$CH_3COCH_2CO_2(CH_2)_9CH_3 \xrightarrow[\substack{NaBH_3CA\\sulpholan/DMF/H^+}]{p\text{-}CH_3C_6H_4SO_2NHNH_2}$$

$$CH_3(CH_2)_2CO_2(CH_2)_9CH_3 \quad (65\%)$$

Hydrogenolysis of ketone dithioketals represents a very mild procedure for the conversion of carbonyl groups into methylene groups. However, since a large excess of Raney nickel is required (7 g/g of substrate), it is normally only of use in small-scale preparations.

8.4.4   Reduction of carboxylic acids and their derivatives

Acids, amides, and esters are resistant to catalytic hydrogenation. Indeed, both ethyl acetate and acetic acid are commonly used solvents in low pressure hydrogenations.

Esters are readily reduced to alcohols by lithium aluminium hydride and by dissolving metal reactions. The latter, the **Bouveault–Blanc** method, only rarely holds advantage over lithium aluminium hydride and has largely been replaced by it. Acids are also reduced to primary alcohols by lithium aluminium hydride. The acyloin reaction of esters of dibasic acids is discussed in section 7.1.5.

$$CH_3(CH_2)_7CH{=}CH(CH_2)_7CO_2C_2H_5 \xrightarrow[\text{ethanol}]{\text{Na}}$$

$$CH_3(CH_2)_7CH{=}CH(CH_2)_8OH \quad (50\%)$$

$$CH_3CH(OCH_3)CH_2CO_2CH_3 \xrightarrow[\text{ether}]{\text{LiAlH}_4} CH_3CH(OCH_3)CH_2CH_2OH \quad (70\%)$$

$$CH_3CH{=}CHCH{=}CHCO_2H \xrightarrow[\text{ether}]{\text{LiAlH}_4}$$

$$CH_3CH{=}CHCH{=}CHCH_2OH \quad (92\%)$$

Acid chlorides can be hydrogenated to aldehydes in presence of the reduced activity **Rosenmund catalyst,** which consists of palladium on barium sulphate to which is added a quinoline–sulphur poison. In many cases an alternative procedure, which can be used in presence of a wide variety of functional groups (see table 8.3), is the use of lithium tri-(t-butoxy)aluminium hydride at low temperatures:

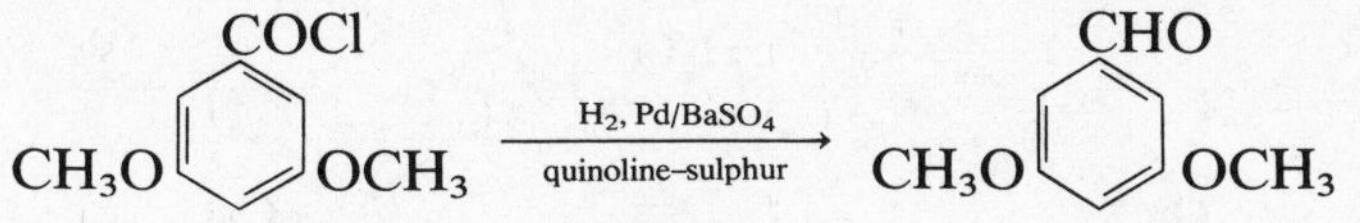

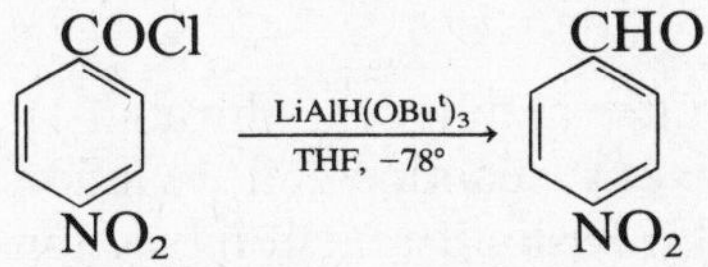

Aldehydes can also be prepared from other acid derivatives by reduction with metal hydrides: *viz.* amides derived from imidazole, carbazole or aziridine with lithium aluminium hydride; simple tertiary amides with lithium triethoxyaluminium hydride; phenyl esters with lithium tri-(t-butoxy)aluminium hydride; and ethyl esters with diisobutylaluminium hydride at low temperatures:

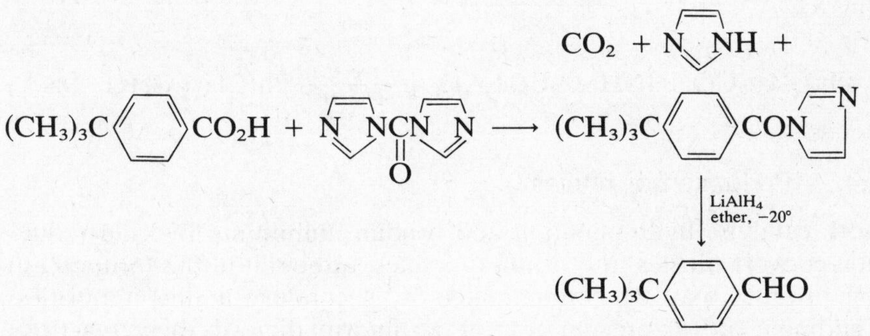

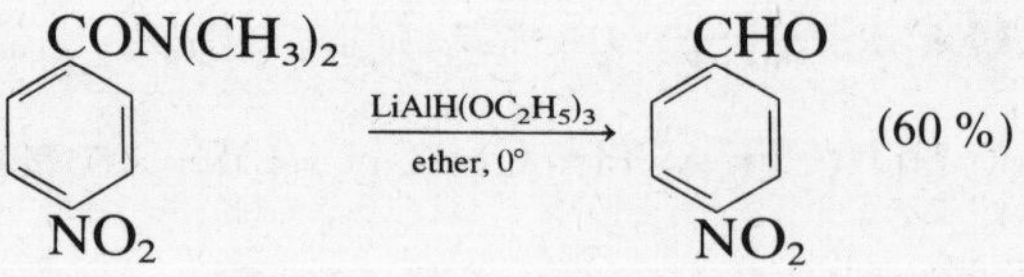

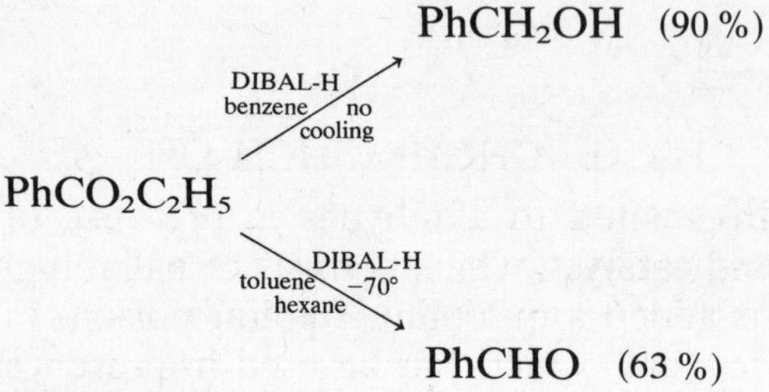

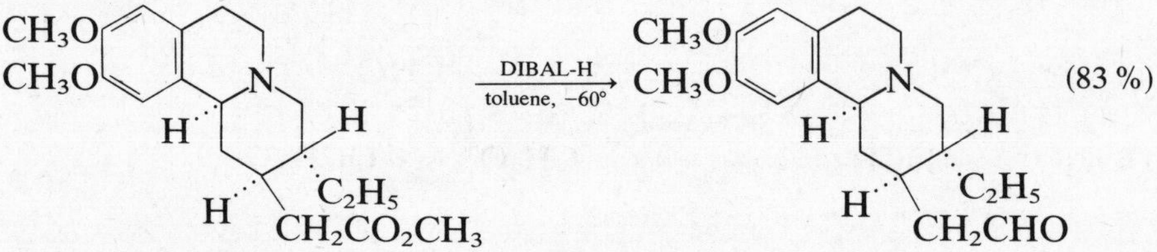

CON(CH₃)₂ [benzene ring with NO₂] →(LiAlH(OC₂H₅)₃, ether, 0°)→ CHO [benzene ring with NO₂]   (60 %)

PhCO₂C₂H₅ →(DIBAL-H, benzene, no cooling)→ PhCH₂OH   (90 %)

PhCO₂C₂H₅ →(DIBAL-H, toluene, −70° hexane)→ PhCHO   (63 %)

[dimethoxy tetracyclic structure with C₂H₅ and CH₂CO₂CH₃] →(DIBAL-H, toluene, −60°)→ [dimethoxy tetracyclic structure with C₂H₅ and CH₂CHO]   (83 %)

Finally, acids may be converted into aldehydes through a sulphonyl-hydrazide (the **McFadyen–Stevens** reaction), a reaction which bears certain similarities to the Wolff–Kishner reaction [section 8.4.3.2]. Yields are, however, often poor.

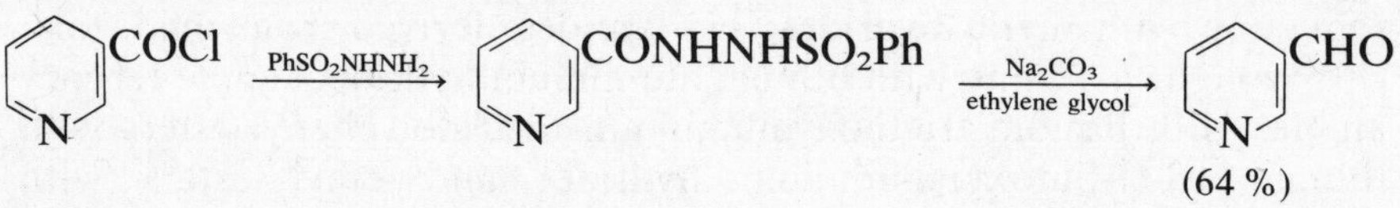

[pyridine]COCl →(PhSO₂NHNH₂)→ [pyridine]CONHNHSO₂Ph →(Na₂CO₃, ethylene glycol)→ [pyridine]CHO   (64 %)

$Ph(CH_2)_2CONHNH_2 \xrightarrow{p-CH_3C_6H_4SO_2Cl}$

$Ph(CH_2)_2CONHNHSO_2C_6H_4CH_3\text{-}p \xrightarrow[\substack{\text{(ii) heat} \\ \textit{in vacuo}}]{\text{(i) NaOCH}_3} Ph(CH_2)_2CHO$ (85 %)

8.4.5 Reduction of nitriles

Both catalytic hydrogenation and lithium aluminium hydride reduction convert nitriles into primary amines, although in the former case the product may be contaminated by secondary amine impurities. The imine (20) is presumed to be an intermediate in these reactions and if the reduction can be stopped at this stage an aldehyde will be formed on hydrolysis. Examples of this are given below:

$$RC\equiv N \longrightarrow [RCH=NH] \longrightarrow RCH_2NH_2$$
$$(20)$$

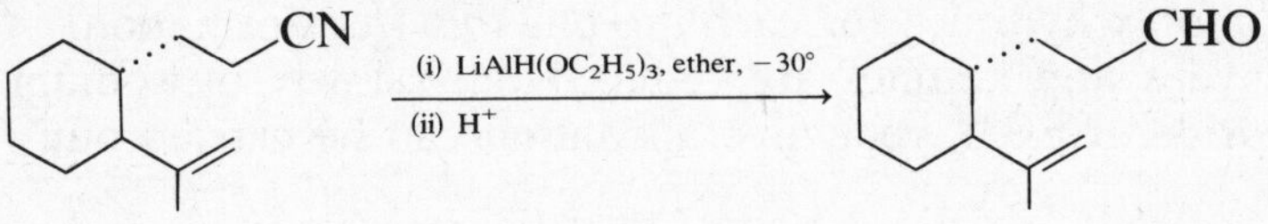

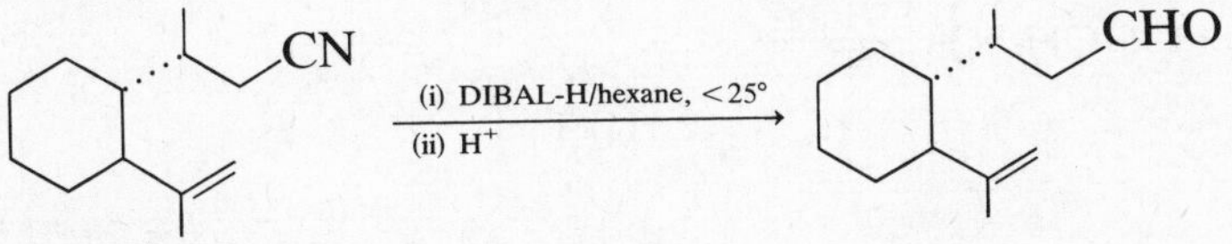

(i) LiAlH(OC$_2$H$_5$)$_3$, ether, $-30°$

(ii) H$^+$

(i) DIBAL-H/hexane, $<25°$

(ii) H$^+$

(96 %)

(i) DIBAL-H/benzene

(ii) methanol (iii) H$^+$

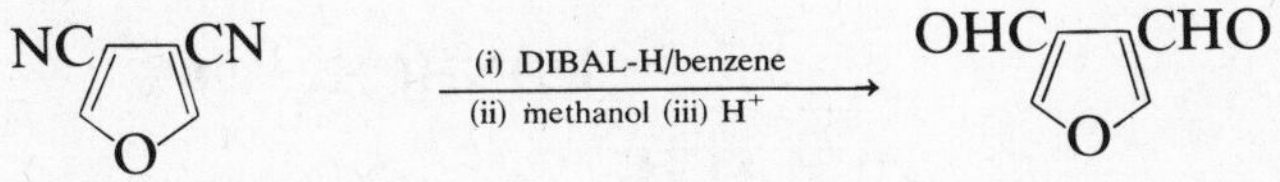

(56 %)

8.4.6 Reduction of imines and oximes, including reductive alkylation

Imines are hydrogenated catalytically to amines. Closely related is the reductive alkylation of amines (including ammonia) and nitro-compounds leading to the formation of primary, secondary and tertiary amines.

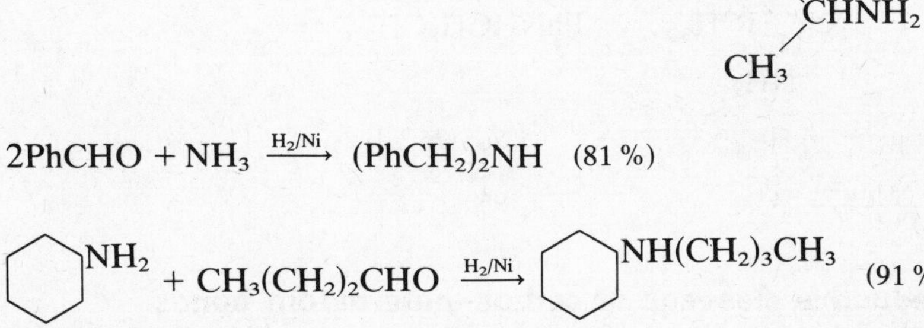

$$CH_3CO(CH_2)_2CH_3 + NH_3 \longrightarrow \left[\begin{array}{c} CH_3(CH_2)_2 \\ \diagdown \\ \diagup C=NH \\ CH_3 \end{array} \right] \xrightarrow{H_2/Ni}$$

$$\begin{array}{c} CH_3(CH_2)_2 \\ \diagdown \\ \diagup CHNH_2 \quad (90\%) \\ CH_3 \end{array}$$

$$2PhCHO + NH_3 \xrightarrow{H_2/Ni} (PhCH_2)_2NH \quad (81\%)$$

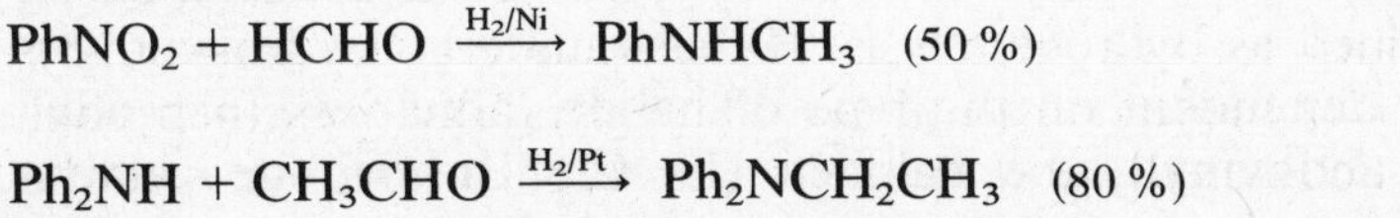

$$+ CH_3(CH_2)_2CHO \xrightarrow{H_2/Ni} \quad (91\%)$$

$$PhNO_2 + HCHO \xrightarrow{H_2/Ni} PhNHCH_3 \quad (50\%)$$

$$Ph_2NH + CH_3CHO \xrightarrow{H_2/Pt} Ph_2NCH_2CH_3 \quad (80\%)$$

Reductions of imines and iminium salts to amines by metal hydrides such as lithium aluminium hydride and sodium borohydride requires neutral or slightly acidic conditions. The greater stability of cyanoborohydrides under such conditions renders them more suitable than other complex hydrides for carrying out this transformation.

Since aldehydes and ketones are reduced only slowly by sodium cyanoborohydride at pH 6, reductive alkylation can be carried out.

$$
\begin{array}{c}
Ph \\
\diagdown \\
C{=}O + CH_3NH_2 \\
\diagup \\
(CH_3)_2CH
\end{array}
\xrightarrow[\substack{CH_3OH}]{pH\,6}
\begin{array}{c}
Ph\diagdown\diagup CH_3 \\
C{=}\overset{+}{N} \\
(CH_3)_2CH\diagup\diagdown H
\end{array}
$$

$$\Big\downarrow \text{NaBH}_3\text{CN}$$

$$
\begin{array}{c}
Ph \\
\diagdown \\
CHNHCH_3 \quad (91\%) \\
\diagup \\
(CH_3)_2CH
\end{array}
$$

Oximes are reduced to primary amines by catalytic hydrogenation over platinum in acetic acid, or by dissolving metal reduction using, for example, sodium in ethanol. Reduction with lithium aluminium hydride gives primary amines almost exclusively from aliphatic oximes. However, the corresponding reduction of aryl ketoximes results in the formation of appreciable amounts of secondary amines. The latter can be the sole product if the reaction is carried out in presence of aluminium chloride, perhaps because of an initial Beckmann rearrangement of the oxime:

$$
\begin{array}{c}
PhCCH_3 \\
\| \\
NOH
\end{array}
\xrightarrow[\text{rearrangement}]{\text{Beckmann}}
PhNHCOCH_3
$$

$$
\begin{array}{ccc}
 & PhCHCH_3 & PhNHCH_2CH_3 \\
 & | & \\
 & NH_2 &
\end{array}
$$

$\xrightarrow[\text{ether}]{\text{LiAlH}_4}$	80	:	20	(75 %)
$\xrightarrow[\text{ether}]{\text{LiAlH}_4/\text{AlCl}_3 (1:4)}$	4	:	96	

8.5 Reductive cleavage of carbon–heteroatom bonds

Reductive cleavage of single bonds by catalytic hydrogenation is usually described as hydrogenolysis. Halides undergo hydrogenolysis with an ease dependent on the type of halide (alkyl less than allyl, aryl, benzyl and vinyl), the halogen (F ≪ Cl < Br < I), the catalyst (palladium catalysts are more effective than Raney nickel which should be the catalyst chosen if hydrogenolysis is undesirable) and solvent (polar solvents and the presence of base favour hydrogenolysis). Halogenoanilines and halogenopyridines are thus very readily hydrogenolysed in other than acidic conditions:

$$\underset{CH_3}{\overset{CO_2C_2H_5}{\bigodot}}\overset{CN}{\underset{N}{Cl}} \xrightarrow[\text{ethanol, R.T., 1 atm}]{H_2,Pd/BaCO_3} \underset{CH_3}{\overset{CO_2C_2H_5}{\bigodot}}\overset{CN}{\underset{N}{}} \quad (95\,\%)$$

$$\underset{Cl}{\overset{NO_2}{\bigodot}} \xrightarrow[\text{ethanol, 100°, 20 atm}]{H_2,Ni} \underset{Cl}{\overset{NH_2}{\bigodot}} \quad (90\,\%)$$

Lithium aluminium hydride and sodium borohydride both reduce primary and secondary alkyl halides to hydrocarbons. However, a wide range of other functional groups in the molecule may also be affected. The reaction appears to involve an S_N2 mechanism with inversion of configuration at the reaction centre. At approximately pH 6, sodium cyanoborohydride reduces few functional groups other than carbon–halogen bonds and is a highly specific reagent for carrying out this transformation:

$$\underset{(21)}{ClCH_2CH_2CO_2H} \xrightarrow[\text{ether}]{LiAlH_4} \underset{(62\,\%)}{CH_3CH_2CH_2OH} + \underset{(21\,\%)}{Cl(CH_2)_3OH}$$

$$\xrightarrow[\text{THF} \quad 0°]{AlH_3}$$

$$\underset{(22)}{Cl(CH_2)_3OH} \quad (61\,\%)$$

$$Br(CH_2)_4CO_2C_2H_5 \xrightarrow{NaBH_3CN} CH_3(CH_2)_3CO_2C_2H_5 \quad (88\,\%)$$

It is also of interest to note that alkyl halides are only slowly attacked by the electrophilic reducing agent, aluminium hydride, and therefore choice of this reagent will minimise unwanted carbon–halogen bond cleavage during reduction of other functional groups [(21) → (22)].

An extension of the cyanoborohydride method (8.11) provides a method for the direct conversion of primary alcohols into hydrocarbons: no improvement in yield is achieved by isolating the intermediate iodo-compound. Secondary and tertiary alcohols undergo elimination:

$$CH_3(CH_2)_9OH \xrightarrow{Ph_3\overset{+}{P}Me\ I^-} CH_3(CH_2)_9I \xrightarrow{NaBH_3CN}$$

$$CH_3(CH_2)_8CH_3 \quad (100\%) \quad (8.11)$$

Aryl halides react only slowly with lithium aluminium hydride, but organotin hydrides, which attack halides in a free-radical process, can be used to cleave aryl halides and other halides which cannot undergo S_N2 reactions:

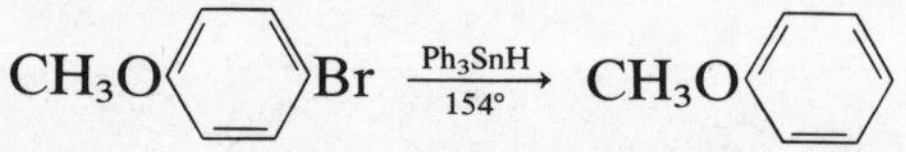

$$CH_3O\langle\rangle Br \xrightarrow[154°]{Ph_3SnH} CH_3O\langle\rangle \quad (90\%)$$

Carbon–halogen bonds are also cleaved by dissolving metal and electrochemical methods. The reaction is of particular significance in partial reduction of *gem*-dihalides where similar stereospecificities are observed:

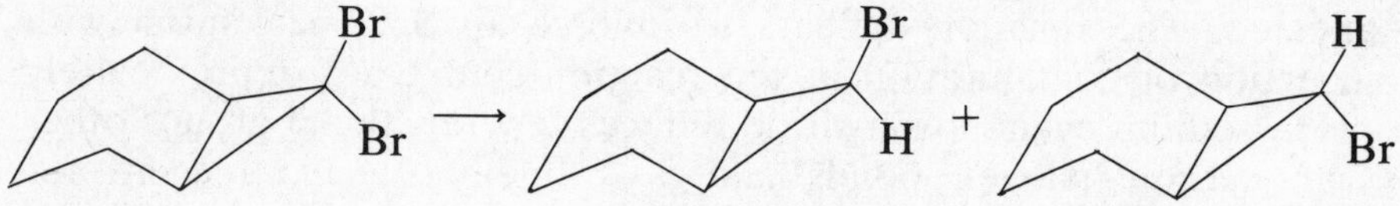

(23)   (24)

Under both electrolytic and dissolving metal conditions (23) and (24) are formed in the ratio of between 1:1 and 5:1.

For heteroatoms other than halogen, the reductive cleavage of greatest synthetic importance occurs at benzylic positions. This can be achieved by catalytic hydrogenation, complex metal hydride, or electron transfer methods. Hydrogenolysis is normally the method of choice, the order of reactivity being $PhCH_2-\overset{+}{N}{\lessgtr} > PhCH_2O— > PhCH_2N{\lessgtr}$. This renders the benzyl group very useful for protecting hydroxyl and amino groups (cf. Ch. 10).

$$HOCH_2CHCO_2H \xrightarrow[ethanol]{H_2,\ Pd/C} HOCH_2CH(NH_2)CO_2H \quad (90\%)$$
$$\underset{NHCH_2Ph}{|}$$

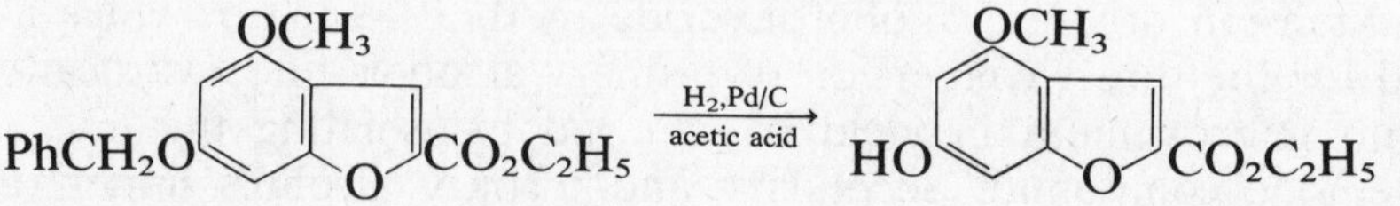

Other reductive cleavage reactions of importance are the Rosenmund reduction of acid chlorides (section 8.4.4) and desulphurisation using Raney nickel (section 8.4.3) already discussed, and reduction of primary and secondary alcohols *via* sulphonate esters. Reductive ring opening of epoxides will be considered in the following section.

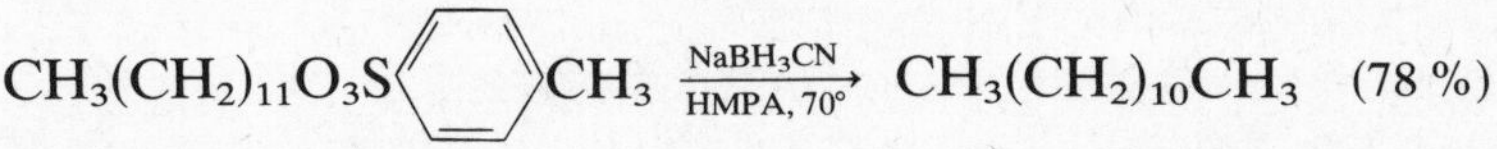
CH3(CH2)11O3S
CH3
NaBH3CN
HMPA, 70°
CH3(CH2)10CH3
(78 %)

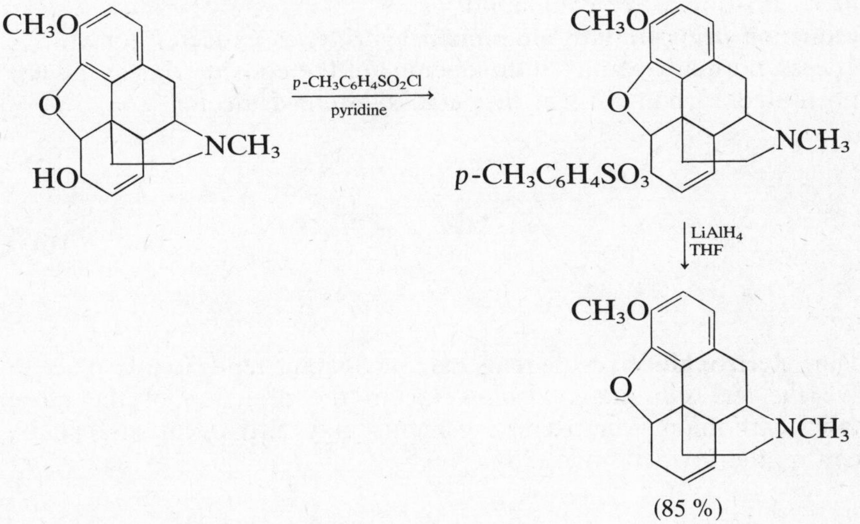
CH3O
O
HO
NCH3
p-CH3C6H4SO2Cl
pyridine
CH3O
O
NCH3
p-CH3C6H4SO3
LiAlH4
THF
CH3O
O
NCH3
(85 %)

8.6 Reductive ring-opening of epoxides

In **hydrogenolysis** of epoxides the following observations have been made:

(i) In acidic solvents hydrogenation takes place rapidly over **platinum** catalysts to give the ring-opened product(s) derived from the *more stable carbonium ion*:

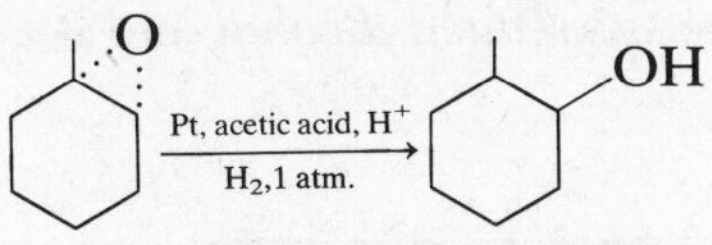
O
Pt, acetic acid, H+
H2,1 atm.
OH

cis and *trans*

(ii) **Palladium** on carbon is the most effective catalyst and *in neutral medium the more substituted alcohol is obtained*:

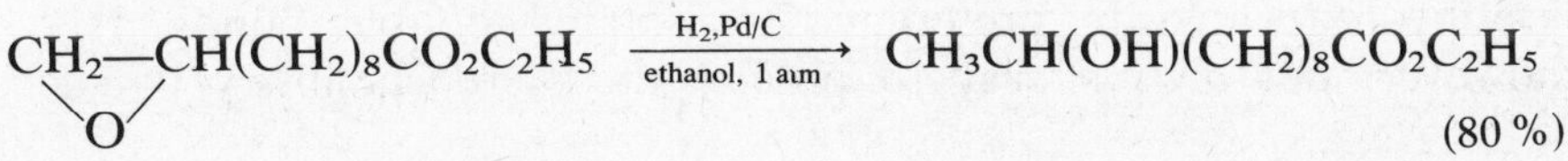
CH2—CH(CH2)8CO2C2H5
O
H2,Pd/C
ethanol, 1 atm
CH3CH(OH)(CH2)8CO2C2H5
(80 %)

(iii) **Raney nickel** requires a high pressure and temperature and *the less substituted alcohol predominates* in neutral solution, but in presence of base *the more substituted alcohol is formed.*

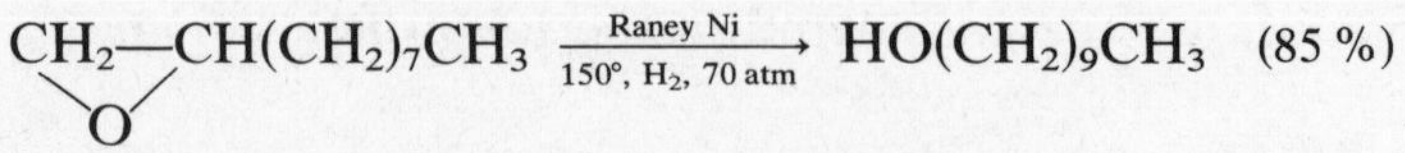

$$CH_2\!\!-\!\!CH(CH_2)_7CH_3 \xrightarrow[150°, H_2, 70\,atm]{Raney\ Ni} HO(CH_2)_9CH_3 \quad (85\,\%)$$

(iv) 1-Aryl-1,2-epoxides are opened under all conditions to give the 1-aryl-2-hydroxy compound.
Reduction using **lithium aluminium hydride,** as expected for an S_N2 process, normally results in the opening of the epoxide ring at the less substituted carbon to give *the more substituted alcohol.*

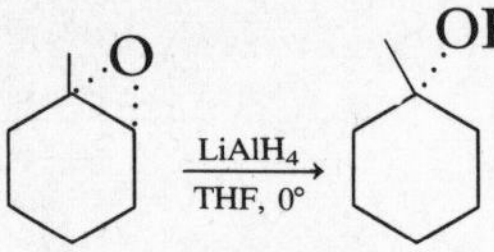

Using **electrophilic hydride reagents** ring opening tends to take place to give *the less substituted alcohol,* i.e. in the direction of the more stable carbonium ion. Rearrangements may also occur, as can be seen by the formation of (25).

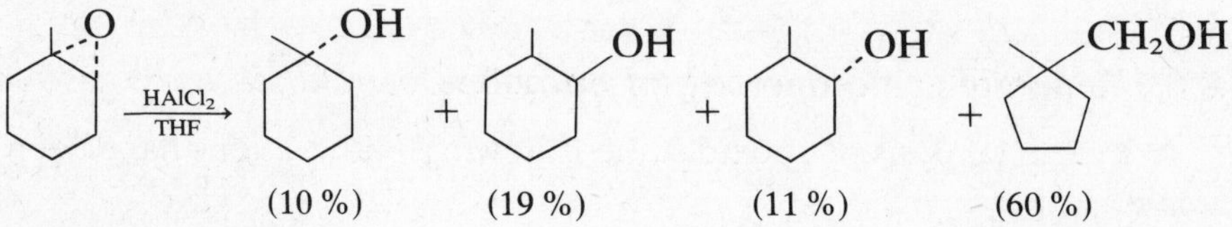

The reductive ring opening of epoxides with **lithium in ethylenediamine** also gives rise to *the more substituted alcohol* and is a superior method for hindered epoxides.

8.7   Reduction of α,β-unsaturated carbonyl compounds

Carbon–carbon double bonds are more readily hydrogenated than carbonyl double bonds or nitriles. Palladium is the preferred catalyst in this case, and in basic media carbonyl-conjugated double bonds are hydrogenated in preference to isolated double bonds. It is, however, not always easy to predict the stereochemistry of such hydrogenations:

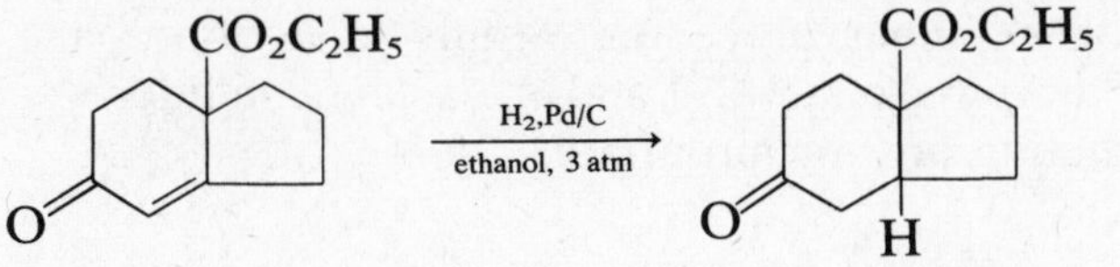

(83 %)

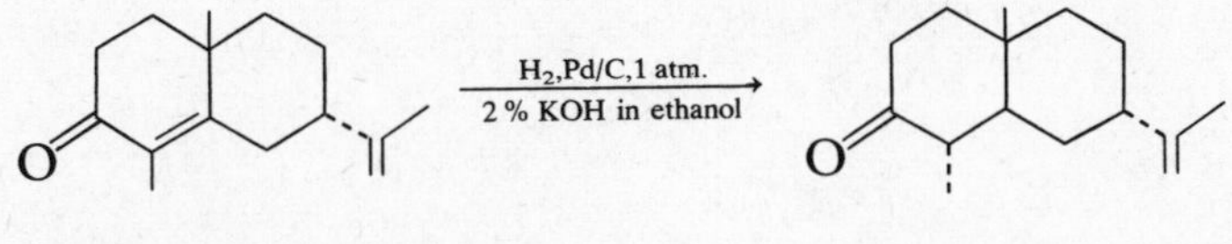

Although techniques such as inverse addition (i.e. addition of hydride to a solution of the α,β-unsaturated compound) at low temperatures have been successful in reducing such compounds to allylic alcohols, a more satisfactory reagent for this transformation is di-isobutylaluminium hydride:

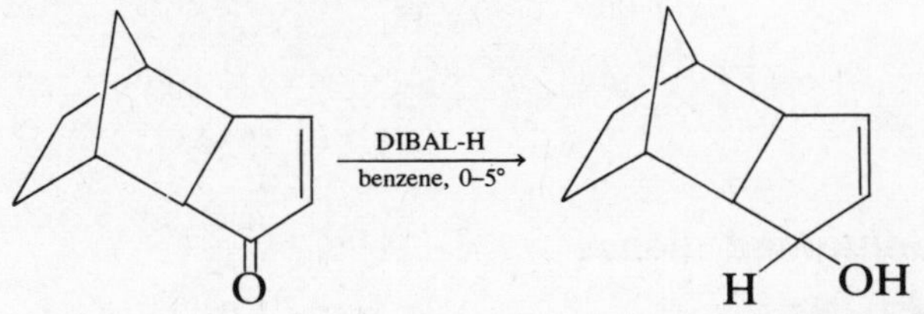

(90 %)

Dissolving metal reduction, on the other hand, leads to reduction of the carbon–carbon double bond. The stereochemistry of the product may, however, differ from that of the product of catalytic hydrogenation:

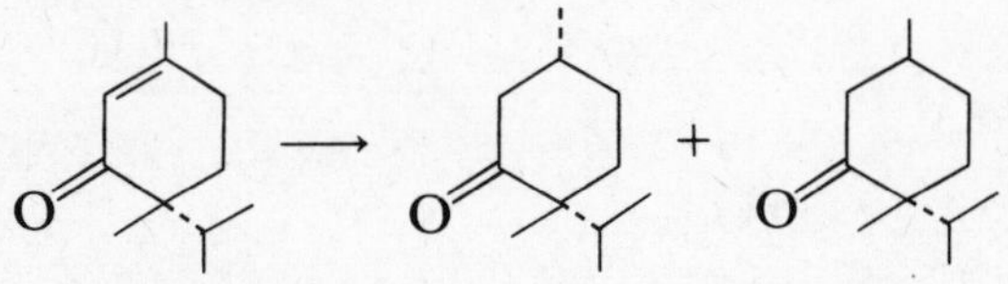

(~80 %)

(26)   (27)   (28)

$$(26) \xrightarrow[\text{methanol}]{\text{H}_2,\text{Pd/C,1 atm.}} (27) + (28)$$
$$(30\%) \quad (70\%)$$

$$(26) \xrightarrow[\text{liq. NH}_3]{\text{Li}} (27) + (28)$$
$$(75\%) \quad (25\%)$$

Reduction of α,β-unsaturated ketones to the alkene is usually effected by desulphurisation of the dithioketal by Raney nickel (8.12), since the Wolff–Kishner procedure results in cyclopropane formation (8.13) *via* the pyrazoline (29) and the acidic conditions of the Clemmensen reduction are inappropriate.

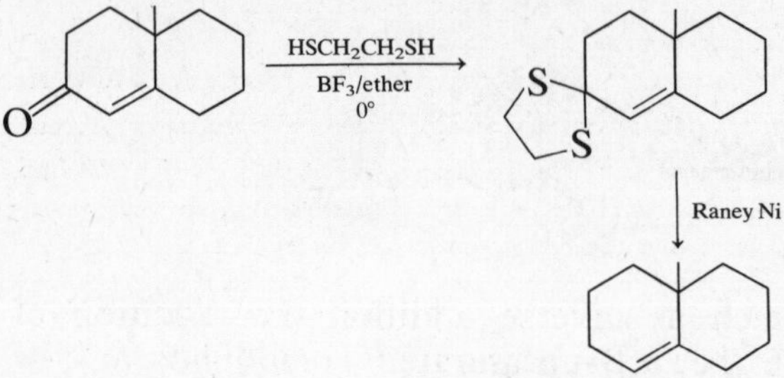

$$(8.12)$$

(60 %)

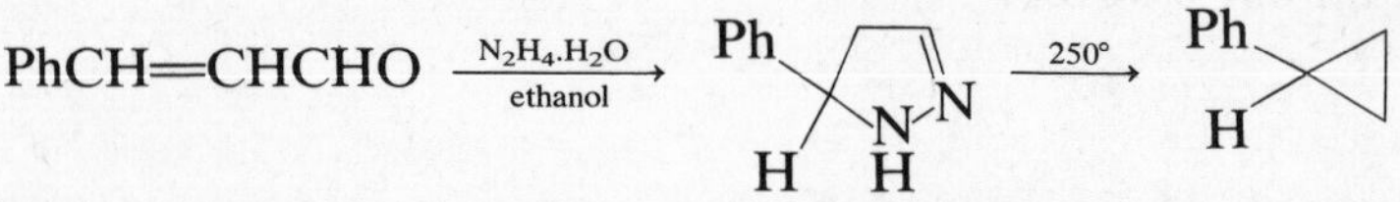

(50 %) (8.13)

(29)

8.8 Reduction of conjugated dienes

Dissolving metal reduction of 1,3-dienes results in 1,4-addition giving a mixture of *cis* and *trans* alkenes, the isomer ratio being temperature-dependent. Trapping experiments indicate that the initial radical anion (30) has the *cis*-configuration shown:

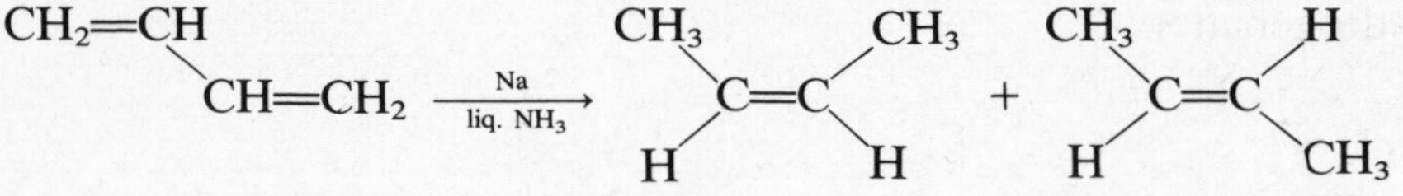

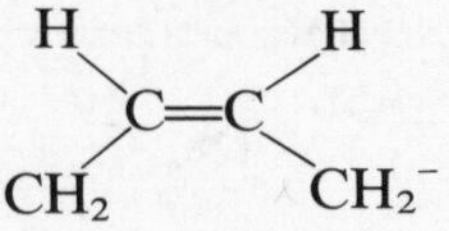

(30)

Conjugated dienes are completely hydrogenated using nickel, platinum or palladium catalysts. Analysis of partially hydrogenated butadiene reveals that but-1-ene, and *cis*- and *trans*-but-2-ene are present in amounts which depend on the catalyst used.

8.9 Reduction of aromatic and heteroaromatic compounds

Catalytic hydrogenation of benzenoid compounds usually requires high-pressure conditions and in these circumstances other groups such as olefinic double bonds and carbonyl groups will also be reduced. Benzenoid compounds except those with one or more electron-withdrawing substituents are not affected by hydride reagents. The only reduction of benzenoid compounds which will be considered here is the dissolving metal reduction (**Birch reduction**) using lithium or sodium in liquid ammonia. The product of this reduction in the case of benzene is cyclohexa-1,4-diene. As expected, electron-withdrawing substituents facilitate the reaction whereas anisole is only reduced with difficulty.

In the case of compounds with electron-withdrawing substituents, reduction takes place at the carbon bearing the substituent (8.14), but with electron-donating substituents reduction occurs at an *ortho*-carbon (8.15). This can be rationalised in terms of the relative stabilities of the intermediate anion radicals (31) and (32). In bicyclic compounds the ring of lower electron density is reduced (8.16):

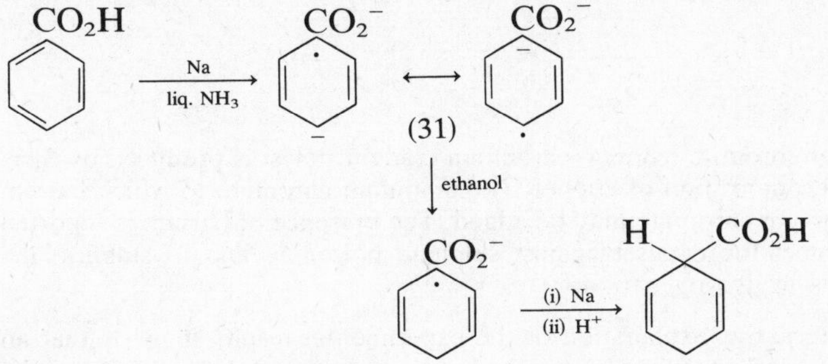

$$(90\,\%) \quad (8.14)$$

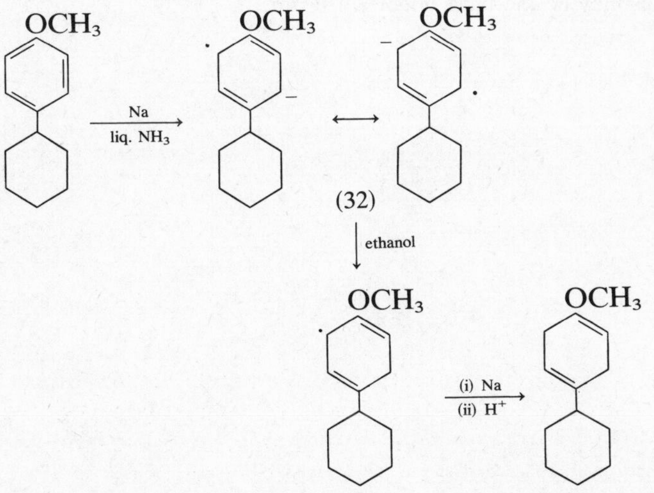

$$(85\,\%) \quad (8.15)$$

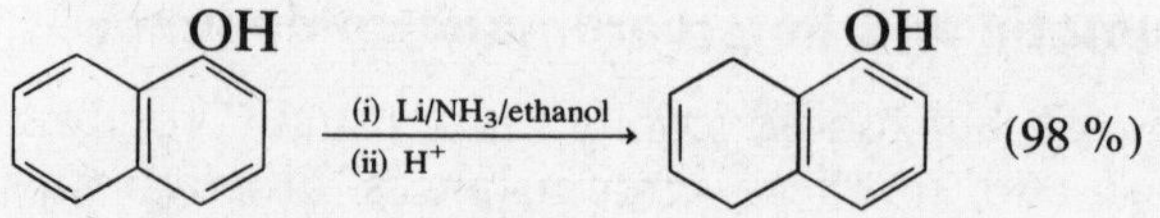

$$(8.16)$$

High-pressure hydrogenation of pyridinium salts results in the formation of piperidines. These can also be formed by reduction using sodium borohydride. 1,2-Dihydro- and 1,2,5,6-tetrahydropyridines can also be isolated depending on the reaction conditions used. Dissolving metal reduction of pyridines in presence of a proton donor also leads to the formation of piperidines: in this case, however, partial reduction leads to the formation of 1,4-dihydropyridines which are converted on hydrolysis into 1,5-dicarbonyl compounds. This is a reaction of considerable synthetic significance:

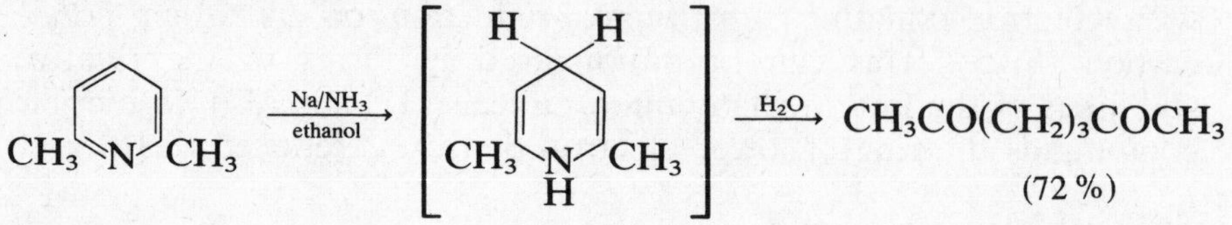

Notes

1.   Copper chromite (copper–chromium oxide) catalyst is produced by thermal decomposition of copper(II) ammonium chromate to which barium ammonium chromate may be added. The presence of barium is reported to protect the catalyst against sulphate poisoning and to stabilise the catalyst to hydrogenation.

2.   An alternative explanation of the experimental results suggests that an eclipsing effect between the hydride reagent and the adjacent axial hydrogens may be the most important factor.

9   Oxidation

Oxidation is, of course, the opposite of reduction, and so the reactions to be described in this chapter should be, in principle at least, the reverse of those discussed in Chapter 8. Three distinct types of reduction have been described, *viz.* the addition of hydrogen to multiple bonds (using catalytic and non-catalytic methods); the substitution of a functional group by hydrogen; and one-electron addition to an electrophilic centre. The opposite of these processes should therefore constitute oxidations: elimination of hydrogen to form multiple bonds; substitution of hydrogen by a functional group; and one-electron abstraction from a nucleophilic centre.

Examples of all three types are well known, as will be apparent in the sections which follow. To these three types must be added a fourth, the addition of oxygen-containing reagents to multiple bonds (the reductive counterparts of such reactions are, in general, much less common, and have not been considered in Chapter 8), and to heteroatoms such as nitrogen, phosphorus, and sulphur.

9.1 General principles

9.1.1 Dehydrogenation

This heading covers a variety of reactions:

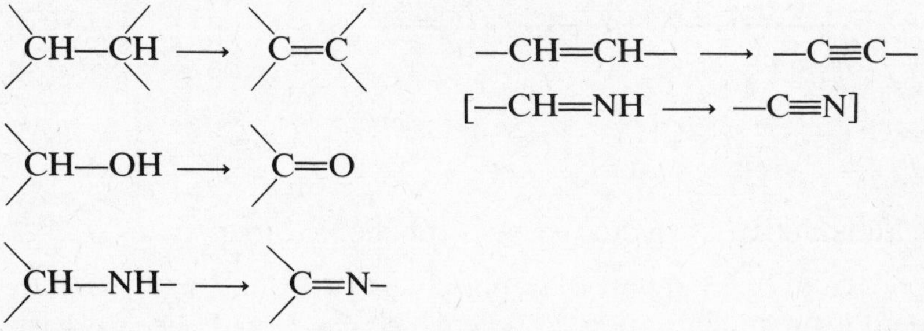

$$\text{CH—CH} \longrightarrow \text{C=C} \qquad\qquad \text{—CH=CH—} \longrightarrow \text{—C≡C—}$$

$$[\text{—CH=NH} \longrightarrow \text{—C≡N}]$$

$$\text{CH—OH} \longrightarrow \text{C=O}$$

$$\text{CH—NH-} \longrightarrow \text{C=N-}$$

It also embraces a variety of reaction types:

(i) *Catalytic dehydrogenation.* The capacity of metals like palladium for adsorption of hydrogen and for co-ordination of alkenes is used most frequently to effect hydrogenation of the alkenes. But *in the absence of added hydrogen* palladium may effect the dehydrogenation of an alkylated alkene in such a way that a conjugated diene is produced [reaction (9.1)]:

$$\begin{matrix} \diagdown \diagup \\ C=C \\ \diagup \quad \diagdown \\ \quad CH-CH- \\ \diagup \quad \diagdown \end{matrix} \xrightarrow{\text{Pd/C}} \begin{matrix} \diagdown \diagup \\ C=C \\ \diagup \quad \diagdown \\ \quad C=C \\ \diagup \quad \diagdown \end{matrix} \tag{9.1}$$

The reaction is particularly successful if the conjugated system produced is aromatic (cf. section 7.1.4.2, example 20; also section 9.2.4.1).

(ii) *Dehydrogenation by successive hydride and proton transfers.* These may be represented schematically by reactions (9.2) and (9.3):

$$\begin{matrix} \diagdown \\ CH-XH \end{matrix} \xrightarrow{-H^-} \begin{matrix} \diagdown \\ \overset{+}{C}\text{—}X\text{—}H \end{matrix} \xrightarrow{-H^+} \begin{matrix} \diagdown \\ C=X \end{matrix} \tag{9.2}$$

$$\begin{matrix} \diagdown \\ CH-XH \end{matrix} \xrightarrow{-H^+} -\overset{\overset{\displaystyle H}{|}}{C}\text{—}X^- \xrightarrow{-H^-} \begin{matrix} \diagdown \\ C=X \end{matrix} \tag{9.3}$$

Clearly these processes will be most effective if the intermediate cation or anion is stabilised, and if a good hydride acceptor (a strong electrophile) is present.

(iii) *Dehydrogenation by substitution-elimination and addition-elimination processes.* These are represented by reactions (9.4) and (9.5), and almost certainly constitute the most common dehydrogenation methods: for example, the vast majority of carbonyl-forming oxidations can be represented by reaction (9.4) [X = O]:

$$\begin{matrix} \diagdown \\ CH-XH \end{matrix} \longrightarrow -\overset{\overset{\displaystyle H}{|}}{C}\text{—}X\text{—}Y \xrightarrow{-HY} \begin{matrix} \diagdown \\ C=X \end{matrix} \tag{9.4}$$

$$-CH=CH- \xrightarrow{Y_2} -CHY-CHY- \xrightarrow{-2HY} -C\equiv C- \tag{9.5}$$

9.1.2   Substitution of hydrogen by a functional group

The opposite of hydrogenolysis (replacement of a functional group by hydrogen) is functionalisation, a topic which has already been discussed briefly in Chapter 2. It is unusual, however, to regard functionalisation in general as an oxidative process, except when hydrogen is replaced by an oxygenated function such as OH. The mechanism of these reactions may be either radical or ionic [reactions (9.6) and (9.7)], e.g.

$$RCH_2R' \xrightarrow{X^\cdot} HX + R\dot{C}HR' \xrightarrow{Y^\cdot \text{(or Y--X)}} R\overset{\overset{\displaystyle Y}{|}}{C}HR' \tag{9.6}$$

$$RCH_2R' \xrightarrow{-H^+} R\bar{C}HR' \xrightarrow{Y^+} R\overset{\overset{\displaystyle Y}{|}}{C}HR' \tag{9.7}$$

9.1.3 One-electron abstraction from a nucleophilic centre

The most common form of this reaction involves abstraction of one electron from an anion to give a radical, and subsequent dimerisation of the latter [reaction (9.8)]. Such a process has already been encountered in Chapter 4 in connection with organocopper derivatives (pp. 51 and 54). If the radical is stabilised by delocalisation, coupling may give an unsymmetrical dimer (cf. section 9.4):

$$2R{-}XH \xrightarrow{-2H^+} 2R{-}X^- \xrightarrow{-e^\cdot} 2R{-}X^\cdot \to R{-}X{-}X{-}R \tag{9.8}$$

9.1.4 Addition of oxygen-containing reagents to multiple bonds and heteroatoms

Two types of reactions are included in this section: the hydroxylation of multiple bonds [e.g. reaction (9.9)] and the addition of oxygen (usually from a peracid) to a heteroatom carrying a donatable lone pair [reaction (9.10)]:

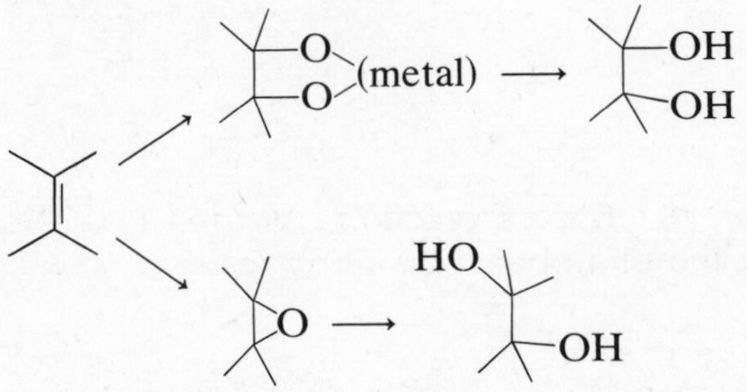

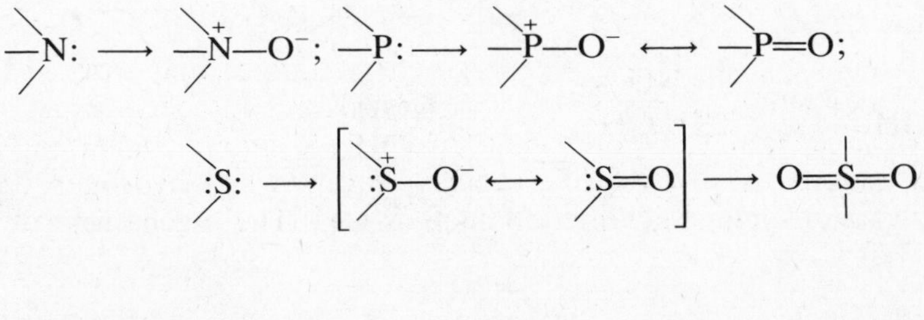

9.2   Oxidation of hydrocarbons

This is, of course, an enormously important area of industrial interest, in relation not only to fuels but also to rubber, plastics, foodstuffs, etc. The purpose of this section, however, is not to discuss these but to concentrate on some synthetically useful oxidations from a 'laboratory' viewpoint.

9.2.1   Alkanes and alkyl groups

Oxidation, like other reactions of alkanes, follows a radical mechanism [reaction (9.6)], in which abstraction of a hydrogen atom is the first step. This is a useful synthetic procedure only if the abstraction occurs specifically at one position. Tertiary hydrogens, for example, are more easily abstracted than secondary or primary hydrogens, and some branched-chain alkanes may be oxidised to tertiary alcohols, e.g.

$$(C_2H_5)_3CH \xrightarrow[\substack{CH_3CO_2H.\\H_2SO_4}]{Na_2Cr_2O_7} (C_2H_5)_3COH \quad (41\%)$$

This generally compares unfavourably, however, with the Grignard method (section 4.1.2) for the preparation of tertiary alcohols.

Much more important from the synthetic standpoint is a group of oxidations in which the radical is generated by *intramolecular* hydrogen abstraction; a radical of the type (1) may rearrange, *via* a six-membered transition state, to (2) [reaction (9.11)]:

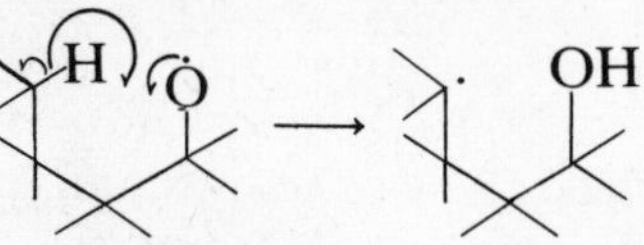

(1)   (2)

(9.11)

In the best-known of these, the **Barton reaction** [reaction (9.11a)], the radical is generated by the photolysis of a nitrite ester:

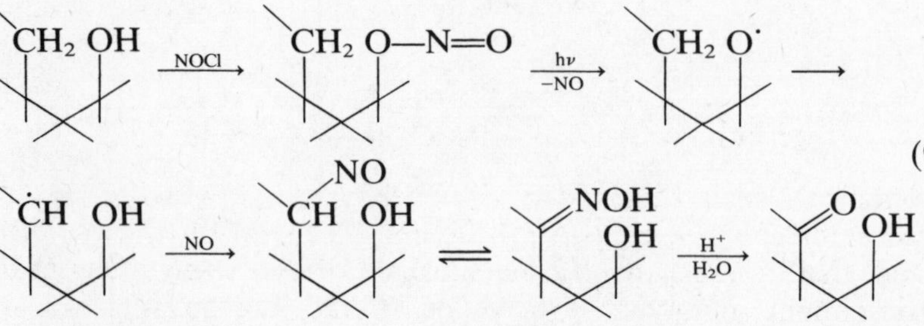

(9.11a)

The reaction has been applied with great success to selective oxidations in steroids, e.g. in the synthesis of the hormone aldosterone (3):

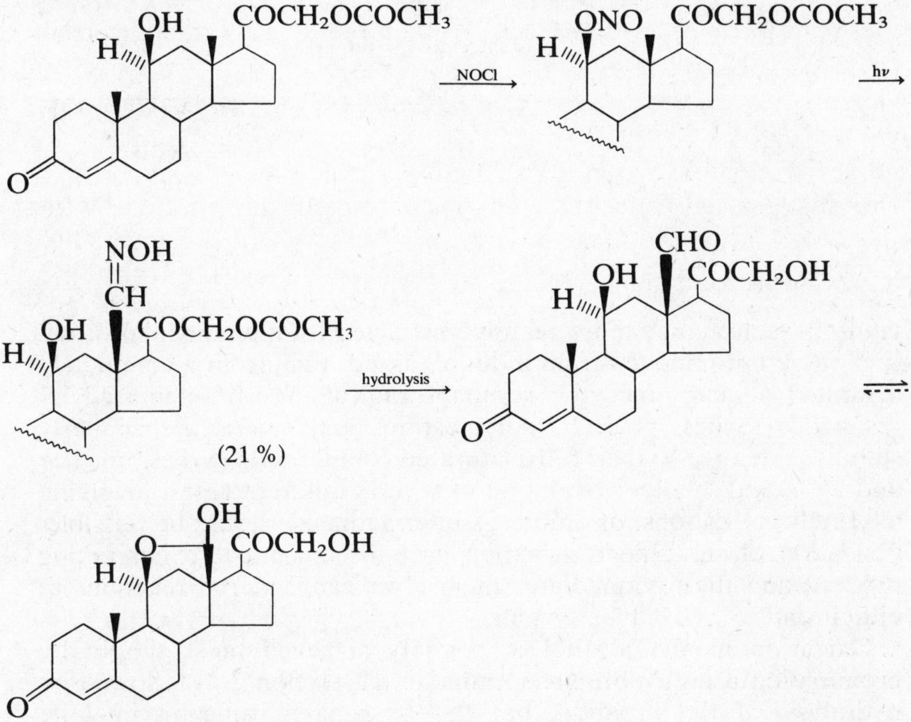

(3)

In the steroid field the intramolecular hydrogen transfer is, presumably, facilitated by the rigidity of the molecular skeleton and the 1,3-diaxial relationship of the interacting groups (cf. structure 4).

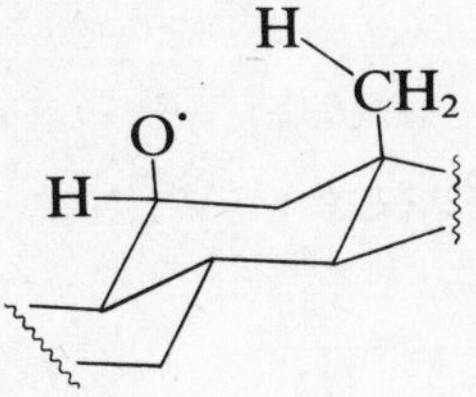

(4)

Oxy-radicals such as (1) can obviously undergo reactions other than intramolecular hydrogen abstraction, and it might therefore be expected that acyclic nitrites, lacking the rigidity of the steroid skeleton, might give very low yields in the Barton reaction. In practice, however, the yields in some cases are surprisingly high, e.g.

$$CH_3(CH_2)_7ONO \xrightarrow[\text{heptane}]{h\nu} CH_3(CH_2)_3\underset{\underset{NO}{|}}{CH}(CH_2)_3OH$$

(isolated as the dimer)

$$\xdashrightarrow{60°, \text{ 2 days}} CH_3(CH_2)_3\underset{\underset{NOH}{\|}}{C}(CH_2)_3OH$$

(26 %)

9.2.2 Allylic oxidation

Hydrogens are much more readily abstracted from an allylic position (i.e. one atom removed from a double bond) than from a completely saturated alkane, since the resultant radical may be stabilised, by resonance (Sykes, p. 302). Allyl cations and anions are similarly stabilised relative to their fully saturated counterparts (Sykes, pp. 104 and 267), and so allylic oxidation of several different types, involving allyl radicals, cations, or anions as intermediates, should be possible. Oxidation of the alkene grouping itself is, of course, a competing process, and the intermediates themselves can undergo reactions at either 'end' of the allylic system.

Oxidation to the alcohol is possibly achieved most simply by bromination using N-bromosuccinimide (cf. section 2.2) followed by hydrolysis of the bromide, but the oxygenated function may be introduced directly by the use of lead tetra-acetate, selenium dioxide, or a perester in presence of a copper(I) salt, e.g.

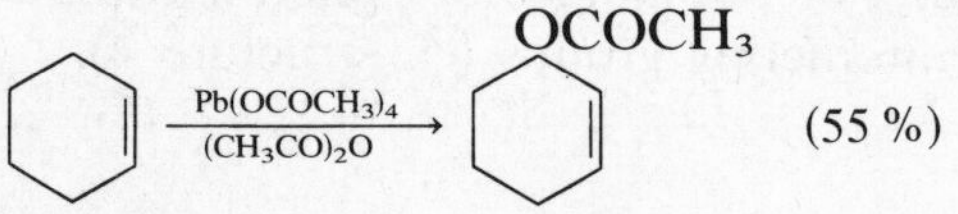

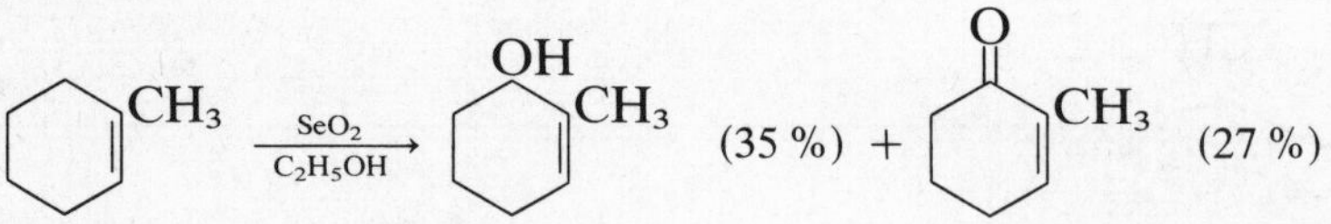

$$CH_3(CH_2)_5CH{=}CH_2 \xrightarrow[\text{Cu}_2\text{Cl}_2,\, 70°]{\text{CH}_3\text{CO}_2\text{OC(CH}_3)_3} CH_3(CH_2)_4\underset{\underset{OCOCH_3}{|}}{CH}CH{=}CH_2 \quad (78\%)$$

$$+ CH_3(CH_2)_4CH{=}CHCH_2OCOCH_3 \quad (11\%)$$

The detailed mechanisms of these reactions need not concern us here; it is sufficient to note that the first and second probably involve initial attack on the double bond and the intermediacy of species such as (5) and (6) or (7). The third example probably involves an allylic radical and then an allylic cation [reaction (9.12)]:

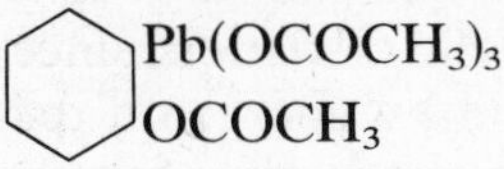

(5)

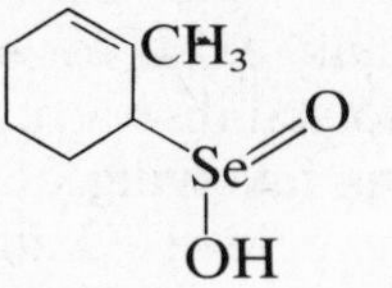

(6)

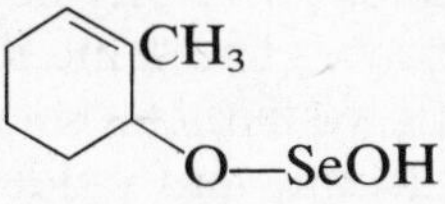

(7)

$$CH_3CO\!-\!O\!-\!OC(CH_3)_3 \xrightarrow{\text{heat}} CH_3CO\!-\!O^{\cdot} + {}^{\cdot}OC(CH_3)_3$$

$$CH_3CO\!-\!O^{\cdot} + Cu(I)^{+} \rightarrow Cu(II)^{2+} + CH_3CO_2{}^{-}$$

$$(CH_3)_3CO^{\cdot} + RCH_2CH\!=\!CH_2 \longrightarrow R\dot{C}HCH\!=\!CH_2 \xrightarrow{Cu(I)^{+}}$$

$$\overset{+}{R}CHCH\!=\!CH_2 \xrightarrow{CH_3CO_2{}^{-}} \overset{\overset{\displaystyle OCOCH_3}{|}}{RCHCH\!=\!CH_2} \quad (9.12)$$

More powerful, and less selective, oxidants carry allylic oxidation beyond the alcohol stage, and may also effect oxidation at the double bond, e.g.

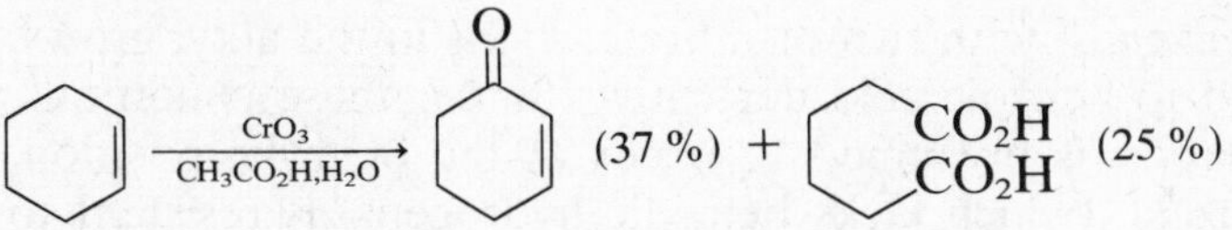

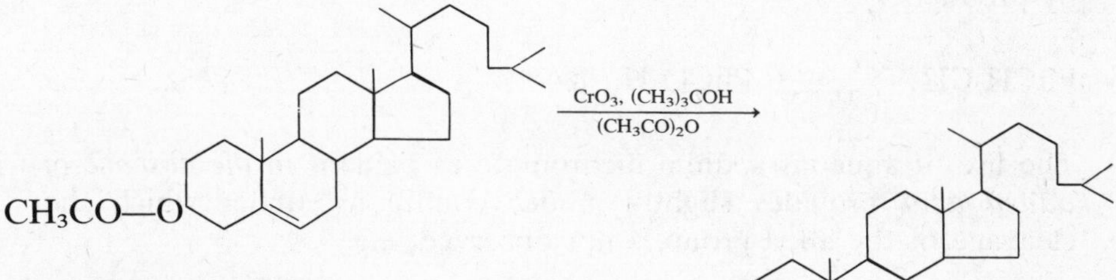

(cholesteryl acetate)

(90%)

Potassium permanganate is generally unsatisfactory for allylic oxidation, since it reacts preferentially with the double bond (cf. section 9.2.5).

9.2.3   Benzylic oxidation

Attention was drawn in the last section to the stabilisation of a radical, a cation, and an anion by an adjacent double bond, and to the consequent diversity of mechanistic routes by which allylic oxidation may occur. The same diversity exists for benzylic oxidation, since an aromatic system can serve to stabilise a radical or a charge on the benzylic carbon, and oxidations involving benzylic radicals, cations, and anions are all known (see below). Oxidation elsewhere in the molecule is not usually a serious problem in these reactions, since aromatic rings are in general oxidised only with difficulty.

Strong oxidants, such as potassium permanganate or chromic acid, oxidise benzylic carbons to the highest degree possible, e.g.

$$Ph_3CH \xrightarrow[CH_3CO_2H]{CrO_3} Ph_3COH \quad (85\%)$$

$$PhCH_2CH_3 \xrightarrow[H_2SO_4, H_2O]{CrO_3} PhCO_2H \quad (80\%)$$

Other alkylbenzenes, with two or more carbons in the alkyl group, are also oxidised to benzoic acid derivatives under these conditions. The initial oxidation is presumed to occur at the benzylic position, since t-butylbenzene (which lacks benzylic hydrogens) is resistant to oxidation, and since aryl alkyl ketones are occasionally isolated as by-products.

$$PhCH_2CH_3 \xrightarrow[H_2SO_4, H_2O]{CrO_3} PhCO_2H \quad (80\%)$$

The use of aqueous sodium dichromate as oxidant *in the absence of added acid* provides slightly milder conditions, under which the cleavage of the alkyl group is not observed, e.g.

$$PhCH_2CH_3 \xrightarrow[250°, pressure]{Na_2Cr_2O_7, H_2O} PhCOCH_3 \quad (50\%)^{[1]}$$

Fused-ring aromatic compounds give different oxidation products according to the reagent used. For example, chromium(VI) reagents oxidise naphthalenes to **naphthoquinones** in acidic media, whereas sodium dichromate in the absence of acid oxidises only substituents.

Potassium permanganate carries the oxidation further still, with ring cleavage and the formation of monocyclic dicarboxylic acids:

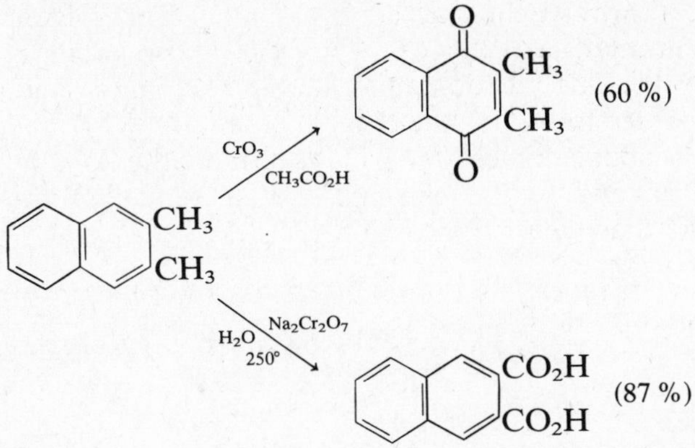

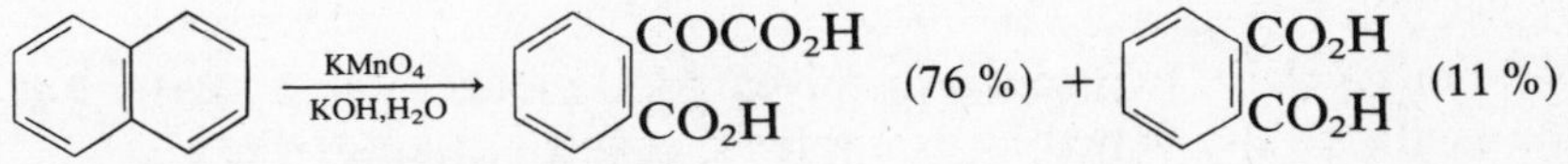

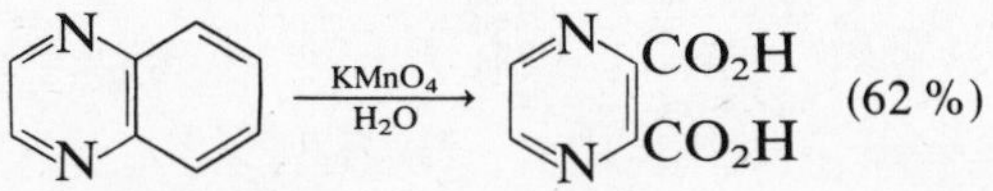

Oxidation of a benzylic centre to a level below the highest attainable presents more difficulties. $ArCH_3 \rightarrow ArCHO$ and $ArCH_2R \rightarrow ArCH(OH)R$, for example, are difficult because the products are more easily oxidised than the starting materials.

Several methods have been developed for the controlled oxidation of methyl groups to aldehydes. The simplest involves free-radical halogenation of the methyl group, and subsequent separation and hydrolysis of the dihalogeno-derivative [reaction (9.13)]:

$$\begin{array}{ccccccc} ArCH_3 & \xrightarrow{X_2} & ArCH_2X & \longrightarrow & ArCHX_2 & \longrightarrow & ArCX_3 \\ & & \downarrow{\scriptstyle \bar{O}H,H_2O} & & \downarrow{\scriptstyle \bar{O}H,H_2O} & & \\ & & ArCH_2OH & & ArCHO & & \end{array} \qquad (9.13)$$

e.g.

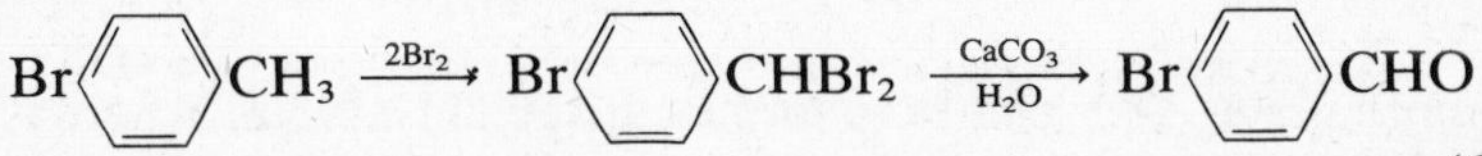

(68 % overall)

A second approach makes use of chromium(VI) reagents, under conditions which ensure that the aldehyde group is generated only in the final work-up. In the best-known of these oxidations, the **Étard reaction**, the oxidant is chromyl chloride, CrO_2Cl_2, in an inert solvent (CCl_4 or CS_2), but chromyl acetate, $CrO_2(OCOCH_3)_2$ (prepared *in situ* from chromium(VI) oxide, acetic anhydride, and sulphuric acid) may also be used successfully. In the Étard reaction (9.14) the intermediate is a 2:1 adduct of chromyl chloride and the toluene derivative, possibly (8), and in the chromyl acetate oxidation the primary product is the diacetate (9) [reaction (9.15)]:

$$ArCH_3 \xrightarrow{2CrO_2Cl_2} \left[\underset{(8)}{ArCH(O\overset{\displaystyle OH}{\underset{\displaystyle |}{C}}rCl_2)_2?} \right] \xrightarrow[\text{(ii) } H^+, H_2O]{\text{(i) } Na_2SO_3} ArCHO \qquad (9.14)$$

$$ArCH_3 \xrightarrow[\text{(CH}_3\text{CO)}_2\text{O}]{CrO_3} \underset{(9)}{ArCH(OCOCH_3)_2} \xrightarrow{H^+, H_2O} ArCHO \qquad (9.15)$$

Neither of these two reactions gives good yields in every case, but many synthetically valuable examples of each are known, e.g.

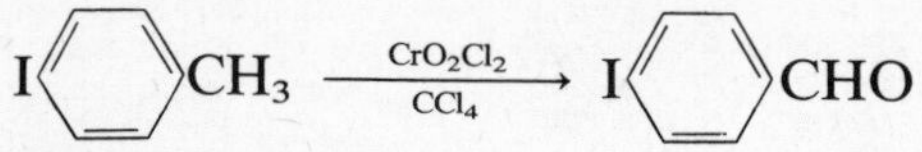

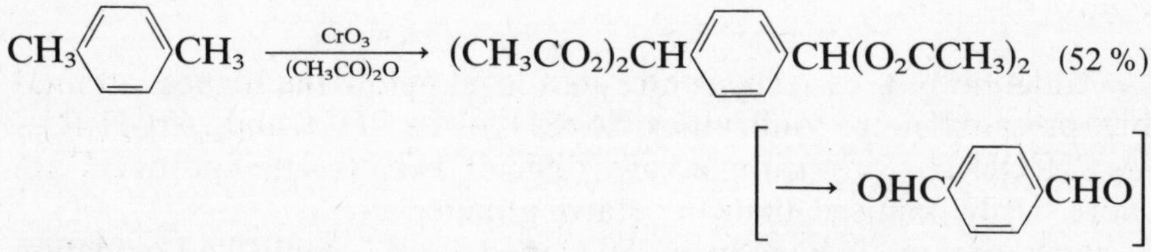

Oxidation of a benzylic centre to the alcohol may be achieved, as in reaction (9.13), by mono-halogenation followed by hydrolysis. Direct oxidation to the alcohol level is also possible using lead tetra-acetate (cf. allylic oxidation: section 9.2.2), e.g.

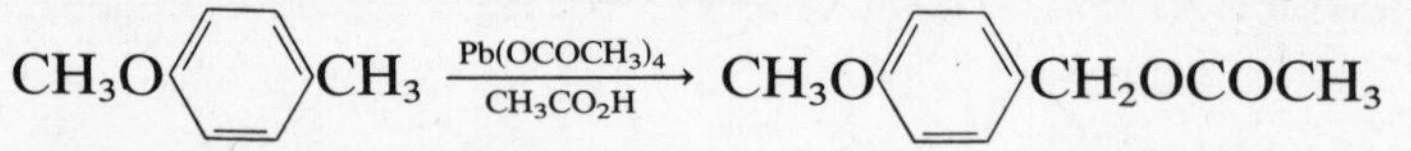

Similarly,

$$Ph_2CH_2 \rightarrow Ph_2CHOCOCH_3 \quad (71\%)$$

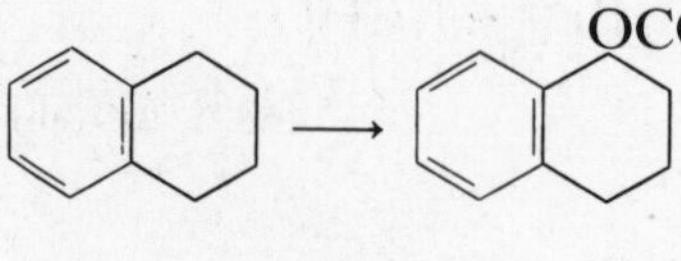

Autoxidation of benzylic compounds, although important industrially (cf. section 2.4.2), is of less value as a 'laboratory' method and is not considered further here.

The mechanisms of many of the reactions described in this section have not been established beyond doubt, but in most cases the initial step is probably abstraction of a *hydrogen radical* or *hydride ion* from the benzylic carbon.

If a benzylic centre can lose a *proton* easily, i.e. if it is a potential carbanion source, it may undergo condensation with a nitrosoarene, and the resulting anil may then be hydrolysed to a carbonyl compound and an arylamine (cf. section 6.3.3), e.g.

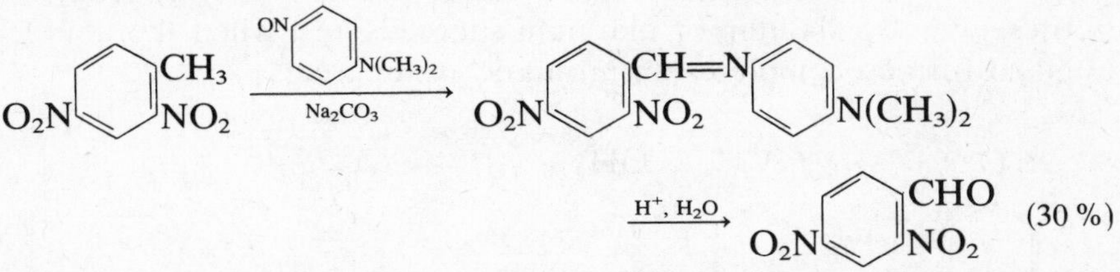

Nitrosation of a benzylic carbanion also leads to oxidation, e.g.

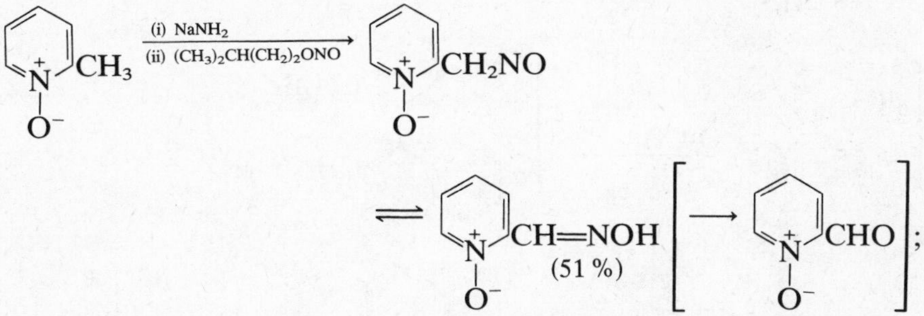

and a deprotonation step, (10) → (11), is the key to the oxidation of 2-methylpyridine to 2-pyridylmethanol[2]:

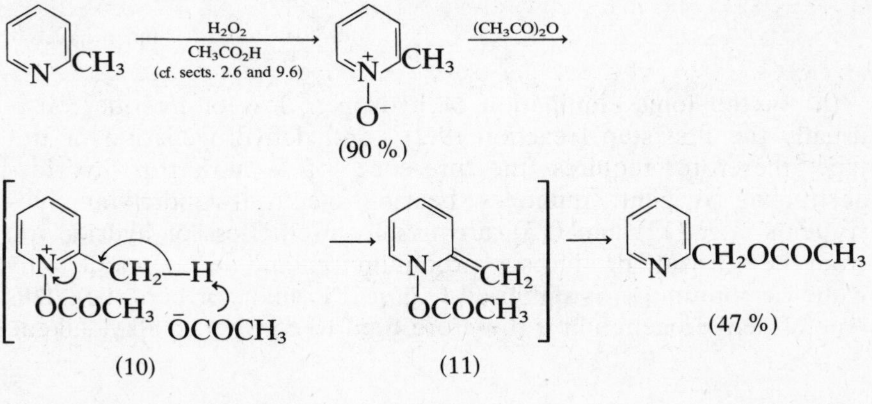

9.2.4 Dehydrogenation of alkanes, alkyl groups, and alkenes

9.2.4.1 Alkanes and alkyl groups

The formation of an alkene by dehydrogenation of the corresponding dihydro-compound cannot be regarded as a general reaction; it succeeds only if the double bond is introduced entirely regiospecifically, and this is possible only if the starting compound possesses the requisite structural features and/or functional groups. Nevertheless the three types of dehydrogenation outlined in section 9.1.1 are all well-known and widely-used methods for the introduction of carbon–carbon double bonds.

(i) As already mentioned (section 9.1.1) catalytic dehydrogenation in presence of palladium or platinum succeeds best when the double bond so formed completes an aromatic system, e.g.

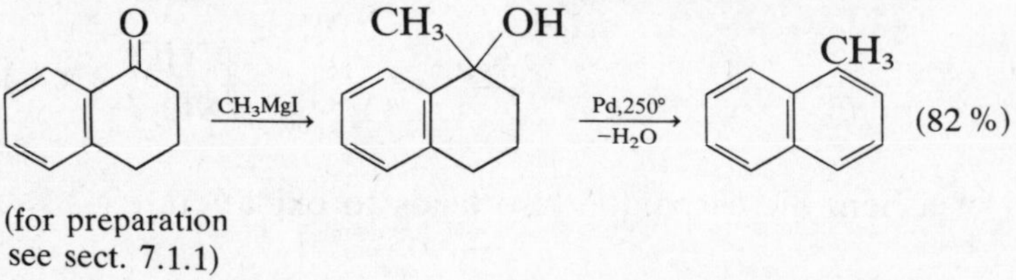

(for preparation
see sect. 7.1.1)

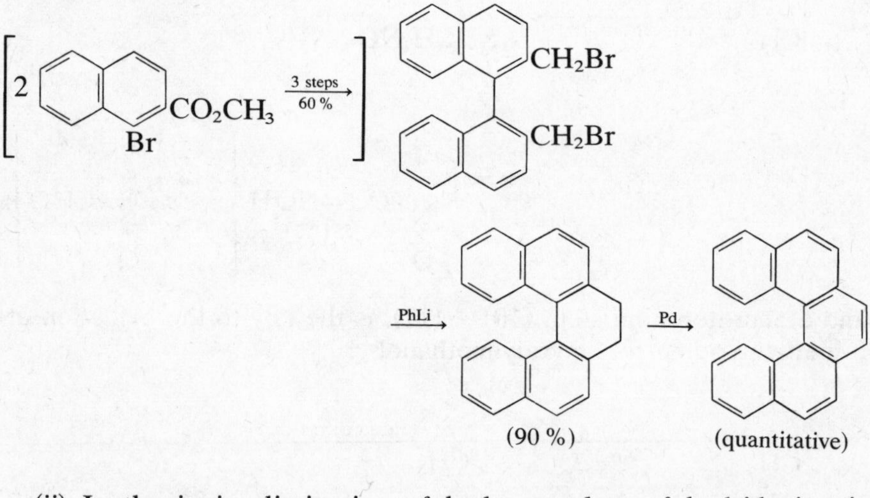

(ii) In the ionic elimination of hydrogen, loss of hydride ion is usually the first step [reaction (9.2)], and dehydrogenation of this type therefore requires the presence of a powerful hydride-abstracting reagent: quinones bearing electron-withdrawing substituents, e.g. (12) and (13), are usually used. Loss of hydride ion from the substrate produces a carbonium ion, and occurs readily only if the carbonium ion is stabilised (e.g. if it is allylic or benzylic). This type of dehydrogenation is therefore used to convert an alkyl–alkene into a conjugated diene, a diene into a triene, an alkylbenzene into a styrene derivative, and so on:

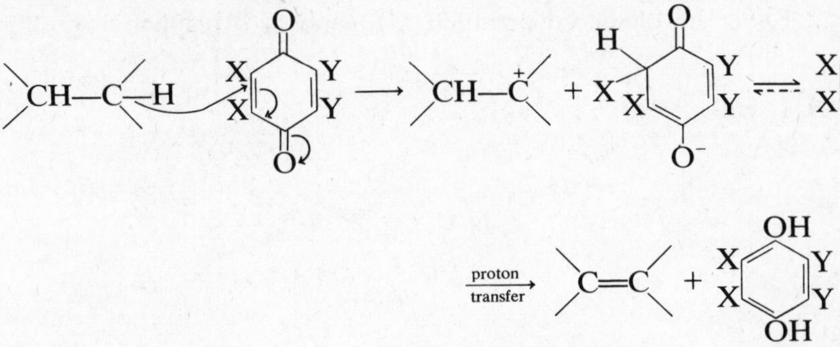

$$\xrightarrow[\text{transfer}]{\text{proton}} \quad \text{C=C} \quad + \quad \text{(9.15)}$$

chloranil   DDQ(**d**ichloro**d**icyanobenzo**q**uinone)

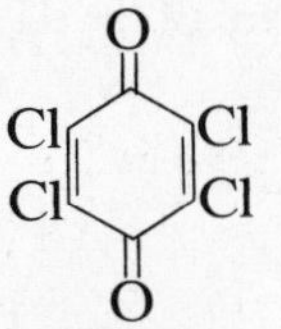

For example,

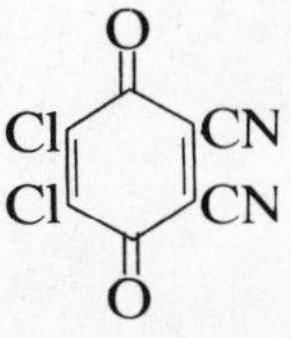

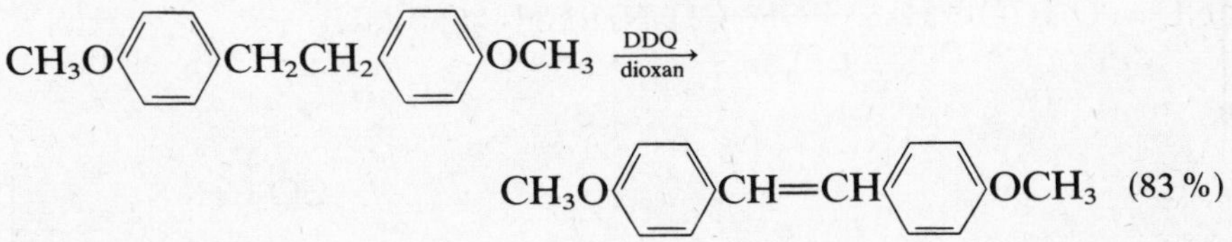

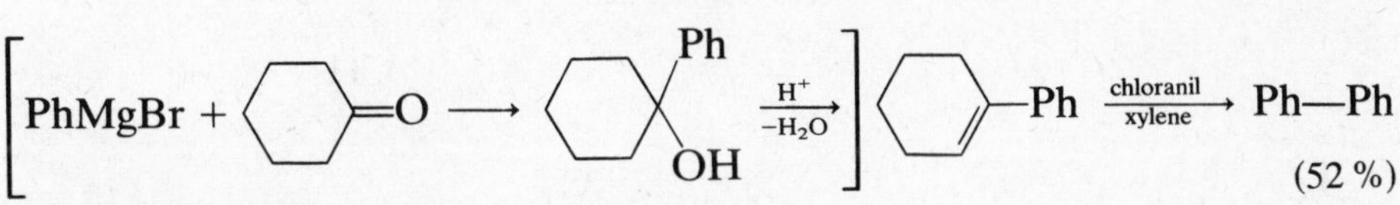

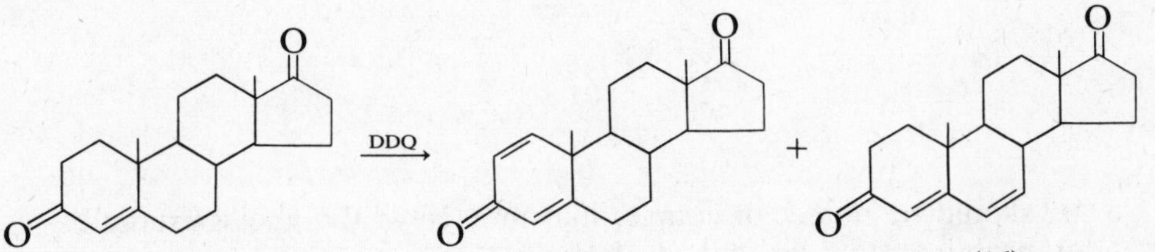

$$CH_3O\text{—}CH_2CH_2\text{—}OCH_3 \xrightarrow[\text{dioxan}]{\text{DDQ}}$$

$$CH_3O\text{—}CH=CH\text{—}OCH_3 \quad (83\%)$$

$$\left[PhMgBr + \bigcirc\text{=O} \longrightarrow \bigcirc\begin{matrix}Ph\\OH\end{matrix} \right] \xrightarrow[-H_2O]{H^+} \bigcirc\text{—Ph} \xrightarrow[\text{xylene}]{\text{chloranil}} Ph\text{—}Ph$$

$$(52\%)$$

[(15):(16) ratio depends on reaction conditions]

This last example is worthy of a little additional comment. Enolisation of (14) is the first step, and this may occur in either of two directions, giving (14a) or (14b). These then lose H^- and H^+ as shown, to give the observed products (15) and (16) respectively:

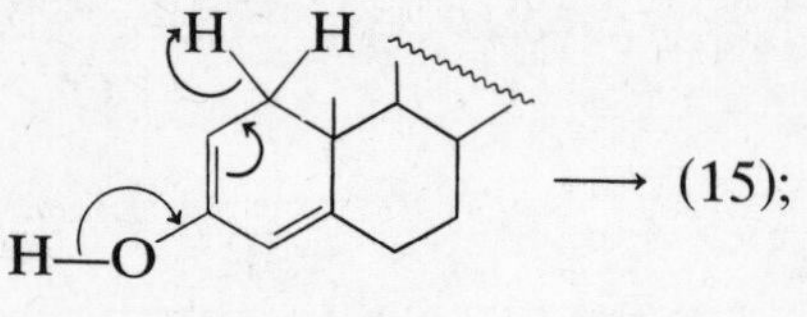
H
H
H—O
→ (15);

(14a)

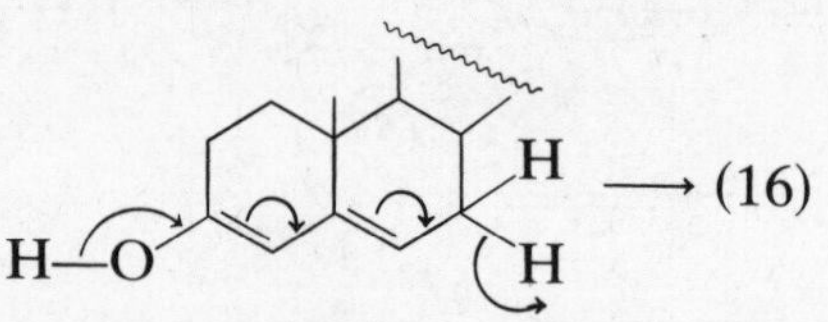
H
H
H—O
→ (16)

(14b)

(iii) The simplest of the substitution-elimination sequences is halogenation (usually bromination) followed by elimination of hydrogen halide, e.g.

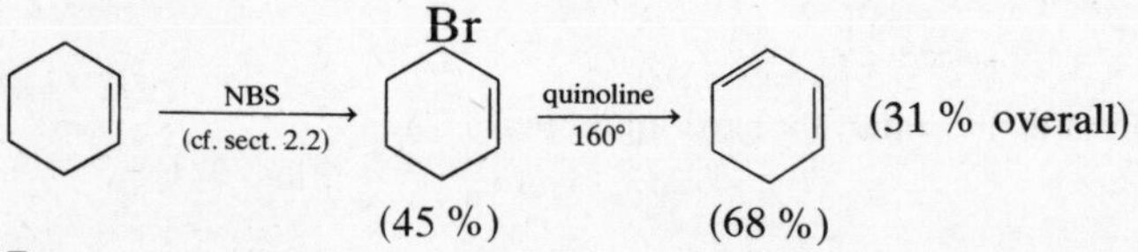
Br
NBS
(cf. sect. 2.2)
quinoline
160°
(31 % overall)
(45 %)
(68 %)

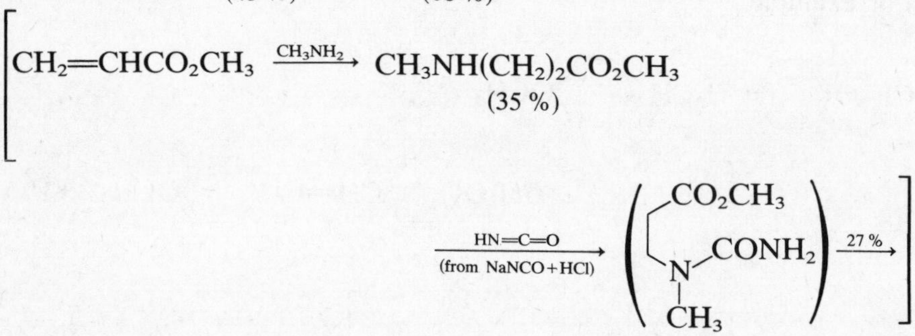
CH2=CHCO2CH3
CH3NH2
CH3NH(CH2)2CO2CH3
(35 %)
HN=C=O
(from NaNCO+HCl)
CO2CH3
CONH2
N
CH3
27 %

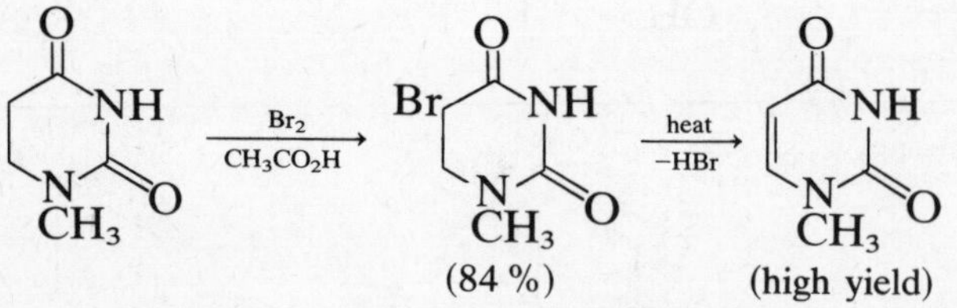
O
NH
N
CH3
O
Br2
CH3CO2H
Br
heat
−HBr
(84 %)
(high yield)

It should be noted, of course, that in each of the above examples, the bromination and dehydrobromination steps lead to a single product, and the success of the method is limited to cases in which such regiospecificity is observed.

9.2.4.2 Alkenes

Bromination followed by dehydrobromination is the usual method for the conversion of an alkene into an alkyne, e.g.

$$CH_2{=}CH(CH_2)_2CH{=}CH_2 \xrightarrow{2Br_2} BrCH_2CHBr(CH_2)_2CHBrCH_2Br \xrightarrow[NH_3(l)]{NaNH_2}$$
$$(95\%)$$

$$HC{\equiv}C(CH_2)_2C{\equiv}CH \quad (53\% \text{ overall})$$
$$(56\%)$$

$$PhCH{=}CH_2 \xrightarrow[CCl_4]{Br_2} PhCHBrCH_2Br \xrightarrow[THF]{NaNH_2} PhC{\equiv}CH \quad (87\% \text{ overall})$$
$$\qquad\qquad (97\%) \qquad\qquad (90\%)$$

9.2.5 Oxidative addition to alkenes

In this section we shall consider two types of addition: the formation of oxirans (epoxides) by addition of an oxygen atom across the double bond, and the formation of 1,2-diols, which is effectively the addition of a hydroxyl group at each end of the double bond.

9.2.5.1 Oxiran formation (epoxidation)

Oxirans (epoxides) are formed by the direct reaction of alkenes with peroxy-acids [reaction (9.16): cf. Sykes, p. 186], the stereochemistry of the alkene being retained in the product:

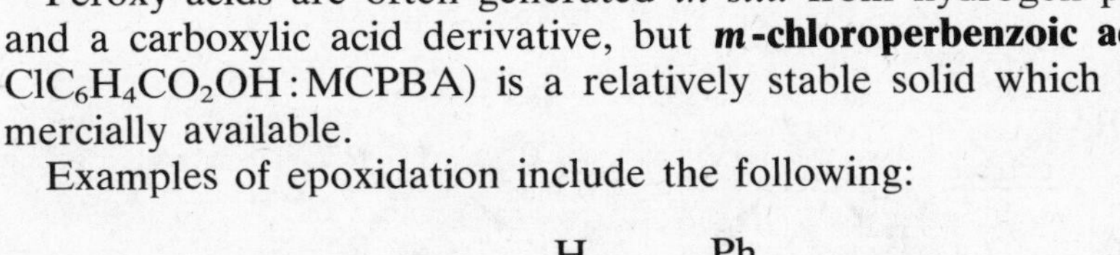

Peroxy-acids are often generated *in situ* from hydrogen peroxide and a carboxylic acid derivative, but **m-chloroperbenzoic acid** (*m*-ClC$_6$H$_4$CO$_2$OH: MCPBA) is a relatively stable solid which is commercially available.

Examples of epoxidation include the following:

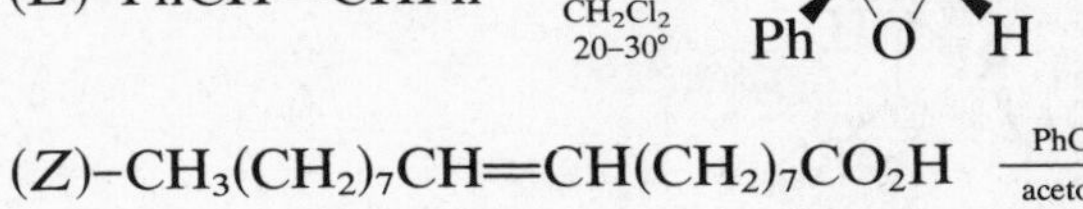

(E)–PhCH=CHPh $\xrightarrow[20-30°]{CH_3CO_2OH \atop CH_2Cl_2}$ (70%)

(Z)–CH$_3$(CH$_2$)$_7$CH=CH(CH$_2$)$_7$CO$_2$H $\xrightarrow[\text{acetone, 0-5°}]{PhCO_2OH^{[3]}}$

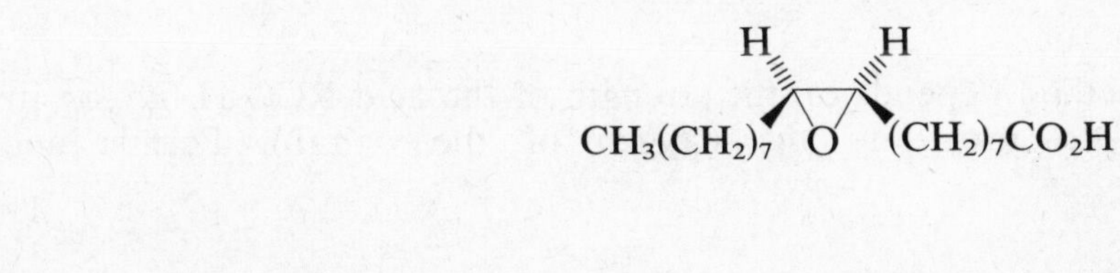

(62%)

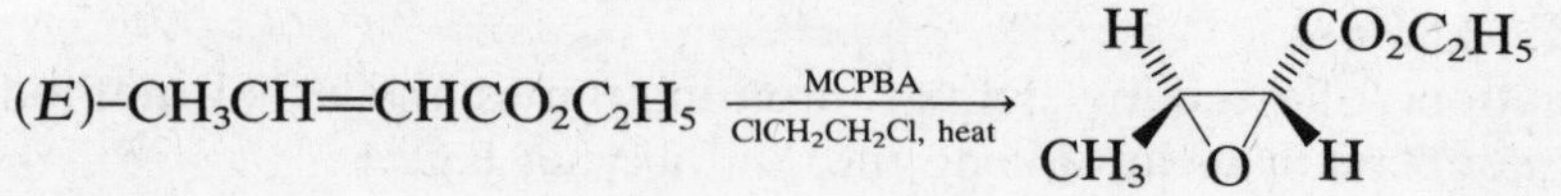

$(E)-CH_3CH{=}CHCO_2C_2H_5 \xrightarrow[\text{ClCH}_2\text{CH}_2\text{Cl, heat}]{\text{MCPBA}}$ (70 %)

This last example illustrates the relative difficulty in oxidising an electron-deficient alkene using a peroxy-acid. Alkenes which are conjugated to a -*M* group are epoxidised by reaction with *alkaline* hydrogen peroxide: this involves Michael-like addition of HO_2^- and subsequent loss of $\bar{O}H$ [reaction (9.17)]. Note that in this case the stereochemistry of the original alkene need not be retained in the product, since free rotation about the C_2—C_3 bond in the intermediate is possible:

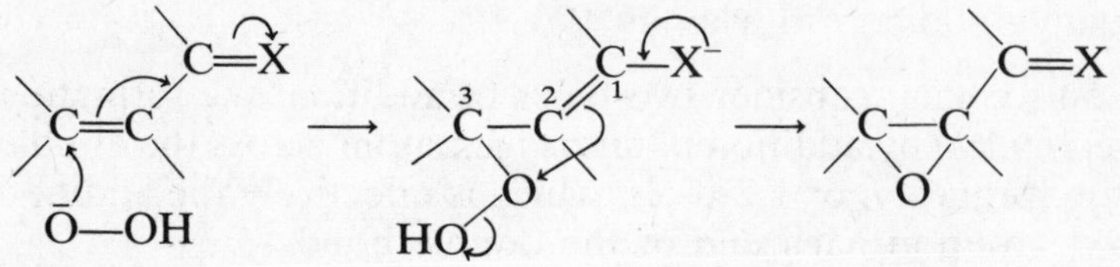

(9.17)

e.g.

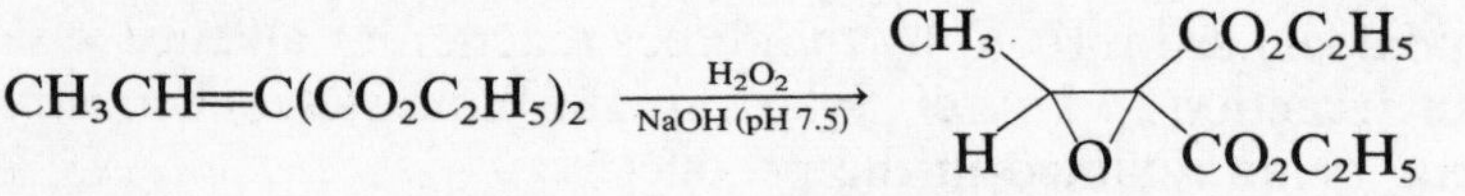

$CH_3CH{=}C(CO_2C_2H_5)_2 \xrightarrow[\text{NaOH (pH 7.5)}]{\text{H}_2\text{O}_2}$ (82 %)

9.2.5.2 1,2-Diol formation (hydroxylation)

Three general methods are commonly used for the conversion of alkenes into 1,2-diols; these complement one another to a certain extent and are thus all worthy of consideration. The first proceeds by way of the oxiran (cf. the preceding section), and leads to the *trans*-adduct [reaction (9.18); cf. Sykes, p. 187].

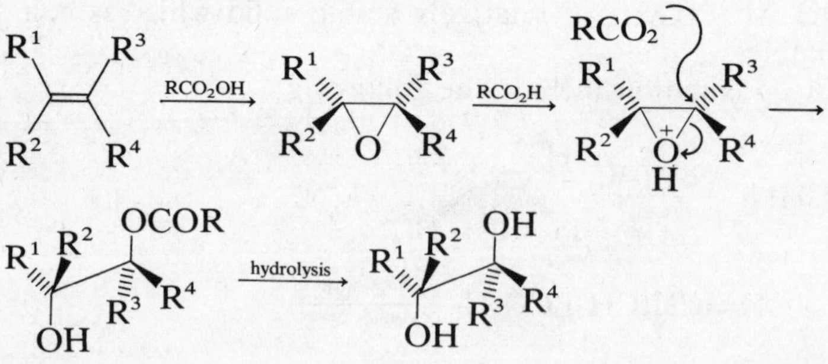

(9.18)

The method depends on the strength of the acid RCO_2H (i.e. on its ability to protonate the oxygen of the oxiran). Formic and trifluoroacetic acids are sufficiently strong to effect the ring-opening, and their peroxy-derivatives are thus commonly used for hydroxylation. Thus, for example,

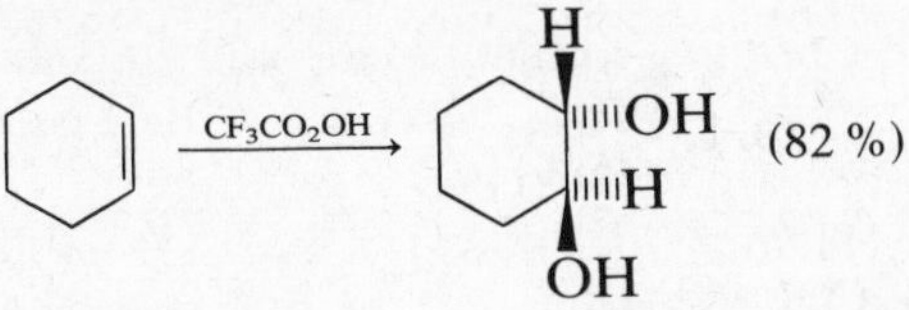

(Z)-$CH_3(CH_2)_7CH{=}CH(CH_2)_7CO_2H \xrightarrow{\ HCO_2OH\ }$

threo-$CH_3(CH_2)_7CHOHCHOH(CH_2)_7CO_2H$ (80 %)

The second involves the formation of a cyclic ester, by reaction of the alkene with potassium permanganate or osmium(VIII) oxide [reaction (9.19): cf. Sykes, pp. 185–6], and subsequent hydrolysis; this leads to the *cis*-adduct.

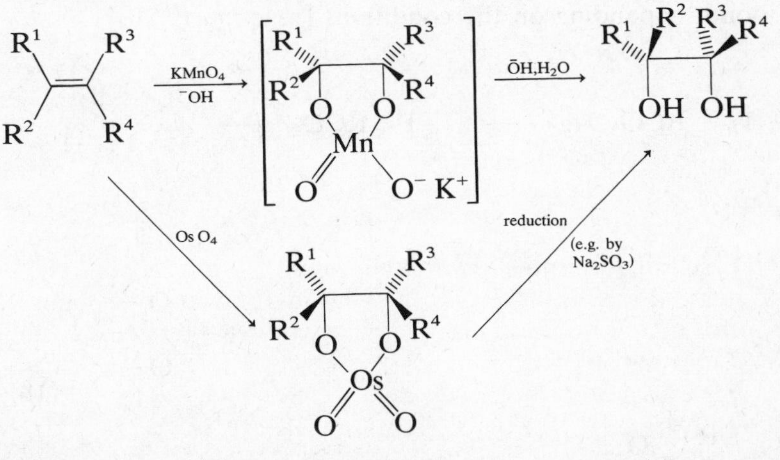

(9.19)

Osmium(VIII) oxide, however, is several hundred times more expensive than potassium permanganate, and is also toxic. It is therefore used only in small-scale reactions where the cost is (relatively) low and high yields are essential. Potassium permanganate, while inexpensive, may bring about further oxidation of the diol (cf. section 9.3.2), and may also oxidise other functional groups in a complex molecule; yields in permanganate hydroxylations, therefore, are not always high. Examples include:

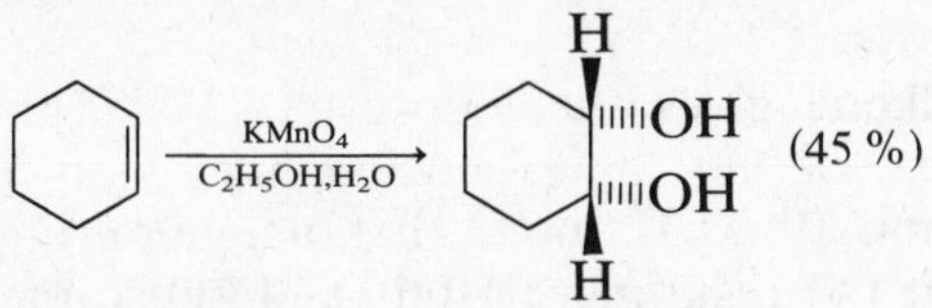

(Z)-$CH_3(CH_2)_7CH{=}CH(CH_2)_7CO_2H$ $\xrightarrow[\text{NaOH,H}_2\text{O}]{\text{KMnO}_4}$

$erythro$-$CH_3(CH_2)_7CHOHCHOH(CH_2)_7CO_2H$ (81%)

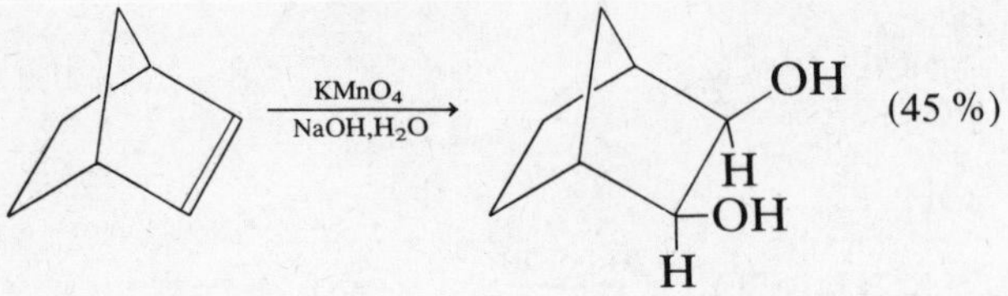

Note, in this last example, that only one of the possible *cis*-diols is isolated: this arises because *the permanganate attacks the alkene on the less hindered side.*

The third hydroxylation method is the **Prévost reaction**, in which the alkene is heated with iodine and a silver salt (usually the benzoate or acetate). This process may lead to *cis*- or *trans*-addition to the double bond, depending on the conditions [reaction (9.20)]:

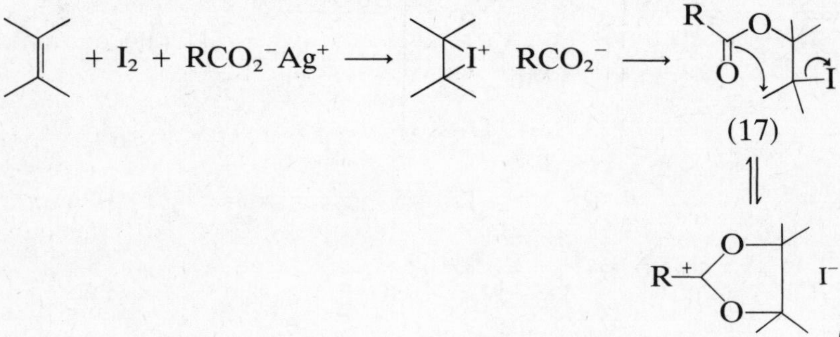

(17)

(18)

$$(18) \xrightarrow{\text{RCO}_2^-} \text{(structure)} \xrightarrow{\text{hydrolysis}} \text{HO} \quad \text{OH} \qquad (9.20a)$$

$$(18) \underset{}{\overset{\text{H}_2\text{O}}{\rightleftharpoons}} \text{(structure)} \longrightarrow \text{(structure)} \xrightarrow{\text{hydrolysis}} \begin{matrix}\text{HO}\\\text{HO}\end{matrix} \qquad (9.20b)$$

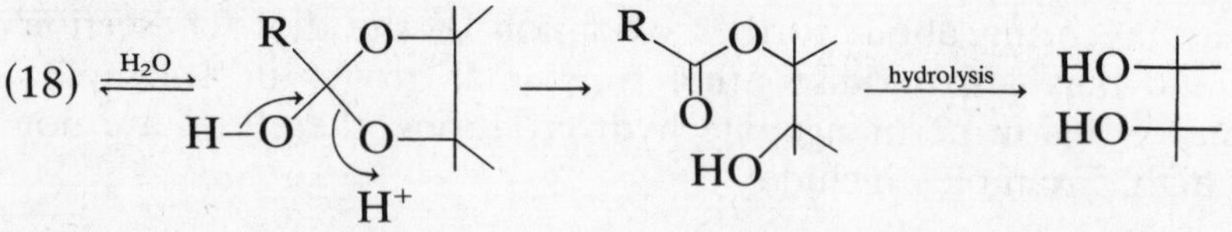

Initial *trans*-addition to the alkene gives the iodo-ester (17). This may then react, in the absence of any other nucleophile, with a second carboxylate ion [reaction (9.20a)]; note that neighbouring group participation ensures that the resulting diester (and hence the diol) retains the *trans* stereochemistry. In presence of water, however, the iodo-ester may be hydrolysed in a different way [reaction (9.20b)] to give the *cis*-diol.

The Prévost reaction under anhydrous conditions thus gives the same diol as the peroxy-acid reaction, e.g.

$$(Z)\text{-}CH_3(CH_2)_7CH{=}CH(CH_2)_7CO_2H \xrightarrow[\text{dry benzene}]{I_2,PhCO_2Ag}$$

$$\textit{threo}\text{-}CH_3(CH_2)_7CHOHCHOH(CH_2)_7CO_2H \quad (75\,\%)$$

and the corresponding reaction in presence of water (the so-called **Woodward modification**) usually gives the same product as the permanganate method, e.g.

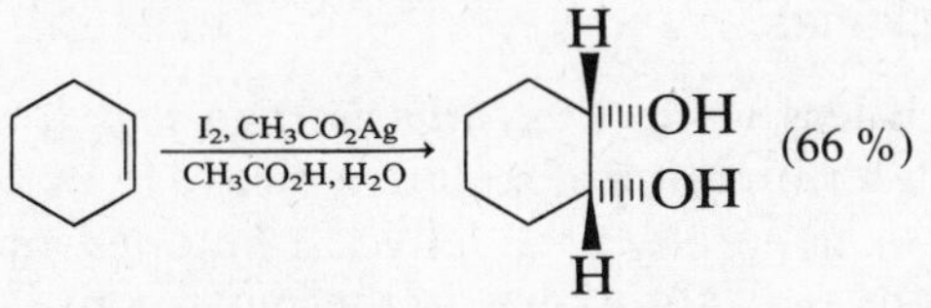

I₂, CH₃CO₂Ag
CH₃CO₂H, H₂O
H
OH
OH
H
(66 %)

The Prévost reaction would thus appear little more than an expensive alternative to other satisfactory hydroxylation methods. However, it holds an obvious advantage over the peroxy-acid method for the *trans*-hydroxylation of acid-sensitive compounds. The Woodward modification is important in cases where a single alkene may produce two *cis*-diols; whereas permanganate oxidation produces the less hindered diol (see above), the Woodward procedure leads to the less hindered iodonium ion and thence [cf. reaction (9.20b)] to the *more hindered diol*, e.g.

Os O₄
HO
HO
I₂,
CH₃CO₂Ag
CH₃CO₂H,H₂O
OH
OH

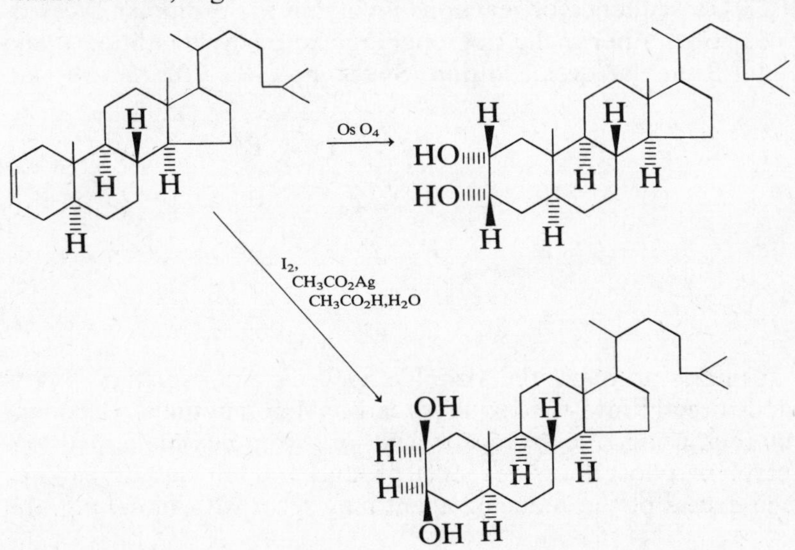

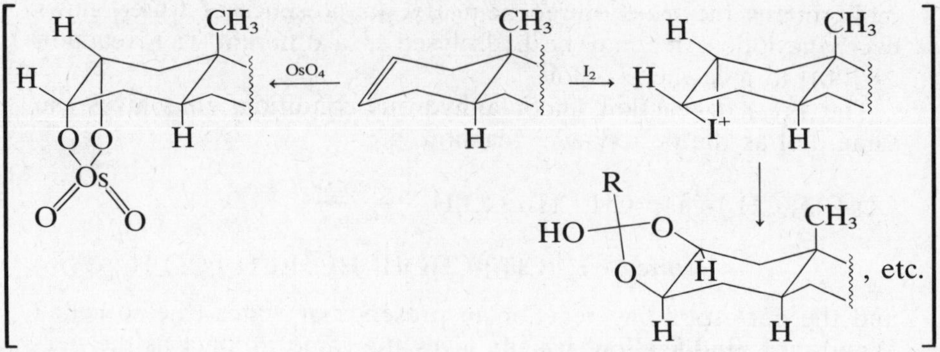

9.2.6 Oxidative cleavage of alkenes

This type of reaction is of much less value in synthesis than it is (or was) in degradative structural determination. Oxidative cleavage of a carbon–carbon double bond leads generally to aldehydes, ketones, or carboxylic acids, and it is seldom that an alkene is the most convenient source of any of these. We have already drawn attention, however (section 7.4.2), to the value of oxidative cleavage as a ring opening procedure, and there are also occasions (cf. section 14.6) when an alkene may be used as a latent carbonyl function in a multi-stage synthesis.

There are two principal methods for bringing about this cleavage. The first involves hydroxylation and subsequent oxidation of the diol (cf. section 9.3.2) using potassium permanganate, lead tetra-acetate, or a periodate. The second involves **ozonisation** of the alkene [reaction (9.21)], a sequence of reactions involving a 1,3-dipolar cycloaddition of ozone, a pericyclic ring opening ('retro-cycloaddition'), and a second 1,3-dipolar cycloaddition (Sykes, pp. 189–90):

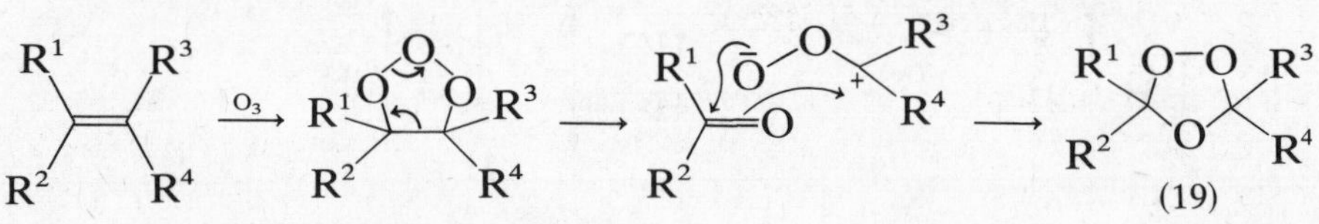

$$(9.21)$$

The primary product, the ozonide (19), is not isolated, but is converted directly into the required carbonyl compounds. *Reductive work-up* (e.g. using zinc and acetic acid, or a complex metal hydride, or a tervalent phosphorus reagent) produces aldehydes or ketones (although excess of the reducing agent may react with these: e.g. the use of lithium aluminium hydride gives alcohols):

(19) 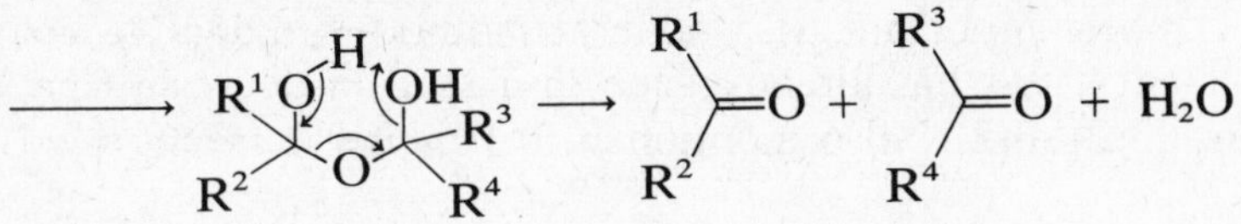 ⟶ ... ⟶ $\displaystyle \begin{array}{c} R^1 \\ \diagdown \\ R^2 \end{array}\!\!=\!O \; + \; \begin{array}{c} R^3 \\ \diagdown \\ R^4 \end{array}\!\!=\!O \; + \; H_2O$

or

$R_3P:$ ... ⟶ $\displaystyle \begin{array}{c} R^1 \\ \diagdown \\ R^2 \end{array}\!\!=\!O \; + \; \begin{array}{c} R^3 \\ \diagdown \\ R^4 \end{array}\!\!=\!O \; + \; R_3PO$

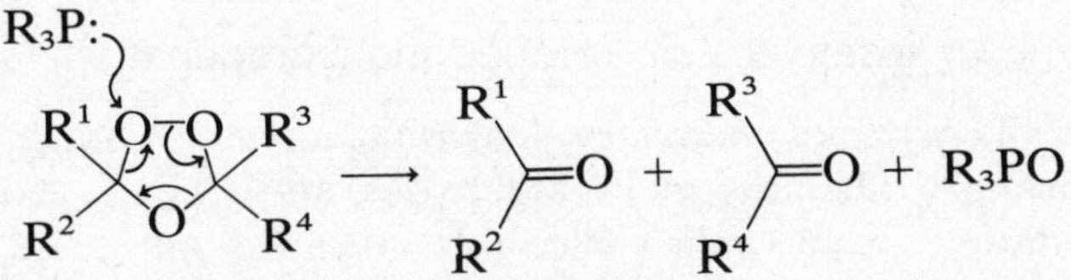

Oxidative work-up usually involves a peroxy-acid, and possibly involves hydrolysis as the first step. Under such conditions the products are carboxylic acids or ketones:

(19) 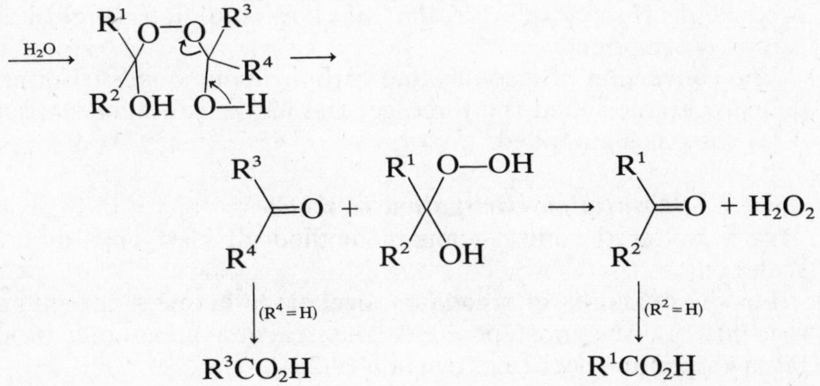 $\xrightarrow{\;H_2O\;}$

$\displaystyle \begin{array}{c} R^3 \\ \diagdown \\ R^4 \end{array}\!\!=\!O \; + \;$ ⟶ $\displaystyle \begin{array}{c} R^1 \\ \diagdown \\ R^2 \end{array}\!\!=\!O \; + \; H_2O_2$

$\Big\downarrow\, (R^4 = H)$ $\Big\downarrow\, (R^2 = H)$

R^3CO_2H R^1CO_2H

9.2.7 Oxidation of alkynes

The oxidation of a carbon–carbon triple bond is a much less-used synthetic procedure than the corresponding oxidation of an alkene. Some examples are known of the formation of 1,2-diketones by hydroxylation methods, e.g.

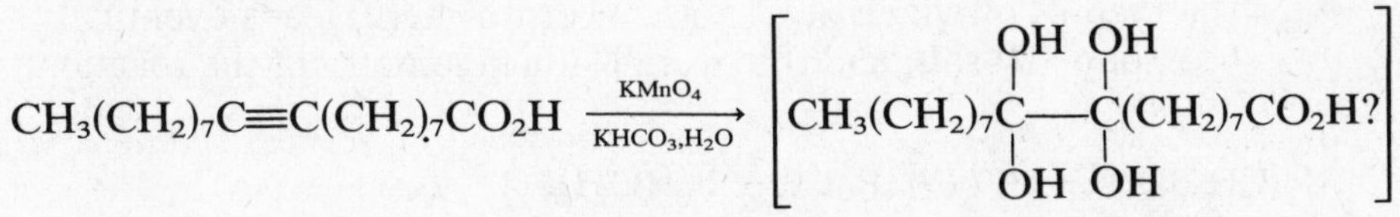

$$CH_3(CH_2)_7C\equiv C(CH_2)_7CO_2H \xrightarrow[KHCO_3,H_2O]{KMnO_4} \left[CH_3(CH_2)_7\underset{\underset{OH}{|}}{\overset{\overset{OH}{|}}{C}}\!\!-\!\!\underset{\underset{OH}{|}}{\overset{\overset{OH}{|}}{C}}(CH_2)_7CO_2H? \right]$$

$$\longrightarrow CH_3(CH_2)_7COCO(CH_2)_7CO_2H \quad (>90\,\%)$$

In many cases, however, complex mixtures result, and hydroxyla-tion of alkynes is thus not a *general* method for diketone preparation.

Much more important as a general method is **oxidative coupling** of alk-1-ynes: this has already been discussed in detail in Chapter 4 (section 4.3.2) and is also mentioned in Chapter 7 (section 7.1.5).

9.3   Oxidation of alcohols and their derivatives

9.3.1   Formation of aldehydes or ketones (dehydrogenation)

Most students of organic chemistry learn at an early stage that oxidation of primary alcohols gives aldehydes and then carboxylic acids, that oxidation of secondary alcohols gives ketones, and that tertiary alcohols are resistant to oxidation unless the conditions are sufficiently vigorous to produce C—C bond cleavage. The oxidations referred to in elementary courses are, as a rule, completely unselec-tive, involving reagents such as hot acidified potassium permanganate or hot chromic acid, but a great deal of work has gone towards the production of methods for the *selective* oxidation of alcohols to carbonyl compounds.

The conversion of alcohols into carbonyl compounds is formally a dehydrogenation, and the three general methods outlined in section 9.1.1 may all be applied.

9.3.1.1   *Substitution-elimination method*
This is by far the most common method, at least on a laboratory scale.

For the **oxidation of secondary alcohols to ketones**, chromium(VI) oxidants are the most popular. The reaction apparently proceeds *via* a chromium ester [e.g. reaction (9.22)]:

$$R_2CHOH \xrightarrow{H_2CrO_4} R_2\overset{H}{\underset{}{C}}{-}O{-}\overset{O}{\underset{O}{Cr}}{-}OH \longrightarrow R_2CO \qquad (9.22)$$

The chromium species produced by this first step [a Cr(IV) deriva-tive] is not the end-product. A further complicated sequence of redox steps (the details of which need not concern us here) leads eventually to a chromium(III) salt, and the overall stoicheiometry of the reaction is:

$$3R_2CHOH + 2CrO_3 \rightarrow 3R_2CO + 2Cr(OH)_3$$

or

$$3R_2CHOH + 2H_2CrO_4 \rightarrow 3R_2CO + 2Cr(OH)_3 + 2H_2O$$

or

$$3R_2CHOH + 2CrO_3 + 6H^+ \rightarrow 3R_2CO + 2Cr^{3+} + 6H_2O$$

A large number of variants of this oxidation are known. If the alcohol contains no other oxidisable functional group, and is not acid-sensitive, chromic acid in aqueous sulphuric or acetic acid is the most convenient, e.g.

$$\underset{\text{OH}}{CH_3CH_2\overset{|}{C}H(CH_2)_3CH_3} \xrightarrow[\text{H}_2\text{SO}_4,\text{H}_2\text{O}]{\text{K}_2\text{Cr}_2\text{O}_7} \underset{\text{O}}{CH_3CH_2\overset{\|}{C}(CH_2)_3CH_3} \quad (70\%)$$

$$\underset{\text{OH}}{Ph\overset{|}{C}HC(CH_3)_3} \xrightarrow[\text{CH}_3\text{CO}_2\text{H},\text{H}_2\text{O}]{\text{CrO}_3} PhCOC(CH_3)_3 \quad (64\%)$$

Alcohols containing double or triple bonds may be selectively oxidised using **Jones' reagent** (an aqueous solution of chromium(VI) oxide and sulphuric acid, in the correct stoicheiometric proportions); the alcohol, dissolved in acetone, is effectively titrated with the reagent at or below room temperature. Under these conditions the alcohol group is selectively oxidised, e.g.

$$\underset{\text{OH}}{Ph\overset{|}{C}HC\equiv CH} \xrightarrow[\text{acetone}]{\text{CrO}_3,\text{H}_2\text{SO}_4} \underset{\text{O}}{Ph\overset{\|}{C}C\equiv CH} \quad (80\%)$$

Similarly

$$\underset{\text{OH}}{CH_3CH=CH\overset{|}{C}HC\equiv CH} \longrightarrow \underset{\text{O}}{CH_3CH=CH\overset{\|}{C}C\equiv CH} \quad (75\%)$$

If acid-sensitivity is a problem, **chromium(VI) oxide in pyridine** may be the oxidant of choice. Alternatively the chromium oxide–pyridine complex may be isolated and used in another organic solvent such as dichloromethane. For example,

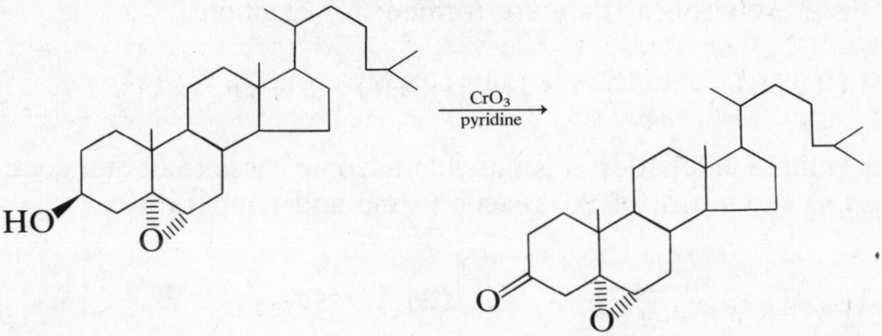

(60%)

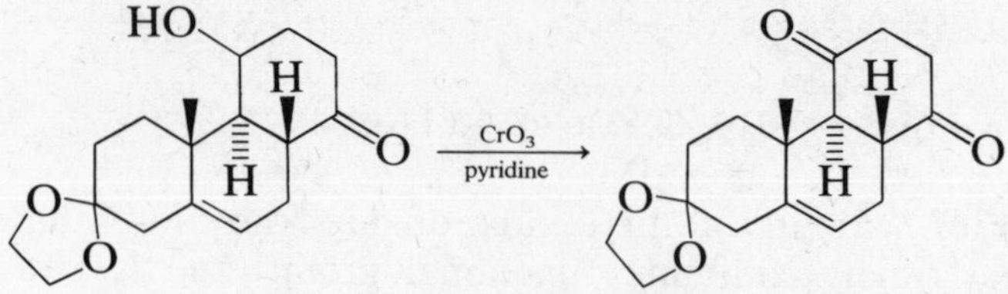

[cf. the synthesis of cortisone: section 14.6]

Among the newer chromium(VI) reagents which can be used for oxidations in organic solvents are **pyridinium chlorochromate**, $(C_5H_5\overset{+}{N}HCrO_3\ \bar{C}l$, PCC), prepared from chromium(VI) oxide, aqueous HCl, and pyridine, and **pyridinium dichromate** $[(C_5H_5NH^+)_2Cr_2O_7{}^{2-}$ PDC], prepared from chromium(VI) oxide, pyridine, and water. Examples of their use include the following:

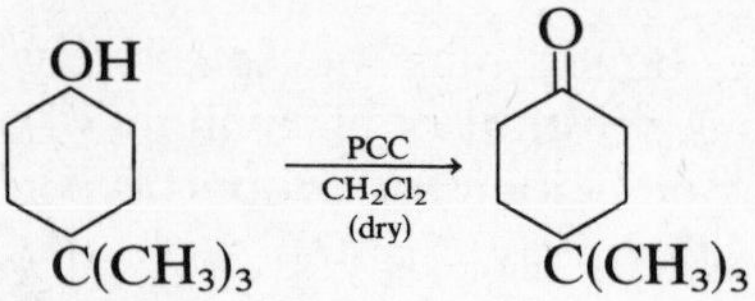

$$CH_3(CH_2)_4\overset{OH}{\underset{|}{C}}HCH{=}CH_2 \xrightarrow[\substack{CH_2Cl_2, \\ 25°}]{PDC} CH_3(CH_2)_4\overset{O}{\overset{\|}{C}}CH{=}CH_2 \ (80\%)$$

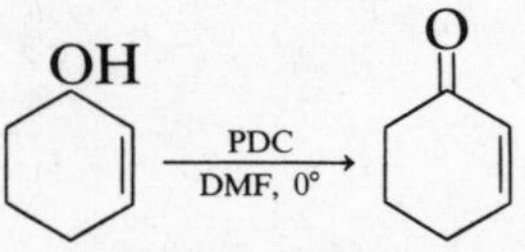

The **oxidation of primary alcohols to aldehydes** requires careful control of the reaction conditions, in order to prevent over-oxidation and the production of carboxylic acids. The use of chromium(VI) reagents for such oxidations is nevertheless widespread. The classical method for preparing the lower aliphatic aldehydes makes use of their relatively low boiling points, the products being distilled out of the oxidising solution as they are formed: for example,

$$CH_3CH_2CH_2OH \xrightarrow[H_2SO_4, H_2O, 95°]{K_2Cr_2O_7} CH_3CH_2CHO \ \ (\text{b.p. } 49°; \ 45\%)$$

For less volatile aldehydes, it is possible in some cases to obtain good yields by strict control of the reaction time and temperature, e.g.

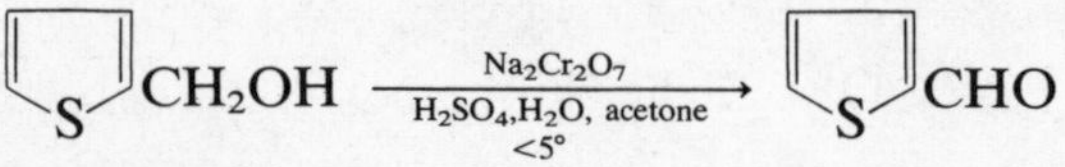

In other cases (particularly those which involve allylic or benzylic oxidation) the CrO_3/pyridine reagent is satisfactory, e.g.

$$CH_3(CH_2)_5CH_2OH \xrightarrow[CH_2Cl_2, 25°]{CrO_3/pyridine\ complex} CH_3(CH_2)_5CHO \quad (77\%)$$

$$PhCH{=}CHCH_2OH \xrightarrow[pyridine(solvent)]{CrO_3, 0°} PhCH{=}CHCHO \quad (81\%)$$

$$+ PhCH{=}CHCO_2H \quad (5\%)$$

PCC and PDC appear to be generally useful for this type of oxidation, e.g.

$$CH_3(CH_2)_4C{\equiv}CCH_2OH \xrightarrow[CH_2Cl_2, 25°]{PCC} CH_3(CH_2)_4C{\equiv}CCHO \quad (84\%)$$

Similarly

$$HO(CH_2)_6OH \rightarrow OHC(CH_2)_4CHO \quad (68\%)$$

$$CH_3(CH_2)_9OH \xrightarrow[CH_2Cl_2, 25°]{PDC} CH_3(CH_2)_8CHO \quad (98\%)$$

Similarly

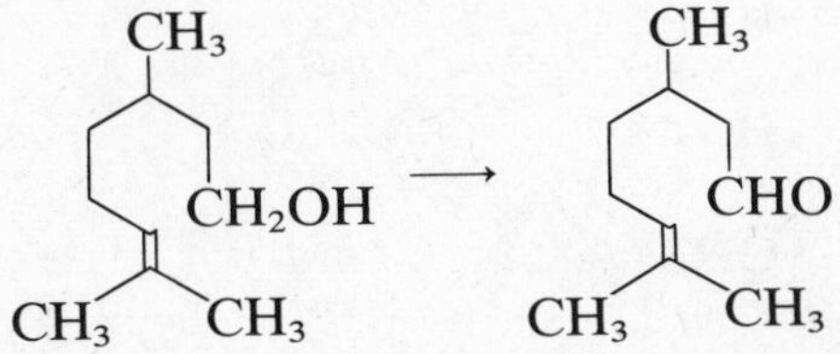

A second useful approach to the selective oxidation of alcohols to aldehydes and ketones makes use of **dimethyl sulphoxide as oxidant**. In principle the reaction is as shown below [reaction (9.23)]:

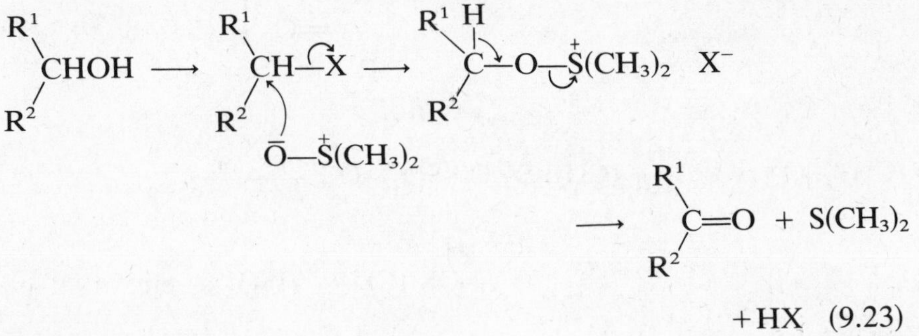

For example,

$$[CH_3(CH_2)_7OH \rightarrow]CH_3(CH_2)_7I \xrightarrow[\substack{150°, 4\ min \\ N_2\ atmosphere}]{DMSO,Na_2CO_3} CH_3(CH_2)_6CHO \quad (74\%)$$

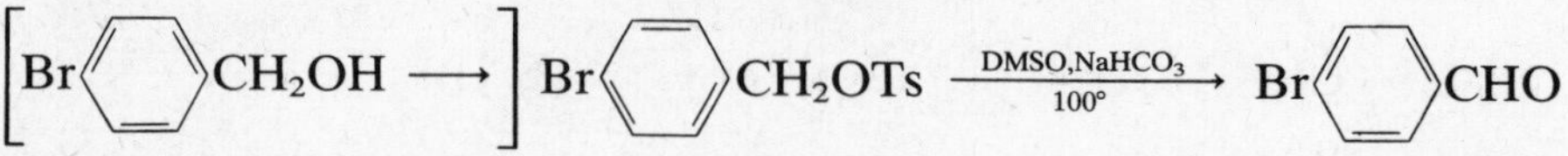

$$\left[Br\langle\bigcirc\rangle CH_2OH \longrightarrow\right] Br\langle\bigcirc\rangle CH_2OTs \xrightarrow[100°]{DMSO,NaHCO_3} Br\langle\bigcirc\rangle CHO$$

$$(65\%)$$

The conversion of the alcohol into the halide or toluene-*p*-sulphonate, however, constitutes an extra step, and the use of unselective reagents. For a multi-step synthesis, therefore, a one-step, selective oxidation is preferable; and this is achieved by converting the dimethyl sulphoxide (a weak *nucleophile*) into a strong *electrophile* which may react directly with the alcohol. Activation of the dimethyl sulphoxide is must commonly achieved using N,N'-dicyclo-hexylcarbodiimide [DCC: reaction (9.24a)], but sulphur trioxide (as its pyridine complex) and acetic anhydride are two other electrophiles which have been used successfully [reactions (9.24b) and (9.24c)]:

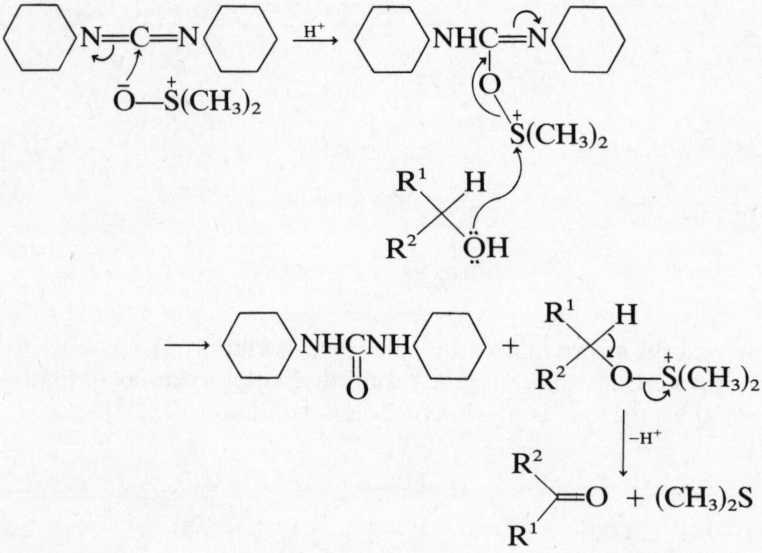

$$R^2 \!\!\!\diagdown\!\!\!=O + (CH_3)_2S \quad (9.24a)$$

$$(CH_3)_2SO + SO_3 \longrightarrow (CH_3)_2\overset{+}{S}-O-SO_2-O^- \xrightarrow{\substack{R^1\ H \\ R^2\ OH}}$$

$$\underset{R^2}{\overset{R^1\ H}{\diagdown}}\!\!-OS^+(CH_3)_2 \ HSO_3^-, \quad \text{etc.} \quad (9.24b)$$

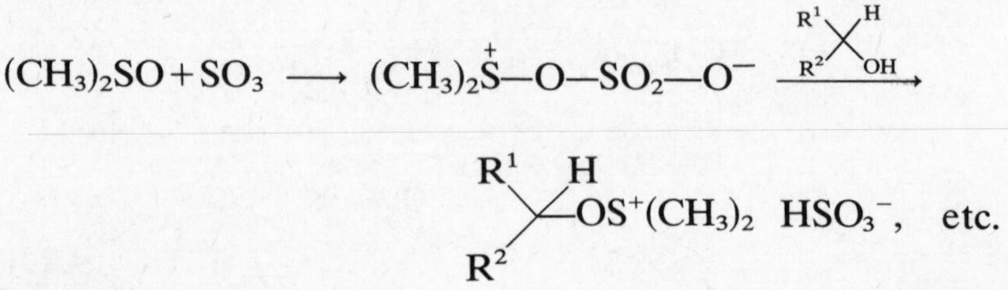

$$(CH_3)_2SO + (CH_3CO)_2O \longrightarrow (CH_3)_2\overset{+}{S}—OCOCH_3 \ \bar{O}COCH_3 \xrightarrow[-CH_3CO_2H]{\begin{smallmatrix}R^1 \\ R^2 \end{smallmatrix}\underset{OH}{\overset{H}{\diagdown}}}$$

$$\underset{R^2}{\overset{R^1}{\diagup}}\underset{O\overset{+}{S}(CH_3)_2}{\overset{H}{\diagdown}} \qquad \bar{O}COCH_3, \text{ etc.} \quad (9.24c)$$

Examples of these processes include:

$$CH_3(CH_2)_7OH \xrightarrow[\substack{CF_3CO_2^- \\ \text{(proton source)}}]{DMSO, DCC} CH_3(CH_2)_6CHO \text{ (quantitative)}$$

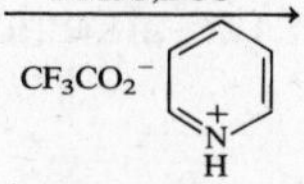

Similarly:

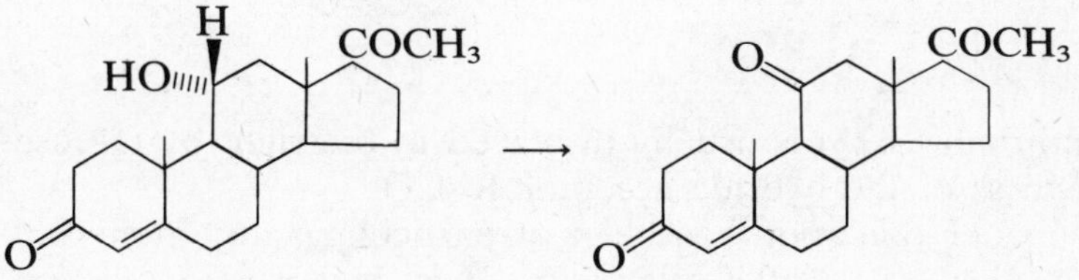

(20) → (21) (99 %)

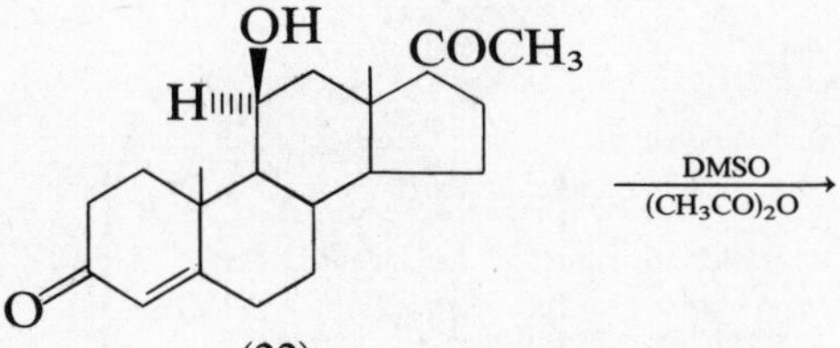

(22) $\xrightarrow[(CH_3CO)_2O]{DMSO}$ (21) (53 %) [4]

$\xrightarrow[SO_3/pyridine/(C_2H_5)_3N]{DMSO}$

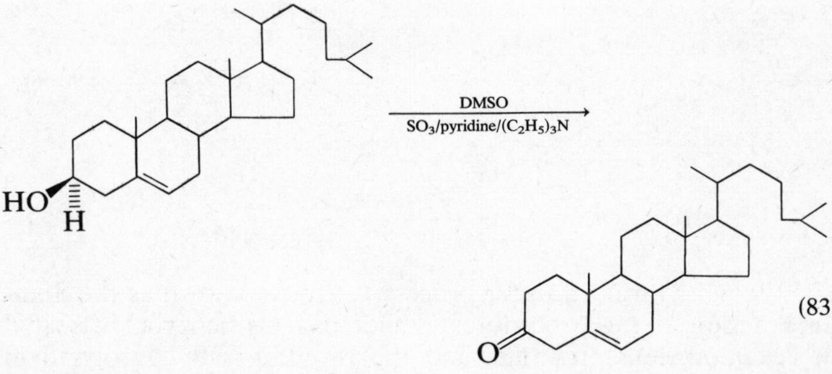

(83 %)

9.3.1.2 Hydride-transfer method

The most important of these is the **Oppenauer oxidation** [reaction (9.25)], in which a secondary alcohol is oxidised to a ketone by another ketone (usually acetone or cyclohexanone) in presence of an aluminium alkoxide (usually isopropoxide or *t*-butoxide):

$$R_2CHOH + R'_2CO \xmrightleftharpoons{Al[OCH(CH_3)_2]_3} R_2CO + R'_2CHOH \tag{9.25}$$

This reaction is the exact opposite of the Meerwein–Ponndorf–Verley reduction [section 8.4.3.1; reaction (8.7)]. It involves deprotonation of the alcohol by equilibration with the alkoxide, followed by hydride transfer to the ketone:

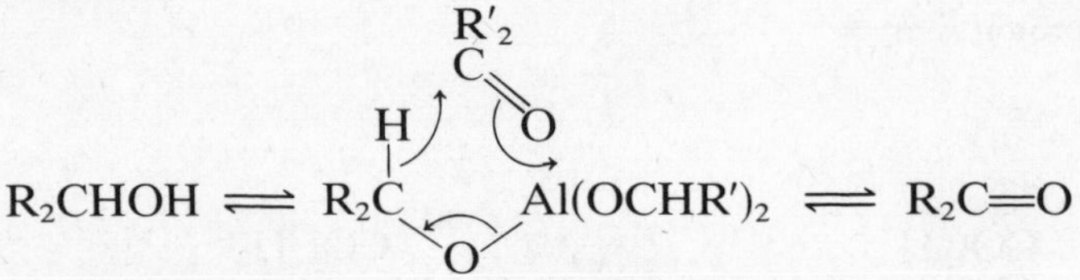

The equilibrium (9.25) is usually displaced to the right by the use of a large excess of the hydride acceptor R'_2CO.

The Oppenauer oxidation has been of particular value in steroid syntheses, in view of its high selectivity. It is, however, a reaction involving a strongly basic medium, and converts β,γ-unsaturated alcohols into α,β-unsaturated ketones, presumably *via* a conjugated enolate ion. Thus, for example,

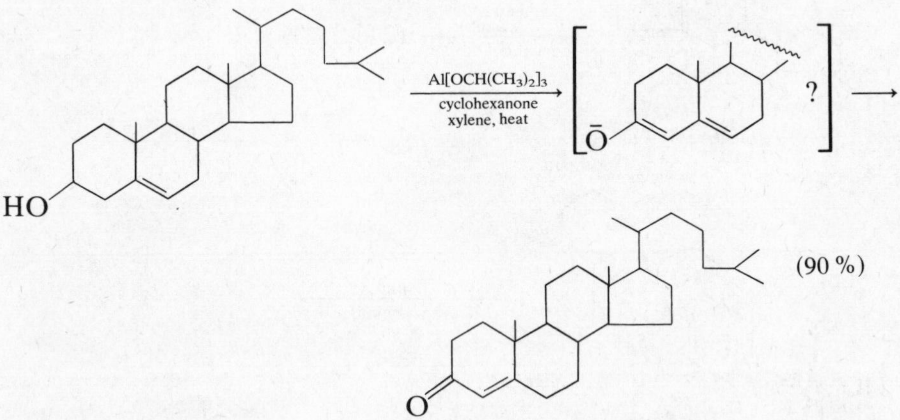

The second hydride-transfer process is usually known as the **Sommelet reaction**. In this procedure, a halide (usually benzylic) is treated with hexamethylenetetramine, and the resulting salt hydrolysed in

presence of an excess of the amine [reaction (9.26)]:

$$ArCH_2 + \underset{\text{(hexamethylenetetramine)}}{\overset{}{\text{N}}} \longrightarrow ArCH_2N^+ \overset{}{\text{N}} \quad X^- \xrightarrow{\text{H}^+,\text{H}_2\text{O}}$$

$$\left[ArCH_2NH_2 + CH_2O + NH_3 \rightleftharpoons \underset{\underset{CH_2=\overset{+}{N}H_2}{}}{\overset{ArCH-\overset{..}{N}H_2}{\underset{H}{|}}} \rightleftharpoons \underset{+ \ CH_3NH_2}{ArCH=\overset{+}{N}H_2} \right]$$

$$\xrightarrow{\text{H}_2\text{O}} ArCHO + NH_3 + CH_3NH_2 \quad (9.26)$$

As with the Oppenauer reaction, the equilibrium is displaced by adding excess of hexamethylenetetramine (i.e. an excess of $CH_2=\overset{+}{N}H_2$). The method gives acceptable yields of aldehydes, as the examples show, but it offers no obvious advantage over the dimethyl sulphoxide method described above (p. 219):

$$\left[\underset{Br}{\overset{CH_2OH}{\bigcirc}} \longrightarrow \right] \underset{Br}{\overset{CH_2Br}{\bigcirc}} \xrightarrow[\text{(ii) } CH_3CO_2H,H_2O]{\text{(i) } (CH_2)_6N_4} \underset{Br}{\overset{CHO}{\bigcirc}} \quad (69\%)$$

Similarly,

$$\underset{}{\overset{CH_2Br}{\bigcirc\bigcirc}} \longrightarrow \underset{}{\overset{CHO}{\bigcirc\bigcirc}} \quad (68\%)$$

9.3.1.3 Other methods

Catalytic dehydrogenation of alcohols, although important industrially, holds no particular advantage on a 'laboratory' scale over the chemical methods described in the previous sections, and so is not considered further here. The oxidation of allylic and benzylic alcohols using manganese(IV) oxide is worthy of mention. It is a heterogeneous reaction, and the detailed mechanism is unknown; its success also depends on the freshness of the oxide used. However, with freshly prepared oxide the yields may be high and the oxidations selective, e.g.

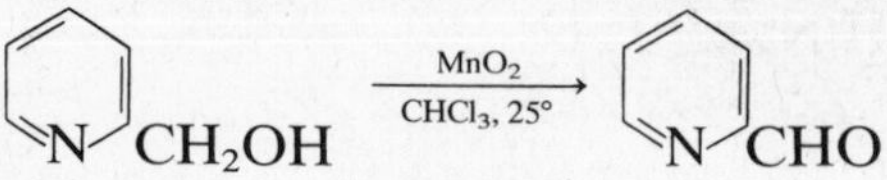

(68 %)

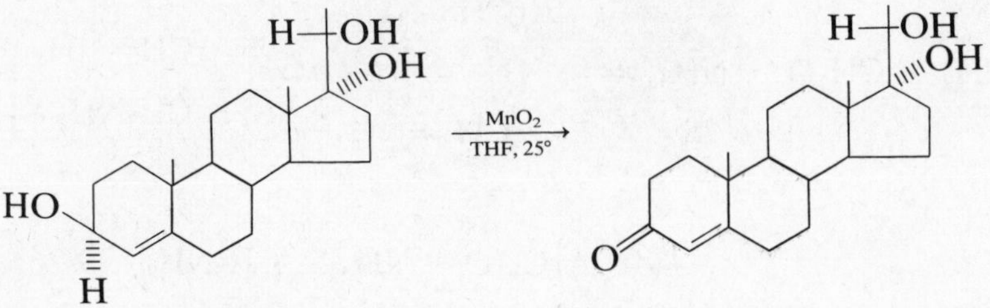

(90 %)

9.3.2   Oxidative cleavage of 1,2-diols

Mention has already been made (section 9.2.6) of the oxidative cleavage of alkenes by ozonisation, and of the (limited) use of this procedure in synthetic work. Alkenes may also undergo oxidative cleavage by hydroxylation to a diol and subsequent cleavage of the latter, although the usefulness of this as a synthetic tool is similarly limited.

The two classical methods for the cleavage of diols involve the use of **lead tetra-acetate** [reaction (9.27)] or **periodic acid** or one of its salts (e.g. **sodium metaperiodate**, $NaIO_4$) [reaction (9.28)]:

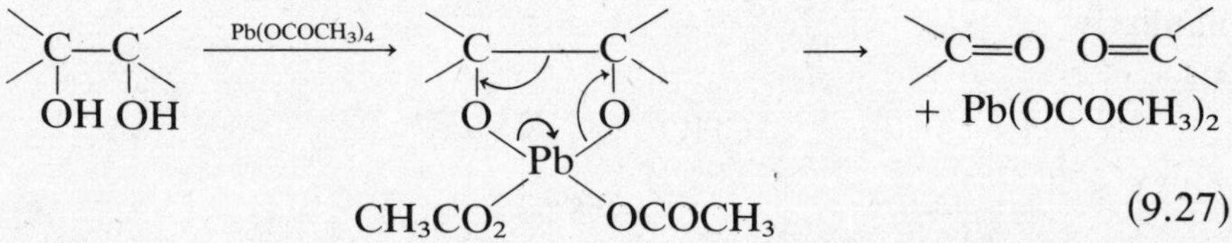

(9.27)

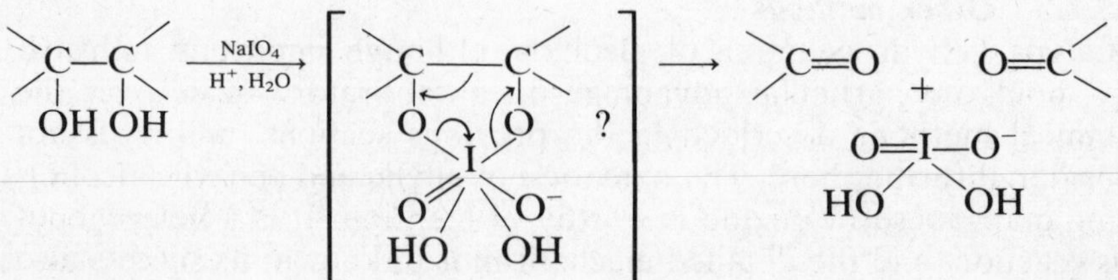

(9.28)

Both reactions involve cyclic intermediates, and thus diols in which the intermediate cannot be formed (e.g. a diaxial *trans*-diol in a ring system) are very resistant to cleavage. For example, the cleavage of (23) to cyclodecane-1,6-dione occurs 300 times faster than that of (24).

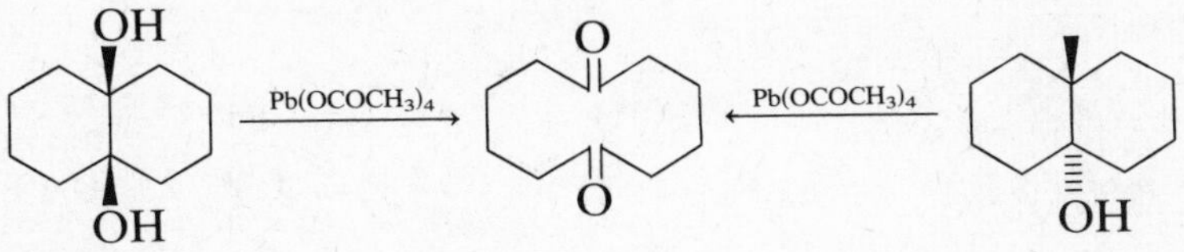

(23)   (24)

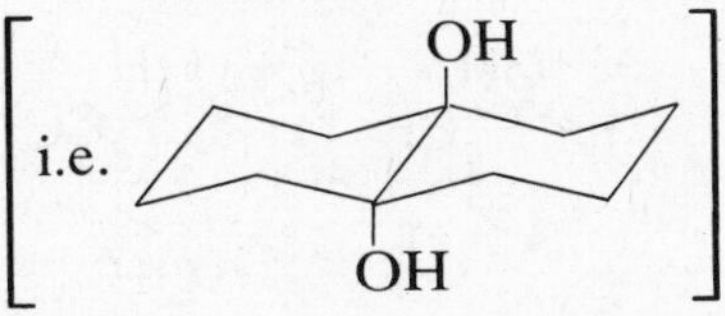

Diol cleavage, like ozonisation, is used synthetically for ring opening (cf. section 7.4.2), to release a carbonyl function from a 'masked' group, and to produce synthetically useful materials from abundant natural products, e.g.

erythro-$CH_3(CH_2)_7CHOHCHOH(CH_2)_7CO_2H \xrightarrow[\text{ethanol}]{KIO_4, H_2SO_4}$

(from oleic acid: cf. p. 212)

$$CH_3(CH_2)_7CHO + OHC(CH_2)_7CO_2H$$
(89 %)   (76 %)

9.4 Oxidation of phenols

Two major features of the chemistry of phenols are the stabilisation of the phenoxide anion (and the phenoxy radical) by the adjacent aromatic ring, and the stabilisation of a positive charge on the ring by the hydroxyl group. The first of these results in the facile removal of the hydroxyl hydrogen (either as a proton or as a radical), and the second results in high reactivity (at the *ortho*- and *para*-positions) towards electrophiles (cf. section 2.5).

Both ionic and radical mechanisms are known for oxidations of phenols. The **Elbs reaction**, for example [reaction (9.29)] is probably an electrophilic substitution, and the oxidation with **Frémy's salt** (a stable free radical) is obviously a radical process [reaction (9.30)]:

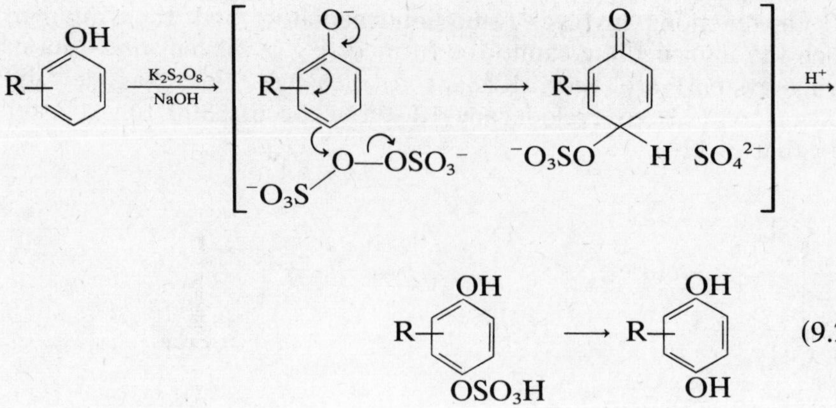

$$(9.29)$$

e.g.

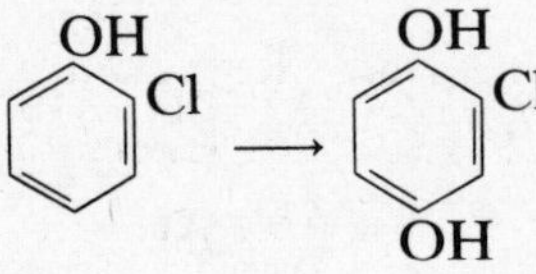

(50 %)

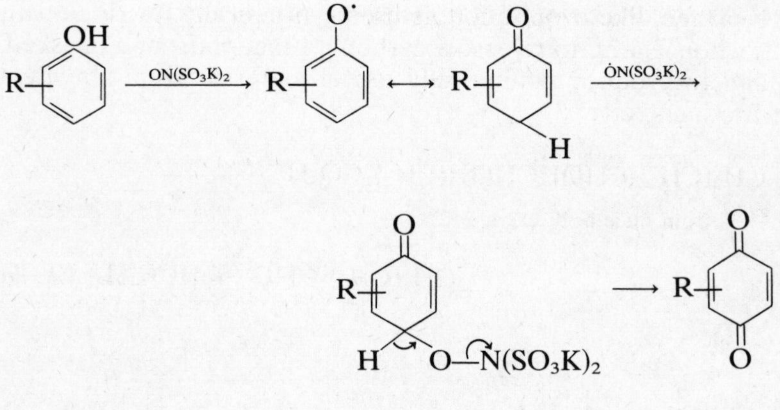

$$(9.30)$$

e.g.

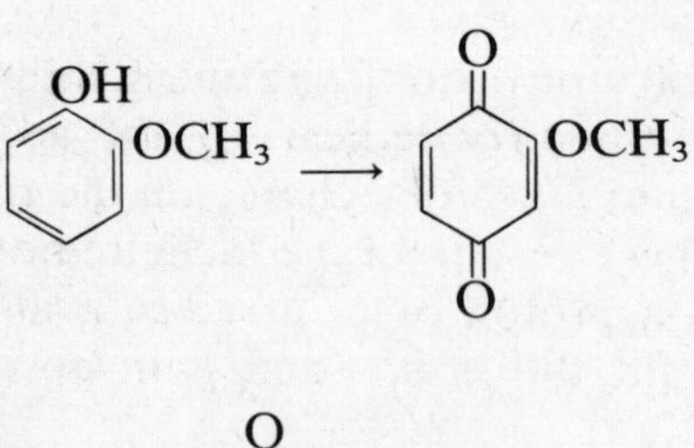

(90 %)

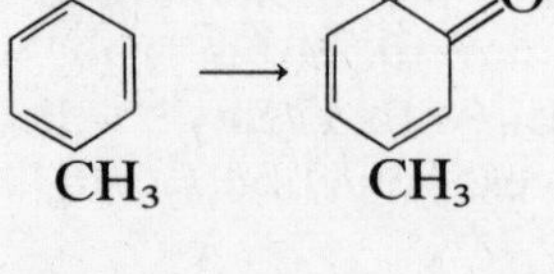

(71 %)

Oxidation of *o*- and *p*-dihydroxybenzenes to quinones is a relatively easy matter, whether by a substitution-elimination sequence or a radical process [reaction (9.31)]:

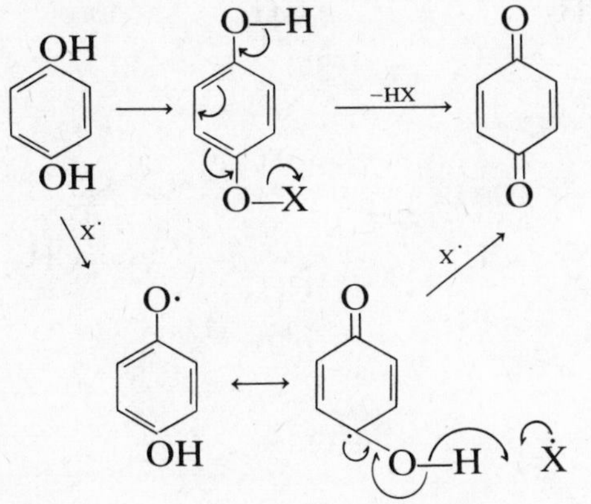

(9.31)

e.g.

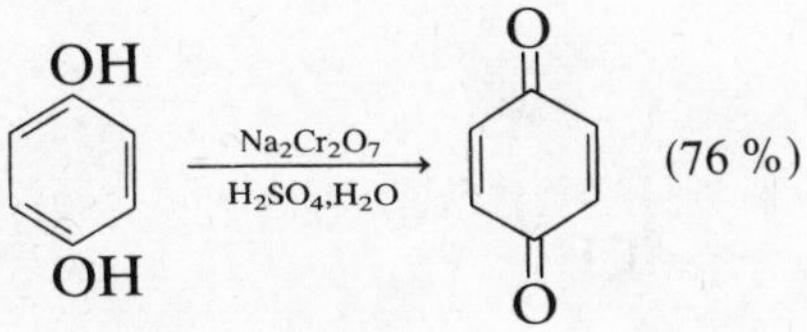

(76 %)

The corresponding oxidations of aminophenols similarly yield quinone-imines and hence quinones and ammonia.

One-electron oxidation of phenoxide ions, very often using an iron(III) compound as oxidant, gives phenoxy radicals which may undergo coupling reactions. **Oxidative coupling**, as it is usually called, is an extremely important biosynthetic procedure,[5] but many *in vitro* examples are also known. It is rare for two phenoxy radicals to give a peroxide dimer; much more common is dimerisation by way of C—C bond formation. Both symmetrical and unsymmetrical dimers are obtainable, but these may themselves be highly reactive and undergo further transformations. For example,

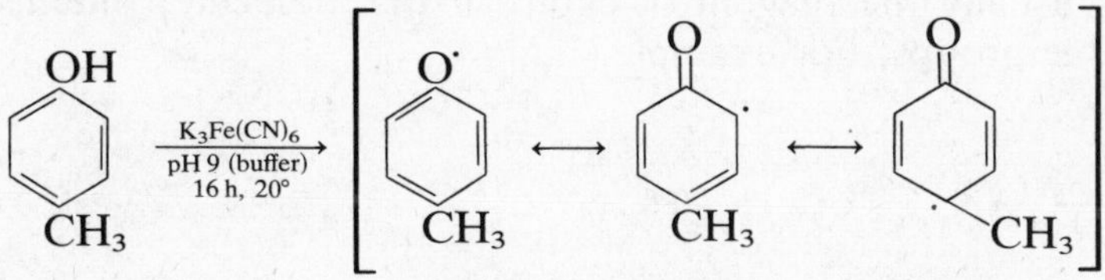

(25a)   (25b)   (25c)

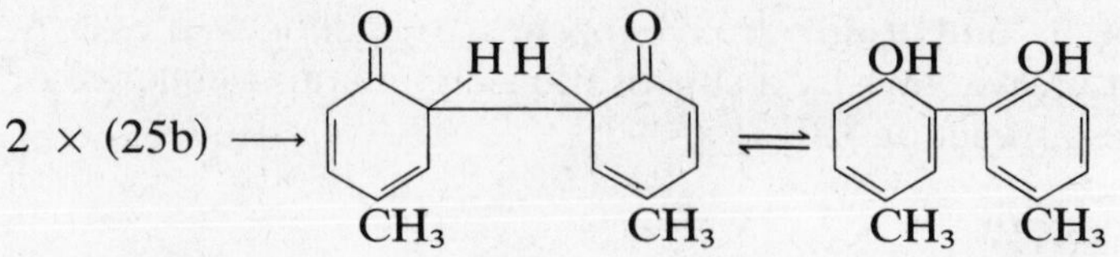

$$2 \times (25b) \longrightarrow$$

(26)

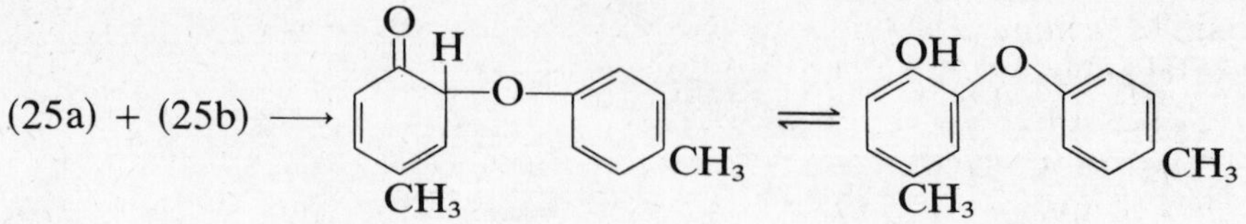

$$(25a) + (25b) \longrightarrow$$

(27)

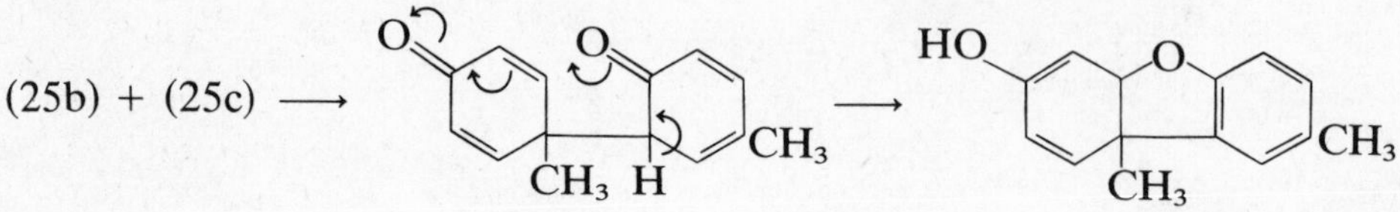

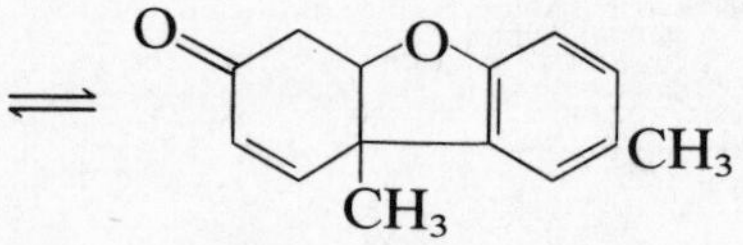

$$(25b) + (25c) \longrightarrow$$

$$\rightleftharpoons$$

$$(26):(27):(28) = 52:7:41$$

(28)

The product ratio depends on temperature, concentration, solvent, and the particular oxidant used.

9.5 Oxidation of aldehydes and ketones

9.5.1 Oxidation to carboxylic acids

We have already referred (section 9.3.1) to the oxidation $RCH_2OH \rightarrow RCHO \rightarrow RCO_2H$ with which most students should be familiar. Aldehydes themselves are very easily oxidised, by chromic acid or potassium permanganate (with or without added acid), by molecular oxygen, or by mild oxidants such as silver oxide: this last-named is usually the reagent of choice if the molecule contains other oxidisable groups. For example,

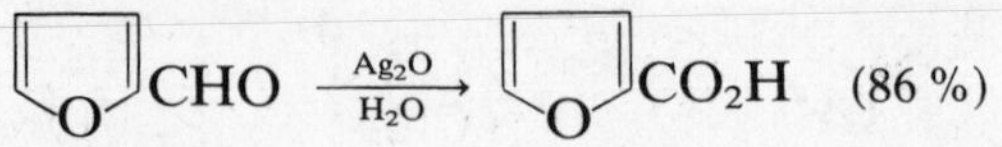

$$CHO \xrightarrow[H_2O]{Ag_2O} CO_2H \quad (86\%)$$

(acid-sensitive)

$$\left[2C_2H_5CHO \xrightarrow{\text{base}} \right] \quad C_2H_5CH\!=\!\overset{\overset{\displaystyle CH_3}{|}}{C}CHO \xrightarrow[H_2O]{Ag_2O} C_2H_5CH\!=\!\overset{\overset{\displaystyle CH_3}{|}}{C}CO_2H$$

(alkene also oxidisable)   (60 %)

Under more vigorous oxidising conditions, side-reactions [especially cleavage: cf. reaction (9.32)] may intervene, and attempts to oxidise primary alcohols to carboxylic acids in a 'one-pot' reaction are similarly subject to side-reactions (e.g. $RCH_2OH \rightarrow RCO_2H$; $RCH_2OH + RCO_2H \rightarrow RCO_2CH_2R$.

Oxidation of ketones to carboxylic acids necessarily involves C—C bond cleavage. In many cases, the reaction apparently involves the enol (or enolate) as intermediate [e.g. reaction (9.32)]:

$$RCH_2COR' \rightleftharpoons RCH\!=\!\overset{\overset{\displaystyle OH}{|}}{C}R' \xrightarrow{KMnO_4} \underset{\underset{\displaystyle OH \ \ OH}{|\ \ \ \ |}}{RCH\!-\!CR'} \xrightarrow{-H_2O} \underset{\underset{\displaystyle OH}{|}}{RCH\!-\!COR'}$$

$$\longrightarrow RCOCOR' \longrightarrow RCO_2H + R'CO_2H \quad (9.32)$$

The synthetic usefulness of this oxidation is restricted to a few particular situations, e.g. ring opening:

⬡=O $\xrightarrow{HNO_3}$ ⬡(CO₂H)(CO₂H)   (60 %)

Methyl ketones, on the other hand, may be converted into carboxylic acids under mild conditions by the **haloform reaction** (Sykes, pp. 287–90). Base-catalysed halogenation is followed by an addition-elimination sequence. The reaction $RCOCH_3 \rightarrow RCO_2H$ [reaction (9.33)] succeeds only if the group R is not itself halogenated under these conditions.

$$RCOCH_3 \xrightarrow[NaOH]{3X_2} \left[RCOCH_2X \longrightarrow RCOCHX_2 \longrightarrow \right] R\!-\!\overset{\overset{\displaystyle O}{\|}}{\underset{\underset{\displaystyle OH}{|}}{C}}\!-\!CX_3$$

$$\longrightarrow R\!-\!\overset{\overset{\displaystyle O^-}{|}}{\underset{\underset{\displaystyle OH}{|}}{C}}\!-\!CX_3 \longrightarrow RCO_2^- + CHX_3 \quad (9.33)$$

e.g.

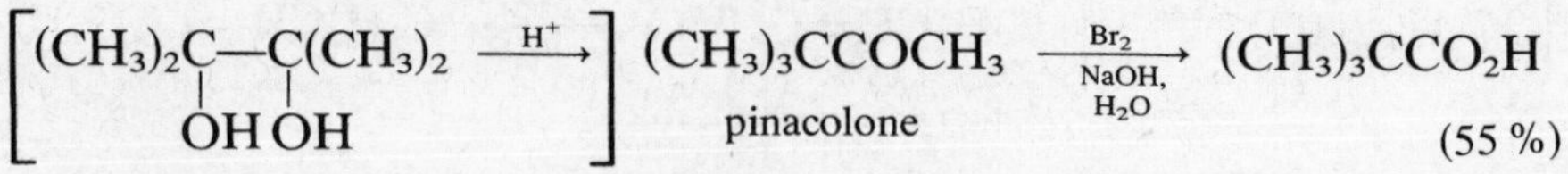

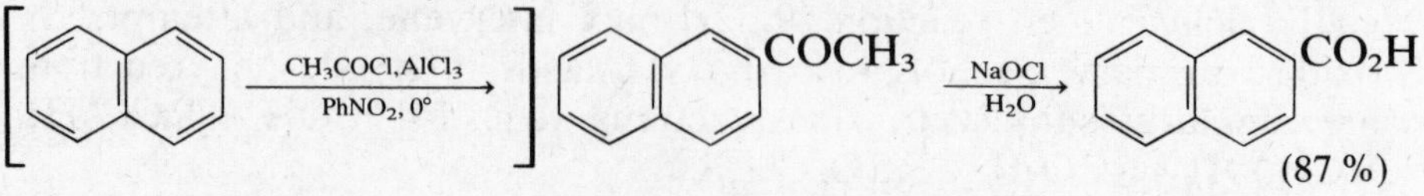

$$\left[\begin{array}{c} (CH_3)_2C-C(CH_3)_2 \\ | \quad | \\ OH \; OH \end{array}\right] \xrightarrow{H^+} \underset{\text{pinacolone}}{(CH_3)_3CCOCH_3} \xrightarrow[\substack{NaOH, \\ H_2O}]{Br_2} (CH_3)_3CCO_2H$$

(55 %)

$$\left[\text{naphthalene} \right] \xrightarrow[PhNO_2, \, 0°]{CH_3COCl, AlCl_3} \text{(naphthalene-COCH}_3) \xrightarrow[H_2O]{NaOCl} \text{(naphthalene-CO}_2H)$$

(87 %)

9.5.2 Oxidation to 1,2-dicarbonyl compounds

Oxidation of an 'active methylene' group to carbonyl is usually carried out by one of three routes; the enol or enolate is again involved as an intermediate in each case.

(i) *Selenium dioxide oxidation* [which probably proceeds *via* the enol selenite (29)]:

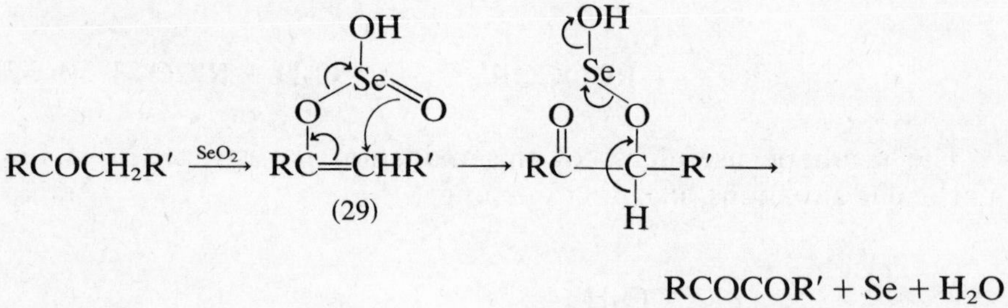

$$RCOCH_2R' \xrightarrow{SeO_2} RC{=}CHR' \longrightarrow RC{-}\overset{|}{\underset{H}{C}}{-}R' \longrightarrow$$
(29)

$$RCOCOR' + Se + H_2O$$

e.g.

$$PhCOCH_3 \rightarrow PhCOCHO \quad (72\%)$$

(60 %)

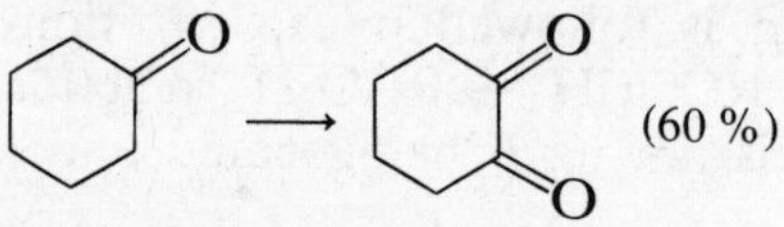

$$CH_3COCH_2COCH_3 \rightarrow CH_3(CO)_3CH_3 \quad (29\%)$$

(ii) *Monohalogenation* followed by *reaction with dimethyl suphoxide* [cf. section 9.3.1.1]:

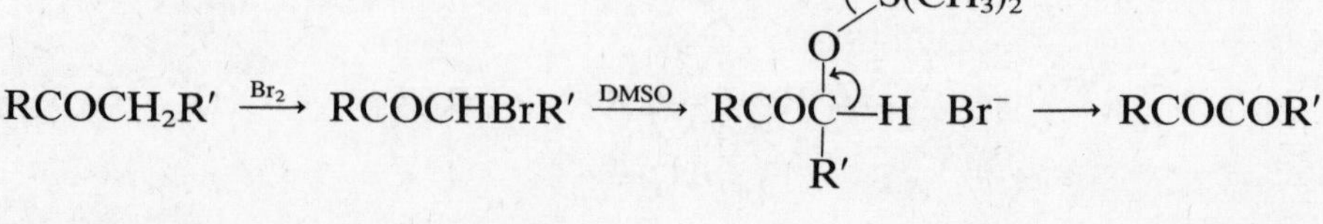

$$RCOCH_2R' \xrightarrow{Br_2} RCOCHBrR' \xrightarrow{DMSO} RCO\overset{+}{\underset{R'}{C}}{-}H \; Br^- \longrightarrow RCOCOR'$$

e.g.

$$O_2N-\!\!\langle\bigcirc\rangle\!-COCH_3 \xrightarrow[CH_3CO_2H]{Br_2} O_2N-\!\!\langle\bigcirc\rangle\!-COCH_2Br \xrightarrow{DMSO}$$

(high yield)

$$O_2N-\!\!\langle\bigcirc\rangle\!-COCHO$$

(72 %)

(iii) *Nitrosation* followed by *hydrolysis* (cf. section 6.3.3):

$$RCOCH_2R' \xrightarrow[\text{or } R''ONO]{NO^+} RCO\overset{NO}{\overset{|}{CH}}R' \rightleftharpoons RCO\overset{NOH}{\overset{||}{C}}R' \xrightarrow[H_2O]{H^+} RCOCOR'$$

e.g.

$$PhCOCH_2CH_3 \xrightarrow[\substack{HCl \\ ether}]{CH_3ONO} PhCO\overset{NOH}{\overset{||}{C}}CH_3 \xrightarrow[H_2O]{H_2SO_4} PhCOCOCH_3.$$

$\qquad\qquad\qquad\qquad\qquad$ (65 %) $\qquad\qquad$ (66 %)

$\qquad\qquad\qquad\qquad\qquad\qquad\qquad\qquad$ (43 % overall)

9.5.3 Oxidation to esters: the Baeyer–Villiger and Dakin reactions

The **Baeyer–Villiger oxidation** involves the reaction of a ketone with hydrogen peroxide or a peroxy-acid [reaction (9.34): cf. Sykes, pp. 126–7] to give an ester:

$$R^1COR^2 \rightleftharpoons R^1\overset{OH}{\underset{+}{C}}R^2 \xrightarrow[\text{(ii) }-H^+]{\text{(i) }RCO_2OH} R^1-\overset{O-H}{\underset{O-OCOR}{C}}-R^2 \longrightarrow \substack{R^1CO-OR^2 \\ + RCO_2H} \quad (9.34)$$

The **Dakin reaction**, although similar, is more restricted in application [reaction (9.35)]:

$$HO-\!\!\langle\bigcirc\rangle\!-COR \xrightarrow[-OH]{H_2O_2} \bar{O}-\!\!\langle\bigcirc\rangle\!-\overset{O^-}{\underset{O-OH}{C}}-R \longrightarrow {}^-O-\!\!\langle\bigcirc\rangle\!-OCOR + \bar{O}H$$

(9.35)

It should be noted that in the Baeyer–Villiger reaction, as in other molecular rearrangements of this type, the group which migrates (R^2 in the equation) is the more nucleophilic of the two; thus:

$$PhCO-\!\!\langle\bigcirc\rangle\!-OCH_3 \xrightarrow{PhCO_2OH} PhCO-O-\!\!\langle\bigcirc\rangle\!-OCH_3 \quad (66\%)$$

($R^2 = p$-methoxyphenyl)

whereas

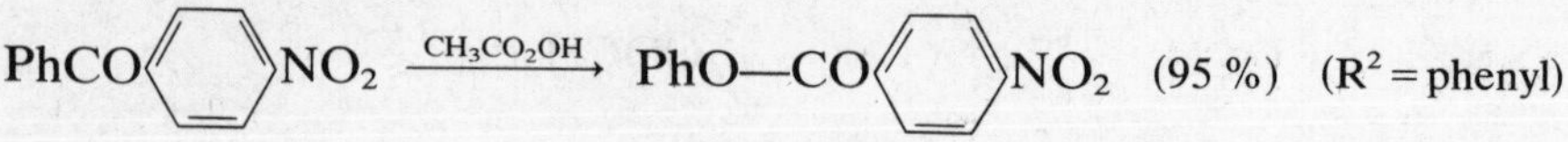

PhCO⟨ ⟩NO$_2$ $\xrightarrow{\text{CH}_3\text{CO}_2\text{OH}}$ PhO—CO⟨ ⟩NO$_2$ (95 %)   (R^2 = phenyl)

Aryl groups migrate more readily than alkyl, e.g.

PhCOC$_2$H$_5$ $\xrightarrow{\text{PhCO}_2\text{OH}}$ PhOCOC$_2$H$_5$ (73 %) (R^2 = phenyl)

and a secondary alkyl more readily than a primary alkyl, e.g.

⟨ ⟩COCH$_3$ $\xrightarrow{\text{PhCO}_2\text{OH}}$ ⟨ ⟩OCOCH$_3$ (61 %)   (R^2 = cyclopentyl)

Examples of the Dakin reaction include:

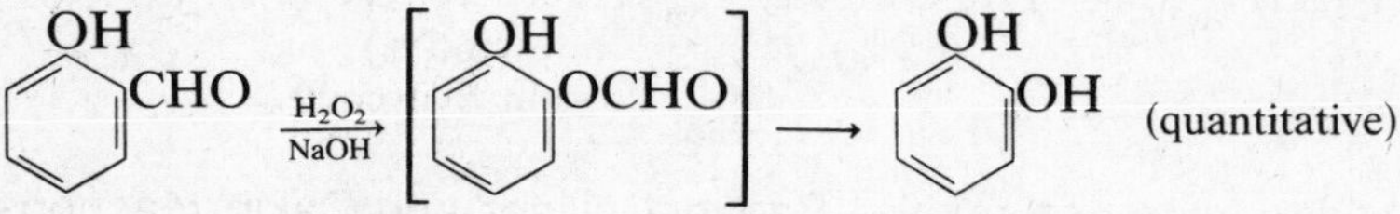

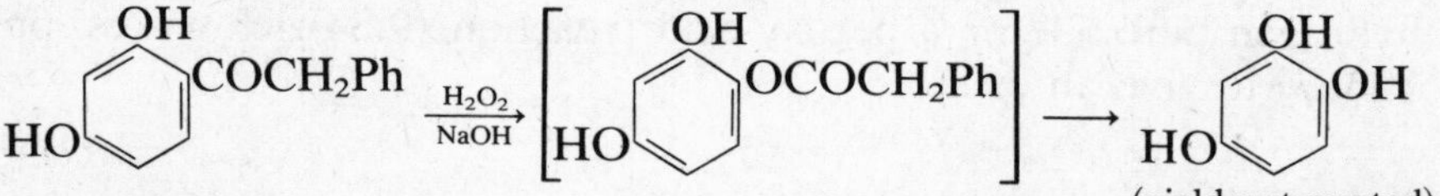

(quantitative)

(yield not quoted)

+ PhCH$_2$CO$_2$H   (95 %)

9.6 Oxidation of functional groups containing nitrogen

9.6.1 Formation of N-oxygenated compounds

Amines, being nucleophilic, react with sources of electrophilic oxygen such as peroxy-acids to produce N-oxygenated compounds. The most familiar examples of such reactions involve the formation of N-oxides from heteroaromatic tertiary amines such as pyridine (cf. section 2.6). N-Oxide formation, however, is a characteristic reaction of tertiary amines in general, e.g.

(CH$_3$)$_3$N $\xrightarrow{\text{H}_2\text{O}_2}_{\text{H}_2\text{O}}$ (CH$_3$)$_3\overset{+}{\text{N}}$—O$^-$ (>90 %)

(C$_2$H$_5$)$_2$NCH$_2$CH$_2$N(C$_2$H$_5$)$_2$ $\xrightarrow{\text{H}_2\text{O}_2}$ (C$_2$H$_5$)$_2\overset{+}{\text{N}}$CH$_2$CH$_2\overset{+}{\text{N}}$(C$_2$H$_5$)$_2$ (92 %)
           | |
           O$^-$ O$^-$

$$PhN(CH_3)_2 \xrightarrow{CH_3CO_2OH} Ph\overset{+}{N}(CH_3)_2 \quad (81\%)$$
$$\underset{O^-}{|}$$

In the case of secondary amines, *N*-oxidation is followed by proton transfer, and the product is a hydroxylamine. Yields are not uniformly high, but in some cases are synthetically acceptable, e.g.

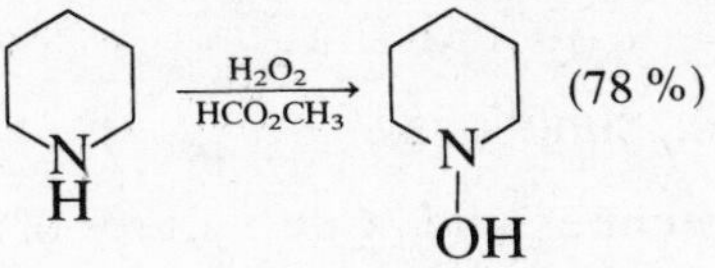

$$\xrightarrow[HCO_2CH_3]{H_2O_2} \quad (78\%)$$

In the case of primary amines, the reaction is more complicated, since the hydroxylamine itself undergoes *N*-oxidation [$RNH_2 \rightarrow RNHOH \rightarrow RN(OH)_2 \xrightarrow{-H_2O} RNO$], e.g.

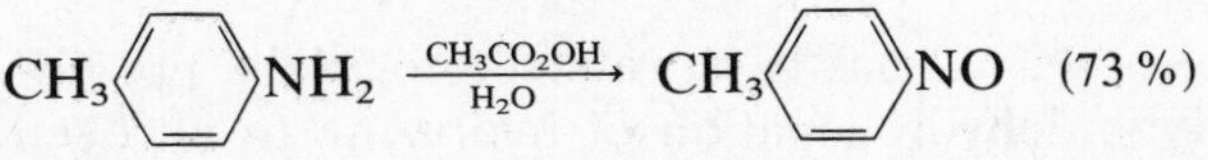

$$\xrightarrow[H_2O]{CH_3CO_2OH} \quad (73\%)$$

$$\xrightarrow[CH_2Cl_2]{CH_3CO_2OH} \quad (44\%: \text{isolated as dimer})$$

If the nitroso-compound contains an α-hydrogen it may undergo tautomerisation to an oxime, e.g.

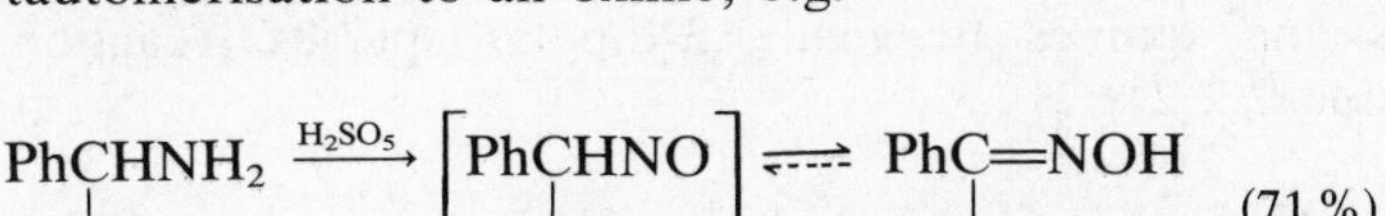

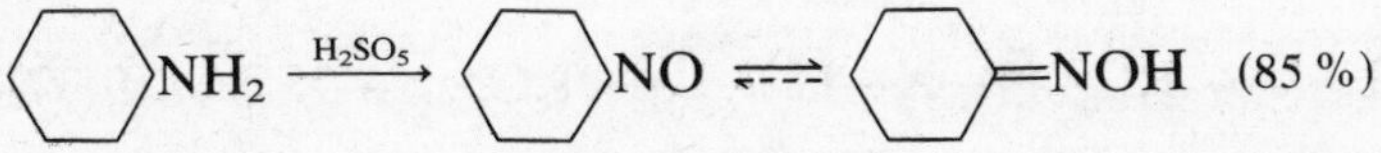

$$\underset{CH_3}{\overset{|}{PhCHNH_2}} \xrightarrow{H_2SO_5} \left[\underset{CH_3}{\overset{|}{PhCHNO}} \right] \rightleftharpoons \underset{CH_3}{\overset{|}{PhC}{=}NOH} \quad (71\%)$$

$$\xrightarrow{H_2SO_5} \quad NO \rightleftharpoons {=}NOH \quad (85\%)$$

The use of peroxytrifluoroacetic acid, or *anhydrous* peroxyacetic acid, can lead to *N*-oxidation even of the nitroso-compound, and to the formation of a nitro-compound. This reaction can be useful for the preparation of unusually substituted nitroarenes, e.g.

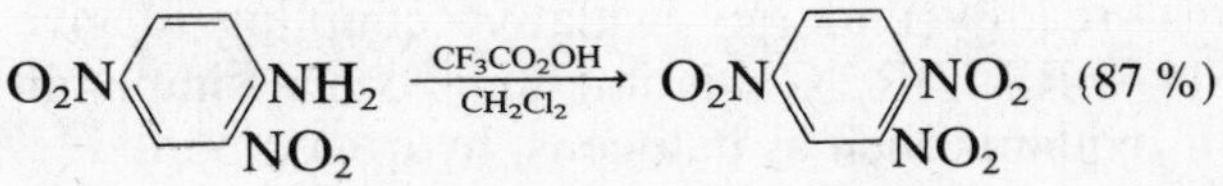

$$\xrightarrow[CH_2Cl_2]{CF_3CO_2OH} \quad (87\%)$$

Also:

$$C_2H_5\underset{\underset{CH_3}{|}}{C}HNH_2 \xrightarrow[\text{ClCH}_2\text{CH}_2\text{Cl}]{\text{CH}_3\text{CO}_2\text{OH, dry}} C_2H_5\underset{\underset{CH_3}{|}}{C}HNO_2 \quad (65\%)$$

[but $CH_3(CH_2)_5NH_2 \xrightarrow{CF_3CO_2OH}$ only $CH_3(CH_2)_5NHCOCF_3$ (80%)]

9.6.2   Dehydrogenation involving nitrogen functions

This general heading embraces a large number and wide variety of reactions. Oxidations of the type $\ce{>CH-NH-} \rightarrow \ce{>C=N-}$ are well known especially if the new double bond forms part of a conjugated system, but these are much less generally used than the corresponding reactions giving $\ce{>C=C<}$ and $\ce{>C=O}$ bonds. Dehydrogenation of the type $-NH-OH \rightarrow -N=O$ has been referred to in the preceding section. The catalytic dehydrogenation of hydrazine to nitrogen may be used to provide hydrogen for catalytic hydrogenation; the intermediate dehydrogenation product, di-imide, also serves as a reducing agent (section 8.4.1). Oxidation of 1,2-disubstituted hydrazines produces azo-compounds ($RN=NR'$).

Hydrazones of the type $R_2C=NNH_2$ are oxidised to diazoalkanes by reagents such as mercury(II) oxide [reaction (9.36)], and substitution-elimination sequences lead to the dehydrogenation of arylhydrazones and oximes to give 1,3-dipolar species [reaction (9.37): cf. section 7.2.2]:

$$R_2C=N-NH_2 + HgO \rightarrow R_2C=\overset{+}{N}=\bar{N} + Hg + H_2O \tag{9.36}$$

$$RCH=N-XH \xrightarrow{Cl_2} R\underset{\underset{}{}}{\overset{\overset{Cl}{|}}{C}}=\ddot{N}-X-H \xrightarrow{-HCl} RC\equiv\overset{+}{N}-X^- \tag{9.37}$$

(X = O or NAr)

9.7   Oxidation of functional groups containing sulphur

9.7.1   Thiols

Thiols, unlike alcohols, readily undergo oxidative coupling to give disulphides, i.e. $2RSH \rightarrow RS-SR$. Oxidation may occur simply in air, or by the action of oxidants such as halogens, hydrogen peroxide, or iron(III) salts. Both radical and electrophile-nuclophile interactions may be involved, e.g.

$$RSH \xrightarrow[\text{or } Fe^{3+}]{O_2} RS^{\cdot}; \quad 2RS^{\cdot} \rightarrow RS\!-\!SR$$

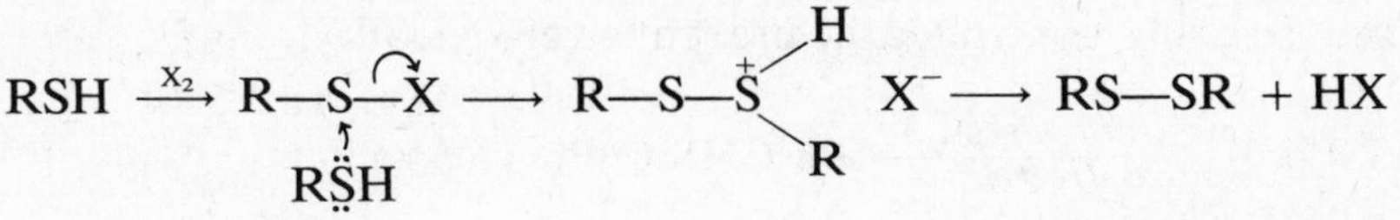

$$RSH \xrightarrow{X_2} R\!-\!\underset{R\overset{\cdot\cdot}{S}H}{\overset{\curvearrowright}{S}\!-\!X} \longrightarrow R\!-\!S\!-\!\overset{+}{\underset{R}{S}}\!\overset{H}{\diagup} \quad X^- \longrightarrow RS\!-\!SR + HX$$

Examples include:

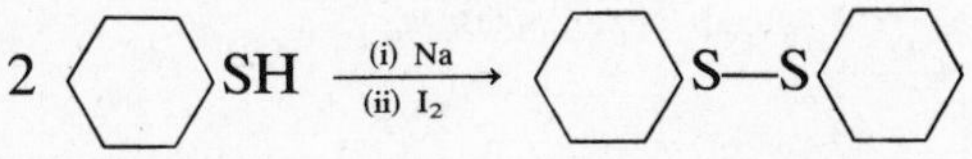

$$2 \langle \rangle SH \xrightarrow[\text{(ii) } I_2]{\text{(i) Na}} \langle \rangle S\!-\!S\langle \rangle \quad \text{(yield not quoted)}$$

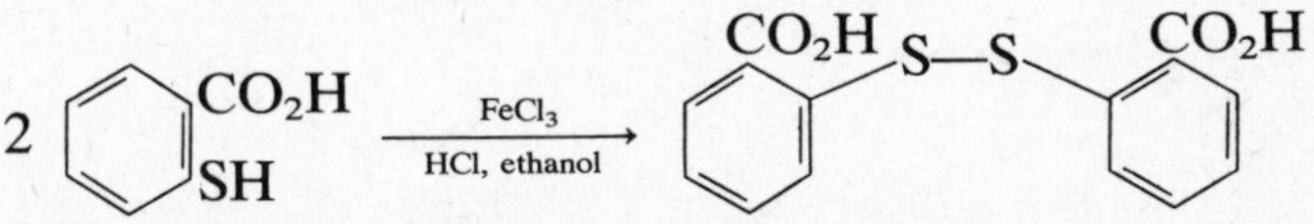

$$2 \underset{\text{SH}}{\overset{CO_2H}{\bigcirc}} \xrightarrow[\text{HCl, ethanol}]{FeCl_3} \overset{CO_2H}{\bigcirc}S\!-\!S\overset{CO_2H}{\bigcirc} \quad \text{(quantitative)}$$

More powerful oxidising agents convert thiols directly into sulphonic acids, e.g.

$$CH_3(CH_2)_{15}SH \xrightarrow[H_2O]{KMnO_4} CH_3(CH_2)_{15}SO_3H \quad (36\%)$$

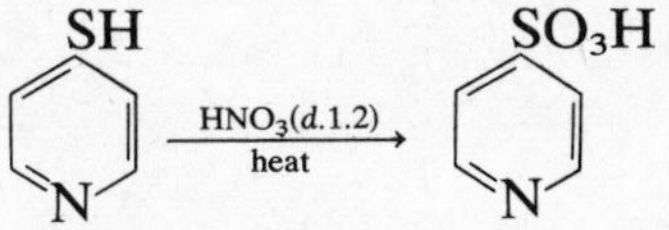

$$\underset{N}{\overset{SH}{\bigcirc}} \xrightarrow[\text{heat}]{HNO_3(d.1.2)} \underset{N}{\overset{SO_3H}{\bigcirc}} \quad (60\%)$$

9.7.2 Sulphides

These react with sources of electrophilic oxygen, e.g. peroxyacids, in the same manner as amines, i.e.

$$R_2S \rightarrow R_2\overset{+}{S}\!-\!\bar{O}(\leftrightarrow R_2S\!=\!O) \rightarrow R_2\overset{+}{\underset{O^-}{S}}\!=\!O \; (\leftrightarrow R_2\underset{O}{\overset{\|}{S}}\!=\!O).$$

The first oxidation step, giving the sulphoxide, is often considerably faster than the second, which gives the sulphone, and many sulphoxides may thus be prepared by this route, e.g.

$$Ph_2S \xrightarrow[4 \text{ days},25^\circ]{H_2O_2,CH_3CO_2H} Ph_2SO \quad \text{(quantitative)}$$

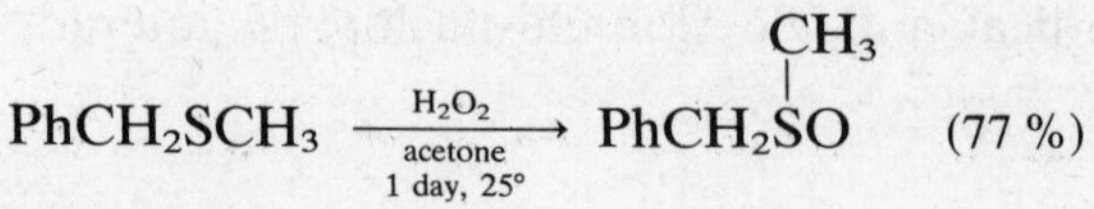

$$PhCH_2SCH_3 \xrightarrow[\substack{acetone \\ 1\ day,\ 25°}]{H_2O_2} PhCH_2\overset{\overset{\displaystyle CH_3}{|}}{S}O \quad (77\%)$$

Other oxidants may also be used, provided that the reaction conditions are carefully controlled to prevent over-oxidation, e.g.

$$Ph_2CHSPh \xrightarrow[CH_3CO_2H, 60-80°, 15\ min]{CrO_3(10\%\ excess), H_2O} Ph_2CHS(O)Ph \quad (96\%)$$

More vigorous reaction conditions lead directly to sulphones, e.g.

$$[CH_3(CH_2)_{15}]_2S \xrightarrow[90]{H_2O_2, CH_3CO_2H} [CH_3(CH_2)_{15}]_2SO_2 \quad (98\%)$$

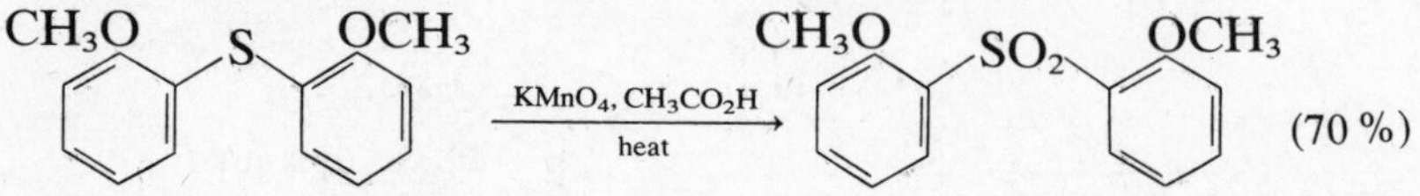

Notes

1.　A literature report that oxidation of ethylbenzene under these conditions gives phenylacetic acid could not subsequently be confirmed [D. G. Lee and U. A. Spitzer, *J. Org. Chem.*, **34**, 1493 (1969)].

2.　The conversion of (11) into the final product may be represented as an analogue of the Cope rearrangement (section 7.4.3):

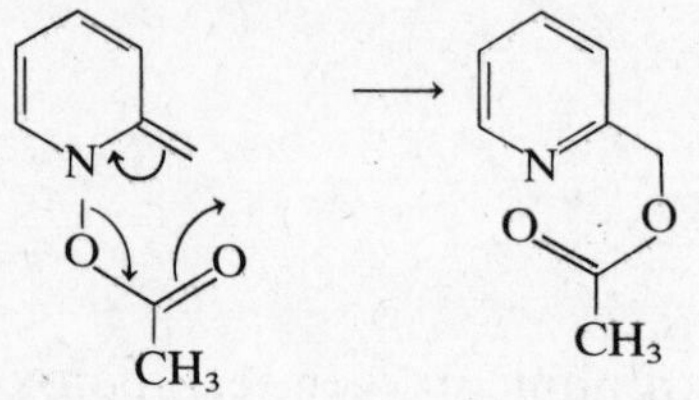

but it is more likely to involve cleavage of the N—O bond in (11) to give an ion-pair or a radical pair:

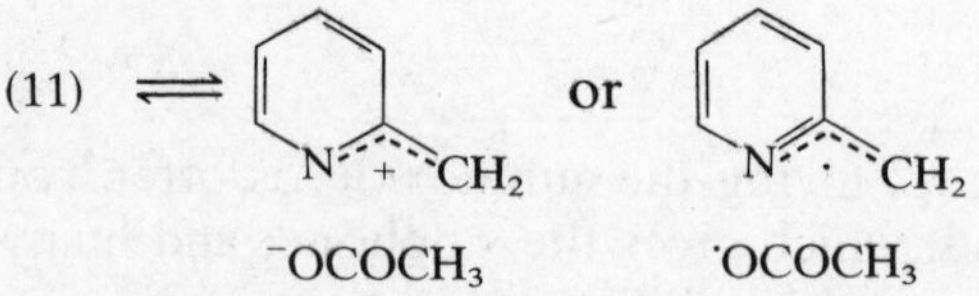

3.　Peroxybenzoic acid is usually made from dibenzoyl peroxide and methanol followed by acid hydrolysis.

4.   The axial alcohol (22) is sterically hindered by the two angular methyl groups: the bulkier DMSO/DCC and DMSO/SO$_3$ reagents do not therefore effect this oxidation.

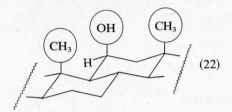

(22)

5.   See, for example, J. Mann, *Secondary Metabolism*, Oxford University Press, 1978, pp. 50–1, 200, and 207–9.

10 Protective groups

This chapter aims to set out the principal features of the use of protective groups in synthetic sequences. It is not intended to be a comprehensive treatment of the topic but rather some illustrative examples will be considered.

10.1 The strategy

In a synthetic sequence, it is frequently necessary to carry out a transformation at one centre while another reactive site remains unchanged. Two principal techniques[1] are used to achieve this purpose. One, to which reference has been made in most, if not all, of the remaining chapters, involves the careful choice of a selective reagent and/or of reaction conditions. The other, which we shall now describe in some detail, involves the temporary modification of the site at which reaction is undesirable in such a manner that it remains intact during reaction at the other site and at the end of the reaction sequence the original group can be easily regenerated. The group modifying the functional group is known as the **protective group.**

Thus we can make the following specification for an ideal protective group:

(i) the group should be introduced under mild conditions,
(ii) the group should be stable under the reaction conditions necessary to carry out transformations at other centres in the compound,
(iii) the group should be removed under mild conditions.

In some instances, this last condition can be relaxed to allow the protected group to be converted directly into another functional group. We shall now show how these specifications can be satisfied by considering the case of protection of hydroxyl groups. Protection of the amino and carbonyl groups will also be dealt with briefly. Further examples will be found in Chapter 14.

10.2 Protection of alcohols

10.2.1 Ether formation

In general, ethers are stable under neutral and alkaline conditions and to most oxidising agents. While methyl and ethyl ethers are readily formed [e.g. scheme 10.1, (1) → (2), (3) → (4)], they are not normally easily removed. Exceptions to this do, however, occur in carbohydrate chemistry. Methyl ethers at C-1, which are in fact *acetals*, are readily hydrolysed [e.g. scheme 10.1, (4) → (5)]. Lewis acids such as boron trichloride can be used to cleave methyl ethers [e.g. scheme 10.1, (10) → (11)].

More commonly used derivatives are benzyl, trityl (triphenyl-methyl), tetrahydropyranyl, and trialkylsilyl ethers. The last mentioned, as the trimethyl derivatives, have been widely used in gas chromatographic and mass spectroscopic applications but are too easily hydrolysed for most synthetic purposes. Recently, however, bulky silyl derivatives such as t-butyldimethyl have been found to be somewhat more stable than the trimethyl derivatives and can be used in situations where selective protection is desired [cf. section 13.4].

Benzyl and trityl ethers are formed by treatment of the alcohol with the appropriate halide in presence of base [e.g. scheme 10.2, (12) → (13)]. The hydroxyl group can be regenerated by hydrogenolysis or, in the case of trityl ethers, by mild acid treatment [e.g. scheme 10.2, (14) → (15), (19) → (20)]. Since tritylation of hindered alcohols is much slower than that of primary alcohols, selective protection is possible [scheme 10.2, (12) → (13)].

Tetrahydropyranyl ethers are formed from the alcohol and 2,3-dihydropyran under acid catalysis. The alcohol is regenerated with dilute sulphuric acid. Alcohols are often protected in this way during reactions involving organometallic compounds when the protective group is often lost in the work-up [scheme 10.3].

10.2.2 Ester formation

Esters, being reasonably stable under acidic conditions, are often used to protect hydroxyl groups during nitration, oxidation and formation of acid chlorides. Acetates and trifluoroacetates are usually formed by treatment of the alcohol with the appropriate anhydride or acid chloride in presence of base [scheme 10.2, (13) → (14)]. Formates are prepared by use of formic acid in presence of perchloric acid. The alcohol is regenerated by treatment with base [scheme 10.2, (17) → (18), scheme 10.4] but in the case of trifluoroacetates, water is often sufficient.

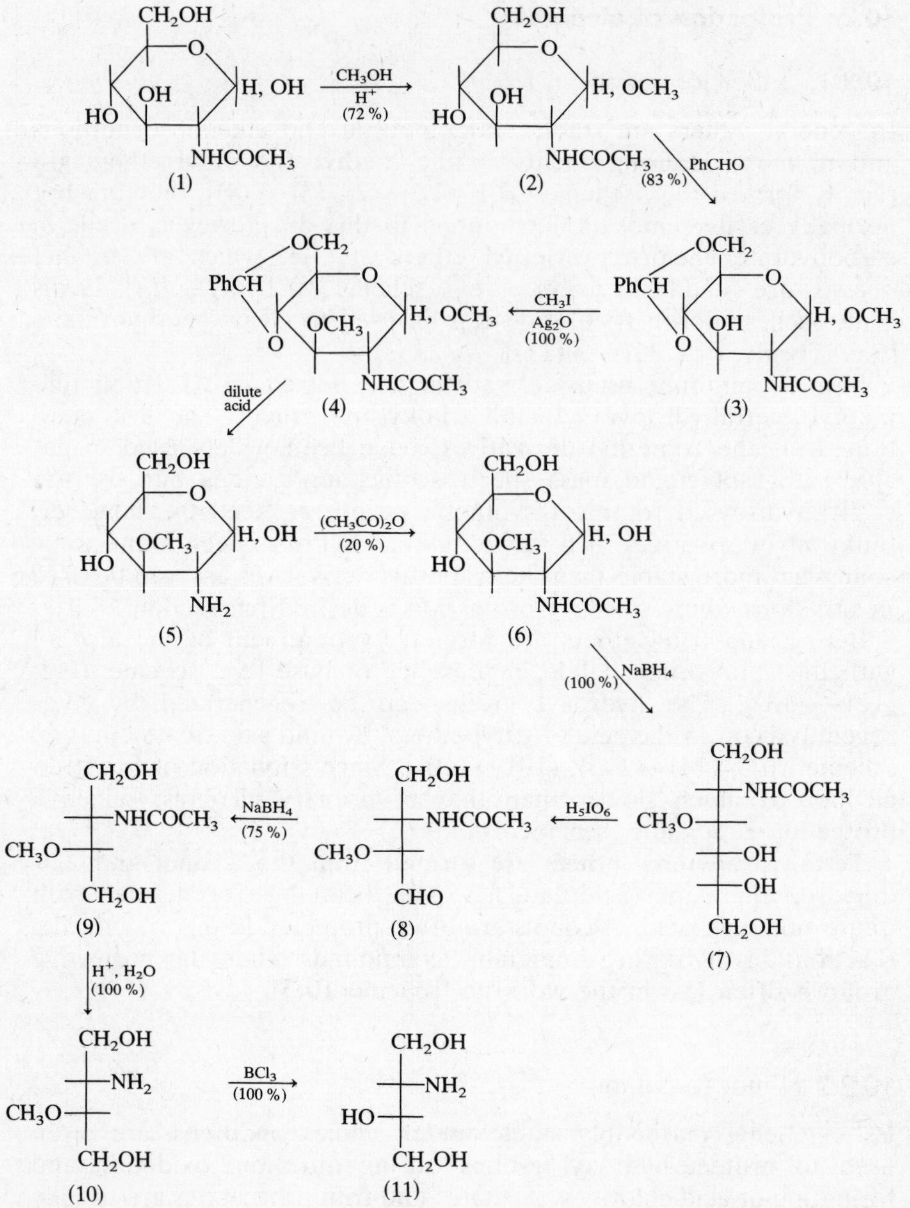

CH$_2$OH · · · O · OH · HO · H, OH · NHCOCH$_3$
(1)

$\xrightarrow[\text{(72 %)}]{\underset{\text{H}^+}{\text{CH}_3\text{OH}}}$

CH$_2$OH · · · O · OH · HO · H, OCH$_3$ · NHCOCH$_3$
(2)

$\xrightarrow[\text{(83 %)}]{\text{PhCHO}}$

OCH$_2$ · PhCH · O · OCH$_3$ · O · H, OCH$_3$ · NHCOCH$_3$
(4)

$\xleftarrow[\text{(100 %)}]{\underset{\text{Ag}_2\text{O}}{\text{CH}_3\text{I}}}$

OCH$_2$ · PhCH · O · OH · O · H, OCH$_3$ · NHCOCH$_3$
(3)

$\xrightarrow[\quad]{\underset{\text{acid}}{\text{dilute}}}$

CH$_2$OH · · · O · OCH$_3$ · HO · H, OH · NH$_2$
(5)

$\xrightarrow[\text{(20 %)}]{\text{(CH}_3\text{CO)}_2\text{O}}$

CH$_2$OH · · · O · OCH$_3$ · HO · H, OH · NHCOCH$_3$
(6)

$\xrightarrow[\text{(100 %)}]{\text{NaBH}_4}$

CH$_2$OH
—NHCOCH$_3$
CH$_3$O—
CH$_2$OH
(9)

$\xleftarrow[\text{(75 %)}]{\text{NaBH}_4}$

CH$_2$OH
—NHCOCH$_3$
CH$_3$O—
CHO
(8)

$\xleftarrow{\text{H}_5\text{IO}_6}$

CH$_2$OH
—NHCOCH$_3$
CH$_3$O—
—OH
—OH
CH$_2$OH
(7)

$\downarrow \underset{\text{(100 %)}}{\text{H}^+,\ \text{H}_2\text{O}}$

CH$_2$OH
—NH$_2$
CH$_3$O—
CH$_2$OH
(10)

$\xrightarrow[\text{(100 %)}]{\text{BCl}_3}$

CH$_2$OH
—NH$_2$
HO—
CH$_2$OH
(11)

Scheme 10.1

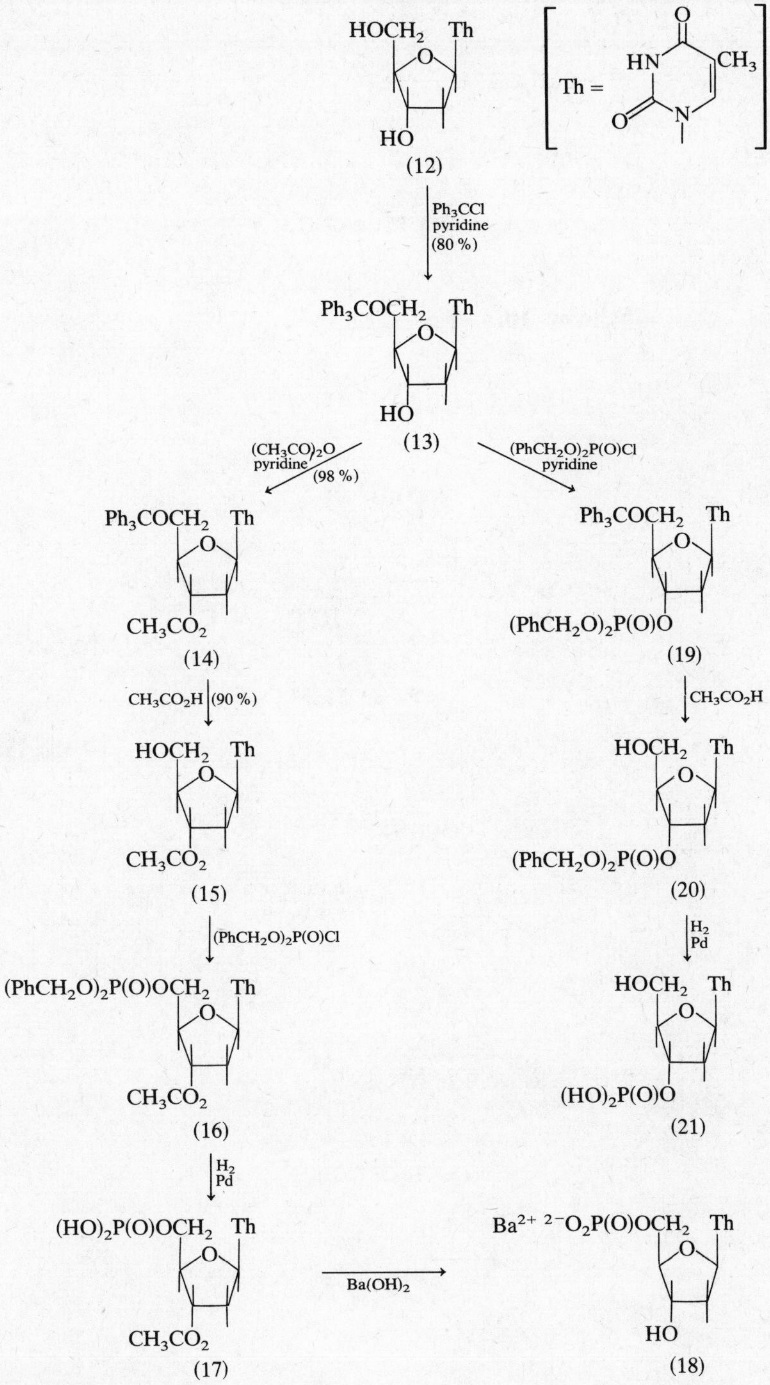
HOCH2 Th
HO
(12)
Th =
HN
CH3
O
N
Ph3CCl
pyridine
(80 %)
Ph3COCH2 Th
HO
(13)
(CH3CO)2O
pyridine
(98 %)
(PhCH2O)2P(O)Cl
pyridine
CH3CO2
(14)
(PhCH2O)2P(O)O
(19)
CH3CO2H (90 %)
CH3CO2H
(15)
(20)
H2
Pd
(PhCH2O)2P(O)OCH2 Th
(16)
(HO)2P(O)O
(21)
(HO)2P(O)OCH2 Th
Ba(OH)2
Ba2+ 2−O2P(O)OCH2 Th
(17)
(18)

Scheme 10.2

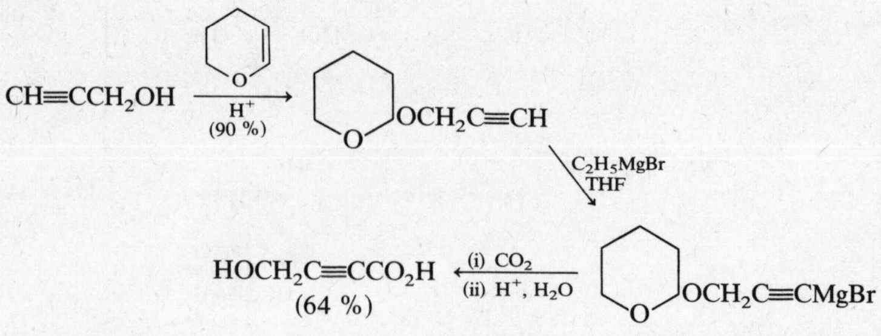

$$CH\equiv CCH_2OH \xrightarrow[\substack{H^+ \\ (90\ \%)}]{} \text{(tetrahydropyranyl)}OCH_2C\equiv CH$$

with reagent $\underset{THF}{\overset{C_2H_5MgBr}{\searrow}}$

$$HOCH_2C\equiv CCO_2H \xleftarrow[\text{(ii) } H^+,\ H_2O]{\text{(i) } CO_2} \text{(tetrahydropyranyl)}OCH_2C\equiv CMgBr$$
$$(64\ \%)$$

Scheme 10.3

$\xrightarrow[\substack{\text{pyridine} \\ (88\ \%)}]{(CF_3CO)_2O}$

$(35\ \%)\ \Big|\ \substack{\text{(i) ClCOCOCl} \\ \text{(ii) } CH_2N_2}$

$\xleftarrow[\substack{(90\ \%)}]{KHCO_3,\ H_2O}$

$\Big|\ \substack{\text{heat} \\ CH_3CO_2H}$

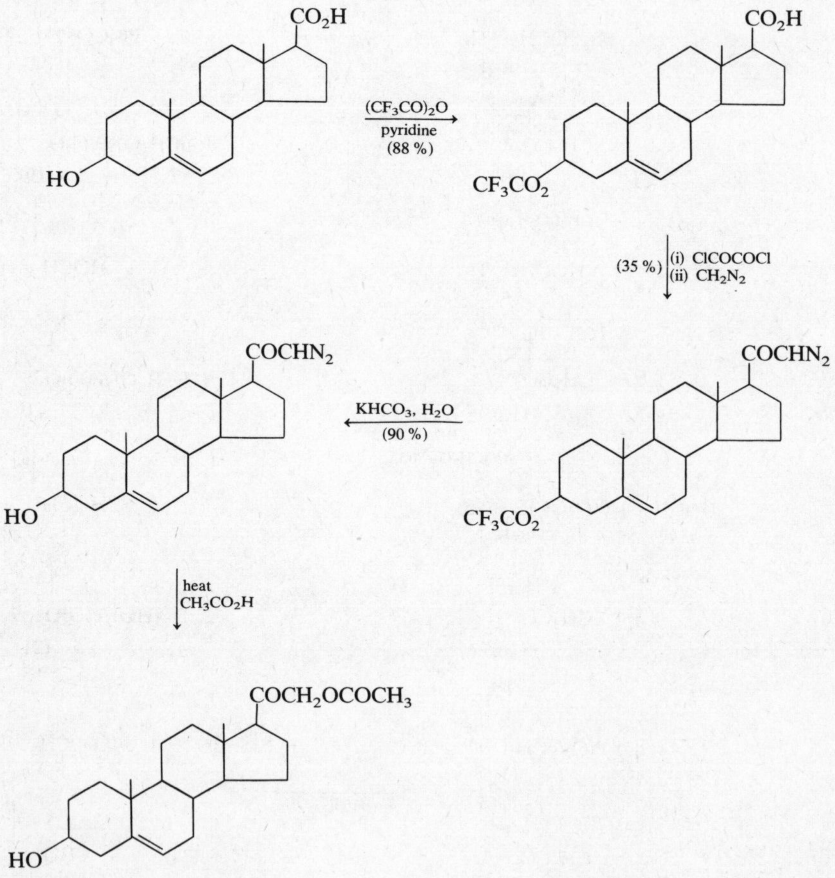

Scheme 10.4

10.3 Protection of diols

It is often convenient to protect hydroxyl groups two at a time in polyhydroxy-compounds. The protective groups can be either acetals, ketals or carbonates.

10.3.1 Acetals and ketals

The commonly used carbonyl compounds in such reactions are acetone and benzaldehyde. Acetone reacts with *cis*-1,2-diols under acid catalysis, and benzaldehyde with 1,3-diols often in presence of zinc chloride [scheme 10.1, (2) → (3)]. The diols are regenerated by treatment with dilute acid [scheme 10.1, (4) → (5)]. Hydrogenolysis can also be used in the case of the benzylidene group. Acetals and ketals are stable under neutral and alkaline conditions and can, therefore, be used to protect diols during alkylation, acylation, oxidation, and reduction provided that the reactions can be carried out under alkaline conditions [scheme 10.5].

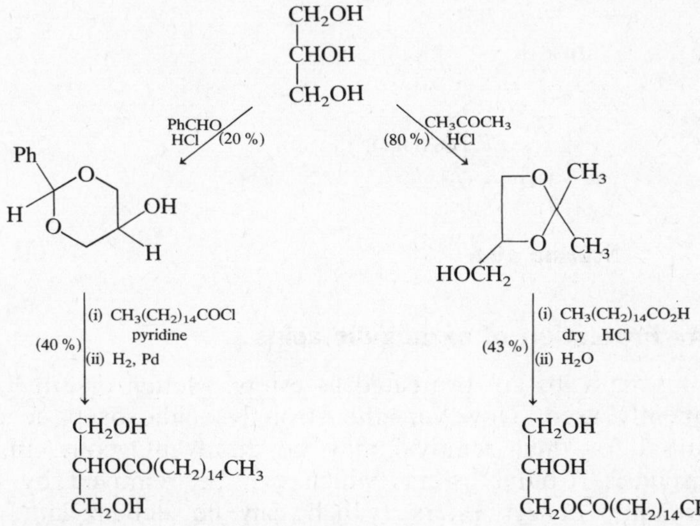

Scheme 10.5

10.3.2 Carbonates

Phosgene reacts with *cis*-1,2-diols in presence of pyridine to give a cyclic carbonate which is stable under neutral and mildly acidic conditions and can protect 1,2-diols during oxidations and reductions carried out under such conditions. Treatment with alkaline reagents regenerates the diol from the carbonate [scheme 10.6].

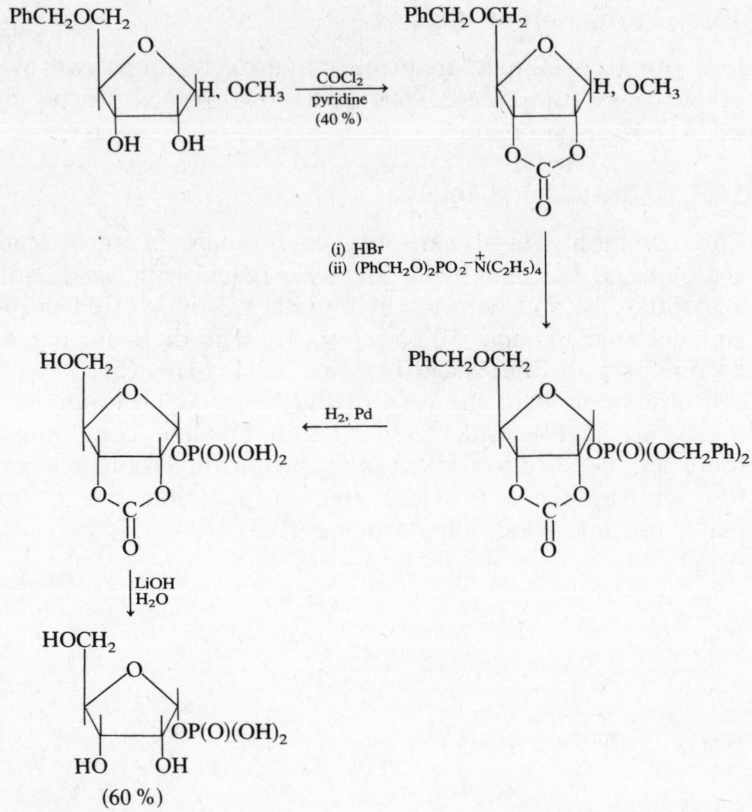

Scheme 10.6

10.4   Protection of carboxylic acids

Carboxylic acids are protected as esters. Methyl or ethyl esters are frequently used. However, the strongly acidic or basic conditions required for their removal may be disadvantageous. In such circumstances, t-butyl esters (which can be removed by mild acid treatment), benzyl esters (which can be debenzylated by hydrogenolysis) or β,β,β-trichloroethyl esters (22) (for which deprotection involves a zinc-induced elimination reaction) may be more useful. Trichloroethyl esters have been used in Woodward's cephalosporin C synthesis where the β-lactam ring in (23) has to be kept intact during removal of the protective groups:

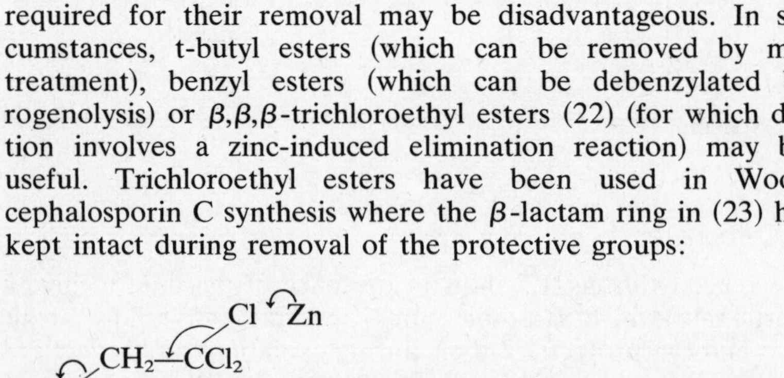

$$CCl_3CH_2O_2CCH(CH_2)_3CONH$$

$$NHCO_2CH_2CCl_3 \qquad (23)$$

with structure containing $CO_2CH_2CCl_3$ and CH_2OCOCH_3 groups

Both benzyl and t-butyl esters are widely used in peptide synthesis (schemes 10.7 and 10.8). Scheme 10.7 shows how the benzyl protective group is used to direct the reaction of the free amino-group with the activated carboxyl group ($-COCOC_2H_5$), $[(26) \rightarrow (27)]$, the

$$\overset{\parallel}{O}\,\overset{\parallel}{O}$$

stability of the group to mild acid treatment $[(27) \rightarrow (28)]$, and its ease of removal by hydrogenolysis $[(28) \rightarrow (29)]$. Scheme 10.8 shows, in addition to the protection of the acid as the t-butyl ester, the stability

$$H_2NCH_2CO_2H \xrightarrow[\text{(C}_2\text{H}_5)_2\text{NH}]{Ph_3CCl} Ph_3CNHCH_2CO_2H$$
$$\qquad (24) \qquad\qquad\qquad (25)$$

$$\searrow \begin{array}{c} ClCO_2C_2H_5 \\ (C_2H_5)_3N \end{array}$$

$$Ph_3CNHCH_2CONHCHCO_2CH_2Ph \xleftarrow[PhCH_2CH(NH_2)CO_2CH_2Ph]{} Ph_3CNHCH_2CO_2CO_2C_2H_5$$
$$(27) \quad CH_2Ph \qquad\qquad\qquad\qquad\qquad\qquad (26)$$

$$\downarrow \begin{array}{c} HCl \\ \text{aqueous} \\ \text{ethanol} \end{array}$$

$$H_2NCH_2CONHCHCO_2CH_2Ph \xrightarrow[(90\%)]{H_2,\ Pd} H_2NCH_2CONHCHCO_2H$$
$$(28) \quad CH_2Ph \qquad\qquad\qquad\qquad (29) \quad CH_2Ph$$

Scheme 10.7

$$H_2NCH_2CO_2H \xrightarrow{PhCH_2OCOCl} PhCH_2OCONHCH_2CO_2H$$
$$(30) \qquad\qquad\qquad (31)$$

$$\searrow \begin{array}{c} PhCH_2CH(NH_2)CO_2C(CH_3)_3 \\ [(C_2H_5O)_2P(O)]_2O \end{array}$$

$$H_2NCH_2CONHCHCO_2C(CH_3)_3 \xleftarrow[(60\%)]{H_2,\ Pd} PhCH_2OCONHCH_2CONHCHCO_2C(CH_3)_3$$
$$(33) \quad CH_2Ph \qquad\qquad\qquad\qquad\qquad (32) \quad CH_2Ph$$

$$(80\%)\downarrow \begin{array}{c} HCl \\ \text{benzene} \end{array}$$

$$H_2NCH_2CONHCHCO_2H$$
$$(34) \quad CH_2Ph$$

Scheme 10.8

of the t-butyl group to catalytic hydrogenation [(32) → (33)] and its facile removal by mild acid treatment [(33) → (34)].

10.5 Protection of the amino group

Schemes 10.7 and 10.8 also show the use of two protective groups frequently used for amines, the benzyloxycarbonyl group and the triphenylmethyl (trityl) group. In scheme 10.8 is shown the formation of the former using benzyl chloroformate [(30) → (31)] and its removal by hydrogenolysis [(32) → (33)]. The trityl derivative is formed by base-catalysed substitution of trityl chloride by the amino group [scheme 10.7, (24) → (25)] and is removed by mild acid treatment [(27) → (28)]. Amines are also protected by acetylation when moderate stability under acidic conditions is required and when removal under strongly basic or acidic conditions can be tolerated. An example is given in scheme 10.1, (7) → (8).

(a)

$(CH_3)_2CClCHO$ $\xrightarrow[\substack{\text{p-toluenesulphonic acid} \\ \text{benzene (86 %)}}]{HOCH_2CH_2OH}$ $(CH_3)_2CClCH$ [dioxolane ring]

$(71\%) \Big| \substack{HOCH_2CH_2OH \\ KF}$

$CH_2{=}C(CH_3)CHO$ $\xleftarrow[\substack{\text{aqueous acetone} \\ (66\%)}]{\text{oxalic acid}}$ $CH_2{=}C(CH_3)CH$ [dioxolane ring]

(b)

[cyclohexane with CO_2CH_3 and ketone] $\xrightarrow[\substack{\text{benzenesulphonic acid} \\ \text{benzene (86 %)}}]{HOCH_2CH_2OH}$ [cyclohexane with CO_2CH_3 and dioxolane]

$(67\%) \Big| LiAlH_4$

[cyclohexane with CH_2OH and ketone] $\xleftarrow[\substack{\text{aqueous} \\ \text{tartaric acid} \\ (95\%)}]{}$ [cyclohexane with CH_2OH and dioxolane]

$\Big| \substack{CH_3COCl \\ pyridine \\ ether}$

[cyclohexane with CH_2OCOCH_3 and ketone] $\xleftarrow[\substack{\text{aqueous} \\ \text{tartaric acid}}]{}$ [cyclohexane with CH_2OCOCH_3 and dioxolane]

Scheme 10.9

10.6 Protection of the carbonyl group

As has been already pointed out, 1,2-diols can be protected as the ketal by reaction with acetone [cf. section 10.3.1]. Aldehydes and ketones are frequently protected as the acetal or the ketal derived from ethylene glycol. Their stability under neutral and alkaline conditions is demonstrated in scheme 10.9.

When slightly greater stability to mildly acidic conditions is required, or when acid sensitive compounds are being used, the monothioketal[2] may be a better protective group, since it is introduced by zinc chloride-catalysed reaction with mercaptoethanol and removed by treatment with Raney nickel [scheme 10.10].

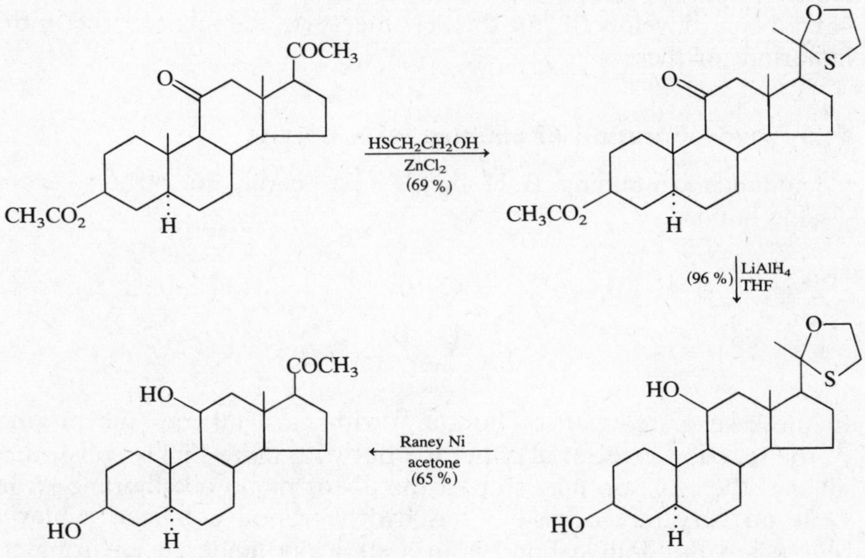

Scheme 10.10

Notes

1.   Another technique, known as latent functionality, involves the carrying through of a reaction scheme, until a late stage, with a relatively inert group which is then converted into the desired labile functional group (cf. section 14.3).

2.   This derivative is also described in the literature as a hemithioketal.

11 Boron reagents

Since the discovery in the 1950s of the facile addition of borane to alkenes, many synthetic applications of boron-containing compounds have been developed. In this chapter, we summarise the more important of these.

11.1 Hydroboration of alkenes with borane

Compounds containing B–H bonds add readily to carbon–carbon double bonds.

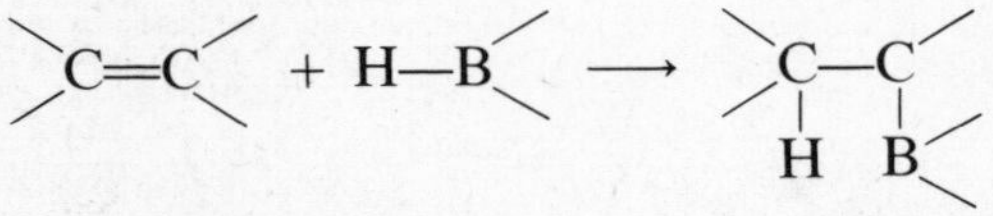

$$\text{\Large $>\!\!C\!\!=\!\!C\!<$} + \text{H--B}< \longrightarrow \text{\Large $>\!\!C\!\!-\!\!C\!<$} \atop \text{\quad\quad H \quad B}<$$

In the case of reaction of borane[1] with most alkenes the product of the reaction is the trialkylborane but with more highly substituted alkenes the reaction may stop at the di- or mono-alkylborane stage. Only on very rare occasions has hydroboration not been achieved [e.g. when the double bond is in a sterically hindered environment such as in (3)]. Hindered boranes, (Sia)$_2$BH (1)[2] and thexylborane (2)[3], are in themselves useful reagents [cf. sections 11.2.1, 11.2.2, 11.3.3, 11.5]:

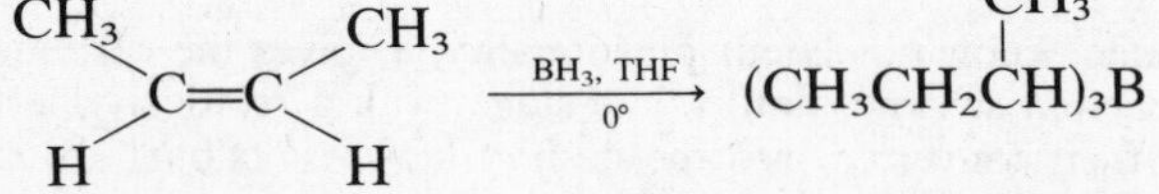

$$\underset{H}{\overset{CH_3}{>}}C\!\!=\!\!C\underset{H}{\overset{CH_3}{<}} \xrightarrow[0°]{BH_3,\ THF} (CH_3CH_2\overset{\displaystyle CH_3}{\overset{|}{C}}H)_3B$$

$$(CH_3)_2C\!\!=\!\!CHCH_3 \xrightarrow[0°]{BH_3,THF} \left((CH_3)_2CH\overset{\displaystyle CH_3}{\overset{|}{C}}H\right)_2BH$$

(1)

$$(CH_3)_2C\!\!=\!\!C(CH_3)_2 \xrightarrow[0°]{BH_3,THF} (CH_3)_2CHC(CH_3)_2BH_2$$

(2)

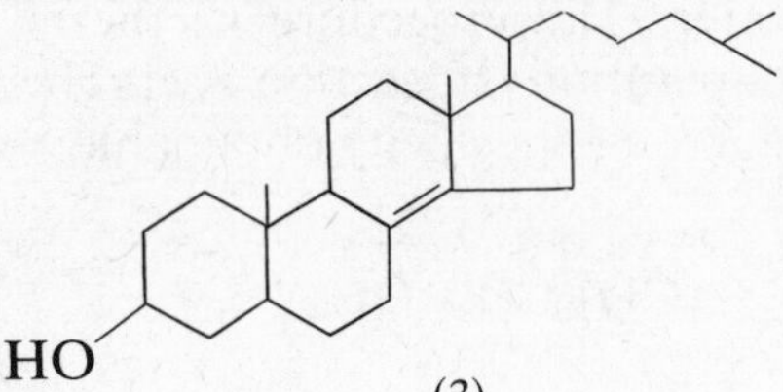

(3)

Of the reactions which will be discussed later [cf. section 11.3], one of the most important is the virtually quantitative oxidation with alkaline hydrogen peroxide (11.1). The overall reaction is addition of water to the carbon–carbon double bond:

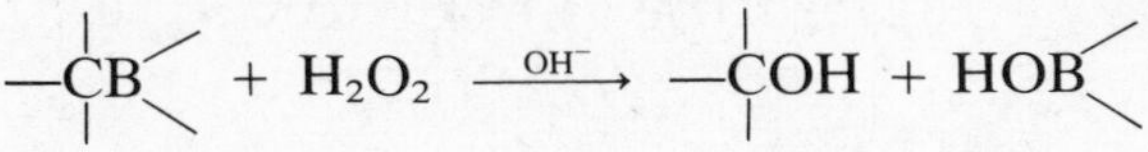

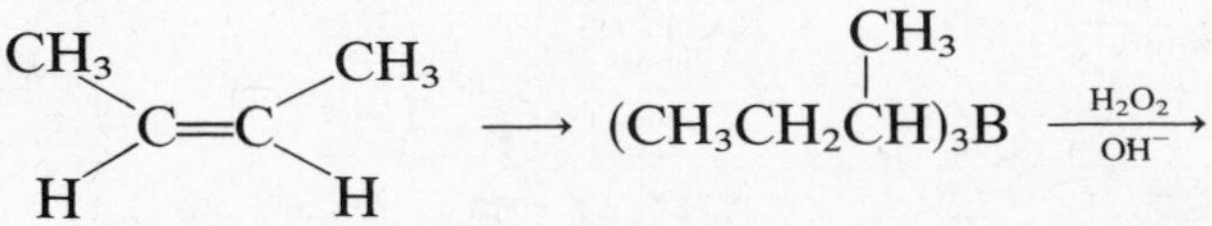

$$CH_3CH_2CH(OH)CH_3 + B(OH)_3 \qquad (11.1)$$

It should be noted that the decomposition of the borane takes place with retention of configuration, and so the stereochemistry of the product is determined by the stereochemistry of the addition of borane to the alkene. The following examples demonstrate that the addition to the double bond is *cis* on the *less hindered side* of the molecule:

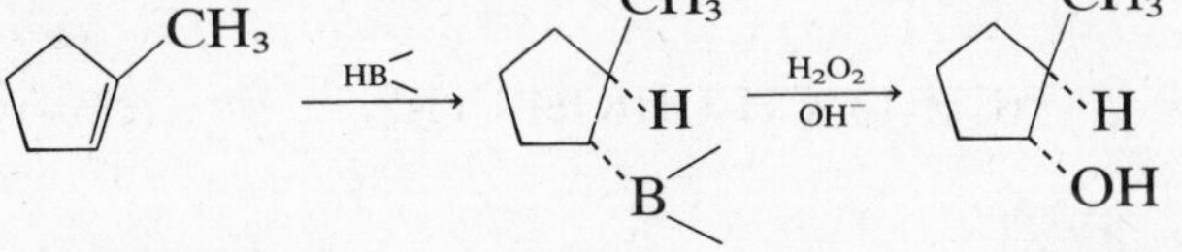

(11.2)

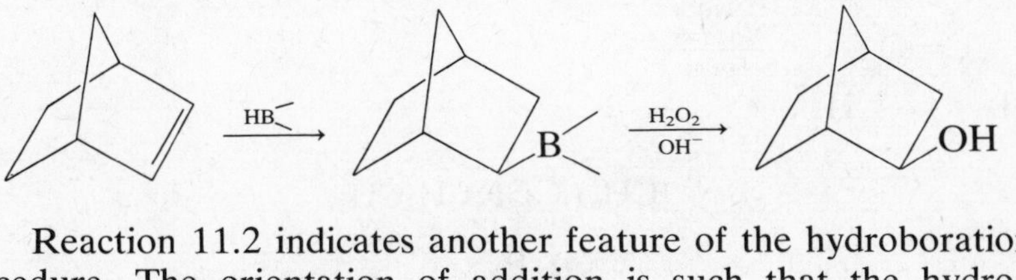

Reaction 11.2 indicates another feature of the hydroboration procedure. The orientation of addition is such that the hydrogen is attached to the more highly substituted carbon, giving, after treatment with alkaline hydrogen peroxide, the product formally derived

by 'anti-Markownikoff' addition of water. The procedure is, therefore, complementary to acid-catalysed hydration [cf. section 2.2]. The
following examples demonstrate the degree of specificity obtainable:

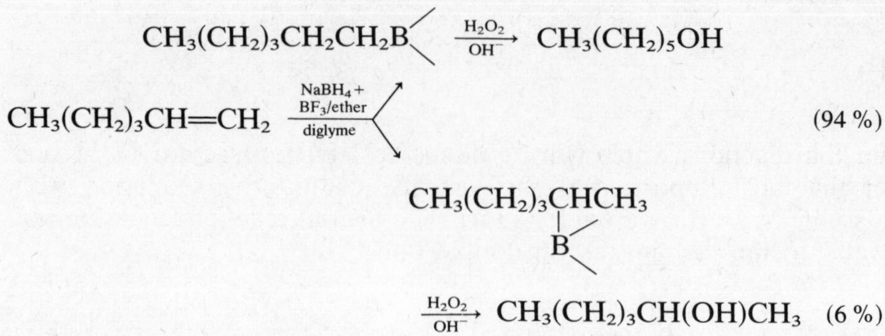

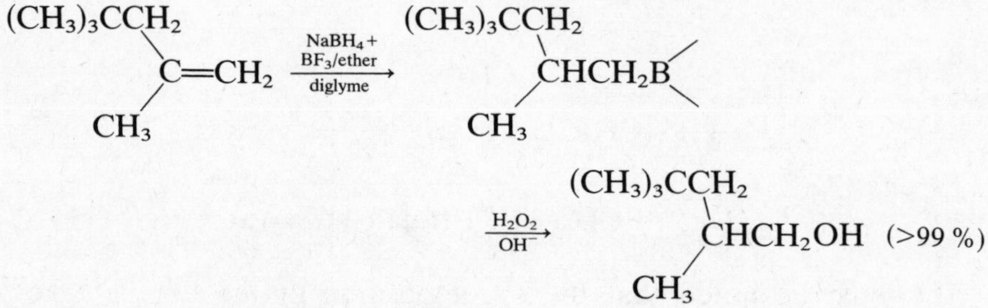

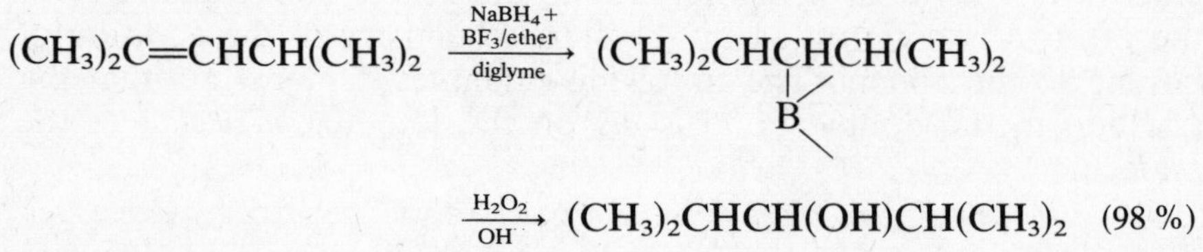

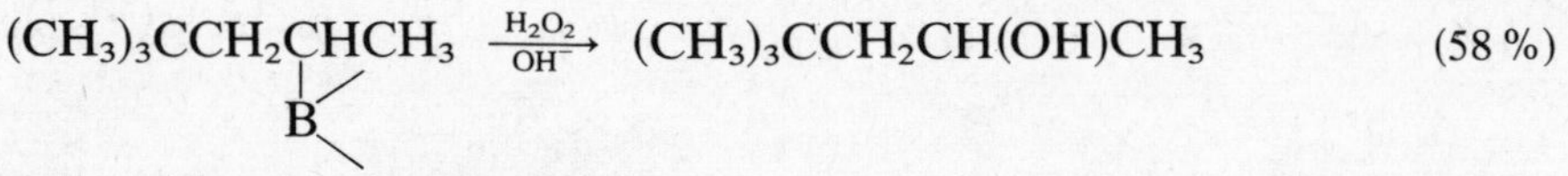

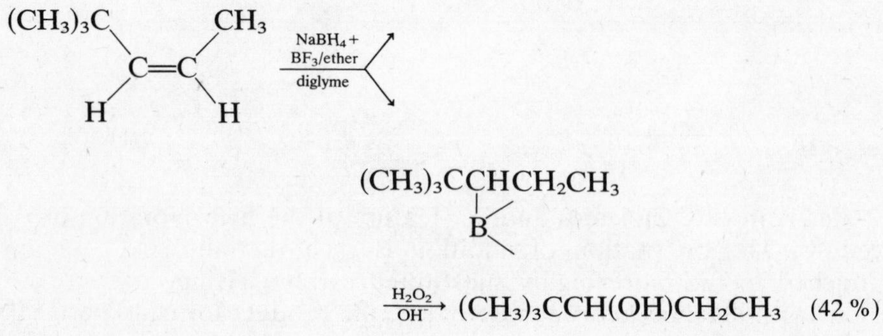

$$CH_3(CH_2)_3CH_2CH_2B\diagup \xrightarrow[OH^-]{H_2O_2} CH_3(CH_2)_5OH$$

$$CH_3(CH_2)_3CH=CH_2 \xrightarrow[diglyme]{\substack{NaBH_4+\\BF_3/ether}}$$

(94 %)

$$CH_3(CH_2)_3CHCH_3$$
$$|$$
$$B\diagup$$

$$\xrightarrow[OH^-]{H_2O_2} CH_3(CH_2)_3CH(OH)CH_3 \quad (6\,\%)$$

$$(CH_3)_3CCH_2$$
$$\diagdown$$
$$C=CH_2 \xrightarrow[diglyme]{\substack{NaBH_4+\\BF_3/ether}}$$
$$\diagup$$
$$CH_3$$

$$(CH_3)_3CCH_2$$
$$\diagdown$$
$$CHCH_2B\diagup$$
$$\diagup$$
$$CH_3$$

$$\xrightarrow[OH^-]{H_2O_2}$$

$$(CH_3)_3CCH_2$$
$$\diagdown$$
$$CHCH_2OH \quad (>99\,\%)$$
$$\diagup$$
$$CH_3$$

$$(CH_3)_2C=CHCH(CH_3)_2 \xrightarrow[diglyme]{\substack{NaBH_4+\\BF_3/ether}} (CH_3)_2CHCHCH(CH_3)_2$$
$$|$$
$$B\diagup$$

$$\xrightarrow[OH]{H_2O_2} (CH_3)_2CHCH(OH)CH(CH_3)_2 \quad (98\,\%)$$

$$(CH_3)_3CCH_2CHCH_3 \xrightarrow[OH^-]{H_2O_2} (CH_3)_3CCH_2CH(OH)CH_3 \quad (58\,\%)$$
$$|$$
$$B\diagup$$

$$(CH_3)_3C \qquad CH_3$$
$$\diagdown \qquad \diagup$$
$$C=C \xrightarrow[diglyme]{\substack{NaBH_4+\\BF_3/ether}}$$
$$\diagup \qquad \diagdown$$
$$H \qquad H$$

$$(CH_3)_3CCHCH_2CH_3$$
$$|$$
$$B\diagup$$

$$\xrightarrow[OH^-]{H_2O_2} (CH_3)_3CCH(OH)CH_2CH_3 \quad (42\,\%)$$

The foregoing reactions demonstrate that in 1-alkyl-, 1,1-dialkyl- and 1,1,2-trialkylethylenes the predominant reaction places the hydrogen on the more substituted carbon, the proportion of isomers formed being largely independent of the alkyl group. Additions to 1,2-dialkylethylenes show very little preference for either orientation. An exception to the small influence of the substituent in orientation of hydroboration is found in the case of p-substituted styrenes (table 11.1).

Table 11.1 Orientation of hydroboration in p-substituted styrenes

$$p\text{-}XC_6H_4CH{=}CH_2 \rightarrow p\text{-}XC_6H_4\overset{\overset{\displaystyle B}{|}}{C}HCH_3 + p\text{-}XC_6H_4CH_2CH_2B{\diagup}$$

X = OCH$_3$	7 %	93 %
X = H	19 %	81 %
X = Cl	27 %	73 %

Boranes react more rapidly with carbon–carbon double bonds than with most other functional groups. So, in many cases, hydroboration can be carried out selectively, e.g.

$$CH_2{=}CH(CH_2)_8CO_2CH_3 \xrightarrow{\text{BH}_3/\text{THF}} {\diagdown}B(CH_2)_{10}CO_2CH_3$$

It is, however, desirable to protect carbonyl groups as acetals or ketals, and acids as esters [cf. sections 10.4, 10.6].

Polar groups attached to the double bond exert a directive effect analogous to that noted previously in the case of p-substituted styrenes, e.g.

$$(CH_3)_2C{=}CHOC_2H_5 \xrightarrow{\text{BH}_3/\text{THF}} (CH_3)_2\overset{\overset{\displaystyle }{|}}{C}CH_2OC_2H_5$$
$$B$$

(≥88 %) (≤12 % of 1-isomer)

but:

$$(CH_3)_2C{=}CHOCOCH_3 \xrightarrow{\text{BH}_3/\text{THF}} (CH_3)_2CH\overset{\overset{\displaystyle }{|}}{C}HOCOCH_3$$
$$B{\diagup}$$

(≥95 %) (≤5 % of 2-isomer)

The effect decreases as the heteroatom is further removed from the double bond, e.g.

$$CH_3CH{=}CHCH_2Cl \xrightarrow{\text{BH}_3/\text{THF}} CH_3CH_2\underset{\underset{\diagdown}{\overset{|}{B}}}{C}HCH_2Cl + CH_3\underset{\underset{\diagdown}{\overset{|}{B}}}{C}HCH_2CH_2Cl$$

$$(>90\,\%) \qquad\qquad (<10\,\%)$$

$$CH_2{=}CHCH_2CH_2Cl \xrightarrow{\text{BH}_3/\text{THF}} \overset{\diagdown}{\underset{\diagup}{B}}(CH_2)_4Cl + CH_3\underset{\underset{\diagdown}{\overset{|}{B}}}{C}HCH_2CH_2Cl$$

$$(80\,\%) \qquad\qquad (20\,\%)$$

11.2 Hydroboration of alkenes with alkylboranes

A number of alkylboranes have found synthetic application and some of these with their uses will now be considered.

11.2.1 Disiamylborane[2]

2-Methylbut-2-ene reacts with borane only slowly beyond the dialkyl-borane stage. The resultant dialkylborane (1), often known as dis-iamylborane or $(Sia)_2BH$, reacts with a very high degree of selectivity with monosubstituted alkenes. In cases such as that of allyl chloride, where addition of borane itself leads to a mixture of adducts (the reverse adduct being formed as the result of the inductive effect of the halogen), the use of the bulkier disiamylborane is necessary to ensure that only 'normal' addition occurs:

$$CH_3(CH_2)_3CH{=}CH_2 + [(CH_3)_2CHCH(CH_3)]_2BH$$

$$(1)$$

$$\to CH_3(CH_2)_5B[CH(CH_3)CH(CH_3)_2]_2$$

$$\overset{\text{(Sia)}_2\text{BH}}{\nearrow} Cl(CH_2)_3B(Sia)_2$$

$$ClCH_2CH{=}CH_2$$

$$\underset{\text{BH}_3/\text{THF}}{\searrow} Cl(CH_2)_3\overset{\diagup}{\underset{\diagdown}{B}} + ClCH_2\underset{\underset{\diagdown}{\overset{|}{B}}}{C}HCH_3$$

$$(60\,\%) \qquad\qquad (40\,\%)$$

The reagent also shows a high degree of selectivity towards the less hindered site in a disubstituted alkene: for example in (4), where reaction with borane produces only a slight excess of the 2-isomer. $(Sia)_2BH$ also reacts preferentially with monosubstituted alkenes:

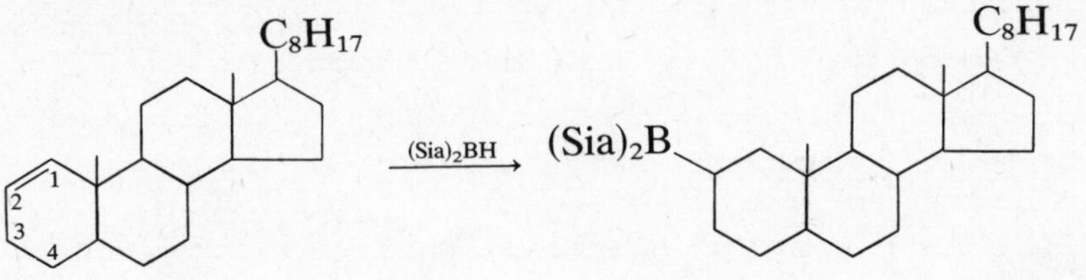

(4)

no detectable amount of 1-isomer

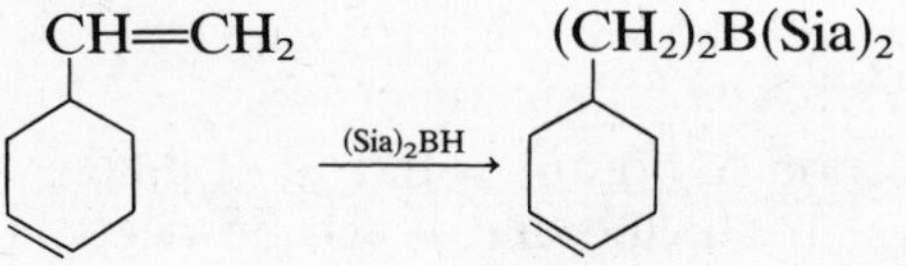

11.2.2 Thexylborane[3]

2,3-Dimethylbut-2-ene reacts with borane to give a monoalkylborane (2) which is often known as thexylborane. It is used mainly when mixed alkylboranes are required (cf. section 11.3.5) and for the alkylation of dienes (cf. section 11.4).

11.2.3 9-Borabicyclo[3.3.1]nonane

As will be seen later [section 11.4], 1.5-dienes react with borane: when cycloocta-1,5-diene is used the product is 9-bora-bicyclo[3.3.1]nonane (5), 9-BBN. This alkylborane reacts more slowly with alkenes than does $(Sia)_2BH$. It has, however, a much greater thermal stability than most dialkylboranes and hence reactions can be carried out in refluxing tetrahydrofuran.

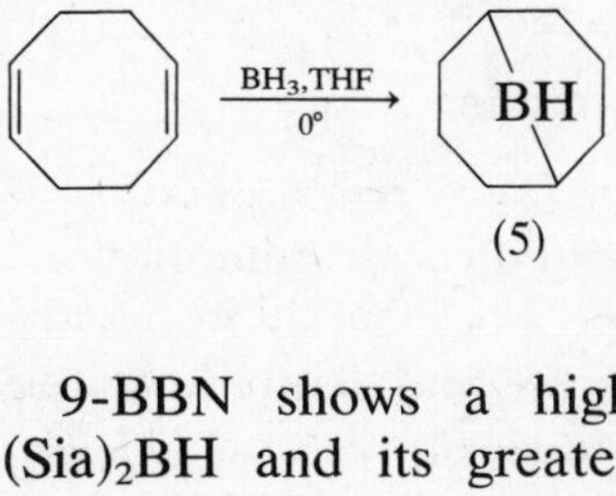

(5)

9-BBN shows a higher degree of regioselectivity than does $(Sia)_2BH$ and its greater stability to heat and towards oxidation renders it the more suitable reagent in most situations:

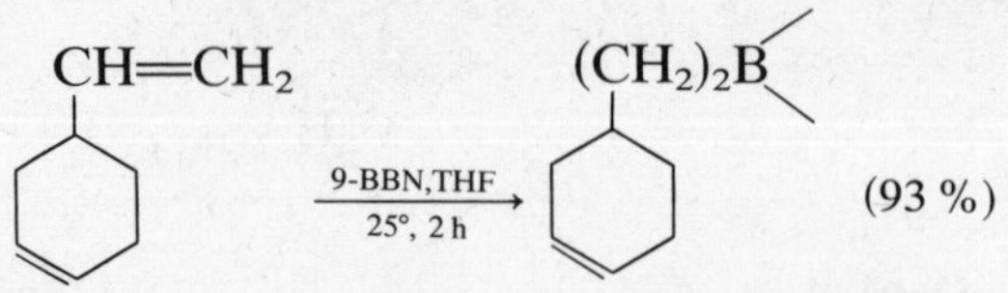

(93 %)

no trace of 2-isomer

(88 %)

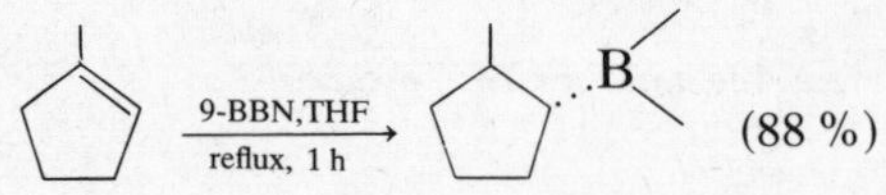

no trace of 1-isomer

11.2.4 (+)-Diisopinocamphenylborane

This dialkylborane, formed by reaction of borane with (+)-α-pinene, reacts with alkenes to give a chiral trialkylborane which on decomposition results in the formation of products of high optical purity[4]:

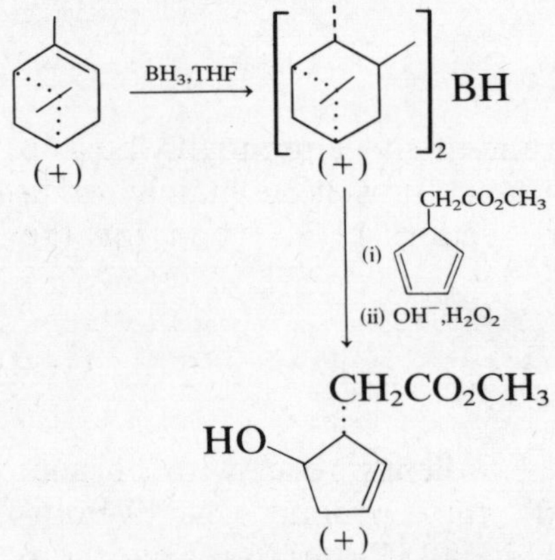

(92 % optically pure)

11.3 Reactions of organoboranes

These are summarised in scheme 11.1 and some of the more important are discussed in the following sections.

11.3.1 Reaction with alkaline hydrogen peroxide

This is probably the single most widely used reaction of organoboranes in which the borane is converted, with retention of configuration (*via* a borate ester), into the alcohol. Thus, the overall reaction of the alkene is its conversion into an alcohol by the *cis*-addition of the elements of water [reactions (11.3) and (11.4)].

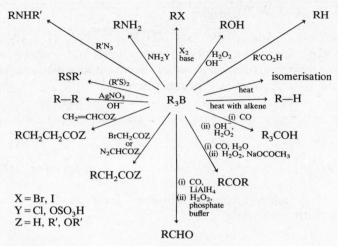

Scheme 11.1

The regiospecificity of the reaction is, of course, determined by that of the hydroboration step [sections 11.1, 11.2].

$$RCH{=}CH_2 \xrightarrow{H-B\diagdown} RCH_2CH_2{-}B\diagdown \xrightarrow{OOH^-} \left[RCH_2CH_2\underset{O-OH}{\overset{|}{B}} \right]$$

$$\longrightarrow RCH_2CH_2O\,B\diagdown + {}^-OH \longrightarrow RCH_2CH_2O^- + HOB\diagdown$$

$$\longrightarrow RCH_2CH_2OH + \bar{O}B\diagdown \quad (11.3)$$

$$\underset{H}{\overset{R}{>}}C{=}C\underset{H}{\overset{R}{<}} \xrightarrow{H-B\diagup} \underset{H\ H}{\overset{R}{\underset{B}{\overset{R}{>}C{-}C}}}H \xrightarrow[H_2O_2]{OH^-} \underset{H\ H}{\overset{R}{\underset{OH}{\overset{R}{>}C{-}C}}}H \qquad (11.4)$$

The reaction is thus in some respects complementary to the acid-catalysed hydration discussed in section 2.2.

$$\text{(cyclohexene)} \xrightarrow[0°]{BH_3,\,THF} \left(\text{(cyclohexyl)}\right)_2 BH \xrightarrow[40-50°]{OH^-,\,H_2O_2} \text{(cyclohexanol)} \quad (85\,\%)$$

$$\text{(4-vinylcyclohexene)} \xrightarrow[0°]{(Sia)_2BH,\,THF} (CH_2)_2B(Sia)_2 \xrightarrow[50°]{OH^-,\,H_2O_2} (CH_2)_2OH \quad (72\,\%)$$

11.3.2   Conversion to amino groups

Boranes react with compounds NH_2X where X is a good leaving group. These are often unstable compounds but hydroxylamine-O-sulphonic acid is a reasonably stable compound which can be used in synthesis. The product is a primary amine:

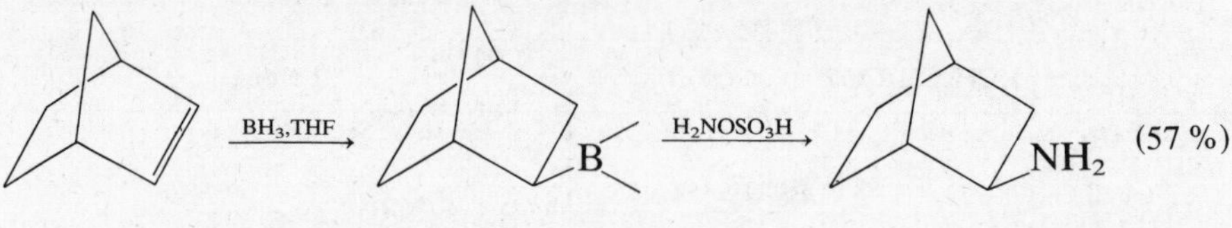

Secondary amines can be obtained from reaction of dichloro-boranes with an azide:

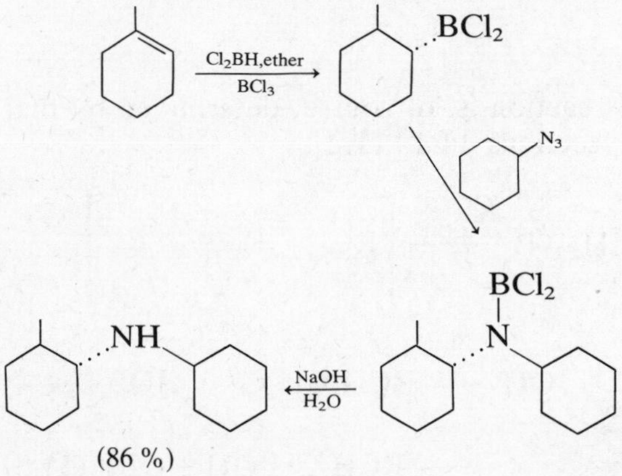

11.3.3   Conversion to halogeno-compounds

Although boranes are quite stable to halogens, rapid reaction follows the addition of alkali. In the case of tri-(primary alkyl)-boranes, only two of the three alkyl groups react; and secondary alkyl groups do not react. Thus, for conversion of terminal alkenes into primary iodides, it is preferable to use Sia_2BH for optimum yield:

$$3RCH\!=\!CH_2 \xrightarrow{BH_3} (RCH_2CH_2)_3B \xrightarrow[2NaOH]{2I_2} 2RCH_2CH_2I$$

$$+\,2NaI+RCH_2CH_2B(OH)_2$$

$$RCH\!=\!CH_2 \xrightarrow{(Sia)_2BH} RCH_2CH_2B(Sia)_2 \xrightarrow[NaOH]{I_2}$$

$$RCH_2CH_2I+NaI+(Sia)_2BOH$$

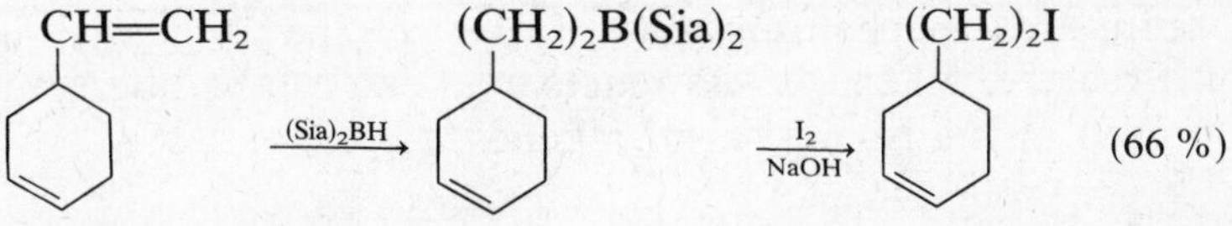

$CH{=}CH_2$ $(CH_2)_2B(Sia)_2$ $(CH_2)_2I$

$\xrightarrow{(Sia)_2BH}$ $\xrightarrow[NaOH]{I_2}$ (66 %)

Conversion into bromides is also observed by use of bromine and sodium methoxide but water must be rigorously excluded (due, possibly, to the formation of hypobromous acid which hydrolyses boranes to alcohols).

It should be noted that in this case (i) secondary alkyl groups react and (ii) the *endo*-bromo-compound is produced from the *exo*-norbornyl borane. The *exo*-isomer is formed by reaction of bromine with the adduct of norbornene with 9-BBN when the reaction proceeds by radical attack of bromine on the α-hydrogen of the alkylborane:

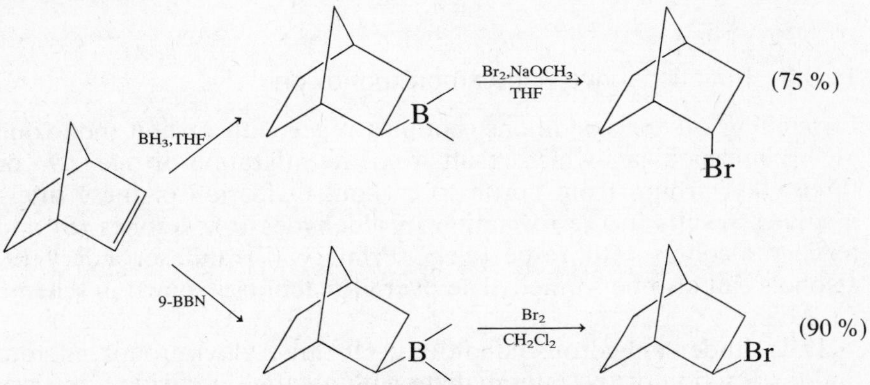

$\xrightarrow{BH_3,THF}$ $\xrightarrow[THF]{Br_2,NaOCH_3}$ (75 %)

$\xrightarrow{9\text{-BBN}}$ $\xrightarrow[CH_2Cl_2]{Br_2}$ (90 %)

11.3.4 Reaction with organic acids

Organic acids convert alkylboranes into alkanes. Propionic acid is often used for this purpose but the reaction has not enjoyed wide synthetic use presumably because of the availability of simpler procedures for reduction of alkenes. The method is more widely used in the decomposition of vinylboranes (cf. section 11.5).

11.3.5 Thermal reactions of alkylboranes

When heated, alkylboranes isomerise so that the boron migrates to the least hindered position of the alkyl group. It is thought that the isomerism takes place by dissociation of the alkylborane followed by hydroboration:

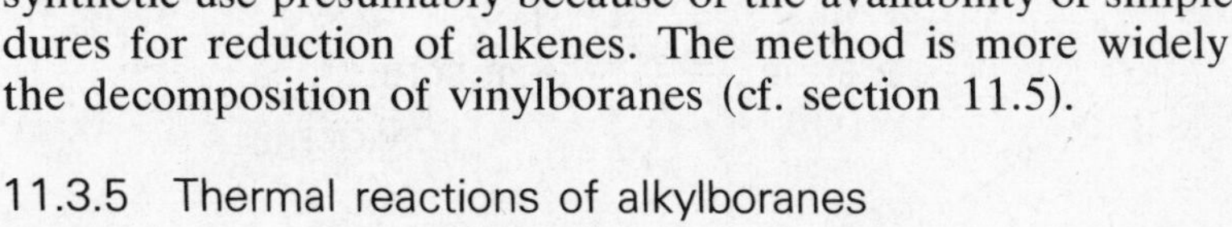

$$-\overset{|}{\underset{H}{C}}-\overset{|}{\underset{B}{C}}-\overset{|}{\underset{H}{C}}- \;\rightleftharpoons\; -\overset{|}{\underset{H}{C}}-\overset{|}{C}{=}C{<} + BH \;\rightleftharpoons\; -\overset{|}{\underset{H}{C}}-\overset{|}{\underset{H}{C}}-\overset{|}{\underset{B}{C}}-$$

It follows from this mechanism that, if the borane is heated in presence of a reactive alkene, a less reactive alkene can be liberated. This can be used in the isomerism of alkenes, for example:

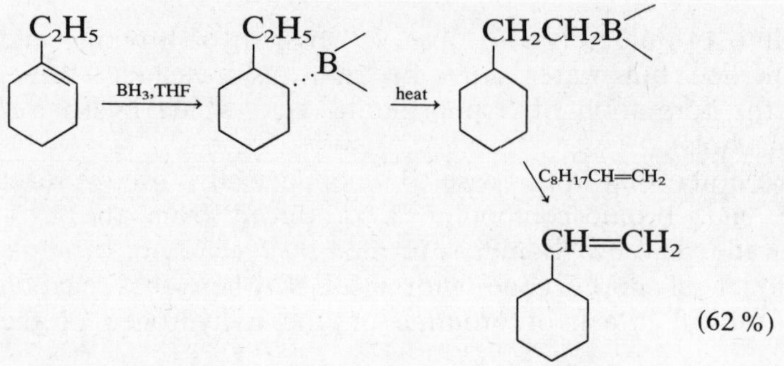

(62 %)

11.3.6   Reactions involving carbon monoxide

Depending on the conditions, boranes react with carbon monoxide giving intermediates which result from the migration of one, two or three alkyl groups from boron to carbon. Oxidation of these intermediates results in the formation of aldehydes (6), ketones (8) and tertiary alcohols (10) respectively. Primary (7) and secondary (9) alcohols can also be formed. The overall reaction is shown in scheme 11.2.

Thus, under anhydrous conditions, all three alkyl groups migrate and on reaction of the intermediate with alkaline peroxide a tertiary

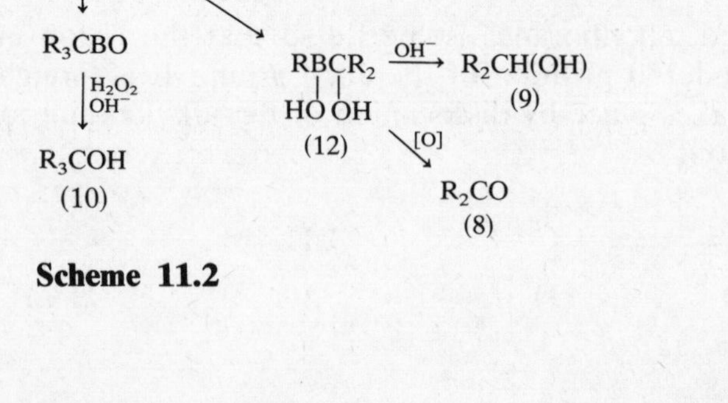

Scheme 11.2

alcohol is formed:

$$\text{cyclopentene} \xrightarrow{\text{BH}_3,\text{THF}} \left(\text{cyclopentyl}\right)_3\!\!B \xrightarrow{\text{CO}} \left(\text{cyclopentyl}\right)_3\!\!CBO$$

$$\downarrow \text{H}_2\text{O}_2,\text{OH}^-$$

$$\left(\text{cyclopentyl}\right)_3\!\!COH \quad (80\%)$$

When a trace of water is present, the reaction is intercepted when only two alkyl groups have migrated and the intermediate boraepoxide (11) is hydrolysed to a boraglycol (12). The boraglycol can be hydrolysed to the secondary alcohol or oxidised to the ketone. With a trialkylborane, one alkyl group is lost but, when hydroboration is performed using thexylborane, the thexyl group usually shows least susceptibility to migration. Unsymmetrical ketones can be prepared in cases where thexylborane can be monoalkylated by hindered alkenes:

$$\text{thexyl}{-}\text{BH}_2 + (\text{CH}_3)_2\text{C}{=}\text{CH}_2 \longrightarrow \text{thexyl}{-}\text{BHCH}_2\text{CH}(\text{CH}_3)_2$$

(2)

$$\downarrow \text{CH}_2{=}\text{CHCH}_2\text{CO}_2\text{C}_2\text{H}_5$$

$$\text{thexyl}{-}\text{B}{-}\text{C}\underset{\text{OH OH}}{\overset{\displaystyle \text{CH}_2\text{CH}(\text{CH}_3)_2}{\diagup}}\!\!(\text{CH}_2)_3\text{CO}_2\text{C}_2\text{H}_5 \xleftarrow[\text{H}_2\text{O}]{\text{CO}} \text{thexyl}{-}\text{B}\underset{\text{CH}_2\text{CH}_2\text{CH}_2\text{CO}_2\text{C}_2\text{H}_5}{\overset{\displaystyle \text{CH}_2\text{CH}(\text{CH}_3)_2}{\diagup}}$$

$$\downarrow \begin{array}{c}\text{H}_2\text{O}_2\\\text{NaO}_2\text{CCH}_3\end{array}$$

$$(\text{CH}_3)_2\text{CHCH}_2\text{CO}(\text{CH}_2)_3\text{CO}_2\text{C}_2\text{H}_5$$

(84%)

If the reaction is carried out in presence of a reactive hydride reducing agent, the boraketone (13) formed on migration of the first alkyl group is reduced. The reduction product is hydrolysed to the primary alcohol and oxidised to the aldehyde. In this case, two of the alkyl groups of the trialkylborane are lost, but use of 9-BBN as hydroboration reagent circumvents the loss of valuable alkyl groups:

$$\text{cyclopentene} \xrightarrow{\text{9-BBN}} \text{cyclopentyl-B} \xrightarrow[\substack{\text{(ii) H}_2\text{O}_2, \text{ phosphate}\\ \text{buffer}}]{\text{(i) CO,LiAlH(OCH}_3)_3} \text{cyclopentyl-CHO} \quad (79\%)$$

11.4 Hydroboration of dienes

Hydroboration of 1,3-, 1,4-, and 1,5-dienes leads to the formation of boracycloalkanes. Use of borane itself may lead to considerable polymerisation and, in most cases, use of thexylborane is preferred. An exception is, of course, the preparation of 9-BBN (cf. section 11.2.3).

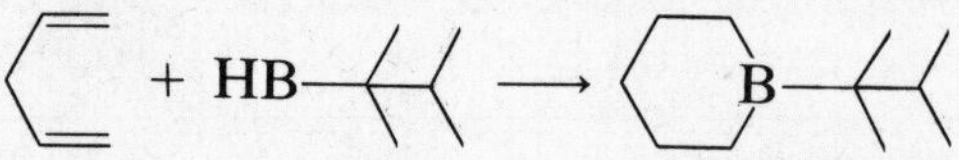

One of the principal synthetic applications of boracycloalkanes is in the preparation of cyclic ketones. This can be achieved by use of the carbonylation reaction (cf. section 11.3.5) but, because, of the high pressure of carbon monoxide (70 atm) required to cause reaction with thexylborocycloalkanes, an alternative procedure involving acylation of cyanoborates [e.g. (14)] represents a more convenient synthetic method. Acylation of cyanoborates results in the migration of two alkyl groups to form the borodihydrooxazole intermediate (15) which can be decomposed in the usual way:

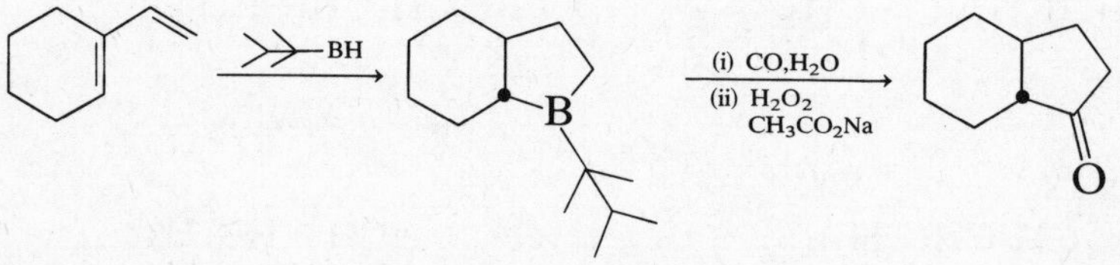

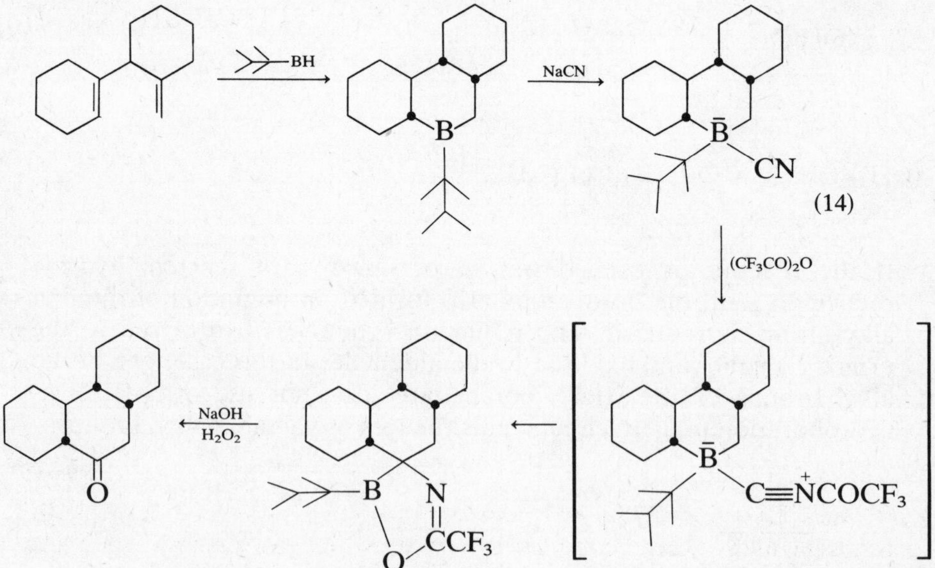

11.5 Hydroboration of alkynes

Monohydroboration of non-terminal alkynes can be achieved by use of controlled amounts of borane, but use of a hindered borane such as disiamylborane prevents further hydroboration. Disiamylborane also yields monohydroboration products on reaction with terminal alkynes which give only dihydroboration products with borane itself. Boranes formed by monohydroboration of alkynes undergo many of the transformations described in section 11.3. Some representative examples having synthetic utility include the following:

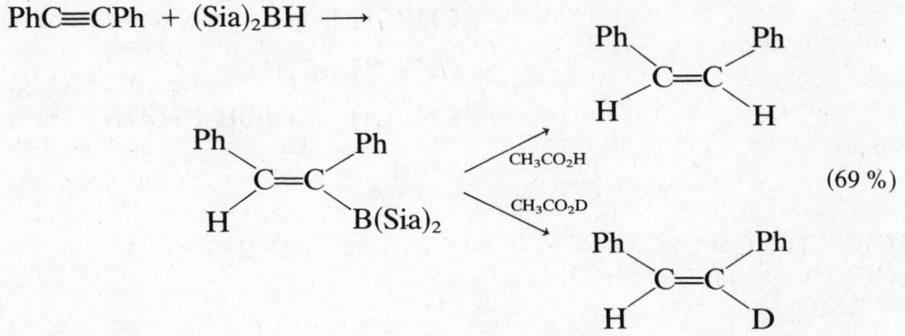
PhC≡CPh + (Sia)2BH
Ph
Ph
C=C
H
H
Ph
Ph
C=C
H
B(Sia)2
CH3CO2H
CH3CO2D
(69 %)
Ph
Ph
C=C
H
D

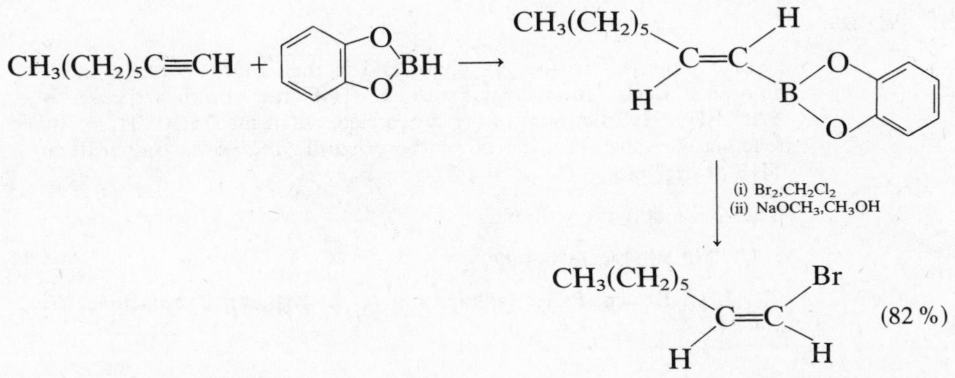
CH3(CH2)5C≡CH +
O
BH
O
CH3(CH2)5
H
C=C
H
O
B
O
(i) Br2,CH2Cl2
(ii) NaOCH3,CH3OH
CH3(CH2)5
Br
C=C
H
H
(82 %)

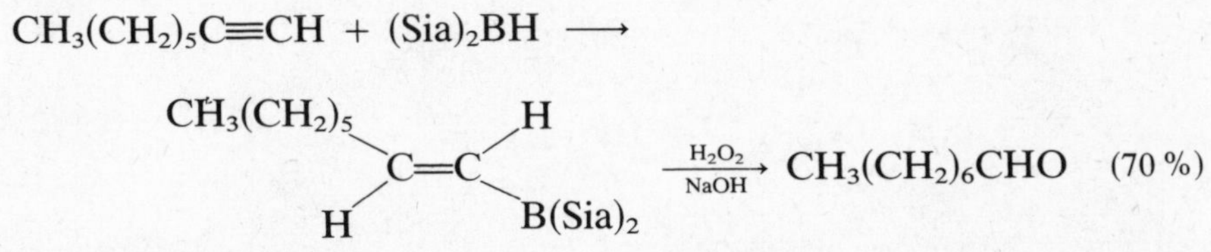
CH3(CH2)5C≡CH + (Sia)2BH
CH3(CH2)5
H
C=C
H
B(Sia)2
H2O2
NaOH
CH3(CH2)6CHO
(70 %)

The dihydroboration products formed by reaction of borane with terminal alkynes are complex polymers which on reaction with alkaline hydrogen peroxide yield mixtures of products. However a synthetically useful procedure is the oxidation of the dihydroboration product of dicyclohexylborane and terminal alkynes to alkanoic acids:

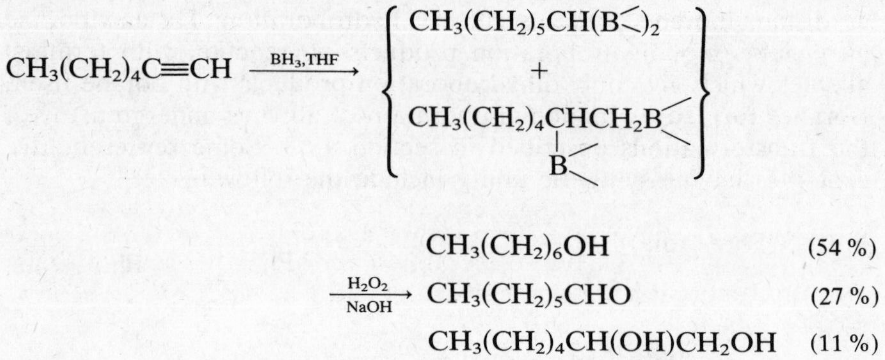

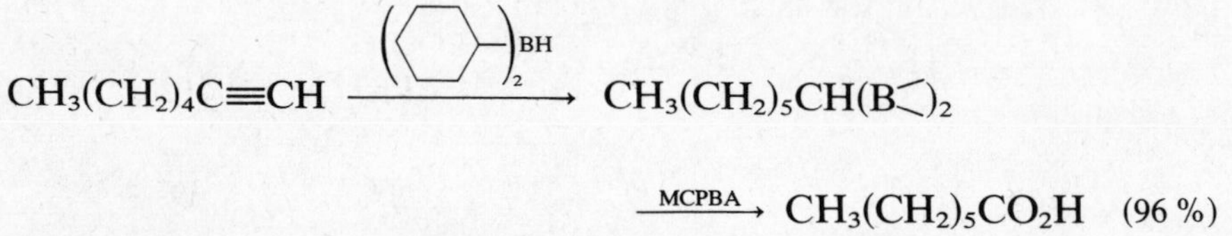

$$CH_3(CH_2)_4C \equiv CH \xrightarrow{BH_3, THF} \begin{Bmatrix} CH_3(CH_2)_5CH(B\diagdown)_2 \\ + \\ CH_3(CH_2)_4\underset{\underset{B\diagdown}{|}}{C}HCH_2B\diagdown \end{Bmatrix}$$

$$\xrightarrow[NaOH]{H_2O_2} \begin{matrix} CH_3(CH_2)_6OH & (54\%) \\ CH_3(CH_2)_5CHO & (27\%) \\ CH_3(CH_2)_4CH(OH)CH_2OH & (11\%) \end{matrix}$$

$$CH_3(CH_2)_4C \equiv CH \xrightarrow{\left(\langle\diamondsuit\rangle\right)_2 BH} CH_3(CH_2)_5CH(B\diagdown)_2$$

$$\xrightarrow{MCPBA} CH_3(CH_2)_5CO_2H \quad (96\%)$$

Notes

1. Borane dimerises in the gas phase and is then more correctly called diborane, B_2H_6. In solvents such as THF the reactive species is $THF:BH_3$. Hydroboration may be carried out using $THF:BH_3$ or by reacting the substrate with sodium borohydride and boron trifluoride in THF or diglyme.

2. Bis(1,2-dimethylpropyl)borane

3. 1,1,2-Trimethylpropylborane

4. Cf. H. C. Brown, P. K. Jadhav, and A. K. Mandal, *Tetrahedron*, **37**, 3547 (1981).

12   Phosphorus reagents

In recent years, there has been an increasing interest in the application of phosphorus reagents in organic synthesis. In the space available in this book, it is possible only to describe some highlights of this field.

12.1 Introduction to organophosphorus chemistry

The versatility of phosphorus is due in large part to several aspects of its chemistry, e.g.:

- (i) phosphorus exists as tri-, tetra-, penta-, and hexa-co-ordinate species and many interconversions of these are known;
- (ii) tervalent phosphorus compounds are weakly basic and highly nucleophilic species, and they react by nucleophilic attack at a variety of sites (e.g. nitrogen, oxygen, sulphur, halogen and electrophilic carbon);
- (iii) phosphorus forms strong bonds with many other elements including carbon, nitrogen, halogen, sulphur and oxygen, with the $P=O$ bond being of particular strength and importance;
- (iv) phosphorus is capable of stabilising adjacent anions.

The highly nucleophilic character of trialkyl- or triarylphosphines is exemplified by their ready reaction with alkyl halides. The quaternary salts formed from triphenylphosphine are the precursors of the familiar Wittig reagents [cf. section 5.3.1]: for example,

$$Ph_3P + CH_3I \rightarrow Ph_3\overset{+}{P}CH_3\ I^- \xrightarrow[\text{base}]{} PH_3\overset{+}{P}{-}\overset{-}{C}H_2$$

The stabilisation of the carbanion in these reagents is due to the adjacent phosphorus.

In the case of phosphites, the reaction takes a different course. In this, the **Michaelis–Arbusov reaction,** the alkoxyphosphonium salts (1) formed undergo further reaction resulting in the formation of phosphonate esters (2):

$$(RO_3)P \xrightarrow{R'Cl} (RO)_2\overset{+}{P}R' \qquad \longrightarrow RCl + (RO)_2\overset{O}{\overset{\|}{P}}R'$$

(1)   (2)

A range of functional groups can be accommodated in the halide, and phosphonate esters of the type $(RO)_2P(O)CH_2R^2$, where R^2 is an electron accepting ($-M$) group are of particular synthetic utility [cf. section 12.2]. Reactions of phosphines and phosphites with α-halogenoketones, which might have been expected to yield ketophosphonium salts (3) and ketophosphonates (4), are more complex.

$$R_3\overset{+}{P}CH_2COR' \ X^- \qquad (RO)_2\overset{O}{\overset{\|}{P}}CH_2COR'$$

(3)   (4)

A trialkyl phosphite can react with an α-halogenoketone in two ways resulting in the formation of a ketophosphonate (4) (the Michaelis–Arbusov product) or an enol phosphate (5) (the Perkow

$$(RO)_3P \xrightarrow{XCH_2COR'} (RO)_2\overset{O}{\overset{\|}{P}}CH_2COR' + (RO)_2\overset{O}{\overset{\|}{P}}O\overset{R'}{\overset{|}{C}}=CH_2$$

(4)   (5)

product). Depending on the structure of the ketone, either or both products may be obtained. The reaction with phosphines can also take two paths. That resulting in the formation of the ketophosphonium salt (3) involves S_N2 reaction at the electrophilic carbon, while that resulting in the formation of the enol phosphonium salt (6) involves attack of the phosphine on halogen. Halogenophosphonium salts (7) have many synthetic applications [cf. section 12.3.1]:

$$R'COCH_2X \xrightarrow{R_3P} R_3\overset{+}{P}CH_2COR' \ X^-$$

(3)

$$R'COCH_2X \xrightarrow{R_3P} R_3\overset{+}{P}X \ CH_2=\overset{R'}{\underset{O^-}{\overset{|}{C}}} \longrightarrow R_3\overset{+}{P}O\overset{R'}{\overset{|}{C}}=CH_2 \ X^-$$

(7)   (6)

The size and polarisability of phosphorus enable it to react more easily at sulphur than at the first row elements oxygen and nitrogen. Indeed, phosphites and phosphines react with sulphur in air to give thionophosphates, $(RO)_3P=S$, and phosphine sulphides, $R_3P=S$, respectively, rather than the oxygen analogues.

12.2 Formation of carbon–carbon double bonds

The Wittig reaction has now become one of the most familiar reactions to the synthetic chemist, and it has been discussed in some detail in section 5.3.1. An alternative procedure which may have certain advantages over the Wittig reaction was developed by Horner and by Wadsworth and Emmons among others. This involves reaction of aldehydes and ketones with stabilised carbanions derived from phosphonate esters. Like Wittig reactions involving stabilised ylides, the E-(or *trans*-)alkene usually predominates, but significant amounts

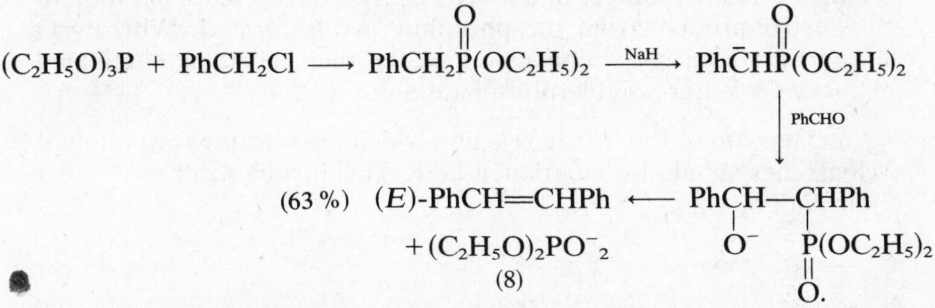

$$(C_2H_5O)_3P + PhCH_2Cl \longrightarrow PhCH_2\overset{\overset{\displaystyle O}{\|}}{P}(OC_2H_5)_2 \xrightarrow{\text{NaH}} Ph\bar{C}H\overset{\overset{\displaystyle O}{\|}}{P}(OC_2H_5)_2$$

$$\Big\downarrow \text{PhCHO}$$

$$(63\%) \quad (E)\text{-PhCH}{=}\text{CHPh} \longleftarrow \underset{\substack{| \\ O^-}}{PhCH}{-}\underset{\substack{| \\ \overset{\displaystyle P(OC_2H_5)_2}{\underset{\displaystyle O.}{\|}}}}{CHPh}$$
$$+ (C_2H_5O)_2PO^-_2$$
$$(8)$$

of Z-(or *cis*-)alkenes are formed in some instances, for example

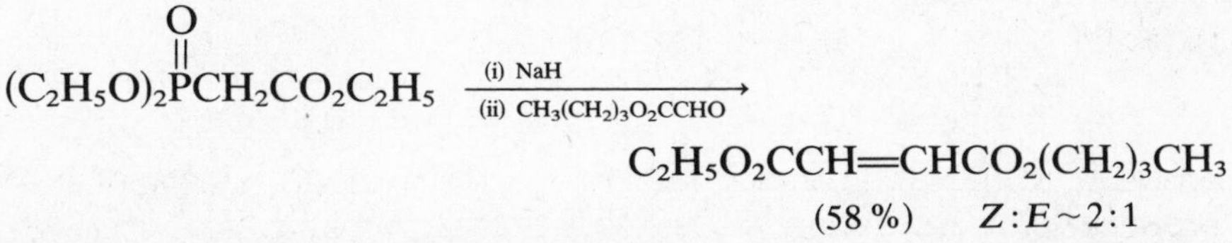

$$(C_2H_5O)_2\overset{\overset{\displaystyle O}{\|}}{P}CH_2CO_2C_2H_5 \xrightarrow[\text{(ii) } CH_3(CH_2)_3O_2CCHO]{\text{(i) NaH}}$$

$$C_2H_5O_2CCH{=}CHCO_2(CH_2)_3CH_3$$
$$(58\%) \quad Z:E \sim 2:1$$

As in the case of the Wittig reaction, the stereochemistry of the product is determined by the rates of formation and decomposition of the *threo*- and *erythro*-forms of the intermediate.

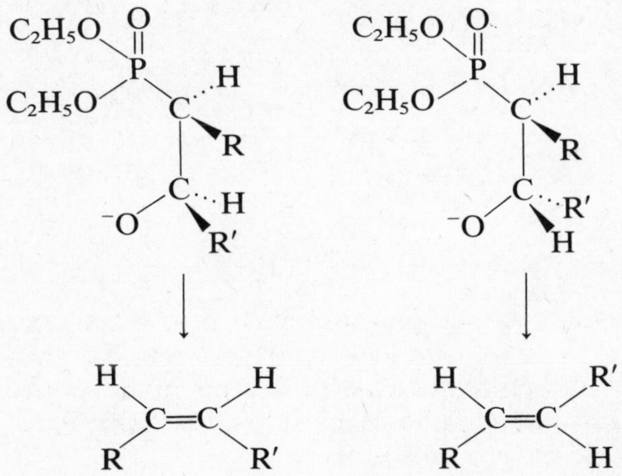

Attempts to control the stereochemistry of the product have been made. However, unlike the situation pertaining to the Wittig reaction, the effect of changing the cation is less predictable.

Advantages claimed for this reaction over the Wittig procedure include the following:

(i) Wittig reactions involving stable ylides are slow and, since additional stabilisation by an electron-withdrawing group is required in almost all successful P=O stabilised carbanion reactions, the latter is the preferred procedure in such cases.

(ii) A major problem in the Wittig procedure is the separation of the product from the phosphine oxide formed. With P=O stabilised carbanion reactions the phosphorus is eliminated as a water-soluble phosphate anion (8).

An extension of the Wittig reaction which leads to the formation of cycloalkenes should be mentioned here. This involves intramolecular C-alkylation, e.g.

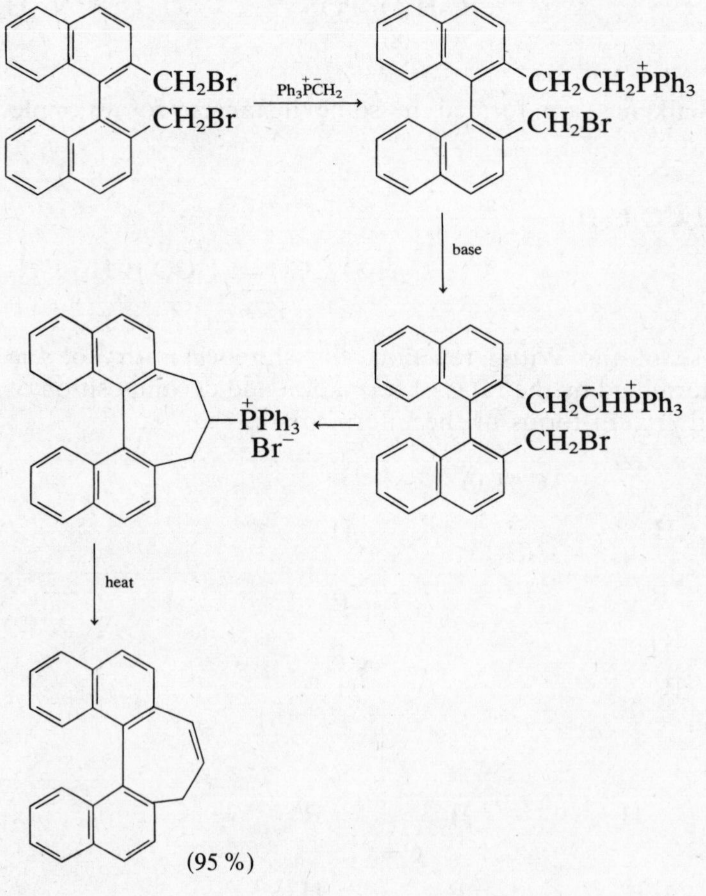

The ready reaction of tervalent phosphorus compounds with sulphur has already been noted. Thiirans undergo elimination of sulphur on reaction with phosphines and phosphites, yielding an alkene in which the stereochemistry is retained. This was of little synthetic utility until recently when a new method of preparing thiirans was developed. This involves the base-catalysed reaction of 2-(alkylthio)-oxazolines [e.g. (9)] with aldehydes or ketones, sulphur being extruded from the resultant thiirans by means of trialkyl phosphites. Where $E:Z$ isomerism is possible the E-isomer predominates:

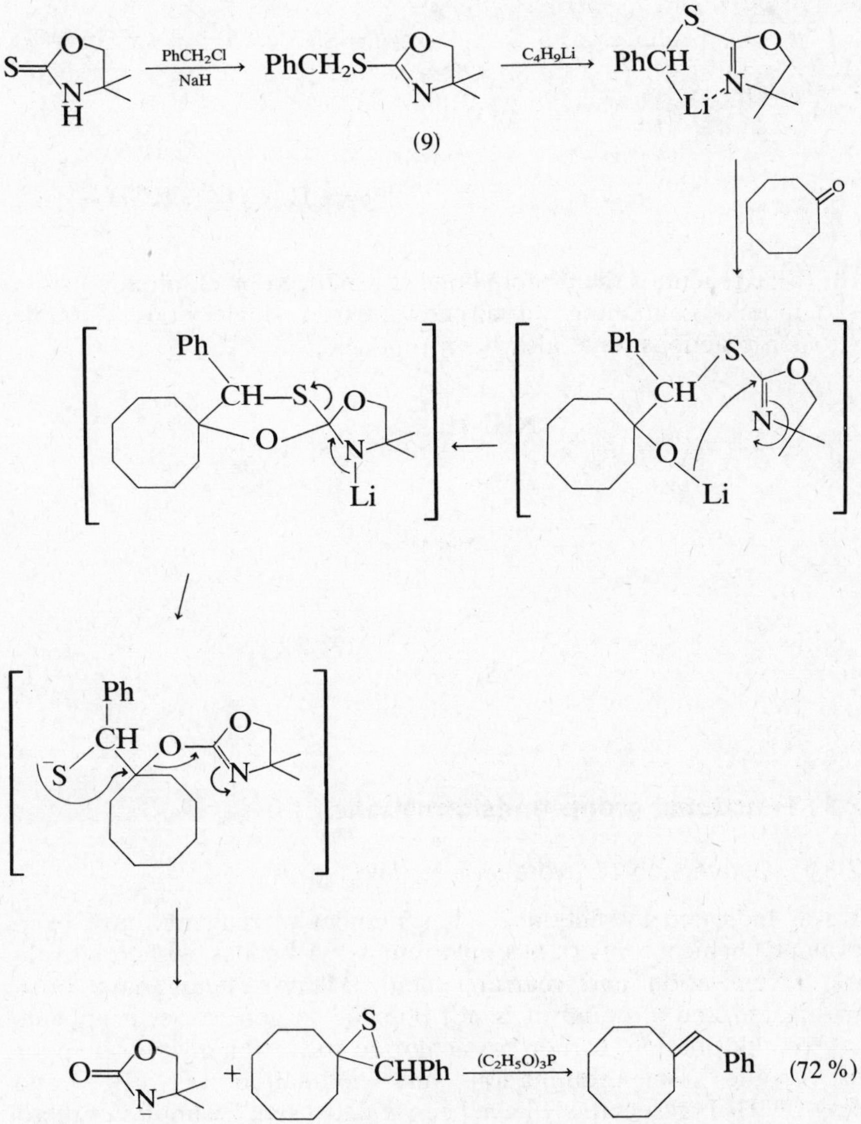

(9)

Another extrusion reaction initiated by reaction of tervalent phosphorus at sulphur is the conversion of 1,2-diols into alkenes *via* an intermediate thionocarbonate. *Z*-Alkenes are derived from *meso*- or *erythro*-diols and *E*-alkenes from racemic or *threo*-diols. This has been used in many reactions including the synthesis of *trans*-cyclooctene shown below:

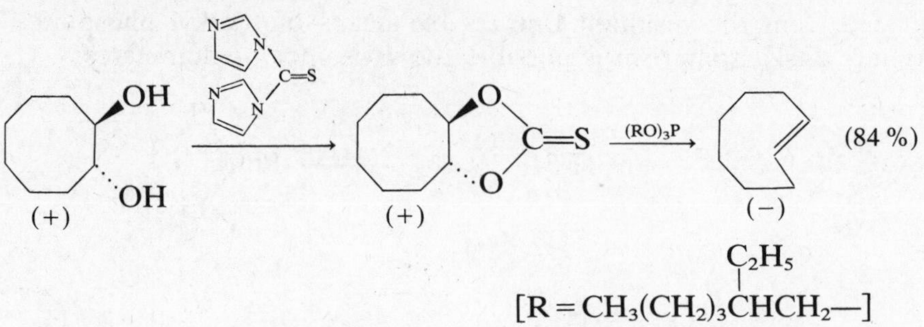

$$[R = CH_3(CH_2)_3\overset{\underset{|}{C_2H_5}}{C}HCH_2-]$$

In these reactions the double bond is produced by elimination from a compound containing an already formed single bond. Double extrusion reactions have also been reported, e.g.

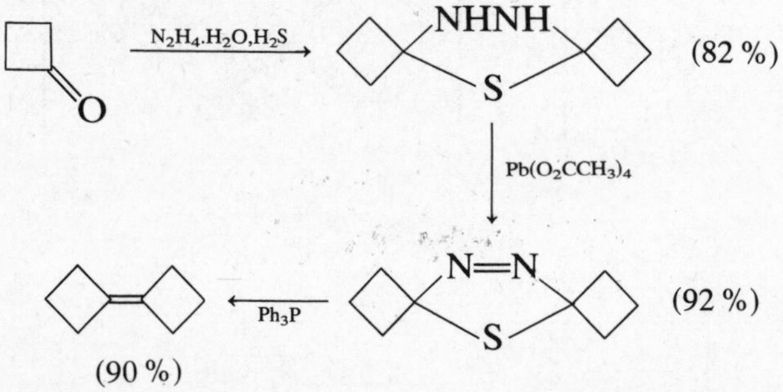

12.3 Functional group transformations

12.3.1 Conversion of hydroxyl into halogen

As was indicated in section 2.7.1, a number of reagents have been developed which convert alcohols into alkyl halides with relatively little racemisation and rearrangement. Many systems have been investigated and although it is not possible to generalise, it appears that for chlorination, carbon tetrachloride with triphenylphosphine is best, while for bromination and iodination, Ph_3PBr_2 and $(PhO)_3\overset{+}{P}CH_3\ I^-$ respectively have been widely used. Examples of these reactions include:

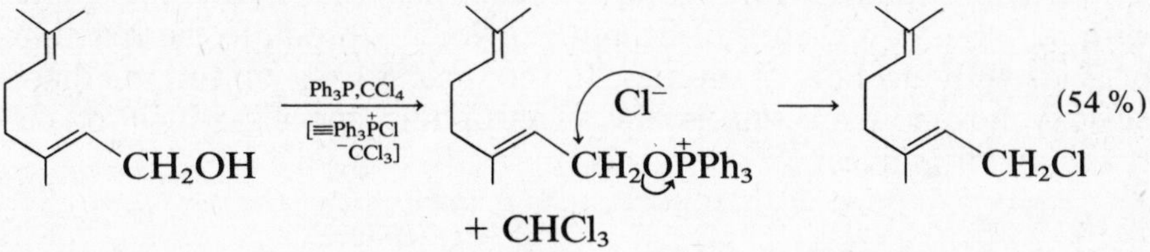

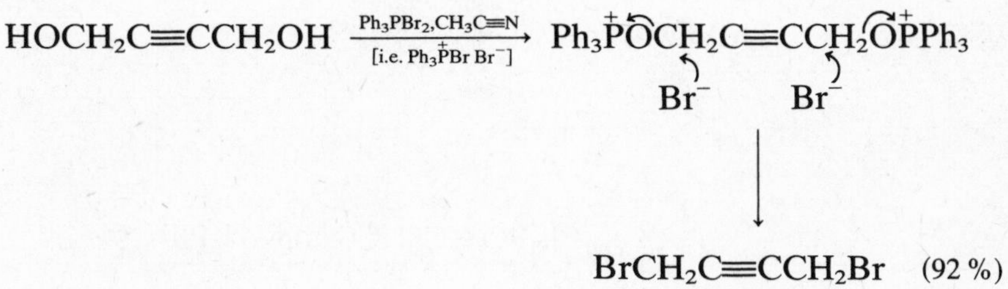

$$\text{HOCH}_2\text{C}{\equiv}\text{CCH}_2\text{OH} \xrightarrow[\text{[i.e. Ph}_3\overset{+}{\text{P}}\text{Br Br}^-]}{\text{Ph}_3\text{PBr}_2,\text{CH}_3\text{C}{\equiv}\text{N}} \text{Ph}_3\overset{+}{\text{P}}\text{OCH}_2\text{C}{\equiv}\text{CCH}_2\overset{+}{\text{O}}\text{PPh}_3$$

Br⁻   Br⁻

$$\text{BrCH}_2\text{C}{\equiv}\text{CCH}_2\text{Br} \quad (92\%)$$

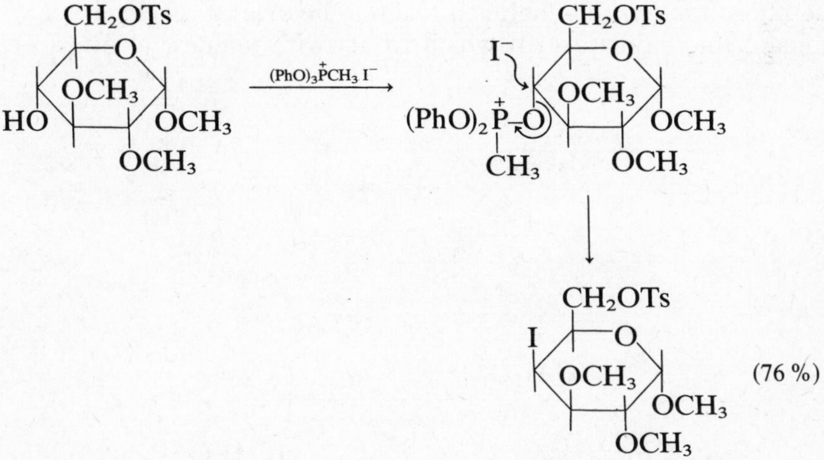

In each case, the reaction is normally of the S_N2 type with inversion at the reaction centre, and in the cases of chlorination and bromination involves initially the attack of a halogeno–phosphonium ion on the hydroxyl group. Reactions of secondary alcohols are slower than those of primary alcohols, and elimination reactions are observed with the former in more polar solvents such as hexamethylphosphoramide in the case of $(\text{PhO})_3\overset{+}{\text{P}}\text{CH}_3\,\text{I}^-$ or dimethylformamide with Ph_3PBr_2. Many other functional groups, except those with acidic hydrogens, are unaffected.

At low temperatures, certain alkoxyphosphonium salts can be isolated and, on reaction with added nucleophiles, can be converted into products such as amines, thiols, nitriles, azides and thiocyanates.

12.3.2 Formation of amides and esters, and related reactions

Triphenylphosphine with carbon tetrachloride, and triethyl phosphite with pyridine in presence of bromine or iodine, promote the reaction of acids with alcohols or amines. In the former case, an intermediate acyloxyphosphonium salt is the electrophilic species which reacts readily with amines:

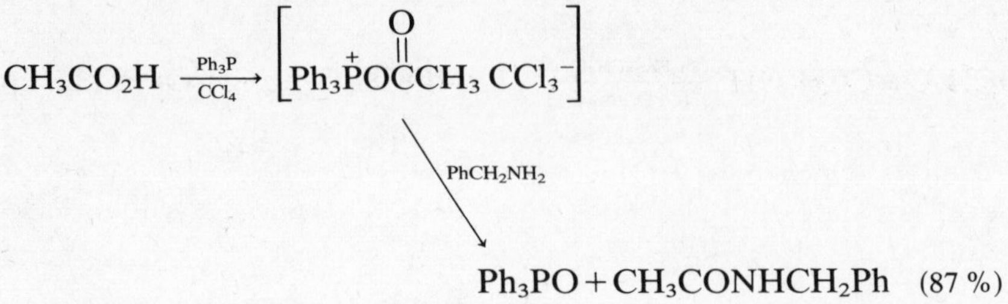

$$CH_3CO_2H \xrightarrow[CCl_4]{Ph_3P} \left[Ph_3\overset{+}{P}O\overset{O}{\overset{\|}{C}}CH_3 \; CCl_3^- \right]$$

$$\downarrow PhCH_2NH_2$$

$$Ph_3PO + CH_3CONHCH_2Ph \quad (87\%)$$

The phosphite/pyridine/halogen reaction involves a reactive penta-coordinate intermediate (10) which reacts with amines, alcohols or acids.

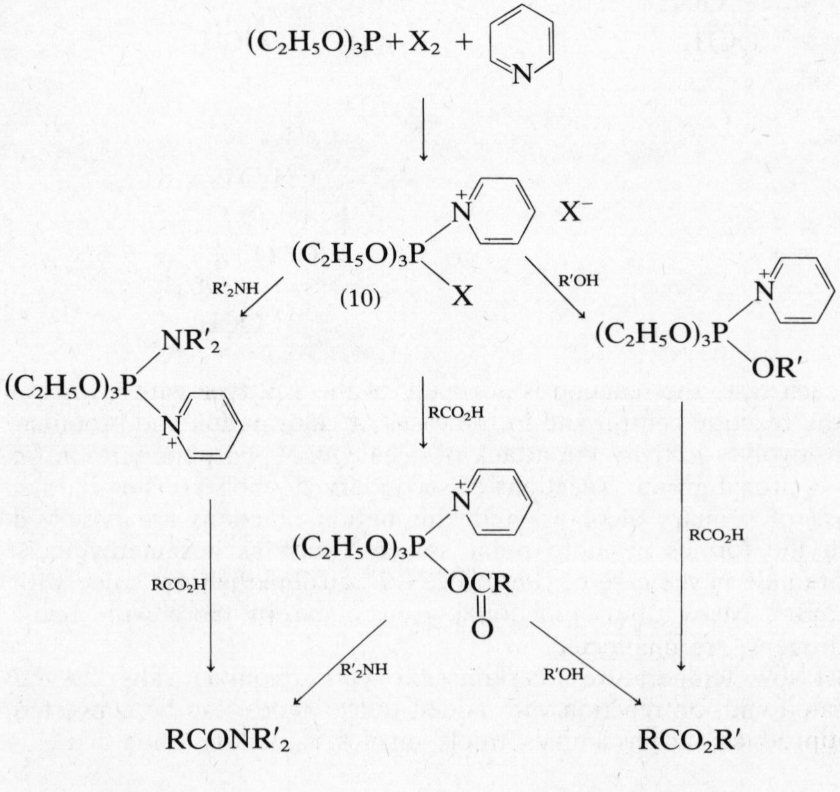

$$(C_2H_5O)_3P + X_2 +$$

An alternative strategy for esterification and amide formation involves an oxidation-reduction process:

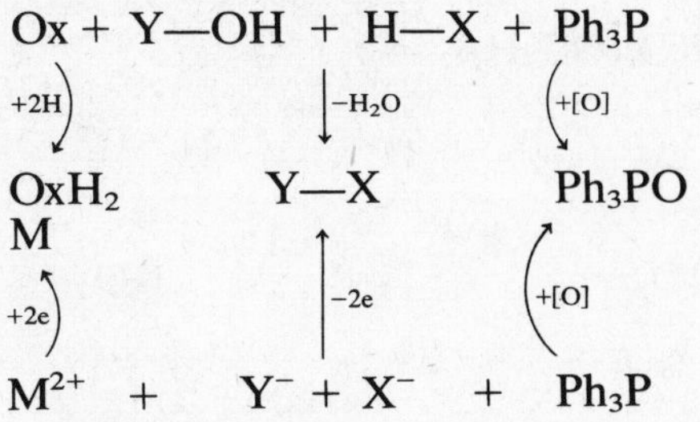

$$Ox + Y-OH + H-X + Ph_3P$$

$$OxH_2 \qquad Y-X \qquad Ph_3PO$$
$$M$$

$$M^{2+} + \quad Y^- + X^- \quad + \quad Ph_3P$$

Oxidants which have been used in these reactions include diethyl azodicarboxylate, sulphenamides and disulphides. Examples include the following[1]:

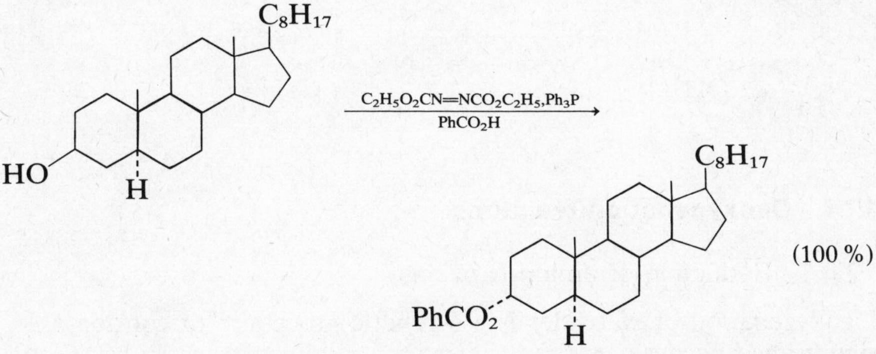

[N.B. inversion of configuration in product]

$$2CH_3(CH_2)_3NHSPh + 2Ph_3P + [CH_3(CH_2)_4CO_2]_2Cu \longrightarrow$$

$$2CH_3(CH_2)_3NHCO(CH_2)_4CH_3 + 2Ph_3PO + (PhS)_2Cu$$

(95 %)

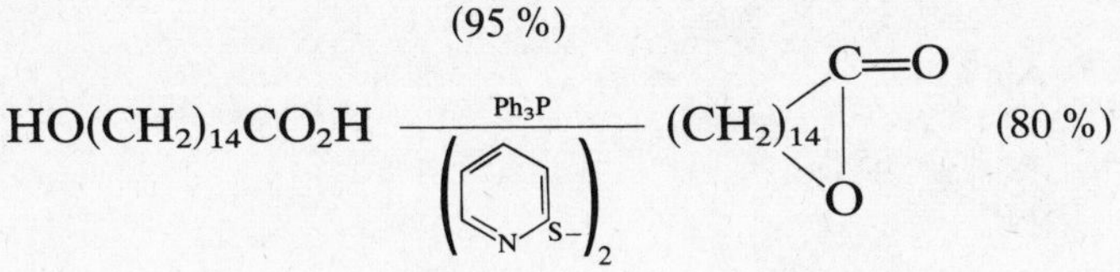

12.3.3 Dehalogenation of aryl halides

Although aryl halides are generally unreactive towards tervalent phosphorus compounds, two dehalogenation reactions are worth noting.

Halogens activated towards nucleophilic attack react with triethyl phosphite to give phosphonates which are cleaved with acids:

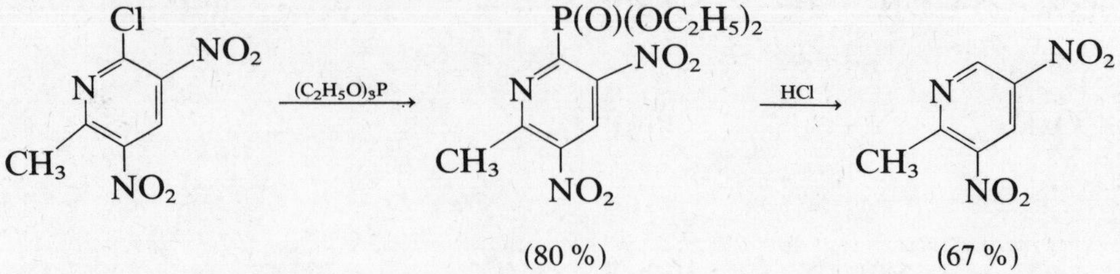

(80 %)   (67 %)

Triphenylphosphine brings about a very rapid debromination of *o*- and *p*-bromophenols, e.g.

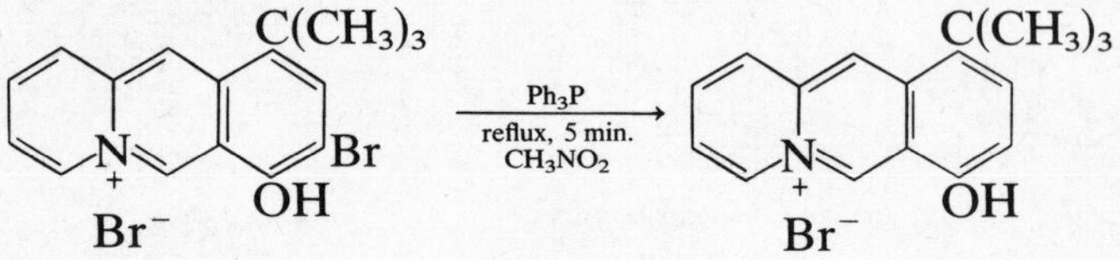

12.4   Deoxygenation reactions

12.4.1   Reduction of amine *N*-oxides

Deoxygenation of *N*-oxides is a synthetic procedure of considerable significance because, in many instances, it is necessary to carry out substitution reactions on electron-deficient heteroaromatic *N*-oxides rather than on the free bases [cf. section 2.6]. Tervalent phosphorus compounds are particularly effective reagents for reduction of *N*-oxides. In general, PCl_3 is the most reactive, but its use may cause side-reactions involving replacement of active nitro groups or halogenation of hydroxyl groups.

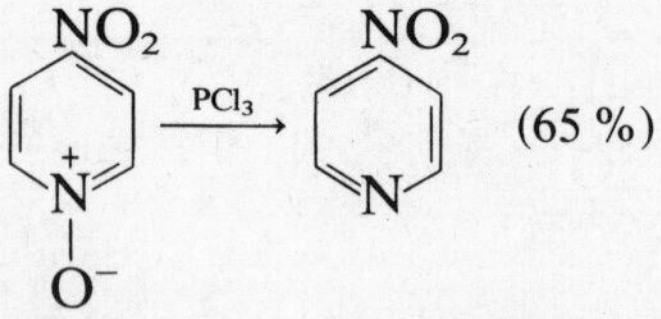

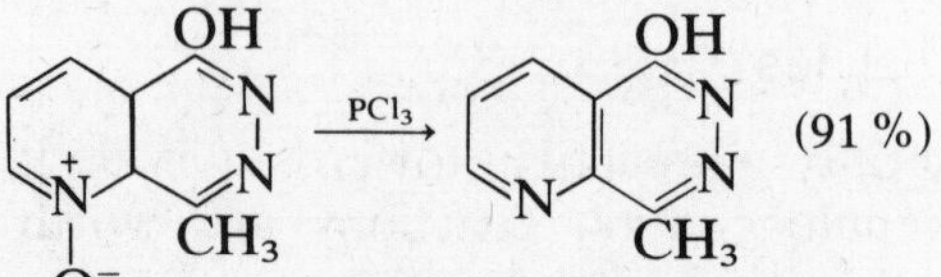

(65 %)

(91 %)

12.4.2 Cyclisation reactions involving nitro- and nitroso-groups

Aryl nitro- and nitroso-compounds react with tervalent phosphorus compounds to give products which may be ascribed to the intermediacy of a nitrene. Thus if the molecule possesses a group appropriately situated which can react with the nitrene, a nitrogen-containing heterocyclic compound is obtained. Some illustrative reactions are given:

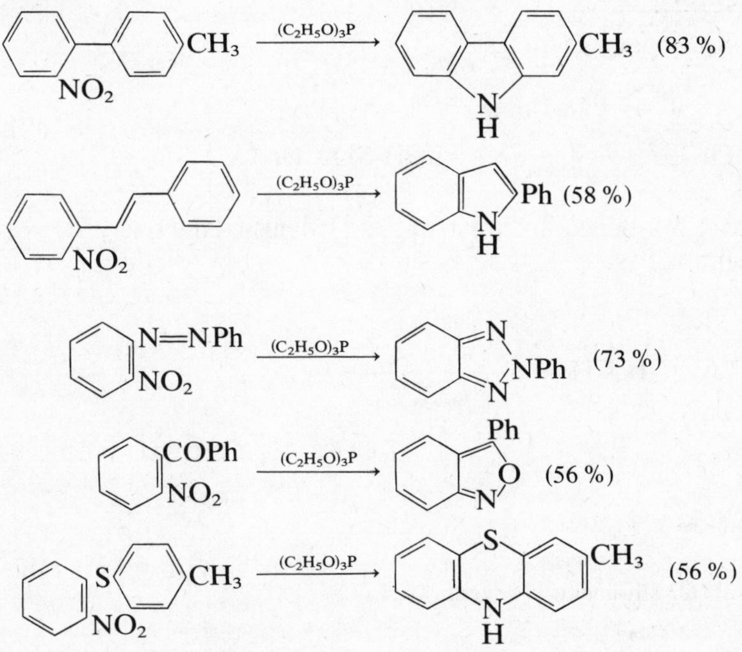

Note that this last reaction involves the initial formation of a five-membered spirodienyl intermediate (11) which undergoes rearrangement:

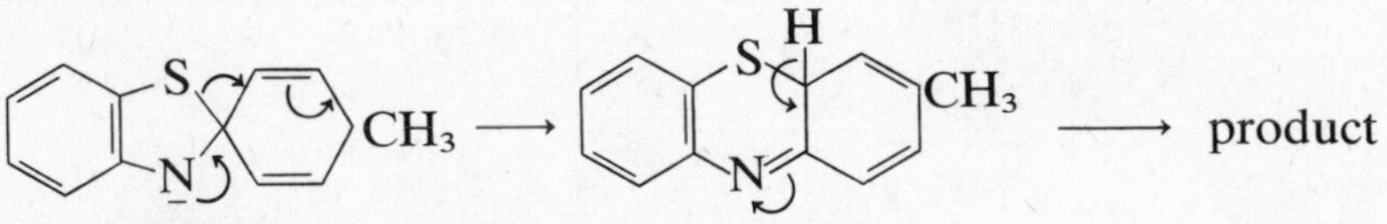

12.4.3 Deoxygenation of sulphoxides

Unlike N-oxides, sulphoxides react only slowly with tervalent phosphorus compounds. However, sulphoxides are readily reduced by the following mixtures: $Ph_3P/I_2/NaI$ and 2-phenoxy-1,3,2-benzodioxa-phosphole (12)/carbon tetrachloride/I_2. Both reactions may involve a diiodide:

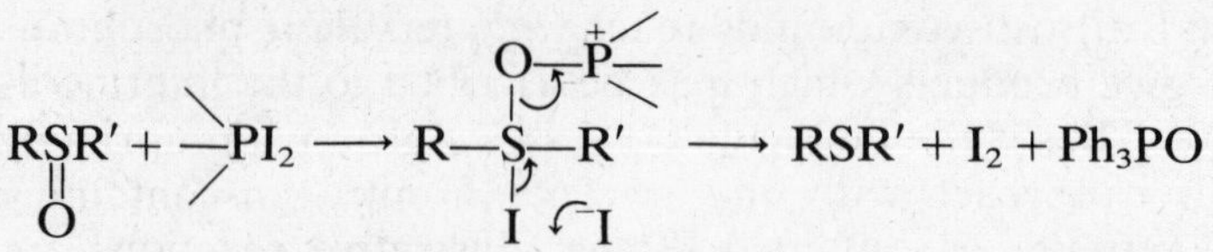

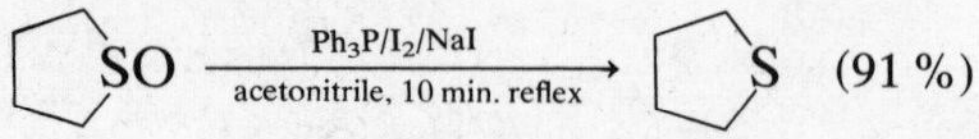

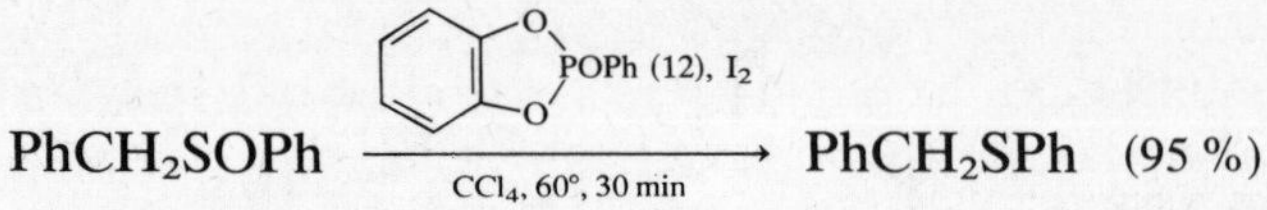

$$RSR' + \underset{O}{\overset{}{\parallel}}PI_2 \longrightarrow R\overset{\overset{\displaystyle O-\overset{+}{P}-}{|}}{\underset{\underset{I\ \curvearrowleft I}{|}}{S}}R' \longrightarrow RSR' + I_2 + Ph_3PO$$

SO $\xrightarrow[\text{acetonitrile, 10 min. reflux}]{\text{Ph}_3\text{P/I}_2\text{/NaI}}$ S (91 %)

$PhCH_2SOPh \xrightarrow[\text{CCl}_4, 60°, 30 \text{ min}]{\text{POPh (12), I}_2}$ $PhCH_2SPh$ (95 %)

2-Chloro-1,3,2-benzodioxaphosphole (13) also effectively deoxygenates sulphoxides:

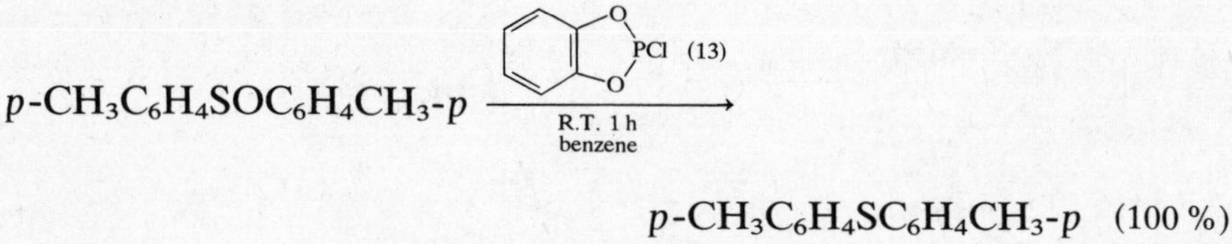

$p\text{-}CH_3C_6H_4SOC_6H_4CH_3\text{-}p \xrightarrow[\substack{\text{R.T. 1 h} \\ \text{benzene}}]{\text{PCl (13)}}$

$p\text{-}CH_3C_6H_4SC_6H_4CH_3\text{-}p$ (100 %)

Notes

1.   Cf. O. Mitsunobu, *Synthesis*, **1981**, 1.

13 Silicon reagents

Over recent years, considerable interest has been shown in the use of organosilicon compounds as intermediates in synthetic sequences. We shall look in some detail at examples which demonstrate certain aspects of the chemistry of such compounds on which the synthetic developments are based.

13.1 Introduction to organosilicon chemistry

Silicon occupies a position below carbon in group IVB of the periodic table. Its electronic configuration, $3s^2 \cdot 3p^2$, indicates quadrivalence but several aspects of its bonding to other elements differ from those of carbon. For example;

(i) Si forms stronger bonds with O and F than does C but weaker bonds with C and H.

(ii) the $3p$-electrons of Si do not overlap effectively with the $2p$-electrons of C or O. Multiple bonds C=Si and O=Si are not, therefore, normally found in stable molecules.

(iii) unlike C, Si can form stable hexaco-ordinate systems, e.g. $SiF_6{}^{2-}$.

In addition to the foregoing, one must remember that silicon is less electronegative than carbon and therefore Si–C bonds are polarised:

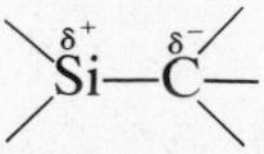

$$\overset{\delta^+}{\diagdown\text{Si}}\!-\!\overset{\delta^-}{\text{C}}\diagdown$$

This results in alkylsilanes being prone to attack by nucleophilic reagents. Silicon also has the ability to stabilise α-carbanions,

$\diagup\bar{\text{C}}\!-\!\text{Si}\diagup$, and β-carbonium ions, $\equiv\!\text{Si}\!-\!\text{C}\!-\!\text{C}^+\diagdown$.

13.2 Synthesis of organosilicon compounds

The most readily available starting materials are the silyl chlorides, $SiCl_4$, $RSiCl_3$, R_2SiCl_2 and R_3SiCl, along with tetraalkylsilanes, R_4Si.

The halides undergo facile nucleophilic substitution reactions in which valuable synthetic intermediates are formed, as illustrated below:

$$SiCl_4 + 2C_2H_5MgBr \xrightarrow[ether]{} (C_2H_5)_2SiCl_2 \quad (75\%)$$

$$(CH_3)_3SiCl + BrMgC\equiv CCH_2OH \xrightarrow[ether]{} (CH_3)_3SiC\equiv CCH_2OH \quad (44\%)$$

$$(CH_3)_3SiCl + NaN_3 \rightarrow (CH_3)_3SiN_3 \quad (87\%)$$

$$(CH_3)_3SiCl + CH_3CONH_2 \xrightarrow[(C_2H_5)_3N]{} (CH_3)_3SiNHCOCH_3 \quad (90\%)$$

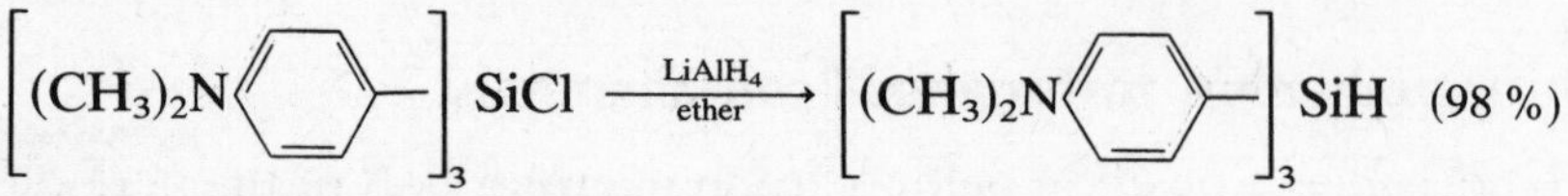

$$\left[(CH_3)_2N\!\!\left\langle\bigcirc\right\rangle\!\! - \right]_3 SiCl \xrightarrow[ether]{LiAlH_4} \left[(CH_3)_2N\!\!\left\langle\bigcirc\right\rangle\!\! - \right]_3 SiH \quad (98\%)$$

13.3   Carbon–carbon bond forming reactions

13.3.1   Reactions involving silicon-stabilised carbanions

When α-silylcarbanions react with carbonyl compounds, the intermediate (1) often decomposes spontaneously to give an alkene. This process (the **Peterson synthesis**) is obviously analogous to the Wittig reaction [cf. section 5.3.1] and the Wadsworth–Emmons–Horner reaction [cf. section 12.2].

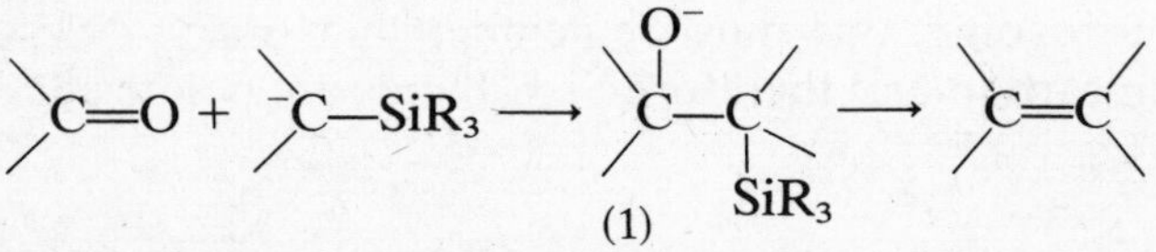

$$\underset{}{{\textstyle\diagdown}C{=}O} + {\textstyle\diagup}\overset{}{\underset{}{C}}{-}SiR_3 \longrightarrow \underset{\underset{(1)}{SiR_3}}{\overset{\overset{O^-}{|}}{C}{-}C} \longrightarrow {\textstyle\diagdown}C{=}C{\textstyle\diagup}$$

However, unlike the Wittig reaction, in most cases where E and Z isomers can be produced both isomers are formed in almost equal proportions: for example,

$$(CH_3)_3SiCH_2\overset{\overset{O}{\parallel}}{S}Ph \xrightarrow[THF, -70°]{(CH_3)_3CLi} (CH_3)_3Si\overset{-}{C}H\overset{\overset{O}{\parallel}}{S}Ph \xrightarrow{PhCHO} PhCH{=}CH\overset{\overset{O}{\parallel}}{S}Ph \quad (87\%)$$
$$E\!:\!Z = 1\!:\!1$$

α-Silylcarbanions can also be prepared by reaction of a polysilylated methane with an alkoxide. The driving force for this reaction is presumed to be the thermodynamically favoured formation of a silicon–oxygen bond. The carbanions so formed have been used in the preparation of alkenes from non-enolisable ketones exemplified below:

$$[(CH_3)_3Si]_2CH_2 + NaOCH_3 \longrightarrow (CH_3)_3SiCH_2^- Na^+ + (CH_3)_3SiOCH_3$$

↓ PhCOPh

$$Ph_2C{=}CH_2 \quad (53\%)$$

$$[(CH_3)_3Si]_3CH + LiOCH_3 \longrightarrow [(CH_3)_3Si]_2CH^- Li^+ + (CH_3)_3SiOCH_3$$

↓ PhCOC(CH₃)₃

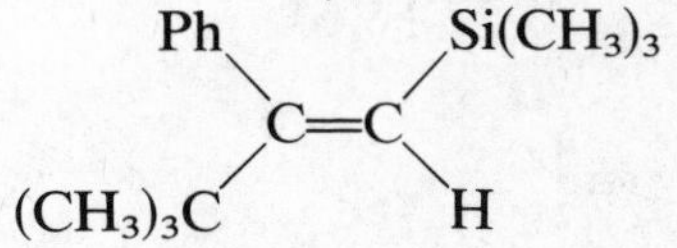

Ph   Si(CH₃)₃

C=C

(CH₃)₃C   H

(13 %)

It should be noted that in the latter case only one geometrical isomer (Z) is formed. This has been explained as resulting from cis-elimination from the more stable eclipsed conformation (1A) of the intermediate anion.

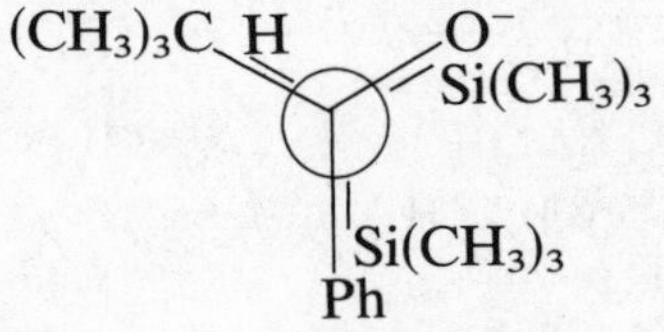

(CH₃)₃C H   O⁻

Si(CH₃)₃

Si(CH₃)₃

Ph

(1A)

α-Silyl Grignards are useful reagents which on reaction with carbonyl groups form alcohols. Elimination results in the formation of alkenes, and the procedure appears to have, in some instances at least, advantage over the Wittig reaction for the conversion of $\diagdown\!C{=}O$ into $\diagdown\!C{=}CH_2$, for example (2) → (3): (2) is unreactive towards $Ph_3P{=}CH_2$:

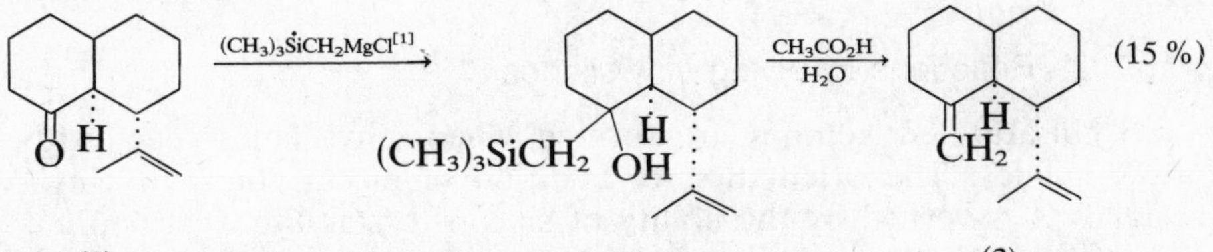

(2)   (3)

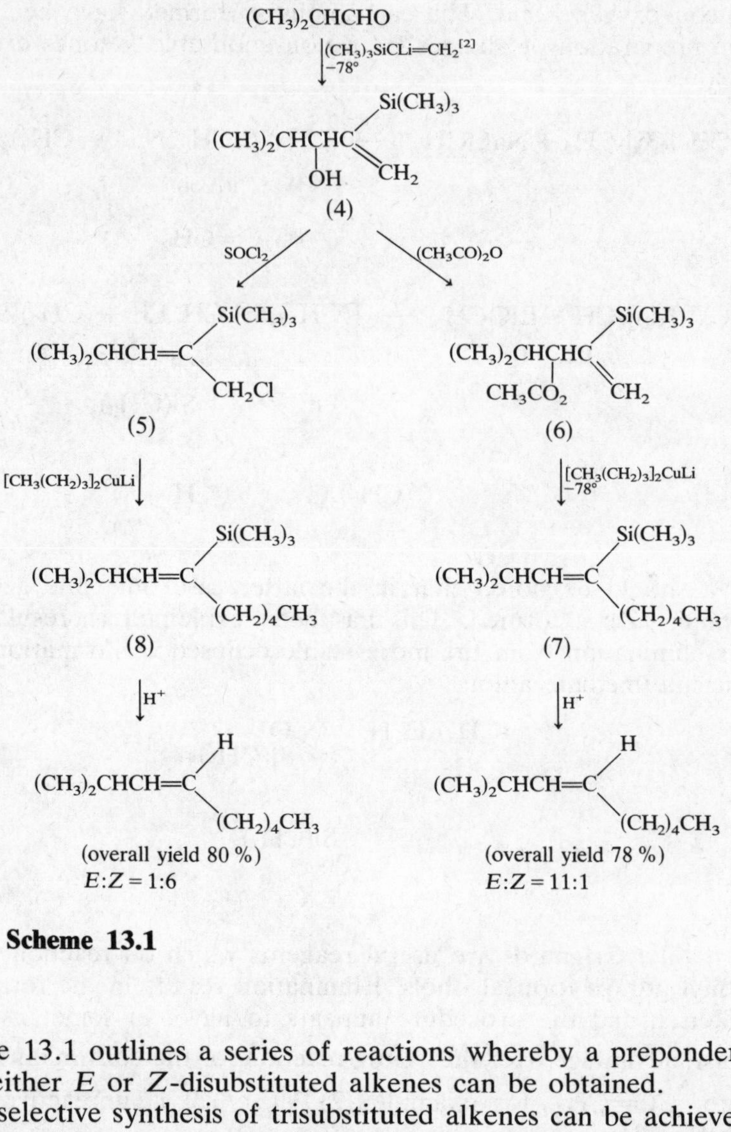

$(CH_3)_2CHCHO$

$\downarrow$ $(CH_3)_3SiCLi{=}CH_2{}^{[2]}$
$-78°$

$\underset{\underset{OH}{|}}{(CH_3)_2CHCHC}\overset{Si(CH_3)_3}{\underset{CH_2}{\diagup}}$

(4)

$SOCl_2\diagdown$ $\diagdown(CH_3CO)_2O$

$(CH_3)_2CHCH{=}C\overset{Si(CH_3)_3}{\underset{CH_2Cl}{\diagup}}$ $\underset{\underset{CH_3CO_2}{|}}{(CH_3)_2CHCHC}\overset{Si(CH_3)_3}{\underset{CH_2}{\diagup}}$

(5)   (6)

$[CH_3(CH_2)_3]_2CuLi\Big|$ $\Big|\begin{matrix}[CH_3(CH_2)_3]_2CuLi\\ -78°\end{matrix}$

$(CH_3)_2CHCH{=}C\overset{Si(CH_3)_3}{\underset{(CH_2)_4CH_3}{\diagup}}$ $(CH_3)_2CHCH{=}C\overset{Si(CH_3)_3}{\underset{(CH_2)_4CH_3}{\diagup}}$

(8)   (7)

$\Big|H^+$ $\Big|H^+$

$(CH_3)_2CHCH{=}C\overset{H}{\underset{(CH_2)_4CH_3}{\diagup}}$ $(CH_3)_2CHCH{=}C\overset{H}{\underset{(CH_2)_4CH_3}{\diagup}}$

(overall yield 80 %)   (overall yield 78 %)
$E:Z = 1:6$ $E:Z = 11:1$

Scheme 13.1

Scheme 13.1 outlines a series of reactions whereby a preponderance of either E or Z-disubstituted alkenes can be obtained.

Stereoselective synthesis of trisubstituted alkenes can be achieved by reaction of the isomeric vinylsilanes (7) and (8) with electrophiles [cf. section 13.3.2].

13.3.2   Reactions involving vinylsilanes

α,β-Unsaturated ketones are formed when vinylsilanes react with acyl halides. The orientation of addition of electrophiles to vinylsilanes is governed by the ability of silicon to stabilise a carbonium ion β to it (13.1):

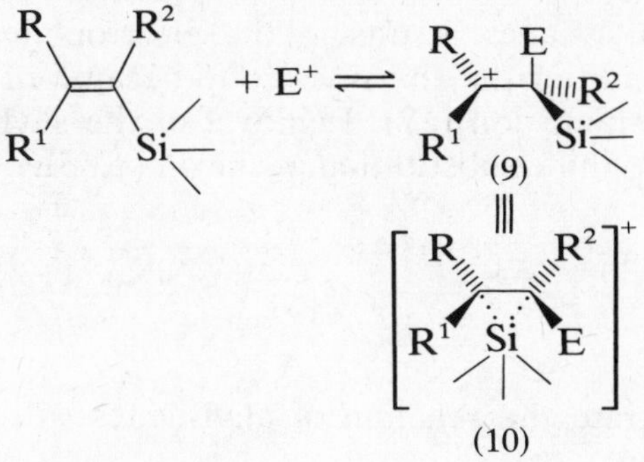

(13.1)

Rotation about the C–C bond of (9) takes place so that the full stabilisation of the carbonium ion by silicon, indicated by (10), can take place. Nucleophilic displacement at silicon can now take place, releasing as the leaving group an alkene in which the silyl group has been replaced stereospecifically by the electrophile (13.2).

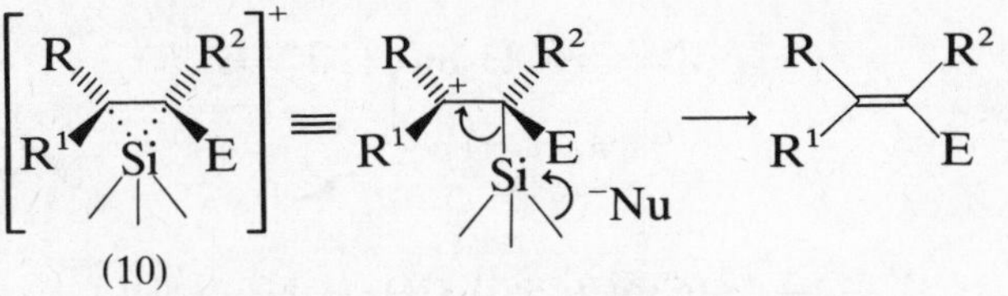

(13.2)

If the electrophile is an acyl halide, an α,β-unsaturated ketone is formed and an α,β-unsaturated aldehyde is produced from α,α-dichlorodimethyl ether[3]:

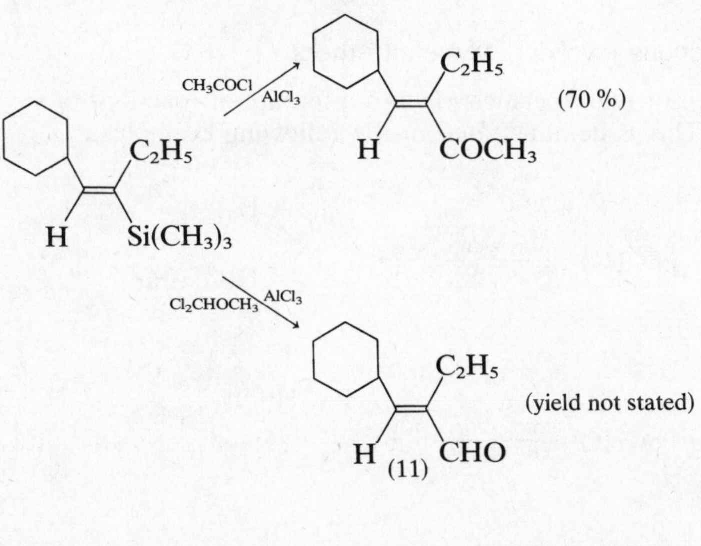

13.3.3   Reactions involving allylsilanes

Addition of electrophiles to allylsilanes results in the electrophile being attached to the carbon remote from the silyl group because of the stability of the β-silyl carbonium ion (12). Removal of the silyl group occurs as a result of nucleophilic substitution at silicon (13.3):

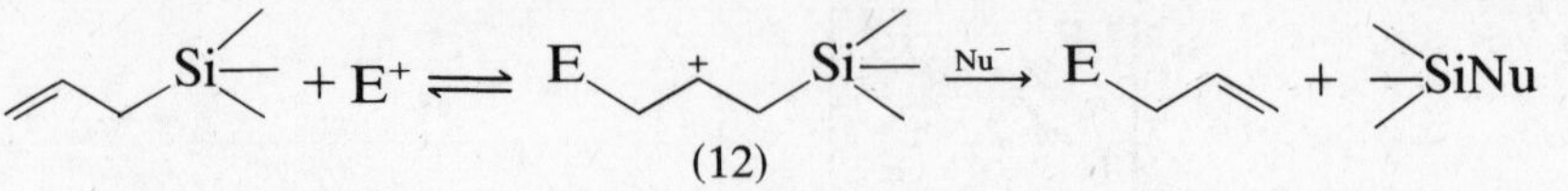

$$\text{allyl-Si} + E^+ \rightleftharpoons E\text{-Si}^+ \xrightarrow{Nu^-} E + \text{Si-Nu} \quad (13.3)$$
$$(12)$$

The following reactions illustrate the reaction of allylsilanes with the electrophilic carbonyl carbon:

$$(CH_3)_3SiCl \xrightarrow[\text{ether}]{CH_2=CHCH_2MgBr} CH_2=CHCH_2Si(CH_3)_3$$

$$\downarrow \begin{array}{l} \text{(i) } CH_3(CH_2)_5CHO, TiCl_4 \\ \text{(ii) } H_2O \end{array}$$

$$CH_3(CH_2)_5CH(OH)CH_2CH=CH_2 \quad (91\%)$$

$$(CH_3)_3SiCl \xrightarrow[\text{etner}]{(CH_3)_2C=CHCH_2MgCl} (CH_3)_2C=CHCH_2Si(CH_3)_3 \quad (58\%)$$

$$\downarrow \begin{array}{l} (CH_3)_2C=CHCOCl \\ AlCl_3, -60° \end{array}$$

$$(CH_3)_2C=CHCOC(CH_3)_2CH=CH_2 \quad (90\%)$$

This procedure is more regiospecific than the corresponding reactions of allylic Grignard reagents, where products of reaction at both ends of the allylic system may be obtained.

13.3.4   Reactions involving silyl enol ethers

A high degree of regiospecificity is also a feature of reactions of silyl enol ethers. This is demonstrated in the following examples:

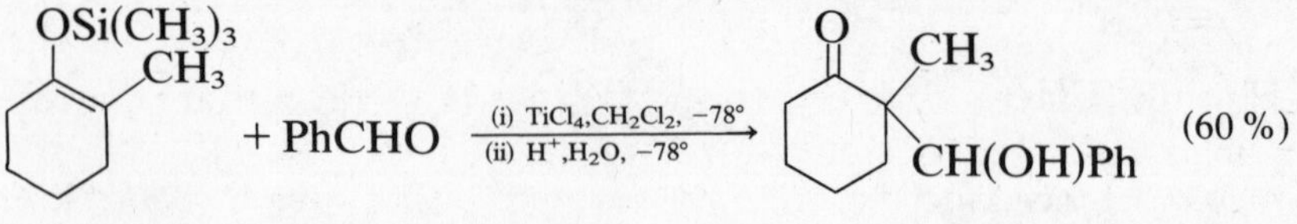

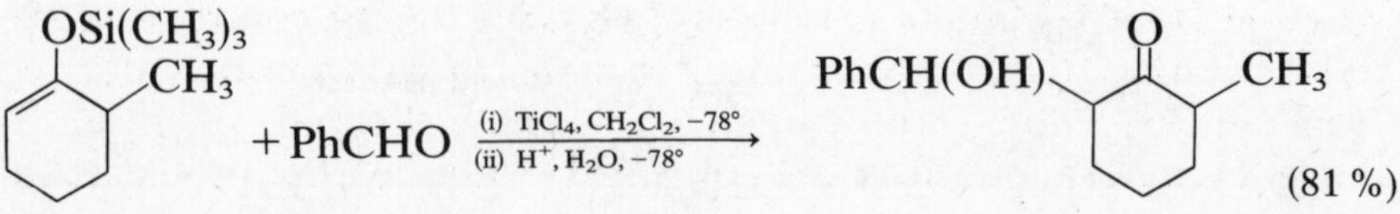

$$+ \text{PhCHO} \xrightarrow[\text{(ii) } H^+, H_2O, -78°]{\text{(i) } TiCl_4, CH_2Cl_2, -78°} \quad (60\%)$$

$$+ \text{PhCHO} \xrightarrow[\text{(ii) } H^+, H_2O, -78°]{\text{(i) } TiCl_4, CH_2Cl_2, -78°} \quad (81\%)$$

These reactions also illustrate the effectiveness of silyl enol ethers in the formation of 'mixed aldol' products without the problems of a mixed condensation reaction [cf. section 5.2.4].

Other reactions of synthetic significance include those with acyl halides, iminium salts and nitroalkenes:

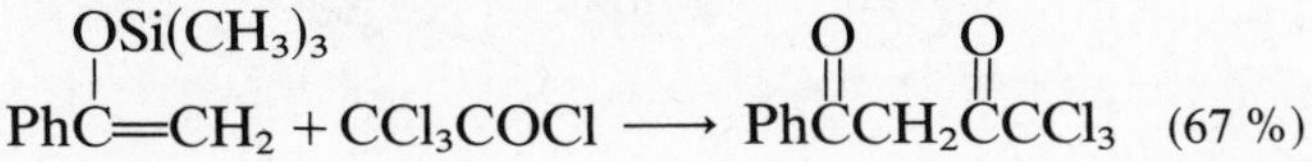

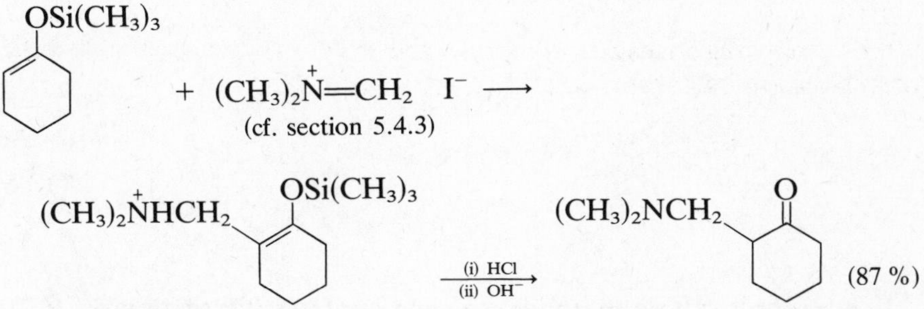

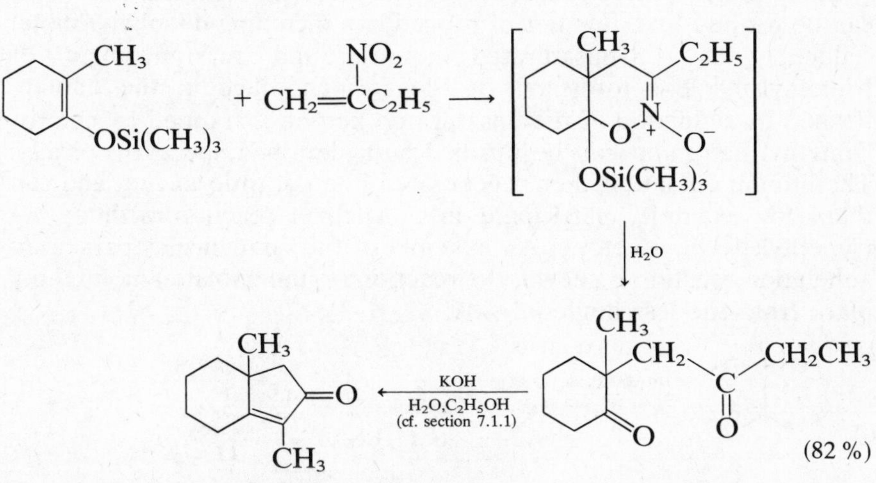

$$\underset{PhC=CH_2}{\overset{OSi(CH_3)_3}{|}} + CCl_3COCl \longrightarrow PhCCH_2CCCl_3 \quad (67\%)$$

(cf. section 5.4.3)

$$+ (CH_3)_2\overset{+}{N}=CH_2 \quad I^- \longrightarrow$$

$$(CH_3)_2\overset{+}{N}HCH_2 \underset{}{\overset{OSi(CH_3)_3}{\diagup}} \quad \xrightarrow[\text{(ii) OH}^-]{\text{(i) HCl}} \quad (CH_3)_2NCH_2 \quad (87\%)$$

$$+ CH_2=CC_2H_5 \longrightarrow$$

$$\Big|\text{H}_2\text{O}$$

$$\xleftarrow[\substack{H_2O,C_2H_5OH \\ (cf.\ section\ 7.1.1)}]{KOH}$$

(14)

(87%)

(13)

(82%)

The 1,4-diketones [e.g. (13)] formed in the last reaction are precursors of partially reduced indenones [e.g. (14)].

Since silyl enol ethers can be formed either under conditions favouring thermodynamic control [e.g. the use of a tertiary amine, reaction (13.4)] or under conditions favouring kinetic control [e.g. the use of lithium diisopropylamide, reaction (13.5)], products derived from either enolate of an unsymmetrical ketone such as 2-methylcyclohexanone can be obtained without the problems of equilibration encountered in reactions involving the use of an excess of strong base (cf. section 5.2.1):

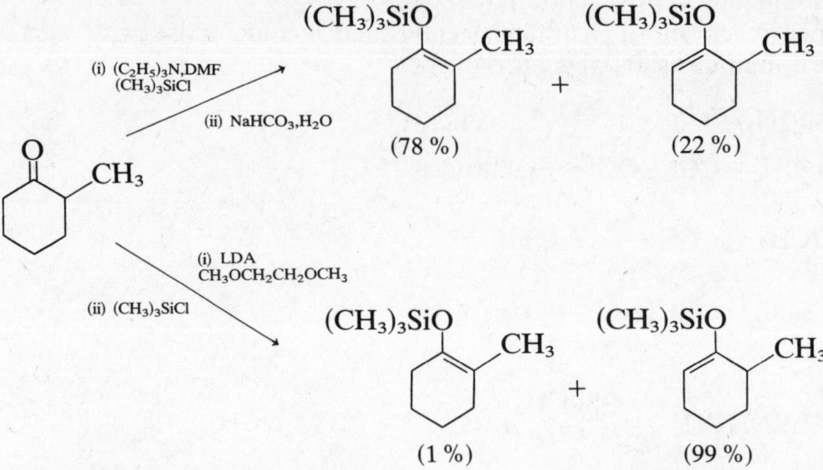

$$(13.4)$$

$$(13.5)$$

In addition to these methods, specific enolates (cf. section 5.2.3.2) can be formed by a number of procedures including dissolving metal reduction of α,β-unsaturated ketones and rearrangement of trimethylsilyl β-keto-esters. In the former reaction, the enolate formed by reduction of the unsaturated ketone is trapped by chloro-trimethylsilane and can be purified and identified spectroscopically. The lithium enolate is then regenerated in an aprotic solvent and can then, for example, participate in a Michael reaction with an α-trimethylsilylvinyl ketone. An example of this sequence as part of an annelation reaction is shown, the reaction on the enolate anion taking place from the less hindered side:

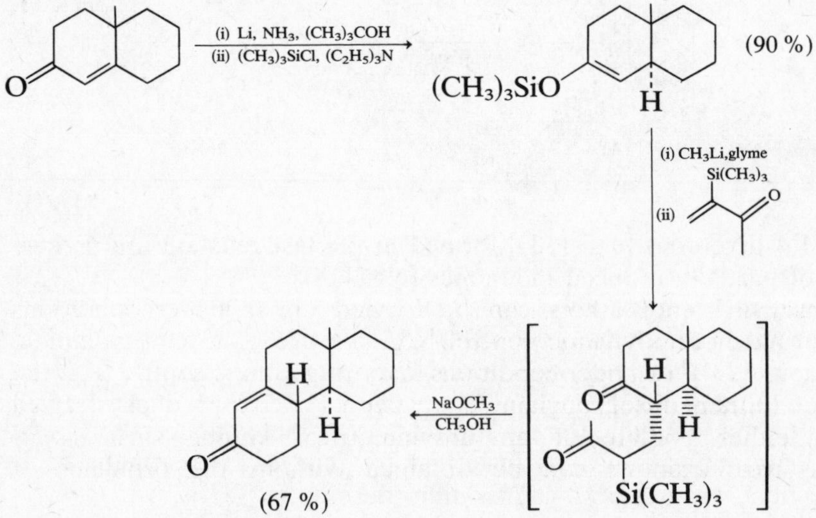

Thermal rearrangement of trimethylsilyl β-keto-esters involves migration of the silyl group with elimination of CO_2 and formation of the silyl enolate in a process analogous to the decarboxylation of β-keto-acids. The following example demonstrates how a silyl enol ether so formed is used in an alkylation reaction:

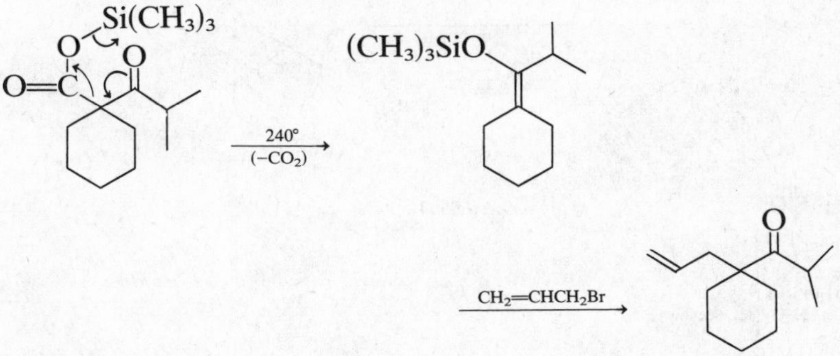

13.4 Silyl groups as protecting groups for alcohols

As has already been noted (cf. section 10.2.1) trimethylsilyl groups have enjoyed much success in derivatisation of hydroxyl groups for gas chromatography and for mass spectrometry. However, the limited stability of trimethylsilyl ethers towards hydrolysis limits their applicability in protection of hydroxyl groups.

Other alkylsilyl ethers are more stable and are formed more selectively than trimethylsilyl ethers. For example,

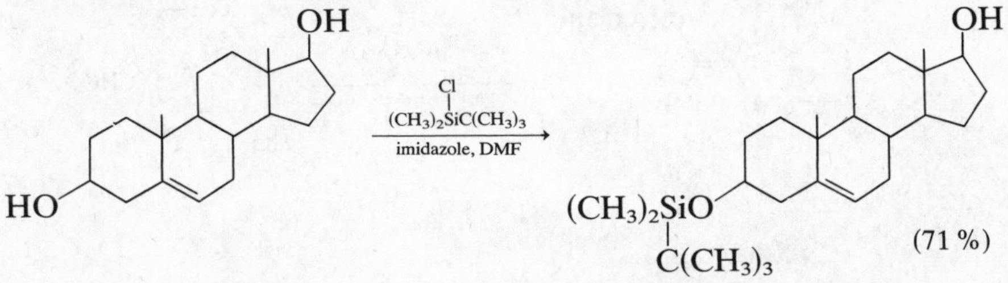

t-Butyldimethylsilyl ethers are more stable than tetrahydropyranyl ethers (cf. section 10.2.1) and have been widely used in, for example, prostaglandin synthesis. The synthesis, shown in scheme 13.2, of prostaglandin $F_{2\alpha}$ and its 15-epimer (15) demonstrates the stability of t-butylsilyl ethers ($\equiv$—Si OR) to strong base, oxidation and metal hydride reduction. The group is also stable to catalytic hydrogenation but is readily removed by mild acid treatment or by treatment with tetrabutylammonium fluoride in THF.

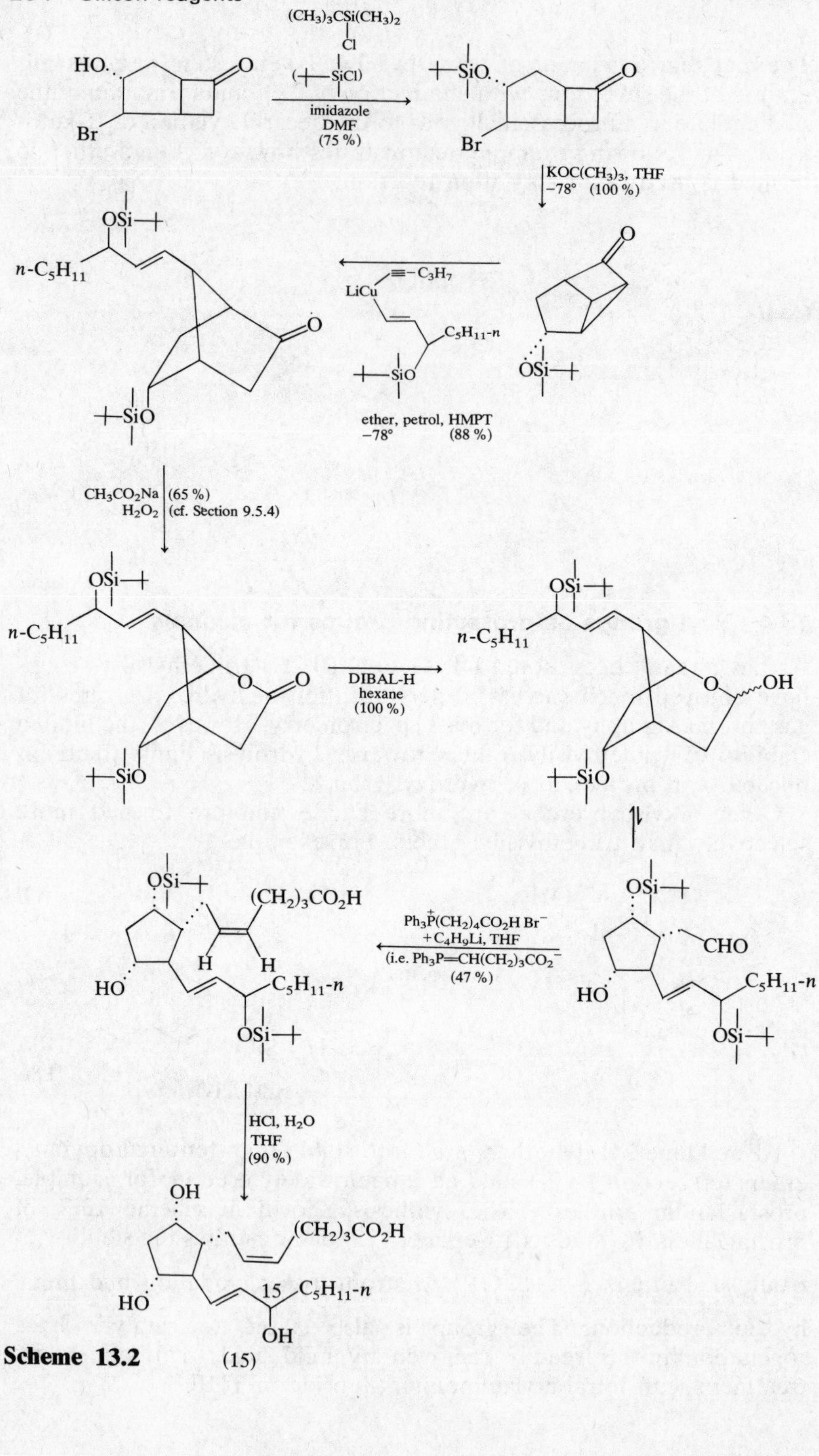

$(CH_3)_3CSi(CH_3)_2$

Cl

$(t-)SiCl$

imidazole
DMF
(75 %)

KOC(CH_3)_3, THF
−78°　(100 %)

n-C$_5$H$_{11}$

$\equiv$—C$_3$H$_7$

LiCu

C$_5$H$_{11}$-n

$+$SiO

ether, petrol, HMPT
−78°　　(88 %)

OSi$+$

CH$_3$CO$_2$Na (65 %)
H$_2$O$_2$ (cf. Section 9.5.4)

n-C$_5$H$_{11}$

$+$—Si O

DIBAL-H
hexane
(100 %)

OSi$+$

n-C$_5$H$_{11}$

$+$SiO

OSi$+$

(CH$_2$)$_3$CO$_2$H

Ph$_3\overset{+}{P}$(CH$_2$)$_4$CO$_2$H Br$^-$
+C$_4$H$_9$Li, THF
(i.e. Ph$_3$P=CH(CH$_2$)$_3$CO$_2^-$)
(47 %)

H　H

HO

C$_5$H$_{11}$-n

OSi$+$

OSi$+$

HO

CHO

C$_5$H$_{11}$-n

OSi$+$

HCl, H$_2$O
THF
(90 %)

OH

(CH$_2$)$_3$CO$_2$H

HO

15　C$_5$H$_{11}$-n

OH

Scheme 13.2　　(15)

Notes

1. The halide used in formation of this reagent is produced by chlorination of tetramethylsilane.

2. The reagent is formed in the following way:

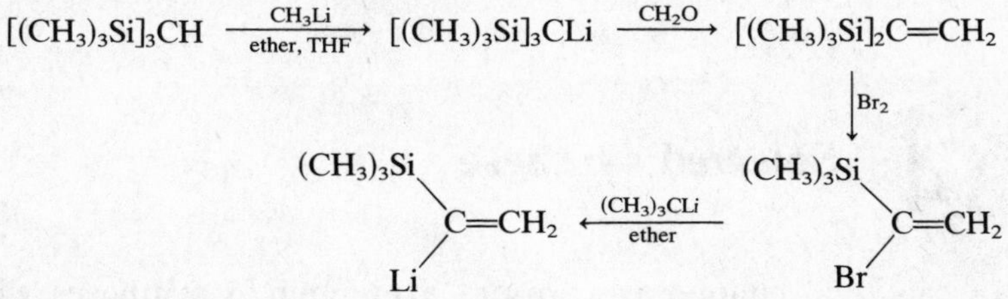

$$[(CH_3)_3Si]_3CH \xrightarrow[\text{ether, THF}]{CH_3Li} [(CH_3)_3Si]_3CLi \xrightarrow{CH_2O} [(CH_3)_3Si]_2C{=}CH_2$$

$$\downarrow Br_2$$

$$\begin{array}{c} (CH_3)_3Si \\ \diagdown \\ C{=}CH_2 \\ / \\ Li \end{array} \xleftarrow[\text{ether}]{(CH_3)_3CLi} \begin{array}{c} (CH_3)_3Si \\ \diagdown \\ C{=}CH_2 \\ / \\ Br \end{array}$$

3. In this case both *E*- and *Z*-vinylsilanes react to give the *E*-isomer of (11) due to equilibrium *via*

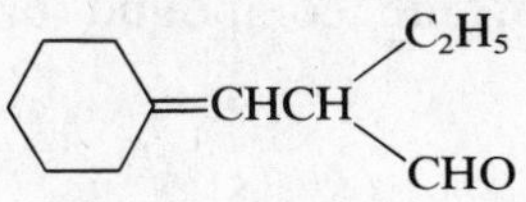

$$\bigcirc{=}CHCH\begin{array}{c} C_2H_5 \\ \diagup \\ \diagdown \\ CHO \end{array}$$

14 Selected syntheses

In this final chapter we present a number of syntheses which, it is hoped, will help to illustrate some of the ideas contained in the earlier parts of the book. At the beginning of each section we give some indication of the importance of the compound or class of compound under discussion.

14.1 Introduction

When an organic chemist is faced with the synthesis of any given molecule, he must plan the synthesis so that (i) readily available starting materials are used, (ii) the smallest number of efficient stages is involved, (iii) reactions involving separation of complex mixtures are avoided, and (iv) the synthesis is unambiguous. To do this may involve a small number of very obvious reactions in simple situations or, in a somewhat more complex case, application of the synthon-disconnection approach described in Chapter 3. However even the latter approach may be too cumbersome and the chemist is forced for example to plan a synthesis by intuitively recognising a key intermediate from which, perhaps by analogy with published syntheses, the target molecule may be obtained.

Two extreme strategies for a multi-stage synthesis can be identified. In one, known as **linear synthesis,** reactions are carried out step by step, each one adding a new part of the target molecule. This approach suffers from two principal drawbacks:

  (i) even if each step proceeds in excellent yield, the overall yield in a multi-stage synthesis can be very low (cf. section 14.6.1);
  (ii) reactive functional groups may have to be carried unchanged through a large number of steps.

The alternative strategy is **convergent synthesis,** in which major parts of the target molecule are synthesised separately and these parts are linked together towards the end of the synthesis. The overall yield may be higher than that obtained in a linear synthesis (cf. section 14.7.1) and the labile features of the target molecule are contained within smaller units.

14.2 *Z*-Heneicos-6-en-11-one

In section 3.1, various ways in which *Z*-heneicos-6-en-11-one could be synthesised from smaller fragments were suggested. We shall now consider further how this pheromone might be synthesised and then look in detail at three published syntheses.

When we consider possible synthetic routes to the target molecule, we should note several points. Firstly, *Z*-alkenes are often prepared by partial hydrogenation of alkynes (cf. section 8.4.2) or by the salt-free Wittig reaction (cf. section 5.3.1.3). Secondly, the functional groups are sufficiently remote from each other to suggest that they can be treated independently. Thirdly, in the case of a synthesis involving alkynes, the alkyne is more stable than the carbonyl group, particularly towards nucleophilic reagents, and so it is preferable to introduce it first.

Looking at possible syntheses involving alkynes we note disconnections for alkynes in table 4.1 and those for ketones in tables 4.1 and 5.1. Synthetic equivalents are found in tables 4.2 and 5.2.

$$RC{\equiv}CR' \Rightarrow RC{\equiv}C^- + R'^+$$

$$RCOR' \Rightarrow R^- + \overset{+}{C}OR'$$

$$RCH_2COR' \Rightarrow R^+ + \bar{C}H_2COR'$$

$$RCOR' \Rightarrow R^+ + \bar{C}OR'$$

Let us now look at our target molecule and consider the possibilities:

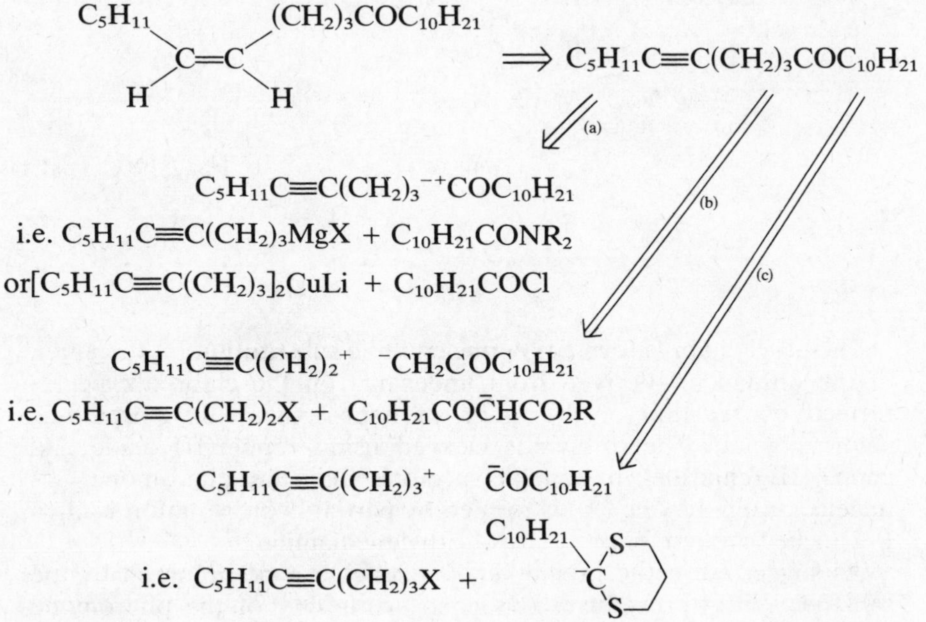

C5H11
(CH2)3COC10H21
C=C
H
H
⇒ C5H11C≡C(CH2)3COC10H21
(a)
(b)
(c)
C5H11C≡C(CH2)3 −+COC10H21
i.e. C5H11C≡C(CH2)3MgX + C10H21CONR2
or[C5H11C≡C(CH2)3]2CuLi + C10H21COCl
C5H11C≡C(CH2)2+ −CH2COC10H21
i.e. C5H11C≡C(CH2)2X + C10H21COC̄HCO2R
C5H11C≡C(CH2)3+ C̄OC10H21
C10H21
S
C
S
i.e. C5H11C≡C(CH2)3X +

Of the synthetic equivalents for the decyl-containing synthons, that from disconnection (c) is readily prepared from commercially available undecanal (syntheses of this compound are given on pp. 47 and 99) and so this disconnection probably offers the best alternative. The electrophilic synthetic equivalent required in (c) is a 1-halogenodec-4-yne. 1-Chlorodec-4-yne could be prepared from the anion of hept-1-yne and 1-chloro-3-bromo- (or iodo)propane when the more reactive halogen (Br or I) will undergo nucleophilic substitution.

The first synthesis of the pheromone, reported in 1975[1], followed the sequence shown below:

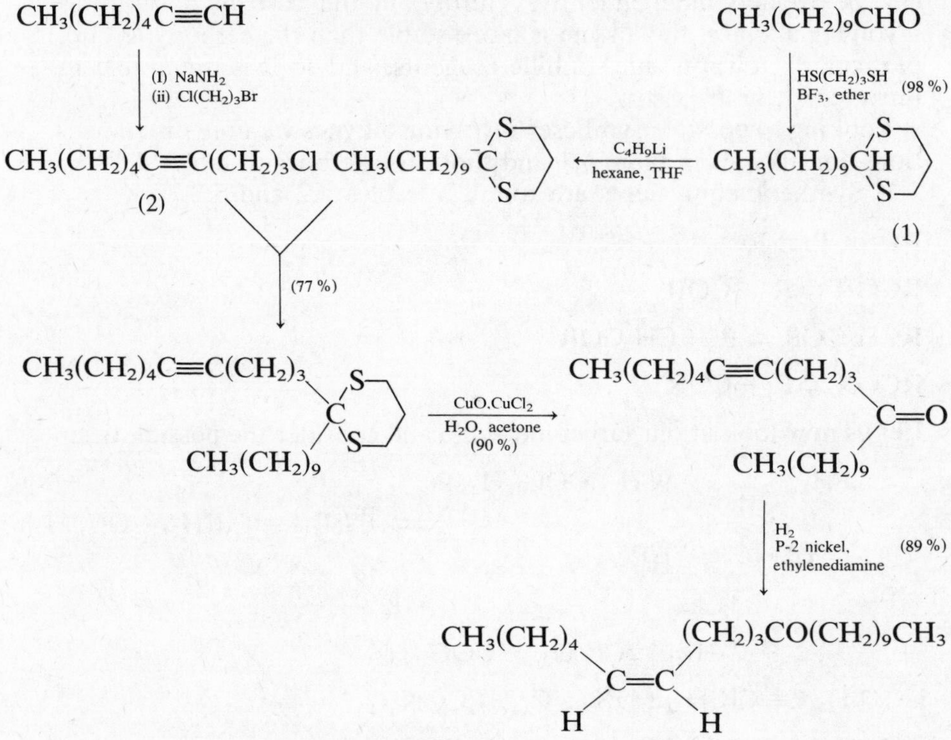

$CH_3(CH_2)_4C\equiv CH$ $CH_3(CH_2)_9CHO$

(I) NaNH$_2$
(ii) Cl(CH$_2$)$_3$Br

HS(CH$_2$)$_3$SH
BF$_3$, ether   (98 %)

$CH_3(CH_2)_4C\equiv C(CH_2)_3Cl$ $CH_3(CH_2)_9\overset{S}{\underset{S}{C}}$ $\xleftarrow[\text{hexane, THF}]{C_4H_9Li}$ $CH_3(CH_2)_9\overset{S}{\underset{S}{CH}}$

(2)   (1)

(77 %)

$CH_3(CH_2)_4C\equiv C(CH_2)_3$
$CH_3(CH_2)_9$ $\overset{S}{\underset{S}{C}}$ $\xrightarrow[\substack{H_2O, \text{ acetone} \\ (90\%)}]{CuO.CuCl_2}$ $CH_3(CH_2)_4C\equiv C(CH_2)_3$
$CH_3(CH_2)_9$ C=O

H$_2$
P-2 nickel,   (89 %)
ethylenediamine

$CH_3(CH_2)_4$ $\quad$ $(CH_2)_3CO(CH_2)_9CH_3$
C=C
H $\qquad$ H

(3)

The alkyl chain is formed by nucleophilic substitution by the anion of the dithian (1) derived from undecanal on the chloroalkyne (2) formed by reaction of the anion of hept-1-yne with 1-bromo-3-chloropropane. The dithian is cleaved using copper(II) oxide and copper(II) chloride in aqueous acetone and the pheromone, Z-heneicos-6-en-11-one (3), is formed by partial hydrogenation using a P-2 nickel catalyst in presence of ethylenediamine.[2]

As suggested in the preamble, Z-alkenes can be prepared by the salt-free Wittig procedure. A second synthesis[3] of the pheromone utilises this method and the *Z*-alkene function is introduced initially. The carbonyl group is prepared *via* the secondary alcohol. The disconnections involved in this route are thus:

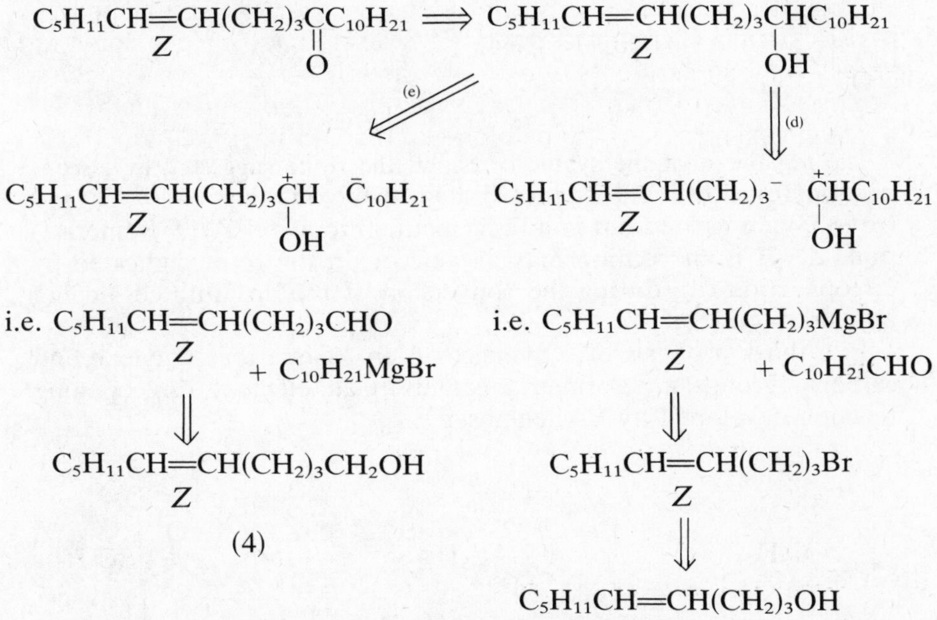

$C_5H_{11}CH{=}CH(CH_2)_3C(O)C_{10}H_{21}$ (Z) ⟹ $C_5H_{11}CH{=}CH(CH_2)_3CH(OH)C_{10}H_{21}$ (Z)
(e)
(d)
$C_5H_{11}CH{=}CH(CH_2)_3\overset{+}{C}H(OH)$ $\bar{C}_{10}H_{21}$ (Z)
$C_5H_{11}CH{=}CH(CH_2)_3^-$ $\overset{+}{C}H(OH)C_{10}H_{21}$ (Z)
i.e. $C_5H_{11}CH{=}CH(CH_2)_3CHO$ (Z) + $C_{10}H_{21}MgBr$
i.e. $C_5H_{11}CH{=}CH(CH_2)_3MgBr$ (Z) + $C_{10}H_{21}CHO$
$C_5H_{11}CH{=}CH(CH_2)_3CH_2OH$ (Z)
(4)
$C_5H_{11}CH{=}CH(CH_2)_3Br$ (Z)
$C_5H_{11}CH{=}CH(CH_2)_3OH$

Both (d) and (e) lead to unsaturated alcohols as the key intermediates. As will be seen below, compound (4) is prepared in a salt-free Wittig reaction from 2-hydroxytetrahydropyran.[4]

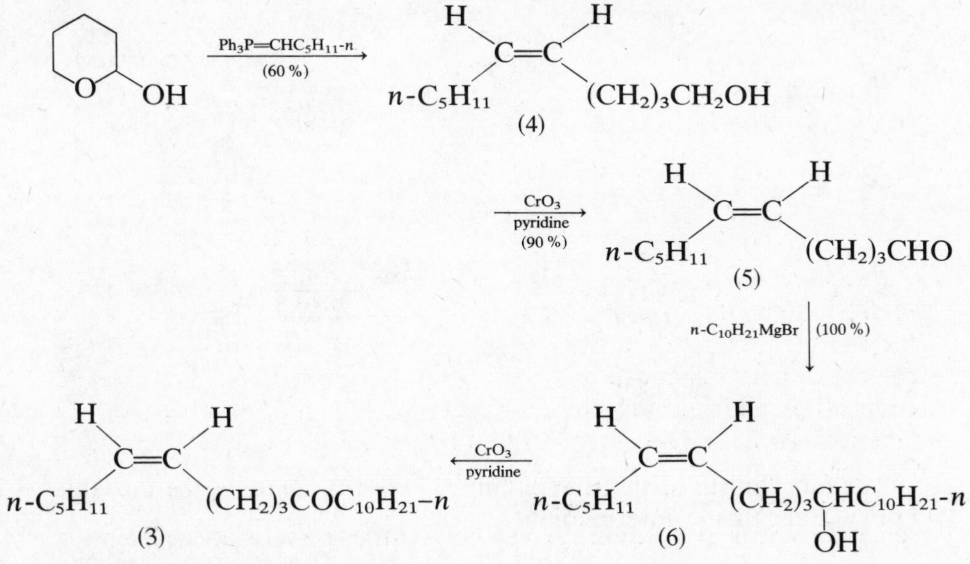

OH
$Ph_3P{=}CHC_5H_{11}$-*n*
(60 %)
H H C=C *n*-C_5H_{11} $(CH_2)_3CH_2OH$
(4)
CrO_3 pyridine (90 %)
H H C=C *n*-C_5H_{11} $(CH_2)_3CHO$
(5)
n-$C_{10}H_{21}MgBr$ (100 %)
H H C=C *n*-C_5H_{11} $(CH_2)_3CH(OH)C_{10}H_{21}$-*n*
(6)
CrO_3 pyridine
H H C=C *n*-C_5H_{11} $(CH_2)_3COC_{10}H_{21}$-*n*
(3)

2-Hydroxytetrahydropyran is the hemiacetal tautomer of 5-hydroxypentanal and its reaction with the Wittig reagent leads to Z-undec-5-en-1-ol (4).

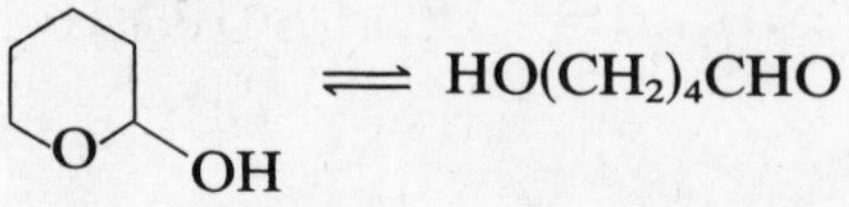

The remainder of the synthesis follows the route suggested by disconnection (e). It should be noted that the oxidations $(4) \rightarrow (5)$ and $(6) \rightarrow (3)$ are carried out in a basic medium to avoid $Z-E$ isomerisation. $Z-E$ isomerisation may also occur on the route indicated by disconnection (d), during the conversion of the unsaturated alcohol into the bromide.

In a third synthesis of Z-heneicos-6-en-11-one the acetylene and carbonyl groups are formed together in an efficient ring opening reaction developed by Eschenmoser[5]:

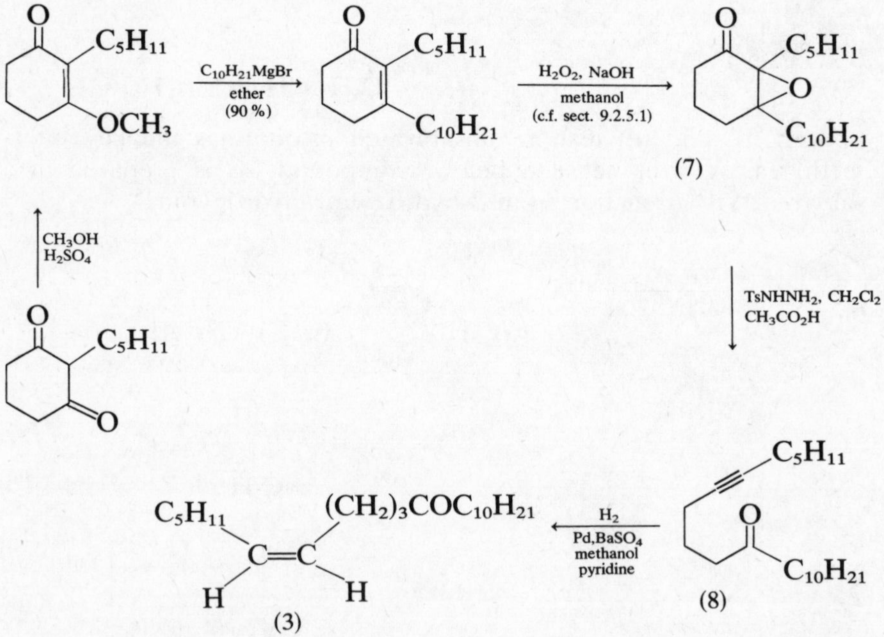

The mechanism of the ring opening $(7) \rightarrow (8)$, involves the unstable epoxy-diazoalkane intermediate (9):

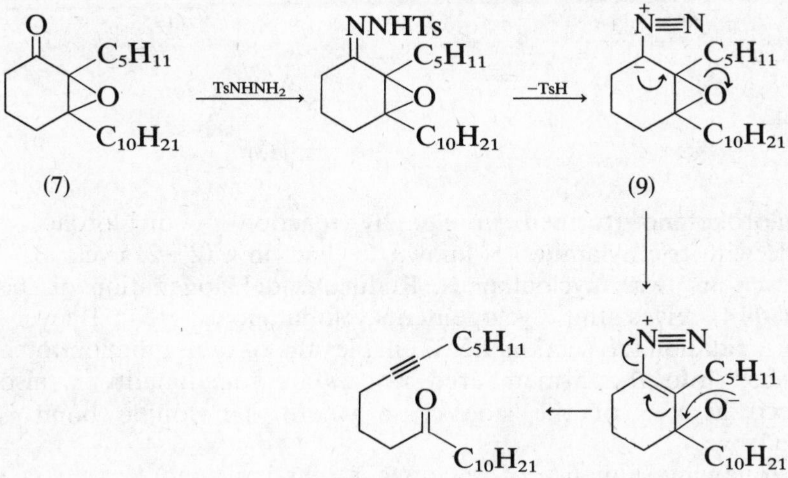

(7)   (9)

(8)

14.3 Synthesis of Z-jasmone

Z-Jasmone (10) is a constituent of jasmine flowers and is widely used in perfumery to reproduce the jasmine fragrance.

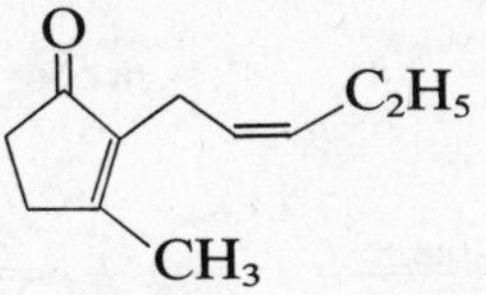

(10)

Synthetic routes might involve a preformed 5-membered ring precursor such as cyclopentadiene or involve cyclisation of a 1,4-diketone to the cyclopent-2-enone (cf. section 7.5.3):

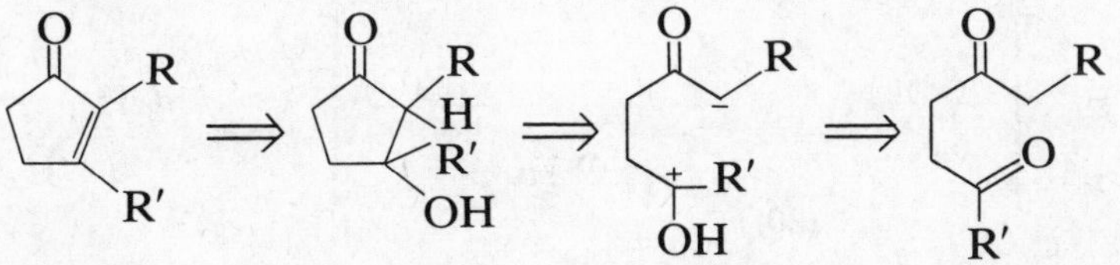

We shall consider three of the syntheses of Z-jasmone described in the literature. Two of these adopt variants of the latter approach[6] and one, which we will describe first, starts from cyclopentadiene.[7]

Cyclopentadiene could, in theory, by suitable reaction at one double bond be converted to (11) which could be isomerised to yield the α,β-unsaturated compound (12), and thence (13):

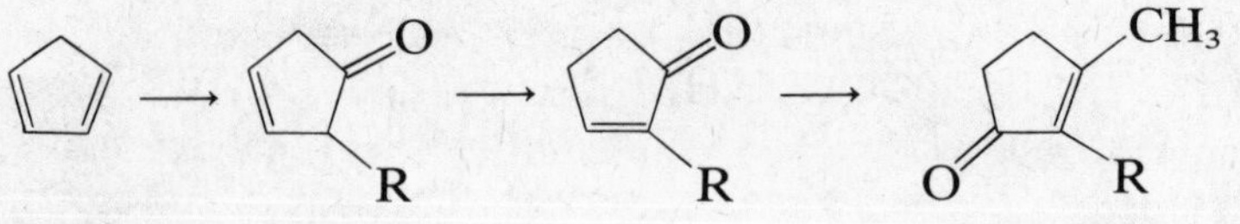

(11)   (12)   (13)

Dichloroketene (formed *in situ* by reaction of dichloroacetyl chloride with triethylamine) is known to react in a (2+2) cycloaddition reaction[8] with cycloalkenes. Reductive dehalogenation of the adduct (14) gives the cyclopentenocyclobutanone (15). Baeyer–Villiger oxidation (cf. section 9.5.3) enables the oxygen function to be introduced into the 5-membered ring while functionality is also produced at C-2 of the side chain where the double bond is required.

The following sequence of reactions is thus indicated:

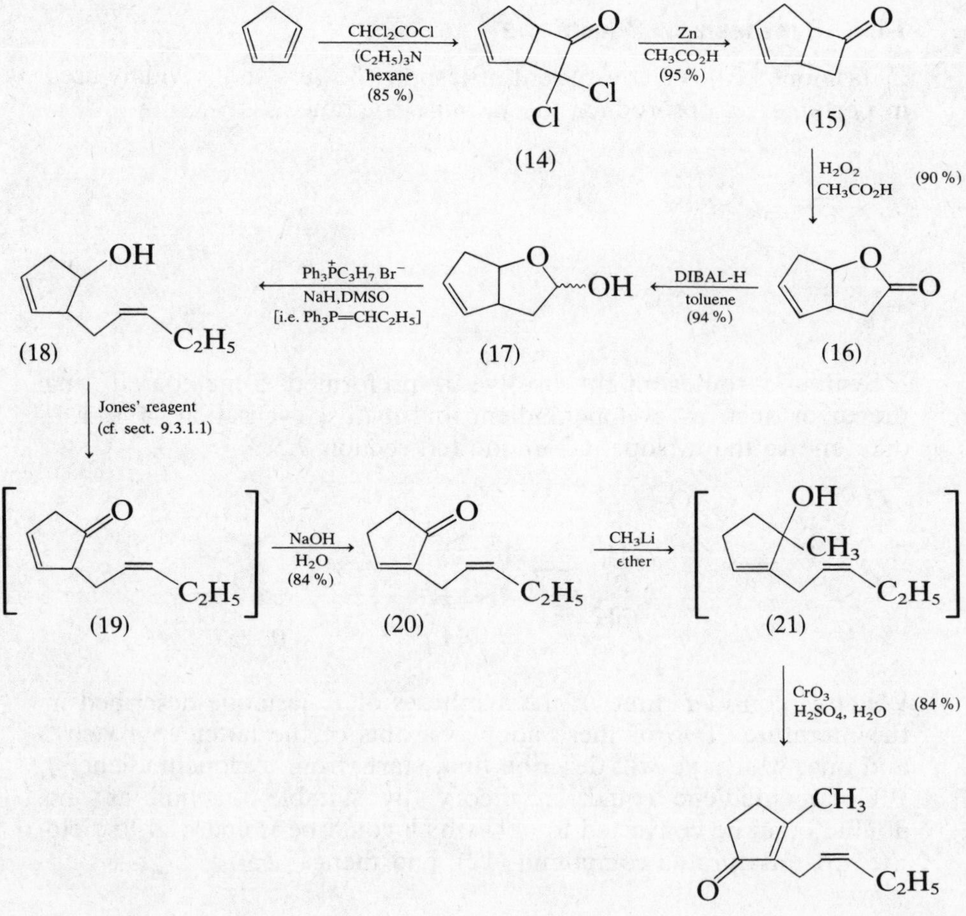

(14)

(15)

(18)   (17)   (16)

(19)   (20)   (21)

The lactone (16) formed in the Baeyer–Villiger oxidation was reduced to the hemiacetal (17) which on reaction with the Wittig reagent gave the cyclopentenol (18) having in the alkyl side chain a Z-double bond. No E-isomer was observed. The secondary alcohol was oxidised to the unsaturated ketone (19) which was converted to the more stable conjugated isomer (20) without isolation of the intermediate. Reaction with methyllithium gave the expected 1,2-addition product (21) (cf. section 4.2.1) which on oxidation underwent allylic rearrangement to Z-jasmone (10). The overall yield is around 40 %.

The other two methods involve cyclisation of the 1,4-diketone (22). It is, of course, undesirable to carry the carbonyl groups through the other steps in the sequence and the two methods differ in their solution of this problem. In the first, the carbonyl groups are present in the form of 1,3-dithians which can be alkylated (cf. section 5.3.2).

$$CH_3CO(CH_2)_2CO(CH_2)_2CH=CHC_2H_5$$

(22)

The starting material for this synthesis is 2,5-dimethoxytetrahydrofuran. This is a bis-acetal and on treatment with acid in presence of propane-1,3-dithiol it is converted to the bis-1,3-dithian of butanedial (23):

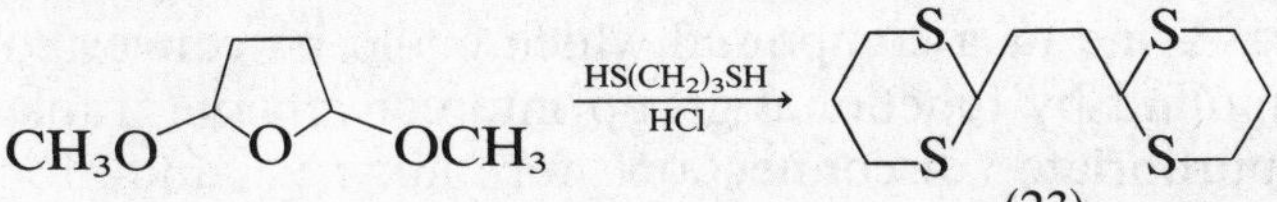

(23)

The next stage is the successive alkylation at C_1 and C_4 by the methyl and Z-hexenyl groups present in jasmone. These steps are carried out in the usual way (metallation using butyllithium followed by reaction with the alkyl halide):

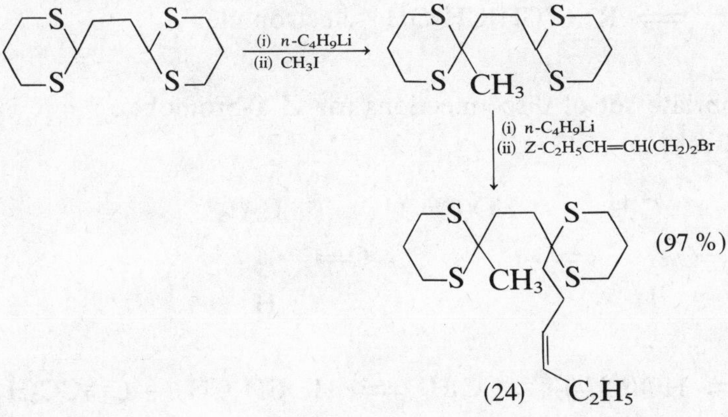

(24)

Hydrolysis of (24) by mercuric chloride and cadmium carbonate resulted in oxymercuration of the double bond. The product had, therefore, to be treated with potassium iodide to isolate the required enedione (22) which yielded jasmone on treatment with alcoholic sodium hydroxide. The overall yield of jasmone prior to purification is 61 %.

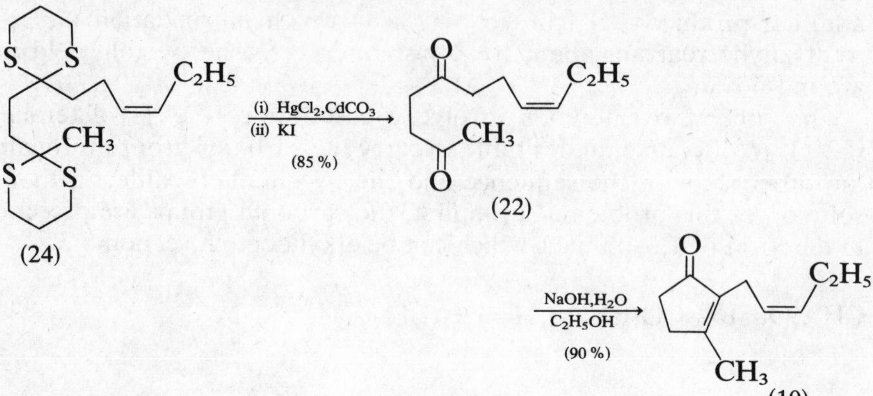

The remaining problem is the synthesis of Z-1-bromohex-3-ene. We recall that Z-alkenes are produced by partial hydrogenation of alkynes (cf. section 8.4.2). The problem is now that of synthesising either 1-bromo-hex-3-yne or a compound which could be converted to the bromo compound by functional group interconversion. Table 4.1 gives the appropriate disconnection for alkynes and for RCH_2CH_2OH ($R—OH \rightarrow R—Br$ being a possible functional group interconversion[9]):

$$RC\equiv CR' \implies RC\equiv C^- + {}^+R' \qquad \text{electrophile } R'–Y$$

$$RCH_2CH_2OH \implies R^- + {}^+CH_2CH_2OH \quad \text{electrophile } \triangle$$

The appropriate set of disconnections for Z-1-bromohex-3-ene is the following:

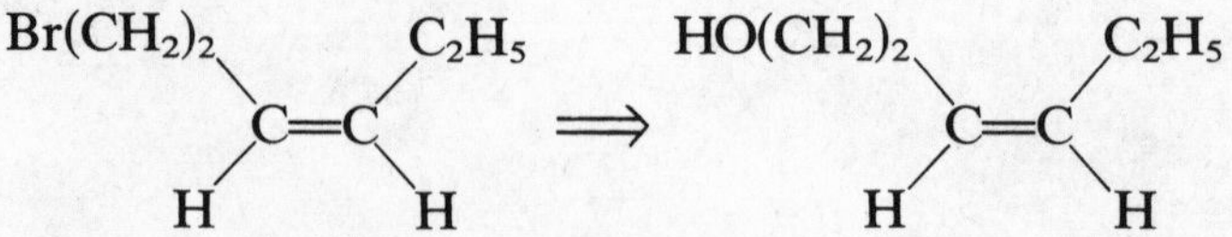

$$\implies HO(CH_2)_2C\equiv CC_2H_5 \implies HOCH_2CH_2{}^+ + \bar{C}\equiv CC_2H_5$$

The synthesis, in fact, starts at acetylene and is shown below:

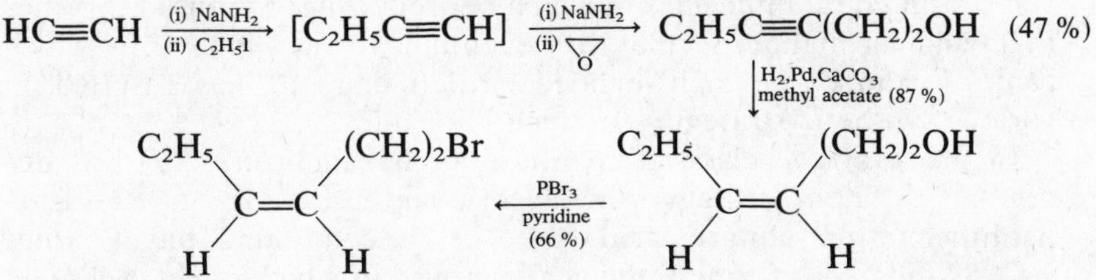
HC≡CH
(i) NaNH2
(ii) C2H5I
[C2H5C≡CH]
(i) NaNH2
(ii)
C2H5C≡C(CH2)2OH (47 %)
H2,Pd,CaCO3
methyl acetate (87 %)
PBr3
pyridine
(66 %)

The second procedure, which eliminates the need to protect the carbonyl groups, is to carry through the reaction sequence with the appropriate 2,5-dialkylfuran. 2,5-Dialkylfurans on hydrolysis are converted into 1,4-diketones. This technique has been given the name 'latent functionality'.[10]

The starting point of this synthesis is 2-methylfuran, which is converted by an acid-catalysed Michael addition with propenal into 3-(5-methyl-2-furyl) propanal (25) in presence of a small amount of hydroquinone as anti-oxidant:

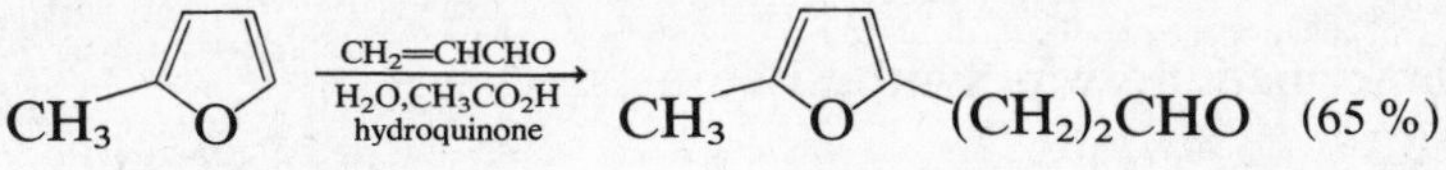
CH2=CHCHO
H2O,CH3CO2H
hydroquinone
(CH2)2CHO (65 %)

(25)

The 'salt-free' Wittig reaction (cf. section 5.3.1.3) on (25) using triphenylpropylphosphonium bromide results mainly in the *Z*-alkene (26) contaminated by about 12 per cent of the *E*-isomer:

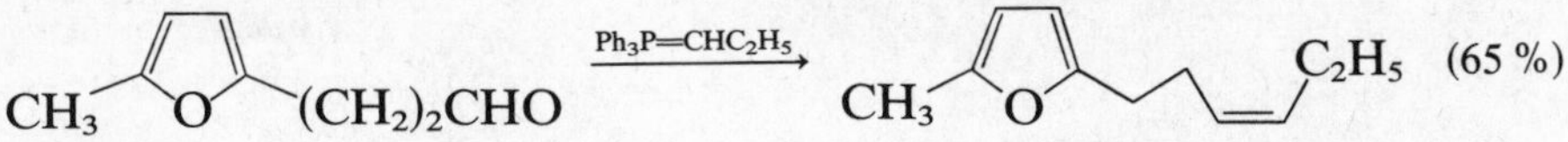
(CH2)2CHO
Ph3P=CHC2H5
C2H5 (65 %)

(26) ($Z:E \sim 8:1$)

Hydrolysis using aqueous acetic acid and cyclisation of the resulting enedione (22) completes the sequence. Jasmone (10) is formed in about 35 % overall yield by this method:

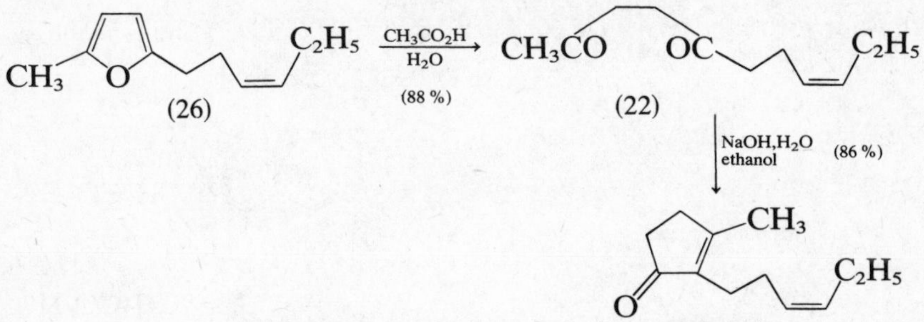
CH3CO2H
H2O
(88 %)
CH3CO
OC
NaOH,H2O
ethanol
(86 %)

(26) (22) (10)

14.4 Helicenes

Interest in chiral molecules in which benzene rings are fused together in an angular manner so that they eventually form a helix (27) [which can, of course, be right-handed or left-handed] has resulted in various synthetic strategies for such molecules.

In the 1950s a 'classical' synthesis of hexahelicene (27) was devised.[11] The strategy involved synthesis of a 3-(di-α-naphthylmethyl)-glutaric acid (28) (i.e. a compound having rings 1, 2, 5, and 6 pre-formed) the acidic groups of which might cyclise on to the naphthyl groups in a Friedel–Crafts type reaction giving a 6-ring system.

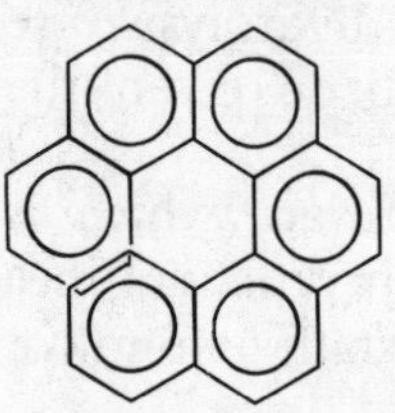

(27)

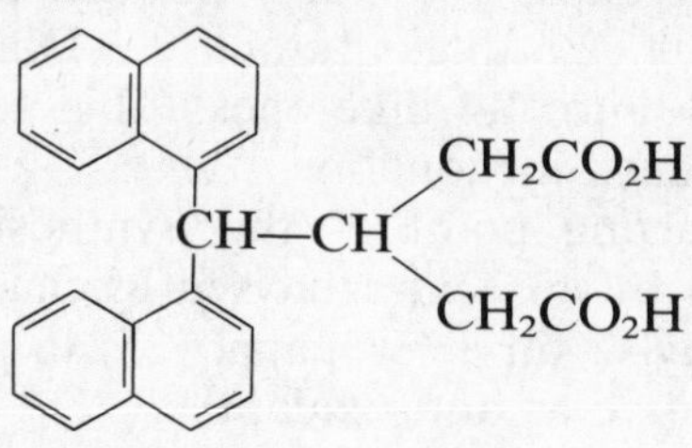

(28)

This key intermediate was synthesised as shown:

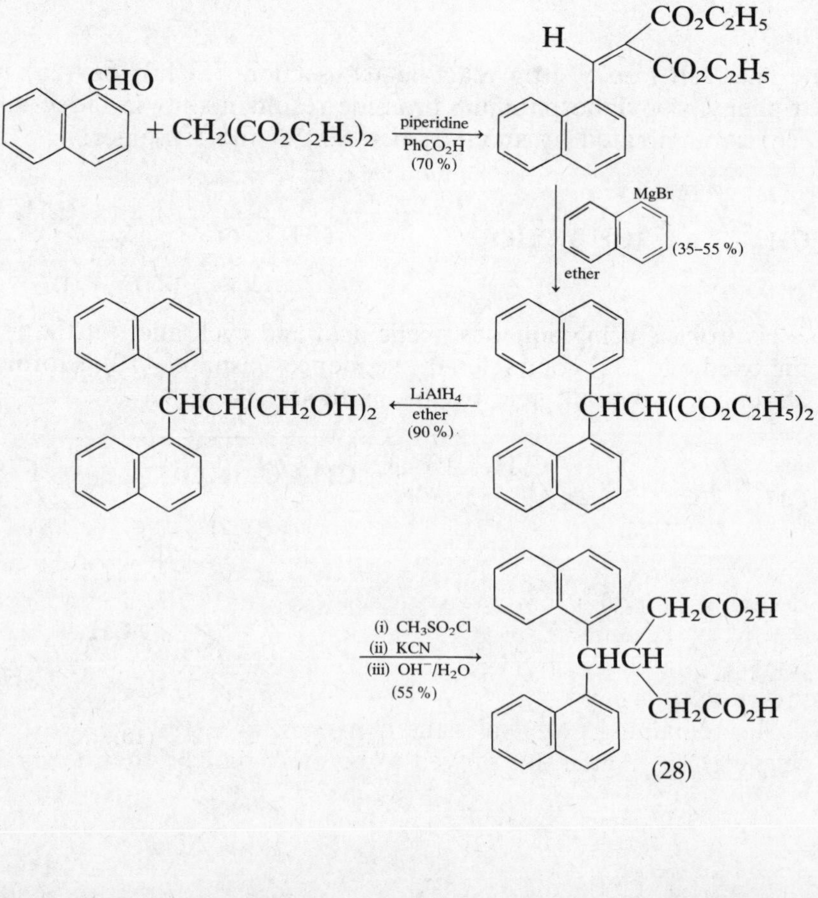

(28)

1-Naphthaldehyde and diethyl malonate undergo a Knoevenagel condensation (cf. section 5.1.4) to give the naphthylidenemalonic ester. Conjugate addition of α-naphthylmagnesium bromide to this α,β-unsaturated ester takes place. After reduction of the ester groups with lithium aluminium hydride, chain extension by the following sequence gives the required glutaric acid (28):

$$ROH + CH_3SO_2Cl \rightarrow ROSO_2CH_3 \xrightarrow{KCN} RCN \xrightarrow[H_2O]{OH^-} RCO_2H$$

Treatment of the glutaric acid (28) with anhydrous HF (Friedel–Crafts acylation) results in only one cyclisation on to a naphthyl group. The fact that the second ring closure does not take place is due to the increased strain. So the two remaining rings must be formed separately:

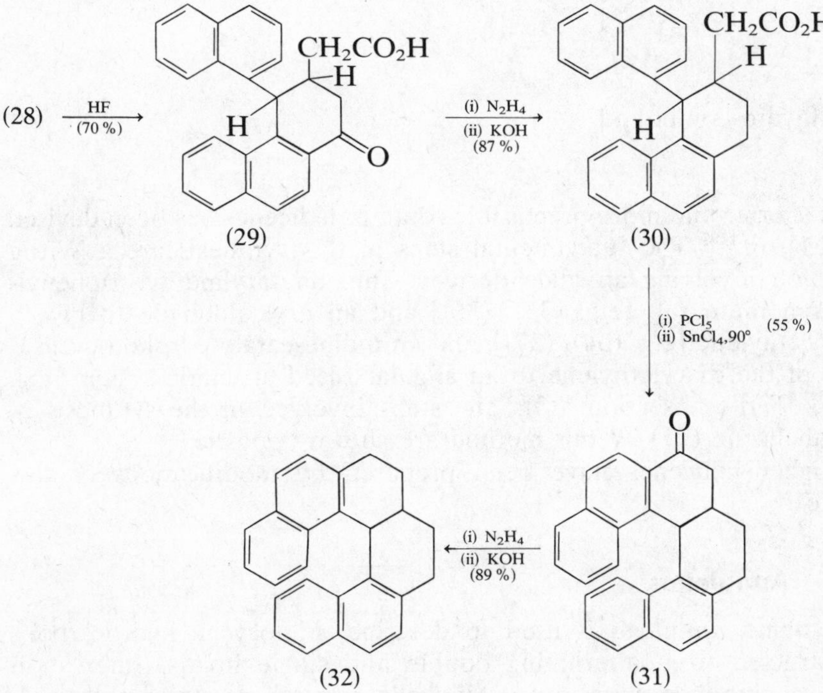

It should be noted that both ketones, (29) and (31), are reduced by the Wolff–Kishner method (cf. section 8.4.3.3) and that the conditions required for the second ring closure (30) → (31) are more severe than usual.

The remaining problem is the dehydrogenation of the two central rings in (32). Again this proved to be more difficult than might have been expected (cf. section 9.2.4) and the only direct method which was successful was a catalytic method using a rhodium catalyst.

One of the standard methods for dehydrogenation, use of DDQ (cf. section 9.2.4) gave a dihydro-compound which proved to be very resistant to further dehydrogenation. It was suggested that the increase in delocalisation energy in this final dehydrogenation is small due to the lack of coplanarity of the rings in hexahelicene and that any increase in delocalisation energy would be offset by an increase in strain:

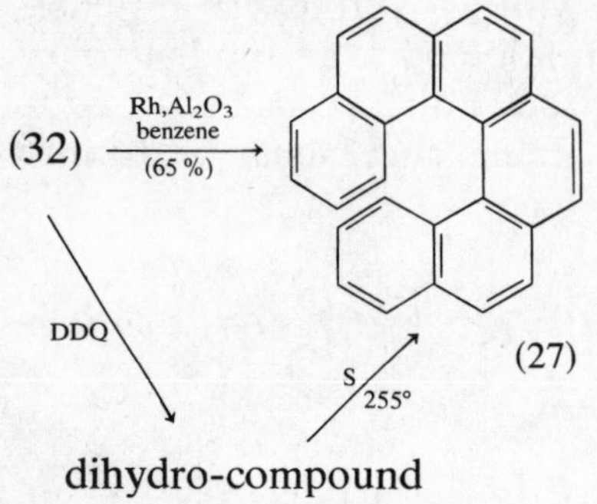

(32) $\xrightarrow[\substack{\text{benzene} \\ (65\%)}]{\text{Rh,Al}_2\text{O}_3}$ (27)

DDQ

S / 255°

dihydro-compound

A shorter and more practicable route to helicenes has been devised by Martin.[12] The fundamental steps in this synthesis are a Wittig reaction involving an ylide derived from an (arylmethyl)triphenyl-phosphonium salt [e.g. (33), (36)] and an aryl aldehyde to give a diarylethylene [e.g. (34), (37)], and an iodine-catalysed photocyclisation of the diarylethylene to an angular fused aromatic system [e.g. (35), (27)] (cf. section 7.3). The steps involved in the synthesis of hexahelicene (27) by this method are shown opposite.

Higher helicenes have been prepared by modifications of this route.

14.5 Annulenes

The name *annulene* is used to describe monocyclic hydrocarbons constructed from alternating double and single bonds. Interest in such compounds arises from Hückel's rule which states that such compounds, provided that the carbon framework can be virtually planar, are aromatic if they contain $(4n+2)$ π-electrons (i.e. those with 10, 14, and 18 carbons in the ring will be aromatic). In the case of 18-membered rings, the structure (38) is stable and its synthesis will be described in section 14.5.1. However, in the case of 10-membered rings, for structure (39) there is too much angle strain for a stable molecule to exist, and in (40) the interactions between the

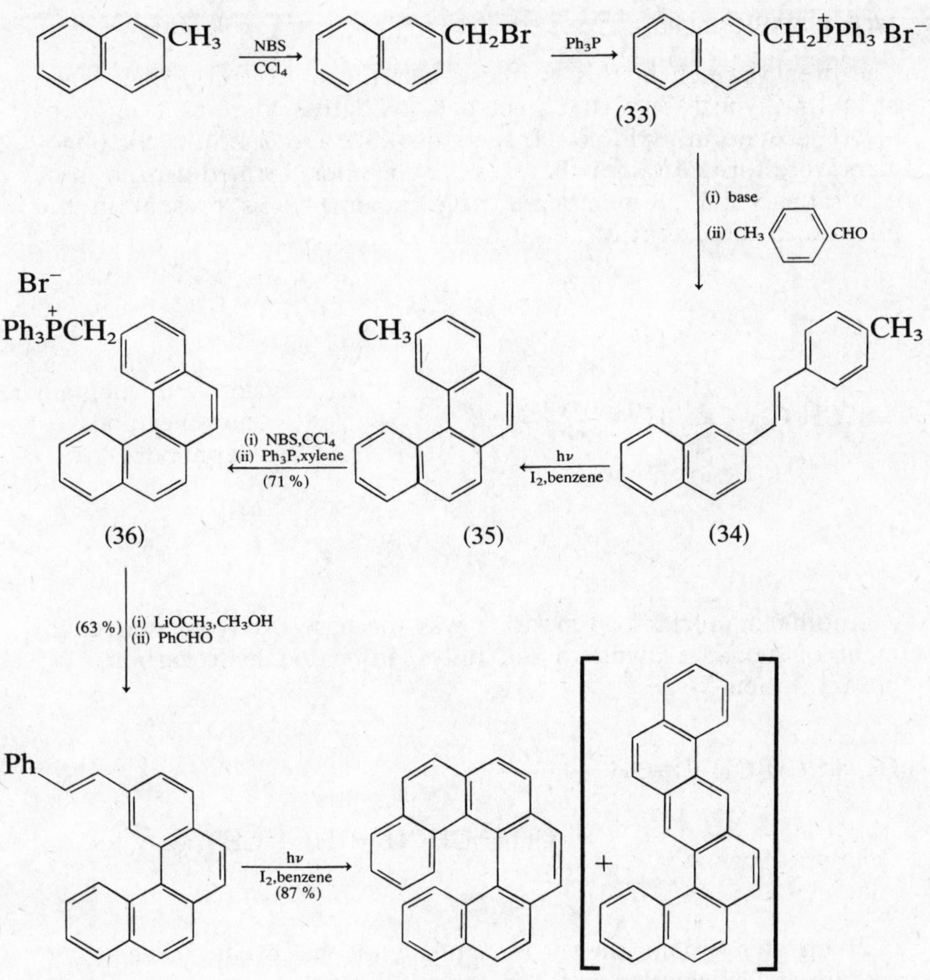
CH3
NBS
CCl4
CH2Br
Ph3P
CH2P̊Ph3 Br⁻
(33)
(i) base
(ii) CH3 CHO
Br⁻
Ph3P̊CH2
CH3
CH3
(i) NBS,CCl4
(ii) Ph3P,xylene
(71 %)
hν
I2,benzene
(36)
(35)
(34)
(63 %)
(i) LiOCH3,CH3OH
(ii) PhCHO
Ph
hν
I2,benzene
(87 %)
+
(37)
(27)

two internal hydrogens prevent the ring carbons achieving coplanarity. Stable aromatic molecules have been made containing a 1,6-bridge [structure (41): Z = CH_2, O, S, etc.]. In section 14.5.2 we shall describe a synthesis of the methano-bridged compound (41, Z = CH_2).

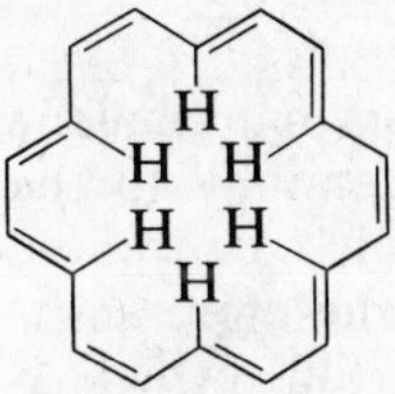
H
H H
H H
H

(38)

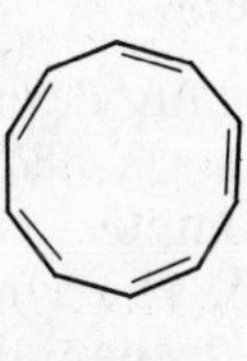

(39)

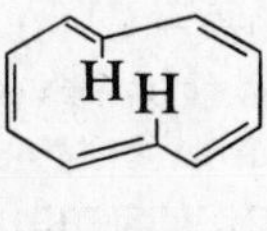
H H

(40)

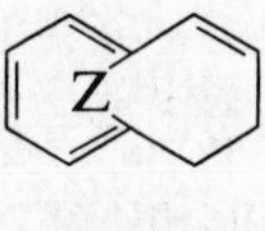
Z

(41)

14.5.1   [18]Annulene[13]

The breakthrough in this synthesis was the observation that, when hexa-1,5-diyne (42) was subjected to oxidative coupling using copper(II) acetate in pyridine (cf. sections 4.3.2 and 7.1.5), cyclic oligomers were formed rather than acyclic oligomers formed using Glaser conditions. The 18-membered ring trimer (43) is present in the mixture to the extent of about 6 %.

$$HC{\equiv}CCH_2CH_2C{\equiv}CH \xrightarrow[\text{pyridine}]{Cu(O_2CCH_3)_2}$$

(42)

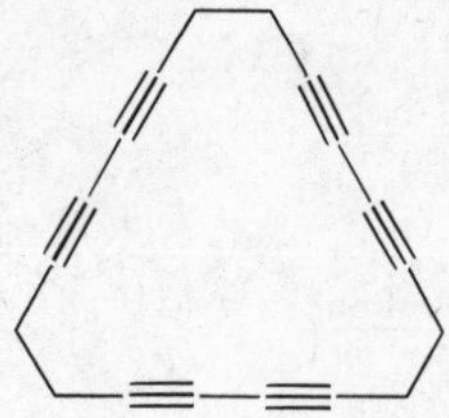

+ tetramer, pentamer, hexamer and heptamer

(43)

Another important observation was the base catalysed rearrangement of hexa-1,5-diyne to the fully conjugated hydrocarbon (44), hexa-1,3-dien-5-yne:

$$HC{\equiv}CCH_2CH_2C{\equiv}CH \xrightarrow[\text{t-butanol}]{KOC(CH_3)_3}$$

$$CH_2{=}CHCH{=}CHC{\equiv}CH \quad (33\% \text{ isolated})$$

(44)

When this isomerisation is applied to the cyclic trimer (43), trisdehydro[18]annulene (45) is formed. If the reaction from hexa-1,5-diyne is carried through to (45) without isolating the trimer the yield is about 3 %. [18]Annulene (38) is formed when (45) is catalytically hydrogenated. It should be noted that in this case a *trans*-double bond is formed from the triple bond and because of the stability of the product (due to its aromatic character) the use of a reduced activity catalyst is not required.

14.5.2   1,6-Methanocyclodecapentaene[14]

This compound (41, Z = CH$_2$) is formally derived from naphthalene by addition of :CH$_2$. However, carbenes add preferentially to the *1,2-bond* in naphthalene to give compounds in which benzenoid stabilisation of one ring is maintained. The synthesis therefore starts with a carbene addition to tetrahydronaphthalene (46) which is

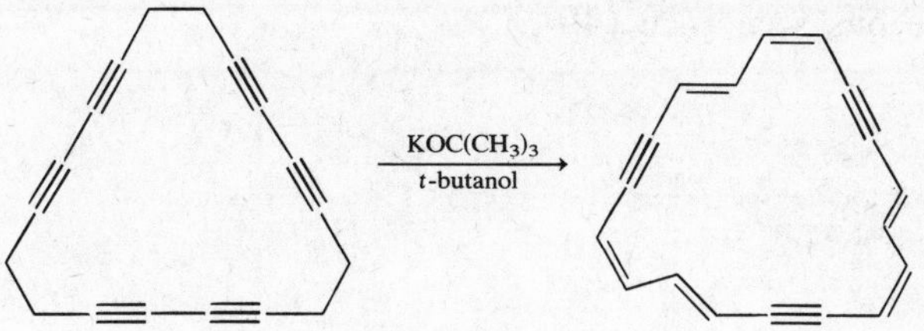

(43) → (45)

KOC(CH₃)₃
t-butanol

[overall yield from (42)
~3 %]

H₂, Pd on C (25 %)
in benzene

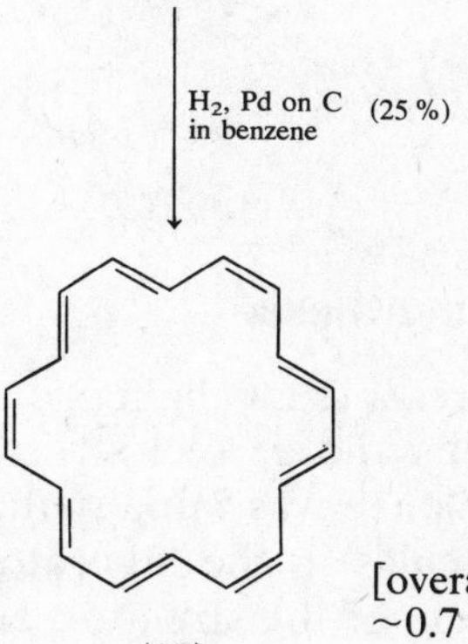

[overall yield from (42)
~0.7 %]

(38)

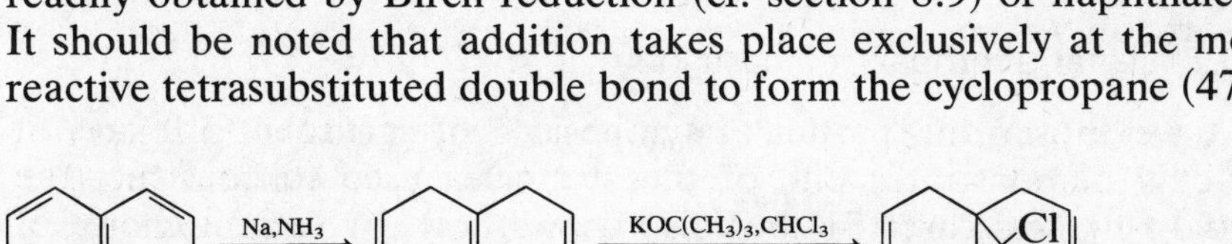

readily obtained by Birch reduction (cf. section 8.9) of naphthalene.
It should be noted that addition takes place exclusively at the more
reactive tetrasubstituted double bond to form the cyclopropane (47):

Na,NH₃
ethanol
ether
(60 %)

(46)

KOC(CH₃)₃,CHCl₃
(i.e.: CCl₂).

(47)

The carbon–halogen bonds in (47) are reductively cleaved by a
dissolving metal method. The next step is to introduce two additional
double bonds [giving (48)] by a bromination-dehydrobromination procedure (cf. section 9.2.4.2). Compound (48) can undergo pericyclic
ring opening (cf. section 7.3) to the desired compound (49). Spectroscopic and chemical properties of the product indicate that it is

aromatic and therefore has structure (49):

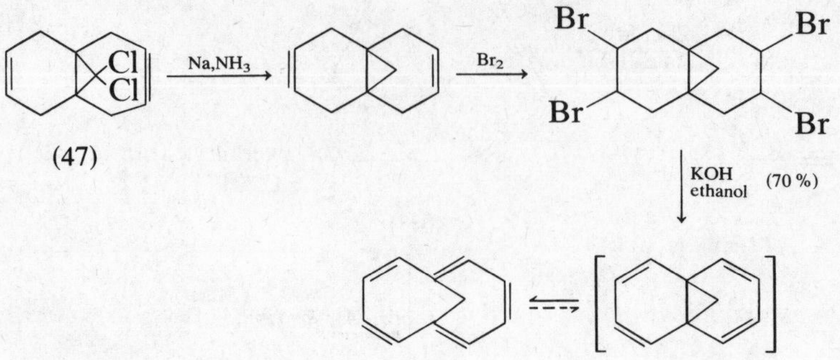

(47)

(49) ⟸ [(48)]

14.6   Steroid synthesis

Steroids have presented a challenge to the synthetic chemist since the early 1930s when the correct structure of cholesterol was first proposed. The challenge was initially that of attempting to produce such important molecules in the laboratory simply as a test of ingenuity and as verification of the structure elucidated mainly by degradation. Later, as the importance of many steroids as pharmaceuticals grew, it became necessary to supplement material isolable from natural sources. In the main, the solution to this last requirement is not to start from simple compounds, as in total synthesis, but to use abundant natural steroids (for example, the sapogenins, derived from plant glycosides), which contain many of the desired functionalities and which can be converted by a relatively small number of high yield steps into medicinally important steroids such as progesterone and cortisone.

14.6.1   Total synthesis of cortisone

We have chosen this particular synthesis[15] of cortisone (50) since it not only demonstrates one of the strategies used to construct the steroid ring structure (51) but also shows how, by careful choice of reagent and of reaction conditions, selective transformations can be carried out.

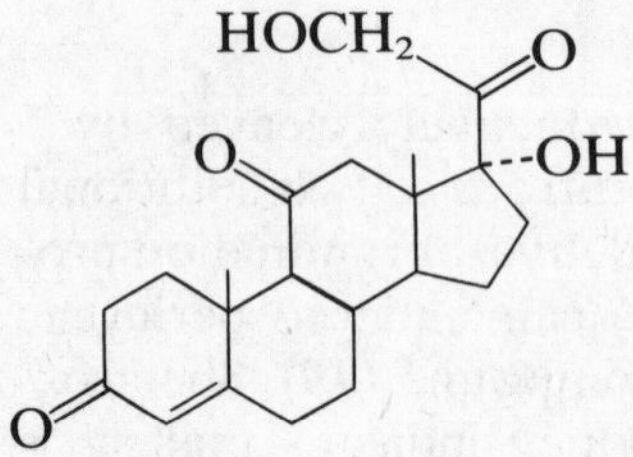

(50) 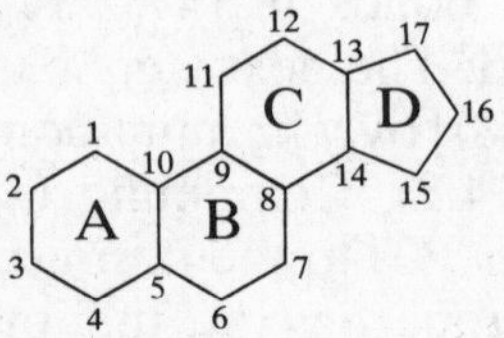   (51)

It would be somewhat difficult to use the disconnection approach to decide which route to cortisone might offer the shortest and most effective synthesis. We shall, therefore, set out the complete synthesis (see scheme 14.1) and then comment on the individual steps and on the strategy of the synthesis as it develops.

The first stage involves a Diels–Alder reaction (cf. section 7.2.1) on 3-ethoxypenta-1,3-diene[16] in which, as expected, the product is that of *syn*-addition *via* the favoured *endo*-transition state. The rings so formed will become rings B and C of the product, and the enol ether part of the molecule will be converted to a ketone which will be used in a Robinson annelation reaction to form ring A. However, before the ketone is released, the double bond and carbonyl groups of the enedione system are reduced. The reduction is carried out in two stages, first the double bond and then the carbonyl groups. [It is important not to affect the stereochemistry of the ring junction at this stage.] So catalytic hydrogenation is chosen to reduce the double bond rather than 'dissolving metal' reduction (cf. section 8.7) and as expected the enol ether double bond is unaffected. Thereafter lithium aluminium hydride is used to reduce the carbonyl groups to give the diol (52) which is extremely sensitive to acid and moisture and with aqueous acetic acid the enol ether is hydrolysed to the ketone (53). Base-catalysed Michael addition to butenone takes place, as expected, *via* the carbanion formed at the more substituted α-carbon atom on the less hindered side of (53) (cf. section 5.1.5) to give (54):

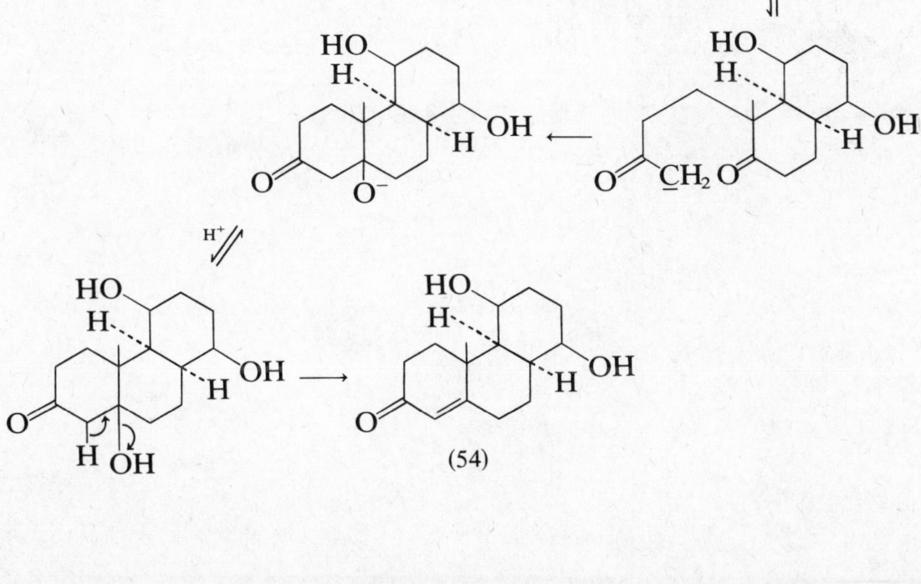

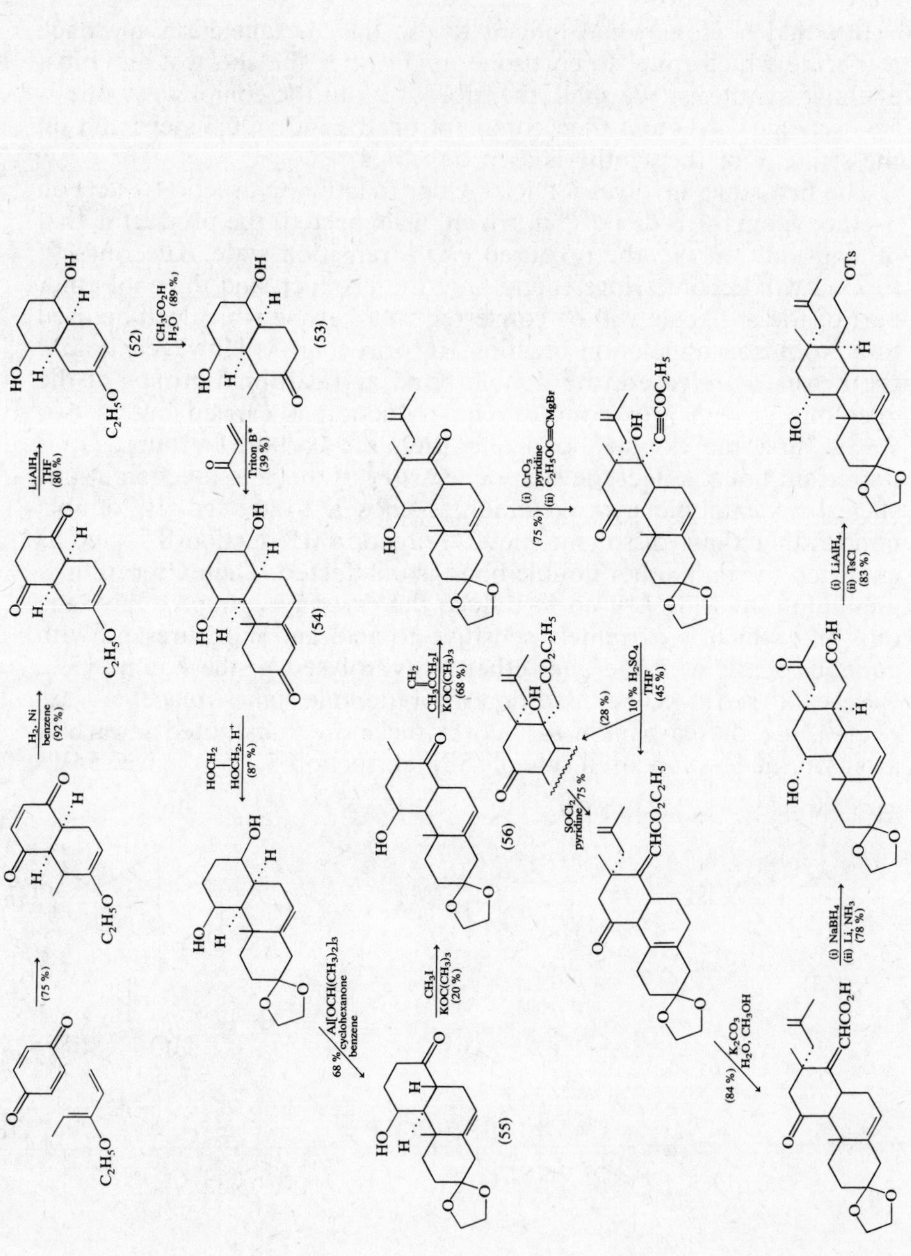

(52)

CH_3CO_2H / H_2O (89%)

(53)

C_2H_5O

$\frac{LiAlH_4}{THF}$ (80%)

Triton B* (39%)

(54)

C_2H_5O

H_2, Ni benzene (92%)

$HOCH_2$ / $HOCH_2$, H^+ (87%)

C_2H_5O

(75%)

C_2H_5O

$\frac{(i)\ CrO_3\ pyridine}{(ii)\ C_2H_5OC\equiv CMgBr}$ (75%)

$\equiv COC_2H_5$

$\frac{CH_2=CHCH_2I}{KOC(CH_3)_3}$ (68%)

(56)

$CO_2C_2H_5$

(28%)

$\frac{10\%\ H_2SO_4}{THF}$ (45%)

$\frac{SOCl_2}{pyridine}$ 75%

$CHCO_2C_2H_5$

$\frac{CH_3I}{KOC(CH_3)_3}$ (20%)

68% / $Al[OCH(CH_3)_2]_3$ cyclohexanone benzene

(55)

$\frac{K_2CO_3}{H_2O,\ CH_3OH}$ (84%)

$CHCO_2H$

OTs

$\frac{(i)\ LiAlH_4}{(ii)\ TsCl}$ (83%)

CO_2H

$\frac{(i)\ NaBH_4}{(ii)\ Li,\ NH_3}$ (78%)

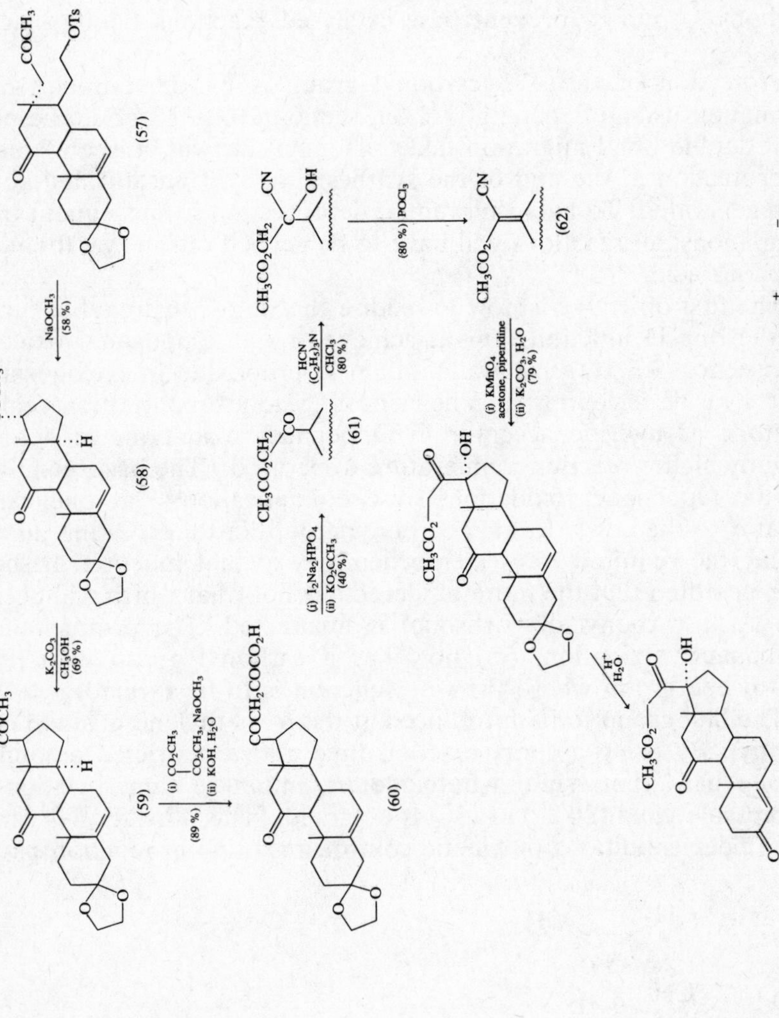

Scheme 14.1

*Triton B = $(CH_3)_3\overset{+}{N}CH_2Ph$ $\bar{O}H$

Let us now review the progress of the synthesis. Rings A, B and C have now been constructed. Rings A and B have the functional groups and unsaturation of the target molecule except that the B/C ring fusion is *cis* rather than *trans* as required in cortisone. Ring C contains an oxygen substituent at the carbon which will become the carbonyl at C-11 in cortisone. The other hydroxyl group in ring C is at one position where the C/D ring junction will be formed. Tactically, it will be simpler to form the D-ring if the hydroxyl group is oxidised to the ketone and it will be necessary to protect the C-3 carbonyl group to prevent base-catalysed reactions taking place in ring A.

Protection of the C-3 carbonyl group is by the familiar acetal formation using ethylene glycol (cf. section 10.6). It should be noted that double bond migration takes place but, as will be seen later, on deprotection at the end of the synthesis the α,β-unsaturated ketone is regenerated. To avoid unwanted deprotection in subsequent transformations, all reactions will have to be carried out in the absence of aqueous acid.

The first objective is now to oxidise the ring-C hydroxyl at the site of the ring-D junction. This is achieved by an Oppenauer oxidation (cf. section 9.3.1.2) using aluminium isopropoxide in cyclohexanone (the hydride acceptor) and benzene. [Cyclohexanone often replaces acetone as hydride acceptor in Oppenauer oxidations to permit a slightly higher reaction temperature to be used.] The basic conditions of the Oppenauer oxidation cause epimerisation[17] at one carbon (that α to the newly formed carbonyl group) of the B/C ring junction giving the required *trans* stereochemistry at that junction. It should also be noted that the more hindered alcohol (that which will become the C-11 carbonyl of cortisone) is unaffected. The product of the Oppenauer oxidation (55) now has a carbonyl group with its α-methylene group where the ring junction is to be formed.

The first group to be introduced at the α-methylene position is the methyl. By using a short reaction time and a restricted amount of strong base (potassium t-butoxide) to minimise dimethylation, an acceptable yield (20 %) of (56) is obtained. [This alkylation is carried out under conditions of kinetic control, resulting in reaction *via* the less substituted enolate.] A second alkylation, using potassium t-butoxide and methallyl iodide, introduces part of the incipient ring D. As we shall see, only the methylene group shown in bold type becomes part of ring D. The three remaining carbons will form the side chain. The italicised C is in fact part of a latent carbonyl group.

The remaining two carbons of ring D will be added at the carbonyl group by a Grignard reaction and it is undesirable to have a hydroxyl group present at this stage. This is oxidised to a carbonyl group by the chromium(VI) oxide/pyridine complex (cf. section 9.3.1.1), a move which might at first sight appear to introduce an undesirable complication. However, this carbonyl is so hindered that it is un-affected by the Grignard reagent. Most of the product of the Grig-nard reaction has the stereochemistry shown (i.e. axial OH) and only a small amount of the equatorial isomer. Extremely mild acid treat-ment converts the hydroxy acetylenic ether into the α,β-unsaturated ester while leaving the protecting group intact. In addition, some β-hydroxyester was formed and this was dehydrated to the α,β-unsaturated ester:

$$\underset{\overset{|}{OH}}{\overset{\diagdown}{C}}-C\equiv COC_2H_5 + H^+ \rightleftharpoons \underset{\overset{|}{OH}}{\overset{\diagdown}{C}}H=\overset{+}{C}OC_2H_5$$

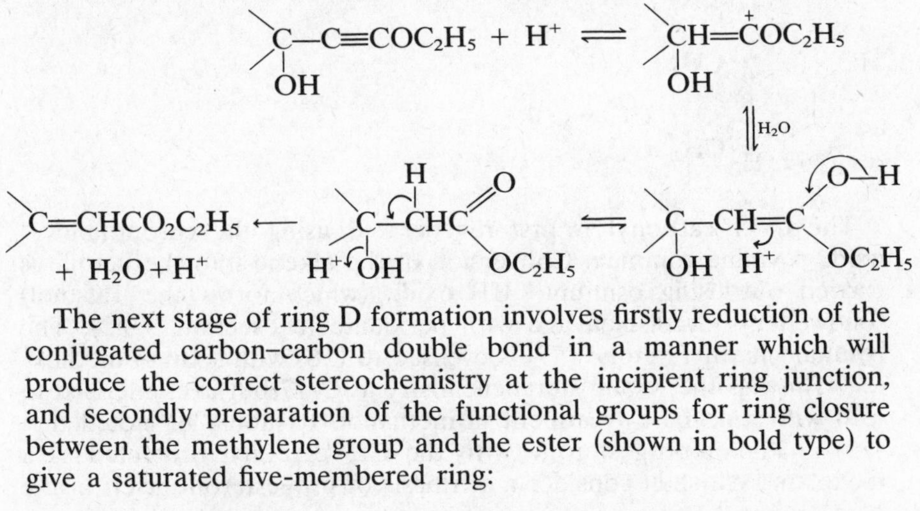

The next stage of ring D formation involves firstly reduction of the conjugated carbon–carbon double bond in a manner which will produce the correct stereochemistry at the incipient ring junction, and secondly preparation of the functional groups for ring closure between the methylene group and the ester (shown in bold type) to give a saturated five-membered ring:

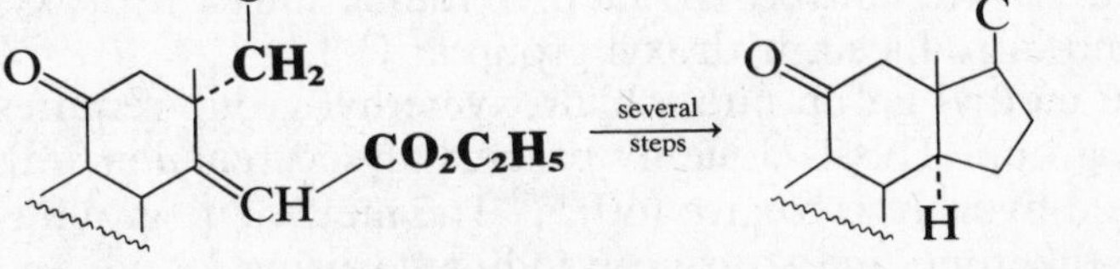

The ester is first hydrolysed under mild conditions and the subse-quent reduction is carried out in two stages, a process which is found to give the desired stereochemistry. First the carbonyl group is reduced using sodium borohydride to give, as expected, the more stable isomer (hydroxyl group equatorial) (cf. section 8.4.3.1). The carbon–carbon double bond is reduced by the dissolving metal method, a process which again usually favours formation of the more stable isomer (cf. section 8.7) – in this case with the carboxy-methylene group equatorial. The carboxylic acid is now reduced to the primary alcohol and the latter converted into the tosylate ester. This now provides an electrophilic carbon with a good leaving group and hence a site at which a carbanion will react. It is therefore necessary to generate a carbanion at the carbon indicated in bold

$$\begin{array}{c}\diagdown \\ \diagup\end{array}\!\!C^{-} \quad \overset{\delta+}{-CH_2}\overset{\delta-}{-OTs} \longrightarrow \begin{array}{c}\diagdown \\ \diagup\end{array}\!\!C-CH_2-$$

type. To do this the latent carbonyl group (shown in italic type) must be exposed.

The C-11 carbonyl is first regenerated using the chromium(VI) oxide/pyridine complex. Conversion of the alkene into the ketone is carried out using osmium(VIII) oxide (which forms the 1,2-diol) followed by glycol cleavage with periodate (cf. section 9.3.2). The resultant methyl ketone (57) is cyclised to (58) with sodium methoxide. This has the wrong stereochemistry at C-17 but is epimerised to (59) with mild base treatment in methanol. It should be noted that (59) is 11-ketoprogesterone with the C-3 carbonyl protected as a dioxolan. [We shall consider a synthesis of progesterone itself in the next section.]

Apart from regenerating the carbonyl group at C-3, the remaining steps in the synthesis will convert the methyl ketone into a hydroxy-methyl ketone and introduce a hydroxyl group at C-17.

To convert the methyl group into a hydroxymethyl group requires activation of the group. This is done by base-catalysed reaction with diethyl oxalate and hydrolysis to give (60).[18] The methylene group is now that of a β-diketone and so can be iodinated using iodine and mild alkali. The iodine is then subjected to nucleophilic substitution by potassium acetate. During this reaction loss of CO and CO_2 also takes place giving the acetoxymethyl compound (61). The side chain ketone is then converted into its cyanohydrin which is then dehydrated using phosphorus oxychloride to give (62). [It should again be noted that it is possible to form the cyanohydrin selectively since the C-11 carbonyl group is very unreactive.] Hydroxylation of the double bond of (62) by potassium permanganate on the less hindered side of the molecule gives a hydroxycyanohydrin. Cortisone acetate (63) is formed when this hydroxycyanohydrin is treated with dilute potassium carbonate and the C-3 carbonyl group is liberated with acid. The overall yield in this synthesis is about 0.01 %.

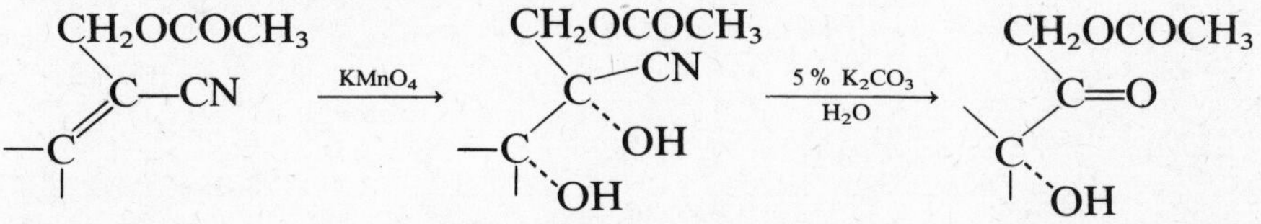

Cortisone (50) can be obtained from cortisone acetate by mild basic hydrolysis ($KHCO_3$ in aqueous methanol).

14.6.2 Conversion of plant steroids to steroid hormones

Total syntheses, such as that described in the previous section, do not normally represent practicable commercial syntheses of pharmaceuticals. Plant steroids provide a source of raw materials which contain many of the features of compounds such as progesterone and cortisone. We will now look in outline at the way in which diosgenin (64)[19] can be converted in a few high-yield steps into progesterone.[20] The conversion is shown in scheme 14.2.

The free hydroxyl group is acetylated and the spiroketal opened to give a dihydrofuran in the first step using acetic anhydride:

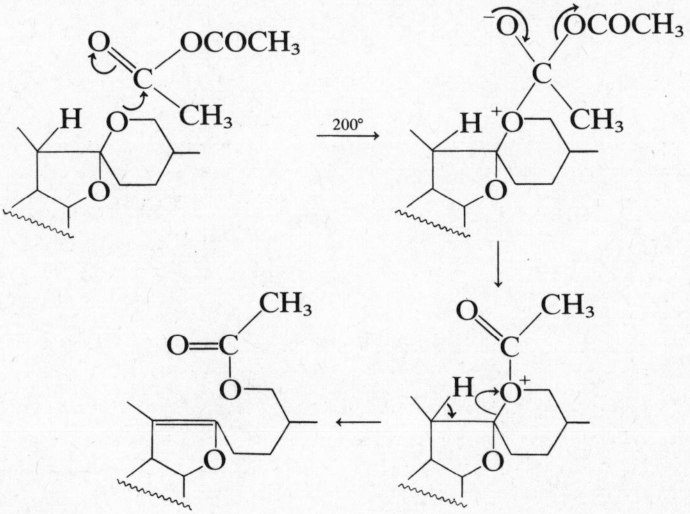

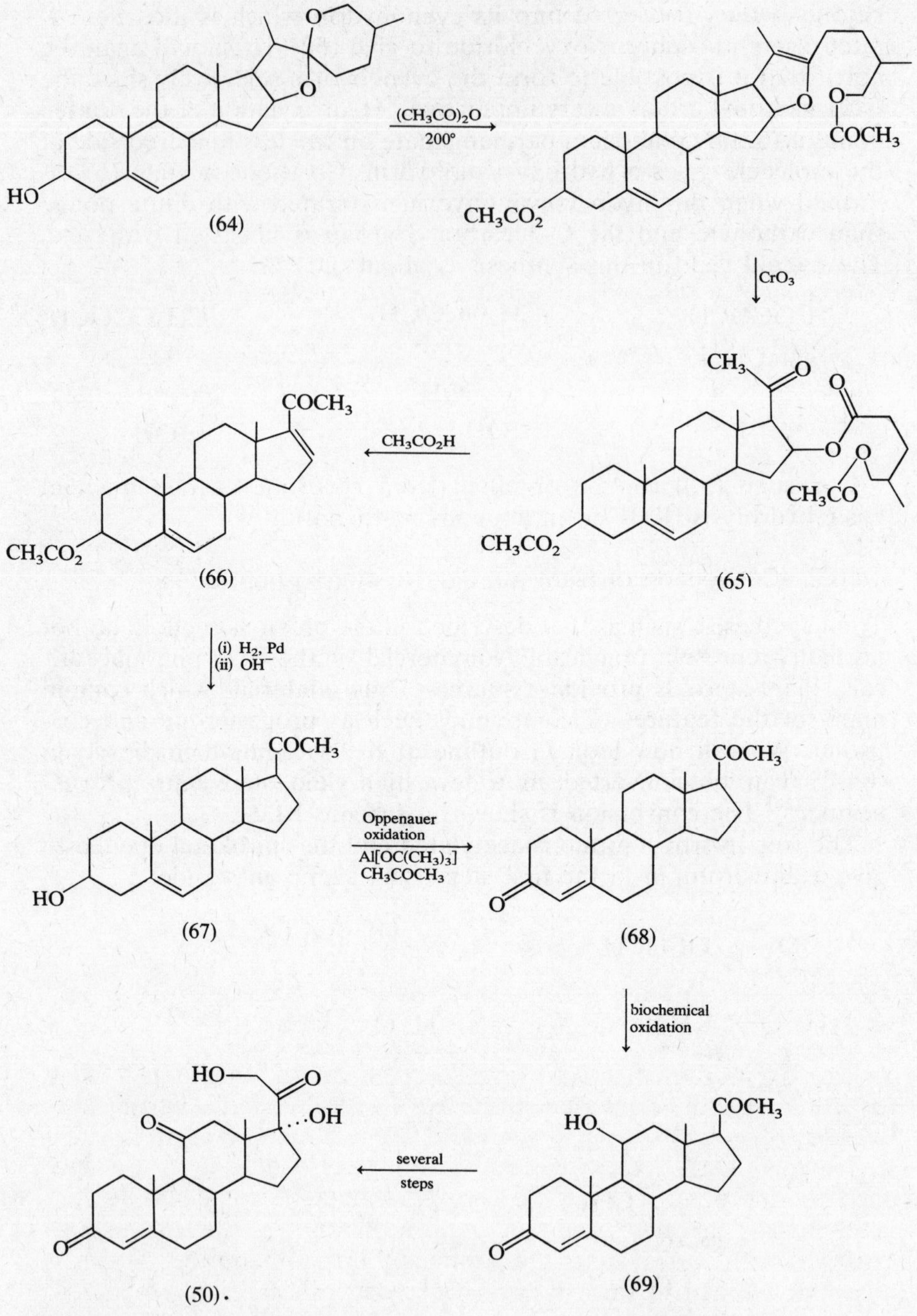
HO
(64)
$(CH_3CO)_2O$
200°
CH_3CO_2
$COCH_3$
CrO_3
CH_3
CH_3CO
CH_3CO_2
(65)
CH_3CO_2H
$COCH_3$
CH_3CO_2
(66)
(i) H_2, Pd
(ii) OH^-
$COCH_3$
HO
(67)
Oppenauer
oxidation
$Al[OC(CH_3)_3]$
CH_3COCH_3
$COCH_3$
(68)
biochemical
oxidation
HO
$COCH_3$
(69)
several
steps
HO
OH
(50).

Scheme 14.2

The product undergoes oxidative ring opening to an ester of a β-ketol (65). Acid-catalysed elimination then gives the α,β-unsaturated ketone (66). The α,β-unsaturated ketone is selectively hydrogenated and the acetate hydrolysed to give pregn-5-en-3β-ol-20-one (67). This can be converted by a number of oxidative procedures including Oppenauer oxidation, into progesterone (68), the overall yield from (64) being of the order of 50 %.

Biochemical oxidation using micro-organisms of the *Rhizopus* species results in the formation of 11α-hydroxyprogesterone (69) in yields as high as 95 %. Thus, cortisone (50) can now be prepared in a number of steps similar to those involved in its formation from (59) (cf. scheme 14.1).

14.7 Peptide synthesis

This section deals with the principles used in synthesis of peptides rather than giving a detailed account of the synthesis of a particular compound. There are several features common to all peptide syntheses:

(i) protection of amino and carboxyl groups not involved in bond formation together with protection of other functional groups (e.g. -OH and $-C\overset{\displaystyle\nearrow\text{NH}}{\underset{\displaystyle\searrow\text{NH}_2}{}}$ which might interfere;

(ii) minimisation of racemisation of the component amino acids, which is a problem particularly during the peptide bond formation step;

(iii) optimisation of yield;

(iv) optimisation of the activity of the product.

14.7.1 Strategy of peptide synthesis

As has been pointed out previously, two extreme possibilities exist for the linking of n amino acid units to form a peptide. In one, the **linear synthesis,** the chain is extended by one amino acid unit in each step involving $n-1$ peptide bond-forming steps (e.g. if $n = 64$ and each peptide bond-forming step proceeds in 90 % yield, the overall yield is $0.9^{63} \times 100$ %, i.e. 0.13 %). In the other, the **convergent synthesis,** amino acid units are connected in pairs and the resultant dipeptides linked to form tetrapeptides, the process being repeated until all amino acid units are linked. In this case, if $n = 64$, 6 sets of reactions depicted below are required, and the overall yield is $0.9^6 \times 100$ %, i.e. 53 %, if each individual step proceeds in 90 % yield.

$$A^1 \quad A^2 \; A^3 \quad A^4 \qquad A^{61} \quad A^{62} \; A^{63} \quad A^{64}$$

$$A^1\text{-}A^2 \quad A^3\text{-}A^4 \qquad A^{61}\text{-}A^{62} \quad A^{63}\text{-}A^{64}$$

$$A^1\text{-}A^2\text{-}A^3\text{-}A^4 \qquad A^{61}\text{-}A^{62}\text{-}A^{63}\text{-}A^{64}$$

$$A^1\text{-}A^2\text{-}A^3\text{-}A^4\text{-} \qquad \text{-}A^{61}\text{-}A^{62}\text{-}A^{63}\text{-}A^{64}$$

$$A^1\text{-}A^2\text{-}A^3\text{-}A^4 \!\!\sim\!\!\sim\!\!\sim\!\!\sim\!\!\sim\!\! A^{61}\text{-}A^{62}\text{-}A^{63}\text{-}A^{64}$$

- 32 reactions
- 16 reactions
- 8 reactions
- 4 reactions
- 2 reactions
- 1 reaction

In practice, however, a combination of these methods, the **block synthesis,** is normally used. The product is divided into small blocks, containing up to, say, 15 amino acid units. These blocks are synthesised often by a linear synthesis using the solid phase technique (cf. section 14.7.2.3). The blocks are then connected to form the required peptide.

The original blocks are chosen so that the final connections are made at sites where racemisation either cannot take place or is liable to be unimportant (cf. section 14.7.3).

14.7.2 Techniques of peptide synthesis

14.7.2.1 Protective groups
In order that amino acids can be linked together in the correct order, it is necessary to protect the amino group in one component and the carboxyl group in the other. The most common protective groups for the amino groups are benzyloxycarbonyl or t-butyloxycarbonyl, because in most cases the use of these groups appears to minimise racemisation at adjacent centres.

Acids are often protected as benzyl esters. In the case of solid-phase synthesis the amino acid is connected to the polymer as a substituted benzyl ester.

Other functional groups may also require protection. In particular the guanidinyl group is nitrated and hydroxyl groups converted into benzyl ethers.

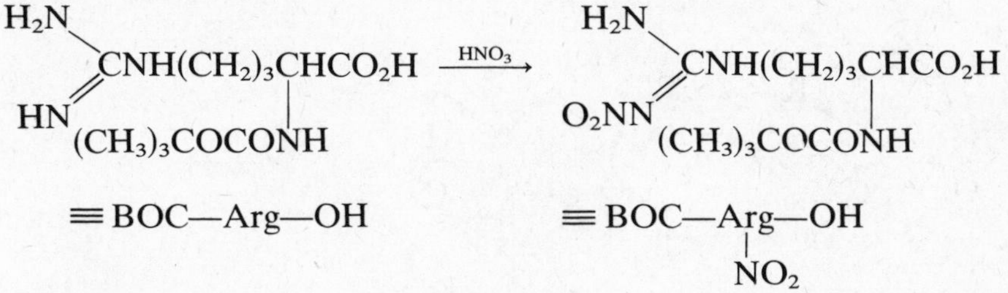

$$H_2N \diagdown$$
$$CNH(CH_2)_3CHCO_2H \xrightarrow{HNO_3}$$
$$HN \diagup \quad (CH_3)_3COCONH$$

$$\equiv BOC{-}Arg{-}OH$$

$$H_2N \diagdown$$
$$O_2NN \diagup CNH(CH_2)_3CHCO_2H$$
$$(CH_3)_3COCONH$$

$$\equiv BOC{-}Arg{-}OH$$
$$\quad\quad\quad\quad | $$
$$\quad\quad\quad\quad NO_2$$

Benzyl groups are removed from the product by treatment with HBr in trifluoroacetic acid and nitro groups by catalytic hydrogenation, as in the final steps of the solid-phase bradykinin synthesis:

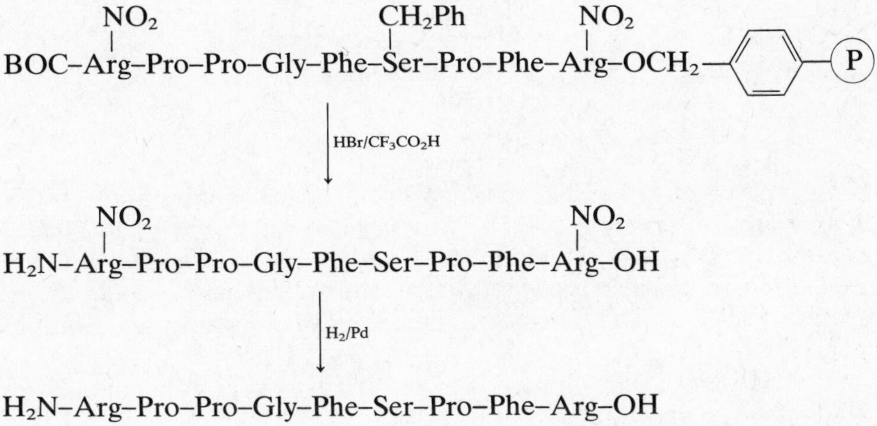

$$\overset{NO_2}{\underset{|}{}} \quad\quad \overset{CH_2Ph}{\underset{|}{}} \quad \overset{NO_2}{\underset{|}{}}$$
BOC–Arg–Pro–Pro–Gly–Phe–Ser–Pro–Phe–Arg–OCH$_2$—⟨ ⟩—Ⓟ

| HBr/CF$_3$CO$_2$H

$$\overset{NO_2}{\underset{|}{}} \quad\quad\quad\quad \overset{NO_2}{\underset{|}{}}$$
H$_2$N–Arg–Pro–Pro–Gly–Phe–Ser–Pro–Phe–Arg–OH

| H$_2$/Pd

H$_2$N–Arg–Pro–Pro–Gly–Phe–Ser–Pro–Phe–Arg–OH

bradykinin

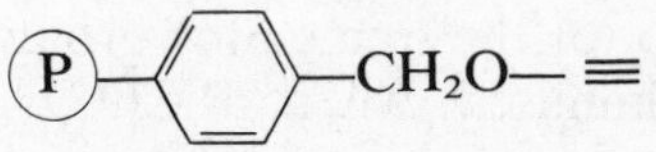

Ⓟ—⟨ ⟩—CH$_2$O— ≡ polymer-bound substituted benzyl group

14.7.2.2 Peptide bond-forming reactions
The basic reaction involved here is the linking of an amino group in one unit with the carboxyl group of another forming an amide (or peptide). Amides are normally prepared by reaction of an amine with an acid chloride, an ester, or an anhydride. Analogues of the last two are commonly used, but the use of an acid chloride leads to extensive racemisation of the amino acid and hence this method is not used.

Racemisation during peptide bond formation results from the formation of an oxazolone (70) which readily racemises [(70) $\rightleftharpoons$ (71)] as shown:

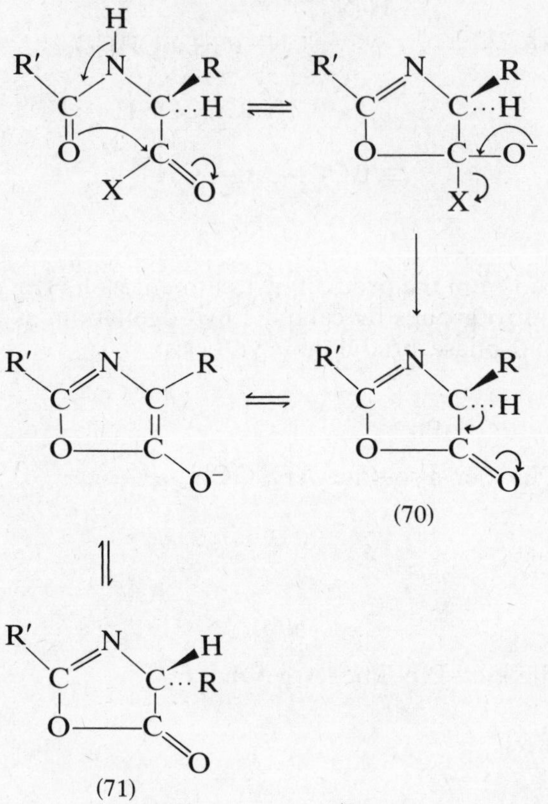

(70)

(71)

The rate of racemisation[22] is dependent on the groups R, R', and X. R'CO may either be the protective group for the free amino group (in which case racemisation may be minimised when R'CO is PhCH₂OCO or (CH₃)₃COCO) or the remainder of the peptide chain (in which case little can be done to affect the degree of racemisation induced). R is determined by the amino acid to be linked to the chain and can, therefore, not be modified. X is the group used to activate the acid and hence a large amount of research has gone into finding the best method of activating the carboxyl group in order to minimise racemisation. We shall now describe a small selection of the more commonly used methods for activation of the carboxyl group.

(a) *Azide method* The reaction sequence involved in this method is outlined below:

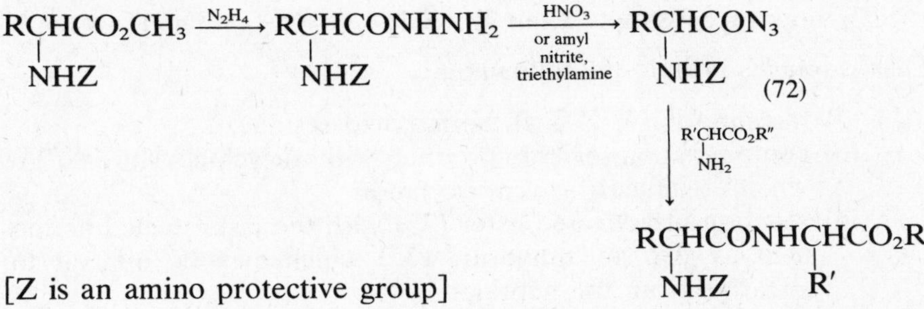

$$RCHCO_2CH_3 \xrightarrow{N_2H_4} RCHCONHNH_2 \xrightarrow[\substack{\text{or amyl} \\ \text{nitrite,} \\ \text{triethylamine}}]{HNO_3} RCHCON_3$$

with NHZ substituents below each, (72)

$$\downarrow \substack{R'CHCO_2R'' \\ NH_2}$$

$$RCHCONHCHCO_2R''$$

[Z is an amino protective group]   with NHZ and R′ substituents

The main advantage of the method is that it results in a very small amount of racemisation but there are two major disadvantages: (i) the number of steps involved in $-CO_2H \rightarrow -CO_2CH_3 \rightarrow -CONHNH_2 \rightarrow -CON_3 \rightarrow -CONHR$; and (ii) rearrangement of the azide (72) to the isocyanate (73) leads to a urea derivative (74) which is difficult to separate from the peptide:

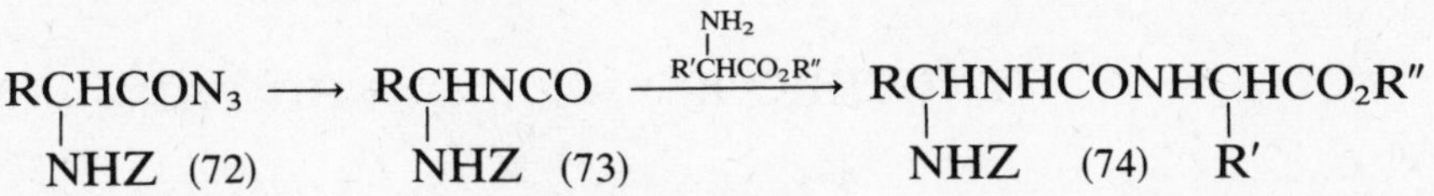

$$RCHCON_3 \longrightarrow RCHNCO \xrightarrow[\substack{R'CHCO_2R'' \\ NH_2}]{} RCHNHCONHCHCO_2R''$$

with NHZ (72)   NHZ (73)   NHZ (74) R′ substituents

This method is often used in block synthesis where it is not feasible to connect blocks at either glycine or proline (cf. section 14.7.3).

(b) *Dicyclohexylcarbodiimide* (DCC)   This involves the conversion of the carboxyl group into a type of activated ester (75):

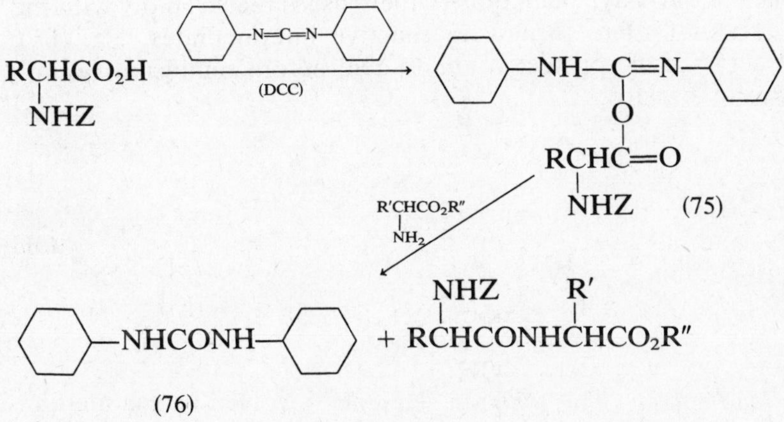

$$RCHCO_2H \xrightarrow{(DCC)}$$

with NHZ substituent

(75)

$$\xrightarrow[\substack{R'CHCO_2R'' \\ NH_2}]{}$$

NHCONH + $RCHCONHCHCO_2R''$

with NHZ and R′ substituents

(76)

The advantages of the DCC method are:

(i) good yields in a short reaction time, and
(ii) low racemisation when $Z = (CH_3)_3COCO$ or $PhCH_2OCO$.

Disadvantages include the following:

(i) racemisation if Z is an amino acid residue,
(ii) contamination of the product with dicyclohexylurea (76) which is difficult to remove, and
(iii) reaction of activated ester (75) with the N-protected amino acid to give an anhydride (77) which may be difficult to separate from the peptide:

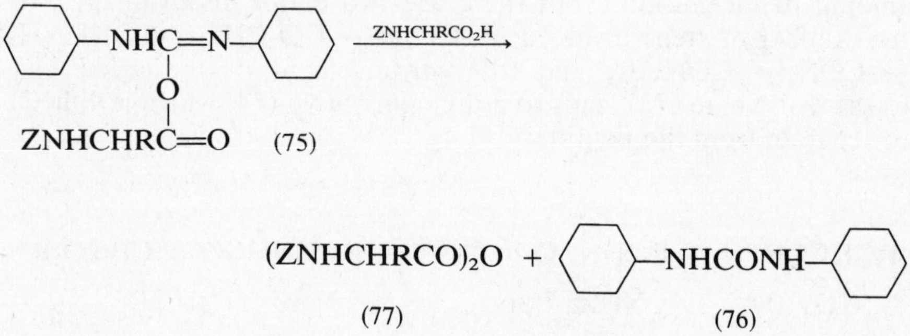

DCC is the condensing agent most commonly used in the solid phase method (cf. section 14.7.2.3).

The use of additives minimises any racemisation encountered in the DCC method. The additives [N-hydroxysuccinimide (78), N-hydroxybenzotriazole (79) and 3-hydroxy-4-oxo-3,4-dihydro-1,2,3-benzotriazine (80) have been quite widely used] react rapidly with the activated ester (75) to form very reactive intermediates, e.g. (81), which react with the protected amino acid before significant racemisation can take place:

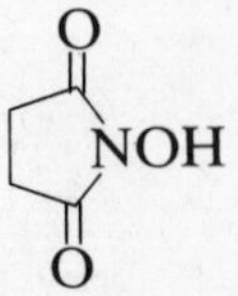

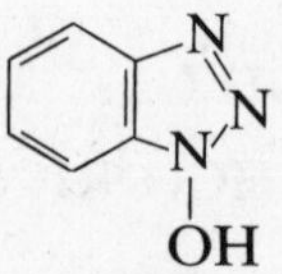

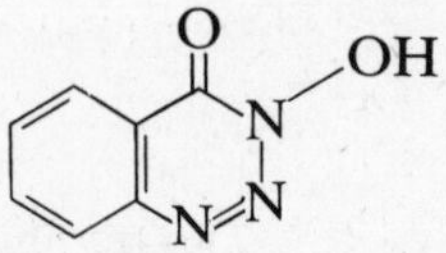

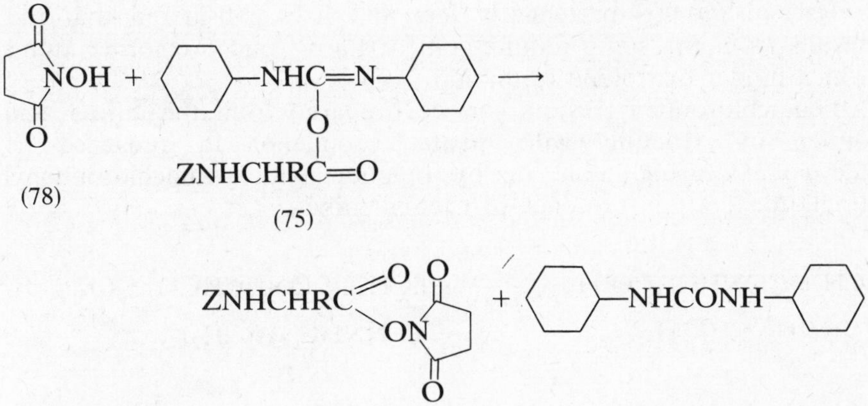

(78)   (75)

(81)

(c) *Active ester method* If the carboxyl group can be readily converted to an ester which reacts rapidly with the free amino group, peptide bond formation would take place in high yield under mild conditions. Many esters have been investigated, including p-nitrophenyl and thiophenyl, but one of the most successful is the pentachlorophenyl ester.

Pentachlorophenyl esters are among the most reactive of all esters, and those of amino acids usually have higher melting points than those of other active esters. In addition, unlike p-nitrophenyl and thiophenyl esters, they are stable to catalytic hydrogenation and hence are suitable when selective removal of, for example, benzyloxycarbonyl groups by catalytic hydrogenation is required:

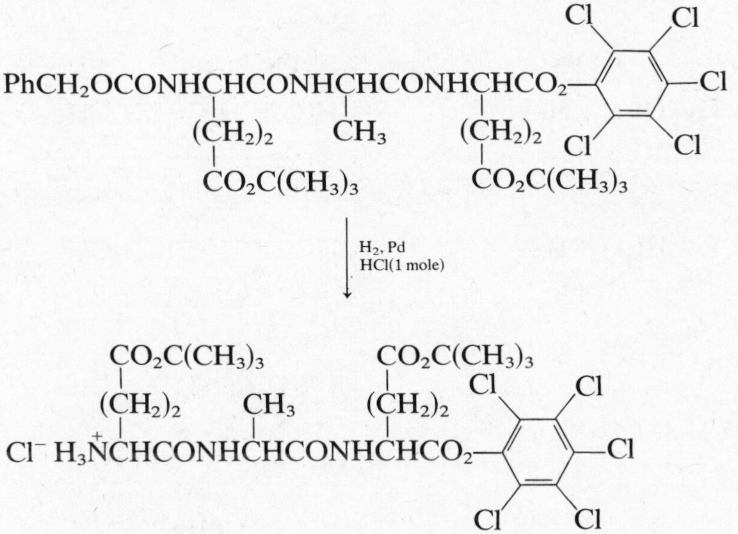

Racemisation is particularly low and it is considered that this results from the steric requirement of the 2- and 6-chlorine atoms which hinder oxazolone formation.

Pentachlorophenyl esters can be prepared from the amino acid either by reaction with pentachlorophenol in presence of dicyclohexylcarbodiimide or by reaction with pentachlorophenyl trichloroacetate in presence of triethylamine:

$$PhCH_2OCONHCHCO_2H \xrightarrow[DCC,CH_2Cl_2]{C_6Cl_5OH} PhCH_2OCONHCHCO_2C_6Cl_5 \quad (70\%)$$

$$\underset{\underset{NH}{\parallel}}{O_2NNHCNH(CH_2)_3} \qquad\qquad\qquad \underset{\underset{NH}{\parallel}}{O_2NNHCNH(CH_2)_3}$$

$$PhCH_2OCONHCH_2CO_2H \xrightarrow[(C_2H_5)N,DMF]{C_6Cl_5O_2CCl_3} PhCH_2OCONHCH_2CO_2C_6Cl_5 \quad (85\%)$$

14.7.2.3 Solid-phase peptide synthesis

One of the major advances in peptide synthesis came in 1962, when Merrifield first described a synthesis of a tetrapeptide, Leu–Ala–Gly–Ala, by a solid-phase technique which is now often associated with his name. The method involves the following steps:

(i) attachment of an N-protected amino acid to a styrene-divinylbenzene co-polymer which has had 5 % of its phenyl groups chloromethylated (82):

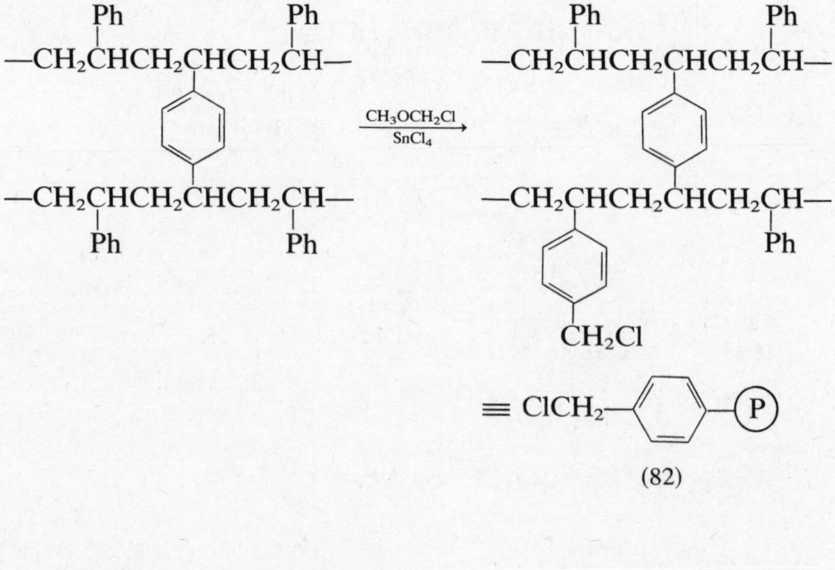

(82)

$$ZNHCHCO_2H + \text{(P)}-\text{\Large\hexagon}-CH_2Cl \xrightarrow{(C_2H_5)_3N}$$
$$\quad\quad R$$

$$ZNHCHCO_2CH_2-\text{\Large\hexagon}-\text{(P)}$$
$$\quad R$$
$$(83)$$

It should be noted that (83) is a substituted benzyl ester and hence this step not only attaches the amino acid to the polymer support but also protects the carboxyl group.

(ii) removal of the *N*-protective group (step A);

(iii) reaction of the free amino group with an *N*-protected amino acid[23] often using DCC (with or without additives) as condensing agent (step B):

$$ZNHCHCO_2CH_2-\text{\Large\hexagon}-\text{(P)} \xrightarrow[\text{step A}]{\text{deprotection}} H_2NCHCO_2CH_2-\text{\Large\hexagon}-\text{(P)}$$
$$\quad R \qquad\qquad\qquad\qquad\qquad\qquad\qquad R$$

$$\text{(step B)} \Big\downarrow \begin{array}{l} ZNHCHCO_2H \\ \quad R' \\ DCC \end{array}$$

$$ZNHCHCONHCHCO_2CH_2-\text{\Large\hexagon}-\text{(P)}$$
$$\quad R' \qquad R$$

(iv) repetition of steps A and B using the required *N*-protected amino acid until all the desired amino acid units have been connected;

(v) removal of the protected peptide from the polymer often using HBr in trifluoroacetic acid. This reagent will also remove many protective groups including *N*-benzyloxycarbonyl, *N*-butyloxycarbonyl, and *O*-benzyl,

(vi) removal of all other protective groups.

$$ZNHCHCONH\cdots\cdots\cdots CONHCHCONHCHCO_2CH_2-\text{\Large\hexagon}-\text{(P)}$$
$$\quad R^n \qquad\qquad\qquad\qquad R' \qquad R$$

$$\Big\downarrow \begin{array}{l} \text{(I) HBr, CF}_3\text{CO}_2\text{H} \\ \text{(ii) complete deprotection} \end{array}$$

$$H_2NCHCONH\cdots\cdots\cdots CONHCHCONHCHCO_2H$$
$$\quad R^n \qquad\qquad\qquad\qquad R' \qquad R$$

It may be undesirable to synthesise a peptide containing more than 15 amino acid units by this method, but **ribonuclease**, which has 124 amino acid units, has been synthesised entirely by solid-phase reactions using the block strategy (cf. section 14.7.1).

Many reviews and books containing critical assessments of the Merrifield technique have been published and there is only space in this work to list some of the main advantages and limitations of the method.

Advantages include the following:

(i) reactions are rapid and high yields are obtained;
(ii) little or no racemisation takes place if Z is t-butyl- or benzyloxycarbonyl, since reaction takes place at the N-protected amino acid;
(iii) purification of intermediates involves nothing more than washing the polymer free of non-polymeric reagents if reaction has gone to completion;
(iv) the procedure can be automated. (The synthesis of a nonapeptide might take about 5 hours.)

Limitations include the following:

(i) incomplete attachment of the C-terminal amino acid to the benzyl groups of the polymer may lead to impurities, e.g. in the synthesis of a pentapeptide A^1–A^2–A^3–A^4–A^5, A^2–A^3–A^4–A^5 might be formed by reaction of the second amino acid with a free benzyl group competing with step B;
(ii) incomplete coupling of the N-protected amino acid with the free amino group (step B) can lead to truncated peptides (e.g. A^1–A^2–A^3) and failure sequences (e.g. A^1–A^2–A^4–A^5);
(iii) lack of analytical techniques for the detection of impurities which do not necessitate removal of the peptide from the polymer. Such removal would be wasteful and time-consuming and might in itself lead to degradation of the product.

14.7.3 Synthesis of a higher molecular weight peptide, Basic Trypsin Inhibitor (BTI) from bovine pancreas

This peptide containing 58 amino acid units (MW ~ 6500) has been shown to have the following sequence of amino acid units:

$$
\begin{array}{l}
\text{Arg}^1\text{–Pro–Asp–Phe–Cys–Leu–Glu–Pro–Pro–Tyr} \\
\qquad\text{Arg–Thr–Cys–Gly}^{56}\text{–Gly–Ala}\qquad\text{Thr} \\
\qquad\text{Met–Cys–Asp}\ \ \text{Ser–Lys}\ \ \text{Lys–Ala}\ \ \mathbf{Gly}^{12} \\
\qquad\text{Leu–Cys}\ \ \text{Glu–Ala}\ \ \text{Phe}\ \ \text{Arg}\ \ \text{Arg}\ \ \text{Pro} \\
\text{Ala–}\mathbf{Gly}^{28}\ \ \text{Gln}\qquad\text{Asn–Asn}\ \ \text{Cys}\ \ \text{Cys} \\
\text{Lys}\qquad\text{Thr–Phe–Val–Tyr–Gly–}\mathbf{Gly}^{37}\text{Lys} \\
\text{Ala–Asn–Tyr–Phe–Tyr–Arg–Ile–Ile–Arg–Ala}\quad (84)
\end{array}
$$

The synthesis of this peptide, described in a series of papers,[24] serves to illustrate a strategy which can be adopted in such cases.

The method used involved synthesising blocks[25] which would eventually be linked at glycine units [indicated in bold type in (84)] in order to eliminate problems of racemisation in the final stages. Thus peptides (1–12), (13–28), (29–56), and (57–78) were synthesised and linked by a solid-phase technique.

Gly–Ala, N-protected with a p-methoxybenzyloxy group [Z(OMe–Gly–Ala) was linked to a bromomethylated styrene–divinylbenzene co-polymer. Each peptide block [(38–56), (29–37), (13–28), and (1–12)] was successively linked to the extending peptide chain using DCC with additive [N-hydroxysuccinimide (NHS)] as condensing agent. The active peptide was removed from the polymer and deprotected by treatment with HF. This is summarised in the following scheme:

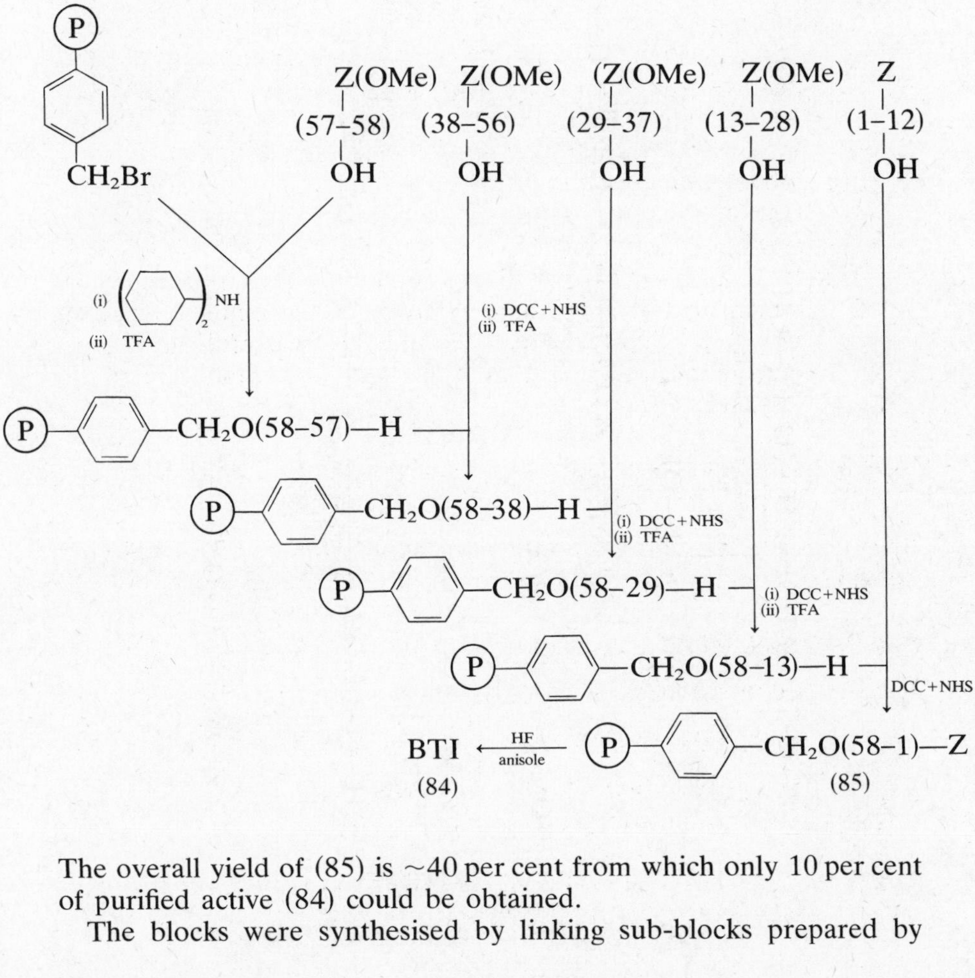

$$
\begin{array}{ccccc}
& Z(OMe) & Z(OMe) & (Z(OMe) & Z(OMe) & Z \\
& | & | & | & | & | \\
& (57\text{–}58) & (38\text{–}56) & (29\text{–}37) & (13\text{–}28) & (1\text{–}12) \\
& | & | & | & | & | \\
& OH & OH & OH & OH & OH
\end{array}
$$

(P)—C₆H₄—CH₂Br

(i) (cyclohexyl)₂NH
(ii) TFA

(i) DCC+NHS
(ii) TFA

(P)—C₆H₄—CH₂O(58–57)—H

(P)—C₆H₄—CH₂O(58–38)—H

(i) DCC+NHS
(ii) TFA

(P)—C₆H₄—CH₂O(58–29)—H

(i) DCC+NHS
(ii) TFA

(P)—C₆H₄—CH₂O(58–13)—H

DCC+NHS

BTI ←—(HF/anisole)— (P)—C₆H₄—CH₂O(58–1)—Z

(84)   (85)

The overall yield of (85) is ~40 per cent from which only 10 per cent of purified active (84) could be obtained.

The blocks were synthesised by linking sub-blocks prepared by

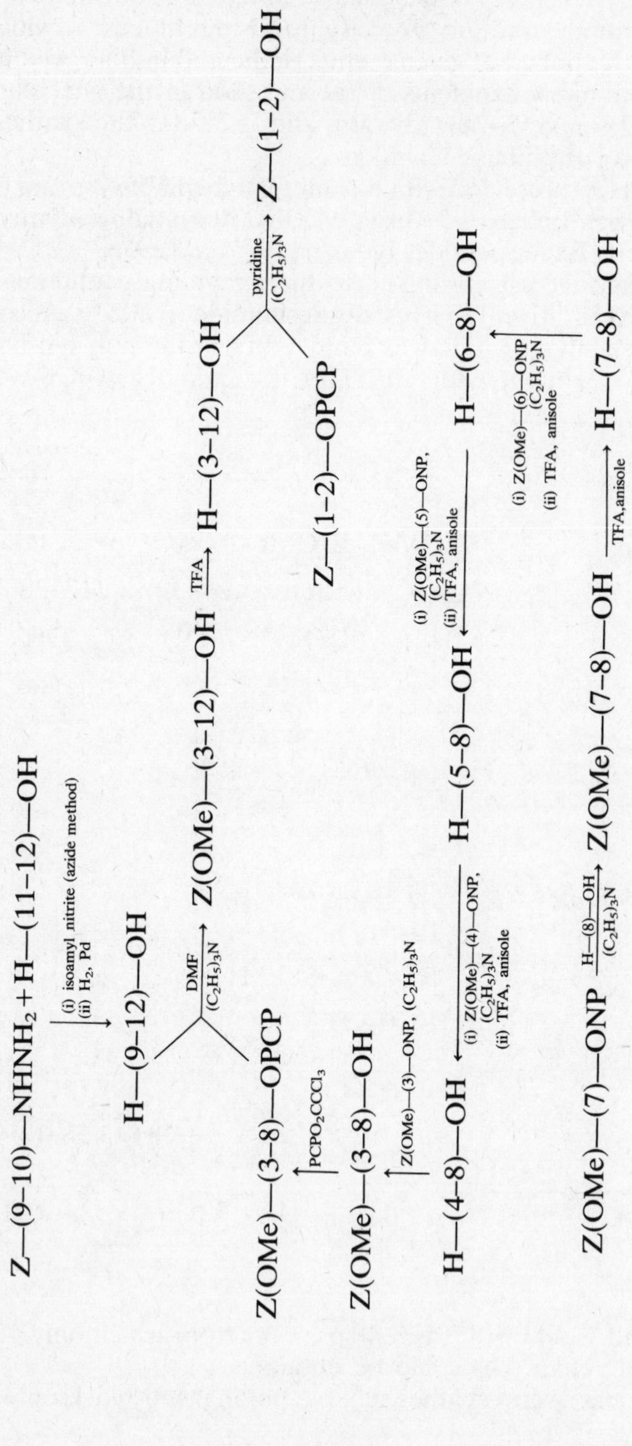

Z—(9-10)—NHNH₂ + H—(11-12)—OH

| (i) isoamyl nitrite (azide method)
| (ii) H_2, Pd
↓

H—(9-12)—OH

Z(OMe)—(3-12)—OH $\xrightarrow{\text{TFA}}$ H—(3-12)—OH $\xrightarrow[(C_2H_5)_3N]{\text{pyridine}}$ Z—(1-2)—OH

⟍ (DMF, $(C_2H_5)_3N$)

Z—(1-2)—OPCP

Z(OMe)—(3-8)—OPCP

↑ PCPO₂CCl₃

Z(OMe)—(3-8)—OH

↑ Z(OMe)—(3)—ONP, $(C_2H_5)_3N$

H—(4-8)—OH $\xrightarrow[\text{(ii) TFA, anisole}]{\text{(i) Z(OMe)—(4)—ONP, }(C_2H_5)_3N}$ H—(5-8)—OH $\xrightarrow[\text{(ii) TFA, anisole}]{\text{(i) Z(OMe)—(5)—ONP, }(C_2H_5)_3N}$ H—(6-8)—OH

↑ (i) Z(OMe)—(6)—ONP $(C_2H_5)_3N$
↑ (ii) TFA, anisole

H—(7-8)—OH

Z(OMe)—(7-8)—OH $\xrightarrow{\text{TFA, anisole}}$

Z(OMe)—(7)—ONP $\xrightarrow[(C_2H_5)_3N]{\text{H—(8)—OH}}$ Z(OMe)—(7-8)—OH either linear or convergent synthesis. In the case of the block consisting of units 1–12, the linkage of sub-blocks was performed at proline units where racemisation is minimal. In most cases, linking was performed by the active ester method using either pentachlorophenyl (PCP) or p-nitrophenyl (NP) esters. The synthesis of the N-benzyloxycarbonyl protected block 1–12 [Z—(1–12)—OH] is outlined on the facing page.

In the other cases, proline units are not available as sites at which to link sub-blocks, and the final linkages of the sub-blocks are made by the azide method which is less liable to induce racemisation.

The overall activity of the product (80 %) is superior to that obtained (30 %) in an alternative stepwise Merrifield synthesis.

Table 14.1 Structures and abbreviations of some common amino acids

Name	Formula	Abbreviation
Glycine	$H_2NCH_2CO_2H$	Gly
Alanine	$CH_3CH(NH_2)CO_2H$	Ala
Valine	$(CH_3)_2CHCH(NH_2)CO_2H$	Val
Leucine	$(CH_3)_2CHCH_2CH(NH_2)CO_2H$	Leu
Iso-Leucine	$CH_3CH_2CH(CH_3)CH(NH_2)CO_2H$	Ile
Phenylalanine	$PhCH_2CH(NH_2)CO_2H$	Phe
Tyrosine	$p\text{-}HOC_6H_4CH_2CH(NH_2)CO_2H$	Tyr
Cysteine	$HSCH_2CH(NH_2)CO_2H$	Cys
Serine	$HOCH_2CH(NH_2)CO_2H$	Ser
Threonine	$CH_3CH(OH)CH(NH_2)CO_2H$	Thr
Methionine	$CH_3SCH_2CH_2CH(NH_2)CO_2H$	Met
Proline	CO_2H	Pro
Aspartic acid	$HO_2CCH_2CH(NH_2)CO_2H$	Asp
Asparagine	$H_2NCOCH_2CH(NH_2)CO_2H$	Asn
Glutamic acid	$HO_2CCH_2CH_2CH(NH_2)CO_2H$	Glu
Glutamine	$H_2NCOCH_2CH_2CH(NH_2)CO_2H$	Gln
Arginine	$\dfrac{H_2N}{HN}{>}CNH(CH_2)_3CH(NH_2)CO_2H$	Arg
Lysine	$H_2N(CH_2)_4CH(NH_2)CO_2H$	Lys

Notes

1. *J. Org. Chem.*, **40,** 1593 (1975).

2. P-2 nickel is prepared by reduction of nickel(II) acetate by sodium

borohydride. When poisoned with ethylenediamine, it can be used as a catalyst for hydrogenation of alkynes to alkenes with $Z:E>97:1$. Yields are high and it is, therefore, an alternative to the Lindlar catalyst for this transformation.

3.   *J. Chem. Soc., Perkin* I, 1978, 842.

4.   2-Hydroxytetrahydropyran is prepared by acid catalysed addition of water to 2,3-dihydropyran.

5.   *J. Org. Chem.*, **41**, 2927 (1976); cf. *Helv. Chim. Acta*, **53**, 1479 (1970).

6.   *J. Chem. Soc. (C)*, 1969, 1024; *J. Chem. Soc. Chem. Comm.*, 1972, 529.

7.   *J. Org. Chem.*, **37**, 2363 (1972).

8.   cf. I. Fleming, *Frontier Orbitals and Organic Chemical Reactions*, Wiley, 1976, p. 143; W. T. Brady, Tetrahedron, **37**, 2949 (1981).

9.   An alternative technique, applicable to the synthesis of 1-chloroalkynes, is described in section 14.2.

10.   *Latent functionality* describes a technique where a labile functional group is carried through a synthesis in the form of a stable group which can easily be transformed into the labile group at an appropriate stage in the synthesis. Examples of latent functionality include the following:

11.   *J. Amer. Chem. Soc.*, **78**, 4765 (1956).

12.   *Tetrahedron*, **28**, 1749 (1972); *Tetrahedron Letters*, 1968, 3507.

13.   *Pure Appl. Chem.*, **7**, 363 (1963).

14.   Chem. Soc. Special Publication No. 21 (1967), p. 113; *Angew. Chem., internat. Edit.*, **3**, 228 (1964).

15.   *J. Amer. Chem. Soc.*, **74**, 1393, 1405, 4974 (1952); **75**, 422, 1707, 2112 (1953); **76**, 1715, 5026, 6031 (1954).

16.   3-Ethoxy-1,3-pentadiene may be prepared as follows:

$$C_2H_5COCH{=}CH_2 \xrightarrow{HC(OC_2H_5)_3} C_2H_5C(OC_2H_5)_2CH_2CH_2OC_2H_5$$

$$\xrightarrow[H^+]{155°} CH_3CH{=}\overset{\overset{\textstyle OC_2H_5}{\textstyle |}}{C}HCH{=}CH_2$$

17.   *Epimerisation* is inversion of configuration at one chiral centre only.

18.   The synthesis in scheme 14.1 gives racemic cortisone. Optically active cortisone is produced by carrying out the remaining steps using optically active (60) obtained by resolution with strychnine.

19.   Diosgenin is a member of the class of compounds known as *sapogenins*. *Saponins* are widely distributed plant glycosides in which the sapogenin is combined with one or more sugar residues. Hydrolysis with acid or enzyme converts the saponin into the sapogenin and the sugars.

20.   *J. Amer. Chem. Soc.*, **62,** 2525, 3349, 3350 (1940); **74,** 5933 (1952); *Rec. Trav. chim. Pays-Bas*, **56,** 137 (1937).

21.   Standard abbreviations for some common amino acids are listed in table 14.1 at the end of this chapter.

22.   It should be noted that glycine is achiral and so cannot be racemised, and that proline has a low tendency to racemisation. So it is common that block synthesis is carried out so that blocks are linked at glycine or at proline units (cf. section 14.7.3).

23.   Other functional groups on the amino acid may also have to be protected (cf. section 14.7.2.1).

24.   *Chem. Pharm. Bull. (Tokyo)*, **22,** 1061, 1067, 1075, 1079, 1087 (1974).

25.   Other functional groups on the constituent amino acids were protected.

Further reading

This list of books and review articles is intended to help readers to locate some more advanced or more detailed accounts of topics referred to in this book. A number of the reviews (especially those from *Organic Reactions*) include experimental instructions.

General

R. O. C. Norman, *Principles of Organic Synthesis*, Second Edition, Chapman and Hall, 1978.

W. Carruthers, *Some Modern Methods of Organic Synthesis*, Second Edition, Cambridge University Press, 1978.

H. O. House, *Modern Synthetic Reactions*, Second Edition, Benjamin, 1972.

F. A. Carey and R. J. Sundberg, *Advanced Organic Chemistry, Part B: Reactions and Synthesis*, Plenum Press, 1977.

Chapter 3

S. Warren, *Designing Organic Syntheses: A Programmed Introduction to the Synthon Approach*, Wiley, 1978.

E. J. Corey, *Pure and Applied Chemistry*, **14,** 19 (1967).

Chapter 4

General

Houben-Weyl, *Methoden der Organischen Chemie*, Georg Thieme Verlag, Vols. 13/1 (1970) and 13/2a (1973) (both in German).

E. Negishi, *Organometallics in Organic Synthesis*, Wiley, Vol. 1 (1979).

Grignard reagents

M. S. Kharasch and O. Reinmuth, *Grignard Reactions of Nonmetallic Substances*, Prentice-Hall, 1954.

B. Blagoev and D. Ivanov, *Synthesis*, **1970,** 615.

B. J. Wakefield, *Chem. and Ind.*, **1972,** 450.

Organolithium reagents

J. M. Brown, *Chem. and Ind.*, **1972,** 454.

D. Ivanov, G. Vassilev, and I. Panayotov, *Synthesis*, **1975,** 83.

Reformatsky reaction

R. L. Shriner, *Org. Reactions*, **1,** 1 (1942).
M. W. Rathke, *Org. Reactions*, **22,** 423 (1975).

Organocadmium reagents

P. R. Jones and P. J. Desio, *Chem. Rev.*, **78,** 491 (1978).

Organocopper reagents

G. H. Posner, *An Introduction to Synthesis using Organocopper Reagents.* Wiley, 1980.
J. F. Normant, *Synthesis*, **1972,** 63.
G. H. Posner, *Org. Reactions*, **19,** 1 (1972); **22,** 253 (1975).
E. C. Ashby *et al*, *J. Org. Chem.*, **42,** 1099, 2805 (1977).

Oxidative coupling

(General) T. Kauffmann, *Angew. Chem., internat. Edit.*, **13,** 291 (1974).
(Alk-1-ynes) G. Eglinton and W. McCrae, *Adv. Org. Chem.*, **4,** 225 (1963).

Chapter 5

Dianions of β-dicarbonyl compounds

T. M. Harris and C. M. Harris, *Org. Reactions*, **17,** 155 (1969).

Knoevenagel condensation

G. Jones, *Org. Reactions*, **15,** 204 (1967).

Michael addition

E. D. Bergmann, D. Ginsburg, and R. Pappo, *Org. Reactions*, **10,** 179 (1959).

Alkylation of aldehydes and ketones *via* metal enolates

D. Caine, in *Carbon–Carbon Bond Formation*, ed. R. L. Augustine, Marcel Dekker, Vol. 1 (1979), Chapter 2.

Claisen acylation ("ester condensation")

C. R. Hauser and B. E. Hudson, *Org. Reactions*, **1,** 266 (1942).

Dihydro-1,3-oxazines

A. I. Meyers, *Heterocycles in Organic Synthesis*, Wiley, 1974, pp. 201–205.
R. R. Schmidt, *Synthesis*, **1972,** 333.

Aldol condensation

A. T. Nielsen and W. J. Houlihan, *Org. Reactions*, **16,** 1 (1968).
Z. G. Hajos, in *Carbon–Carbon Bond Formation*, ed. R. L. Augustine, Marcel Dekker, Vol. 1 (1979), Chapter 1.

Wittig reaction

S. Trippett, *Quart. Rev.*, **17,** 406 (1963)
A. Maercker, *Org. Reactions*, **14,** 270 (1965).
I. Gosney and A. G. Rowley, in *Organophosphorus Reagents in Organic Synthesis,* ed. J. I. G. Cadogan, Academic Press, 1980, Chapter 2.

1,3-Dithians (and 1,3,5-trithians)

D. Seebach, *Synthesis,* **1969,** 17 (in German).
B.-T. Gröbel and D. Seebach, *Synthesis,* **1977,** 357 (in English, but a more advanced review).

Enamines

S. F. Dyke, *The Chemistry of Enamines,* Cambridge University Press, 1973.
P. W. Hickmott and H. Suschitzky, *Chem. and Ind.,* **1970,** 1188.
M. A. Kuehne, *Synthesis,* **1970,** 510.

Gattermann reaction

W. E. Truce, *Org. Reactions*, **9,** 37 (1957).

Gattermann-Koch reaction

N. N. Crounse, *Org. Reactions*, **5,** 290 (1949).

Hoesch reaction

P. E. Spoerri and A. S. DuBois, *Org. Reactions*, **5,** 387 (1949).

Vilsmeier-Haack-Arnold reaction

C. Jutz, *Adv. Org. Chem.*, **9/1,** 225 (1976).

Reimer-Tiemann reaction

H. Wynberg, *Chem. Rev.*, **60,** 169 (1960).

Kolbe-Schmitt reaction

A. S. Lindsey and H. Jeskey, *Chem. Rev.*, **57,** 583 (1957).

Mannich reaction

F. F. Blicke, *Org. Reactions*, **1,** 303 (1942).
M. Tramontini, *Synthesis,* **1973,** 703.

Thermal Michael reaction

G. L. Buchanan *et al.*, *Tetrahedron*, **25,** 5517 (1969), and references therein.

Umpolung (general)

D. Seebach, *Angew. Chem., internat. Edit.*, **18,** 239 (1979).
H. Stetter, *Angew. Chem., internat. Edit.*, **15,** 639 (1976).

Chapter 7

Dieckmann and Thorpe-Ziegler reactions

J. P. Schaefer and J. J. Bloomfield, *Org. Reactions*, **15,** 1 (1967).

Acyloin reaction

K. T. Finley, *Chem. Rev.*, **64,** 573 (1964).

Robinson annelation

R. E. Gawley, *Synthesis*, **1976,** 777.

Pericyclic reactions (general)

I. Fleming, *Frontier Orbitals and Organic Chemical Reactions*, Wiley, 1976.
T. L. Gilchrist and R. C. Storr, *Organic Reactions and Orbital Symmetry*, Second Edition, Cambridge University Press, 1979.

Diels–Alder reaction

J. A. Norton, *Chem. Rev.*, **31,** 319 (1942).
J. G. Martin and R. K. Hill, *Chem. Rev.*, **61,** 537 (1961).
M. Petrzilka and J. I. Grayson, *Synthesis*, **1981,** 753.

1,3-Dipolar cycloaddition

R. Huisgen, *Angew. Chem., internat. Edit.*, **2,** 565, 633 (1963).

Simmons-Smith reaction

H. E. Simmons, T. L. Cairns, S. A. Vladuchick, and C. M. Hoiness, *Org. Reactions*, **20,** 1 (1973).

Cope rearrangement

S. J. Rhoads and N. R. Raulins, *Org. Reactions*, **22,** 1 (1975).

Chapter 8

General

R. L. Augustine (ed.), *Reduction*, Edward Arnold/Marcel Dekker, 1968.
M. M. Baizer, *Organic Electrochemistry*, Marcel Dekker, 1973.

Catalytic hydrogenation

R. L. Augustine, *Catalytic Hydrogenation*, Edward Arnold/Marcel Dekker, 1965.
P. N. Rylander, *Catalytic Hydrogenation over Platinum Metals*, Academic Press, 1967.

Homogeneous catalysis in hydrogenation

F. J. McQuillin, *Prog. Org. Chem.*, **8**, 314 (1973).
A. J. Birch and D. H. Williamson, *Org. Reactions*, **24**, 1 (1976).

Clemmensen reduction

E. L. Martin, *Org. Reactions*, **1**, 155 (1942).
E. Vedejs, *Org. Reactions*, **22**, 401 (1975).

Wolff-Kishner reduction

D. Todd, *Org. Reactions*, **4**, 378 (1948).

Dissolving metal reduction

A. J. Birch and G. Subba Rao, *Adv. Org. Chem.*, **8**, 1 (1972).

Meerwein-Ponndorf-Verley reduction

A. L. Wilds, *Org. Reactions*, **2**, 178 (1944).
G. H. Posner, *Angew. Chem., Intermat. Edit.*, **17**, 487 (1978).

Complex metal hydride reduction

E. R. H. Walker, *Chem. Soc. Rev.*, **5**, 23 (1976).
A. Hajós, *Complex Hydrides, Elsevier*, 1979.

Hydrogenolysis

G. R. Pettit and E. E. Van Tamelen, *Org. Reactions*, **12**, 356 (1962) (desulphurisation).
A. R. Pinder, *Synthesis*, **1980**, 425 (halides).

Rosenmund reduction

E. Mosettig and R. Mozingo, *Org. Reactions*, **4**, 362 (1948).

Chapter 9

General

R. L. Augustine and D. J. Trecker, (eds.), *Oxidation*, Marcel Dekker, Vols. 1 (1969) and 2 (1971).
K. B. Wiberg and W. S. Trahanovsky (eds.), *Oxidation in Organic Chemistry*, Academic Press, Parts A (1965), B (1973), and C (1978).

Barton reaction

R. H. Hesse, *Adv. Free-Rad. Chem.*, **3**, 83 (1969).

Selenium dioxide

M. Rabjohn, *Org. Reactions*, **24**, 261 (1976).

Étard reaction

W. H. Hartford and M. Darrin, *Chem. Rev.*, **58,** 1 (1958).

Dehydrogenation

P. P. Fu and R. G. Harvey, *Chem. Rev.*, **78,** 317 (1978) (general).
D. Walker and J. D. Hiebert, *Chem. Rev.*, **67,** 153 (1967) (DDQ).

Hydroxylation of alkenes (general)

F. D. Gunstone, *Adv. Org. Chem.*, **1,** 103 (1960).

Prévost reaction

C. V. Wilson, *Org. Reactions*, **9,** 332 (1957).

Osmium(VIII) oxide

M. Schröder, *Chem. Rev.*, **80,** 187 (1980).

Pyridinium chlorochromate (PCC) and dichromate (PDC)

E. J. Corey and J. W. Suggs, *Tetrahedron Letters*, **1975,** 2647 (PCC)
E. J. Corey and G. Schmidt, *Tetrahedron Letters*, **1979,** 399 (PDC)

Dimethyl sulphoxide

W. W. Epstein and F. W. Sweat, *Chem. Rev.*, **67,** 247 (1967).
A. J. Mancuso and D. Swern, *Synthesis*, **1981,** 165.

Oppenauer oxidation

C. Djerassi, *Org. Reactions*, **6,** 207 (1951).

Sommelet reaction

N. Blažević *et al.*, *Synthesis*, **1979,** 161, and references therein.

Oxidative coupling of phenols

A. I. Scott, *Quart. Rev.*, **19,** 1 (1965).
D. C. Nonhebel, P. L. Pauson, *et al.*, *J. Chem. Research* (S), **1977,** 12–16.

Baeyer-Villiger reaction

C. H. Hassall, *Org. Reactions*, **9,** 73 (1957).

Chapter 10

J. F. W. McOmie (ed.), *Protective Groups in Organic Chemistry*, Plenum Press, 1973.
J. F. W. McOmie, *Adv. Org. Chem.*, **3,** 191 (1963).
T. W. Greene, *Protective Groups in Organic Synthesis*, Wiley-Interscience, 1981.

Chapter 11

H. C. Brown, *Hydroboration*, Benjamin, 1962.
H. C. Brown, *Boranes in Organic Chemistry*, Cornell University Press, 1972.
H. C. Brown, *Organic Synthesis via Boranes*, Wiley, 1975.
G. M. L. Cragg, *Organoboranes in Organic Synthesis*, Marcel Dekker, 1973.
G. M. L. Cragg and K. R. Koch, *Chem. Soc. Rev.*, **6,** 393 (1977).

Chapter 12

General

A. J. Kirby and S. Warren, *The Organic Chemistry of Phosphorus*, Elsevier, 1967.
B. J. Walker, *Organophosphorus Chemistry*, Penguin Books, 1972.
J. I. G. Cadogan (ed.), *Organophosphorus Reagents in Organic Synthesis*, Academic Press, 1980.

Horner/Wadsworth-Emmons reaction

J. Boutagy and R. Thomas, *Chem. Rev.*, **74,** 87 (1974).

Tervalent phosphorus reagents

J. I. G. Cadogan, *Quart. Rev.*, **16,** 208 (1962) (reducing agents).
J. I. G. Cadogan, *Quart. Rev.*, **22,** 222 (1968) (reduction of -NO and -NO$_2$).
J. I. G. Cadogan and R. K. Mackie, *Chem. Soc. Rev.*, **3,** 87 (1974).

Chapter 13

E. W. Colvin, *Silicon in Organic Synthesis*, Butterworths, 1981.
J. K. Rasmussen, *Synthesis*, **1977,** 91.
E. W. Colvin, *Chem. Soc. Rev.*, **7,** 15 (1978).
T. H. Chan and I. Fleming, *Synthesis*, **1979,** 761.
I. Fleming, *Chem. Soc. Rev.*, **10,** 83 (1981).

Chapter 14

I. Fleming, *Selected Organic Syntheses*, Wiley, 1972.
W. Templeton, *An Introduction to the Chemistry of Terpenoids and Steroids*, Butterworths, 1969.

Insect pheromones

R. Rossi, *Synthesis*, **1977,** 817; **1978,** 413.

Peptides

H. N. Rydon, *Peptide Synthesis*, Royal Inst. of Chem. Lecture Series, 1962, No. 5.
D. T. Elmore, *Peptides and Proteins*, Cambridge University Press, 1968.
M. Fridkin and A. Patchornik, *Ann. Rev. Biochem.*, **43,** 419 (1974).

Index

acetals and ketals
  as protective groups, 17, 243, 247, 304, 306
  formation, 16, 17, 21
acylating agents, 30
acylation
  of alcohols and phenols, 14, 16, 17
  of amines, 17, 19
  of arenes, 8, 9, 102–5
  of enamines, 103, 104
  Friedel–Crafts, 8, 9
  using Grignard reagents, 41–4
  of heteroarenes, 15, 103, 105
  intramolecular (Dieckmann), 128, 129
  of stabilised carbanions, 71–4, 82–4
  of vinylsilanes, 278, 279
acyl halides, formation, 22
acyloin reaction, 141, 142
addition
  to alkenes, 4–6, 174–6, 248–54
  to alkynes, 7, 8, 176–8, 261, 262
  to carbonyl compounds, 30, 31, 178, 180
  to carbonyl compounds, α,β-unsaturated, 31, 32, 190–2
  to cyano compounds, 31, 32, 184, 185
  to imines, 31
alcohols, formation
  from aldehydes and ketones, 178–81
  from alkenes, 4
  from alkyl halides, 19, 20
  from carboxylic acid derivatives, 182–4
  by Grignard synthesis, 40, 41, 44, 45
  via hydroboration, 249, 250, 254, 255, 258, 259, 262
alcohols, reactions, 15–17
  oxidation, 216–24
  protection, 16, 17, 239–42, 283, 284
aldehydes, formation
  by Grignard reaction, 42–3
  via organoboranes, 255, 258–62
  by oxidation of alcohols, 216–24
  by oxidation of methyl groups, 203–5
  by reduction of carboxylic acid derivatives, 183–5
aldehydes, reactions (see also 'aldehydes and

ketones', 'carbonyl compounds')
  alkylation at C=O, 96–8
  α-alkylation, 80, 84–6
aldehydes and ketones, reactions (see also 'aldehydes', carbonyl compounds', 'ketones')
  acetal and ketal formation, 21
  acylation, 82–4
  alkylation, 80–2
  bimolecular reduction (pinacol formation), 180, 181
  condensations, 75–7, 88–92
  with Grignard reagents, 40, 41, 44, 45
  Knoevenagel and Doebner condensations, 75–7
  oxidation, 228–32
  protection, 17, 247, 304, 306
  reduction to alcohols, 178–81
  reduction to methylene, 21, 181, 182
  in Reformatsky reaction, 47, 48
alkanes, reactions
  halogenation, 3, 4
  oxidation, 198–200
alkanes, formation
  from aldehydes and ketones, 21
  from alkenes, 174–6
  from organoboranes, 255, 257
  from organometallic compounds, 49, 51
alkenes, formation
  from alcohols, 16
  from alkanes, 206–8
  from alkyl halides, 19, 20
  from alkynes, 7, 8, 176–8, 261, 262, 288, 290, 295
  by extrusion reactions, 267, 268
  by Peterson reaction, 276–8
  by Wittig and related reactions, 93–8, 265, 266
alkenes, reactions, 4–6
  addition, 4–6
  allylic substitution, 6
  epoxidation, 209, 210
  hydroboration, 248–54
  hydroxylation, 210–14
  as nucleophiles, 36–8

R

I

RETT